ISNM 103
International Series of Numerical Mathematics
Internationale Schriftenreihe zur Numerischen Mathematik
Série Internationale d'Analyse Numérique
Vol. 103

Edited by
K.-H. Hoffmann, Augsburg; H. D. Mittelmann, Tempe;
J. Todd, Pasadena

Birkhäuser Verlag
Basel · Boston · Berlin

ISNM 103:
International Series of Numerical Mathematics
Internationale Schriftenreihe zur Numerischen Mathematik
Série Internationale d'Analyse Numérique
Vol. 103

Edited by
K.-H. Hoffmann, Augsburg; H. D. Mittelmann, Tempe;
J. Todd, Pasadena

Birkhäuser Verlag
Basel · Boston · Berlin

General Inequalities 6

6th International Conference on General Inequalities, Oberwolfach, Dec. 9–15, 1990

Edited by

W. Walter

1992

Birkhäuser Verlag
Basel · Boston · Berlin

Editor

Prof. Dr. Wolfgang Walter
Universität Karlsruhe
Mathematisches Institut I
Kaiserstrasse 12
D–W–7500 Karlsruhe 1
Germany

Deutsche Bibliothek Cataloging-in-Publication Data

General inequalities 6 / 6. International Conference on General Inequalities, Oberwolfach, Dec. 9–15, 1990. Ed. by W. Walter. – Basel ; Boston ; Berlin : Birkhäuser, 1992
(International series of numerical mathematics ; Vol. 103)
ISBN 3-7643-2737-5 (Basel ...)
ISBN 0-8176-2737-5 (Boston)
NE: Walter, Wolfgang [Hrsg.]; Internationale Tagung über Allgemeine Ungleichungen <06, 1990, Oberwolfach>; GT

Printed from the authors' camera-ready manuscripts on acid-free paper in Germany
ISBN 3-7643-2737-5
ISBN 0-8176-2737-5

Foreword

The sixth International Conference on General Inequalities was held from Dec. 9 to Dec. 15, 1990, at the Mathematisches Forschungsinstitut Oberwolfach (Black Forest, Germany). The organizing committee was composed of W.N. Everitt (Birmingham), L. Losonczi (Debrecen) and W. Walter (Karlsruhe). Dr. A. Kovačec (Coimbra) served cheerfully and efficiently as secretary of the meeting. The conference was attended by 44 participants from 20 countries.

Yet again the importance of inequalities in both pure and applied mathematics was made evident from the wide range of interests of the individual participants, and from the wealth of new results announced. New inequalities were presented in the usual spread of the subject areas now expected for these meetings: Classical and functional analysis, existence and boundary value problems for both ordinary and partial differential equations, with special contributions to computer science, quantum holography and error analysis.

More strongly than ever, the role played by modern electronic computers was made clear in testing out and probing into the validity and structure of certain inequalities. Here the computer acts not only for numerical calculations of great complexity, but also in symbolic manipulation of complex finite structures. Problems in inequalities which even a few years ago were intractable, now fall to solution or receive direct and positive guidance as a result of computer applications. The interface between finite and infinite structures in mathematics and the versatility of modern computers is well developed in the subject of general inequalities.

Once again the problems and remarks sessions were filled to overflowing with "Blitzbeweise" and "Blitzgegenbeispiele".

During one of the sessions J.D. Aczél gave a short but moving address in memory of István Fenyő who died in 1987 and who attended many of the previous General Inequalities meetings.

Again the Institute and all its facilities provided a unique environment for the success of the meeting and the well-being of the participants.

The opening remarks for the conference were made by W. Walter, and W.N. Everitt offered comments on the last day of the meeting. Both payed tribute to the high standards set and maintained, firstly by the participants in contributing to the programme of the meeting, and then to the staff of the Institute for their quiet but remarkably efficient service and hospitality.

W.N. Everitt

L. Losonczi

W. Walter

Participants

J. Aczél, University of Waterloo, Waterloo, Ontario, Canada
R.P. Agarwal, National University of Singapore, Kent Ridge, Singapore
C.D. Ahlbrandt, University of Missouri, Columbia, Missouri, USA
C. Alsina, Universitat Politècnica de Barcelona, Barcelona, Spain
H. Alzer, University of South Africa, Pretoria, South Africa
C. Bandle, Universität Basel, Basel, Switzerland
C. Bennewitz, Uppsala University, Uppsala, Sweden
B.M. Brown, University of Wales, Cardiff, Wales, U.K.
R.C. Brown, University of Alabama, Tuscaloosa, Alabama, USA
A. Clausing, Westfälische Wilhelms-Universität, Münster, Germany
M. Essén, Uppsala University, Uppsala, Sweden
W.D. Evans, University of Wales, Cardiff, Wales, U.K.
W.N. Everitt, University of Birmingham, Birmingham, England, U.K.
R. Ger, Silesian University, Katowice, Poland
M. Goldberg, Technion – Israel Institute of Technology, Haifa, Israel
W. Haussmann, Gesamthochschule Duisburg, Duisburg, Germany
H.P. Heinig, McMaster University, Hamilton, Ontario, Canada
C.S. Hönig, University of Sao Paulo, Sao Paulo, Brasil
H. König, Universität des Saarlandes, Saarbrücken, Germany
H. König, Universität Kiel, Kiel, Germany
A. Kovačec, Universidade de Coimbra, Coimbra, Portugal
S. Kremp, Universität des Saarlandes, Saarbrücken, Germany
A. Kufner, Československá Akademie Véd, Praha, CSFR
G.S. Ladde, University of Texas, Arlington, Texas, USA
R. Lemmert, Universität Karlsruhe, Karlsruhe, Germany
L. Losonczi, L. Kossuth University, Debrecen, Hungary
H. Miller, University of Sarajevo, Sarajevo, Yugoslavia
R.N. Mohapatra, University of Central Florida, Orlando, Florida, USA

Participants (Continued)

R.J. Nessel, Rheinisch-Westfälische Technische Hochschule, Aachen, Germany
Zs. Páles, L. Kossuth University, Debrecen, Hungary
J.E. Pečarić, Faculty of Technology, Zagreb, Yugoslavia
M. Plum, Universität Köln, Köln, Germany
Th.M. Rassias, Paradissos Amaroussion, Athens, Greece
D.C. Russell, York University, North York, Ontario, Canada
B. Saffari, Université de Paris-Orsay, Orsay, France
S. Saitoh, Gunma University, Kiryu, Japan
W. Schempp, Universität Siegen, Siegen, Germany
R. Sperb, Eidgenössische Technische Hochschule Zürich, Zürich, Switzerland
G. Talenti, Università degli Studi, Firenze, Italia
P. Volkmann, Universität Karlsruhe, Karlsruhe, Germany
W. Walter, Universität Karlsruhe, Karlsruhe, Germany
J.R.L. Webb, University of Glasgow, Glasgow, Scotland, U.K.
K. Zeller, Universität Tübingen, Tübingen, Germany
A. Zettl, Northern Illinois University, DeKalb, Illinois, USA

Scientific Program of the Conference

Monday, December 10

Opening of the conference — W. Walter

Early morning session — Chairman: W. Walter

A. Zettl: Landau's inequality

A. Kufner: Hardy-type inequalities with applications to differential equations

Late morning session: — Chairman: A. Zettl

H.P. Heinig: Weighted Hardy type inequalities for monotone functions

R.N. Mohapatra: Copson's inequalities and their variations

Problems and remarks

Early afternoon session — Chairman: Heinz König

Hermann König: On Grothendieck's inequality

P. Volkmann: Ein konstruktiver Beweis des Existenzsatzes von Peano

R. Lemmert: Existenzsätze für Differentialgleichungen in geordneten Banachräumen

Late afternoon session — Chairman: M. Essén

M. Goldberg: Multiplicativity factors for seminorms

H. Miller: On several classes of sets III

* * * * **

Tuesday, December 11

Early morning session — Chairman: D.C. Russell

Th.M. Rassias:	Properties of polynomials and extremal problems
R.P. Agarwal:	Sharp inequalities in polynomial interpolation
A. Kovačec:	Systems of overlapping sets and polynomial inequalities

Late morning session: — Chairman: H.P. Heinig

L. Losonci:	Eigenvalues and eigenvectors of multidiagonal matrices
B. Saffari:	The new proofs of the Schaake-van der Corput inequality

Problems and remarks

Early afternoon session — Chairman: W.D. Evans

C. Bennewitz:	Optimal bounds for the ground state of a singular Sturm-Liouville problem
G. Talenti:	An integral inequality for monotonic functions
S. Saitoh:	Norm inequalities derived from the theory of reproducing kernels

Late afternoon session — Chairman: C.D. Ahlbrandt

W. Schempp:	The uncertainty principle in quantum holography
K. Zeller:	Inequalities for summability factors

* * * * **

Wednesday, December 12

Early morning session Chairman: H. Miller

W.N. Everitt: Recent developments in the HELP inequalities

W.D. Evans: HELP inequalities for integrals and sums

B.M. Brown: Numerical estimates for the best constant in the HELP inequality

Late morning session: Chairman: B. Saffari

C.D. Ahlbrandt: Discrete variational inequalities

A. Clausing: Review of cyclic inequalities

Problems and remarks

Afternoon excursion and discussion

* * * * **

Thursday, December 13

Early morning session Chairman: G. Talenti

C. Bandle: Inequalities for maximal solutions of nonlinear elliptic problems

R. Sperb: An isoperimetric inequality for the first eigenvalue of the Laplacian under Robin boundary conditions

W. Walter: The maximum principle for elliptic systems

Late morning session: Chairman: C. Bandle

J.R.L. Webb: An L^p inequality

M. Plum: Existence proofs for nonlinear boundary value problems by error estimates of approximate solutions

Problems and remarks

Early afternoon session Chairman: C. Alsina

J. Aczél: The convex solution of higher order of a functional equation

J.E. Pečarić: Notes on convex functions

R. Ger: Functional inequalities stemming from stability questions

Late afternoon session Chairman: J. Aczél

H. Alzer: Recent results on inequalities

D.C. Russell: Remarks on trigonometric inequalities

Problems and remarks

******

Friday, December 14

Early morning session

M. Essén: Some best constant inequalities for conjugate functions

R.C. Brown: Weighted inequalities of product type on $\mathbb{R}^+$

Heinz König: Non-equivalent complete norms and would-be continuous functionals

Problems and remarks

Late morning session:

C.S. Hönig:	Gronwall-Bellman linear Stieltjes integral inequalities
G.S. Ladde:	Comparison theorems and its applications

Problems and remarks

Early afternoon session

C. Alsina:	On 3-positive semidefinite strict t-norms and semimetrics
R.J. Nessel:	On a problem of best constants for the midpoint and trapezoidal rule
Zs. Páles:	Comparison of two variable homogeneous means

Problems and remarks

Closing of the conference W.N. Everitt

Preface

The present volume contains research articles which were presented at GENERAL INEQUALITIES 6, the sixth Oberwolfach Conference on General Inequalities; it was held from Dec. 9 to Dec. 15, 1990, at the Mathematisches Forschungsinstitut Oberwolfach/Germany. Included are also some contributions by other authors (some were unable to attend). These GI conferences — the first one was held in May 1976 — are distinguished from most of the other Oberwolfach conferences by the special nature of the field of inequalities. The whole body of mathematics is now subdivided in 63 special subjects (1991 Mathematics Subject Classification), but inequalities is not one of them. On the other hand, inequalities occur in almost all of these fields, and they play a significant role in many of them. Our GI conferences have always combined mathematicians from a wide array of interests and research areas, giving the conference a very special flavor. Problems from applications which lead to inequalities are particularly impressing.

Following tradition, the volume has been subdivided into several chapters. Simply looking at the chapter headings, a significant change in emphasis in the time between GI 1 (in 1976) and GI 6 becomes apparent. The most dramatic development has occurred in inequalities connected with differential operators. In GI 1 there were two articles in this field, and now in GI 6 the largest chapter is devoted to this subject. In what follows, a brief description of the content of the book is given.

The volume starts with an article in memoriam István Fenyő (1917–1987). It was from this remarkable man that the editor learned of Leonardo da Vinci's dictum that mechanics is the paradise of mathematicians, where they can see (and harvest) the fruits of mathematics.

The first chapter INEQUALITIES FOR SUMS, SERIES AND INTEGRALS deals with the classical field of inequalities. Kufner and Wannebo consider Hardy type inequalities of the form $||u||_q^w \leq C||u^{(k)}||_p^v$, where the norms are weighted L_p norms over an interval (a,b), $||u||_p^v = (\int_a^b |u(x)|^p v(x)\,dx)^{1/p}$, and u is subject to k boundary conditions, some at a, some at b. Extending earlier results for the case $(a,b) = (0,\infty)$, they consider bounded intervals and prove theorems of the if-and-only-if type depending on the weight functions v, w and p, q with $1 < p \leq q < \infty$.

In 1927 Copson found several inequalities, some of which generalize classical inequalities such as Hardy's inequality $||(A_n)||_p \leq (p/(p-1))||(a_n)||_p$, where $A_n = (a_1 + \cdots + a_n)/n$ (norm in l_p, $a_n \geq 0$). Recently, Bennett has obtained several variants and generalizations of those inequalities. In their article *Copson's inequalities for series and their variations*, R.N. Mohapatra, D.C. Russell and K. Vajravelu derive two new general inequalities which belong to this field. They give corollaries and use them to prove inclusions between sequence spaces.

Inequalities between different means belong to the core of our subject. Among them are the celebrated Ky Fan inequality on quotients of means, a difference analogue due to Alzer, and reverse inequalities. H. Alzer proves counterparts of these inequalities for quasiarithmetic means in his paper *Inequalities between weighted arithmetic and geometric means. Comparison of two variable homogeneous means* is an old subject with quite a bibliography. Zs. Páles derives inequalities for difference means and sum means.

In *A review of Shapiro's cyclic inequality* A. Clausing gives a rather complete account of an eventful development of some 35 years which started in 1954 with the following *Monthly* problem proposed by Shapiro: Establish

$$\sum_{i=1}^{n} \frac{x_i}{(x_{i+1} + x_{i+2})} \geq \frac{n}{2}$$

for $x_i \geq 0$, $x_i + x_{i+1} > 0$, $x_i = x_{n+i}$ with equality only if all denominators are equal. Some highlights: the first counterexample for $n = 20$ by Lighthill in 1956, a proof of a weaker inequality with the sharp lower bound $0.989133 \ldots \cdot \frac{n}{2}$ by Drinfel'd in 1971. The story came to an end after almost 20 more years when Troesch in 1989 closed the last gap in the proof of the theorem that Shapiro's inequality is true for even $n \leq 12$ and for odd $n \leq 23$ and false for all other values of n.

In the second chapter INEQUALITIES IN ANALYSIS AND APPROXIMATION a variety of topics is addressed. M. Essén in his article *Some best constant inequalities for conjugate functions* gives a new approach to inequalities of the type

$$||F||_p^p \leq C_p ||f||_p^p, \quad 1 < p < \infty,$$

and corresponding inequalities of the $L \log L$-type for $p = 1$, where $|| \cdot ||_p$ is the norm in the Hardy space $H^p(U)$ (U unit disc) or $H^p(D)$ (D upper half plane) and $F = f + i\tilde{f}$. His new approach to these basic inequalities relies heavily on inequalities for subharmonic functions. In the case $p = 1$ in D there are still open questions.

G. Talenti derives *An inequality between* u^* *and* $|\text{grad}\, u|^*$, namely

$$u^*(s) \leq n^{-1} \kappa_n^{-1/n} \int_0^{+\infty} |\text{grad}\, u|^*(t)(s+t)^{-1+1/n}\, dt.$$

Here, $u \in L_{\text{loc}}(\mathbb{R}^n)$ with $|\text{grad } u| \in L(\mathbb{R}^n)$, κ_n is the volume of the unit ball and u^* the decreasing rearrangement of u. Rearrangement (or symmetrization) techniques are used in partial differential equations (viscosity solutions of Hamilton-Jacobi equations, eigenvalue problems,...), and they are closely related to Sobolev embedding theorems. As an application, an embedding theorem between Lorentz spaces is proved, which improves a classical result significantly.

Trigonometric functions are involved in the work of L. Losonczi *On an extremal property of nonnegative trigonometric polynomials* and L. Lorch and D.C. Russell *Remarks on trigonometric inequalities.* Extending an old inequality of Egerváry and Szász (1928), Losonczi gives bounds for the coefficients of a *nonnegative* trigonometric polynomial $T_n(x) = \frac{1}{2}a_0 + \sum_{k=1}^{n}(a_k \cos kx + b_k \sin kx)$. A special case reads

$$(a_k^2 + b_k^2)^{1/2} + (a_l^2 + b_l^2)^{1/2} \leq a_0 \cos \frac{\pi}{2\left[\frac{n-k}{l-k}\right] + 3},$$

where $(n+1)/2 \leq k < l \leq n$.

Motivated by the work of Turán and Halász on power sums of the form $g(\nu) = \sum_{j=1}^{n} b_j z_j^{\nu}$ ($\nu > 0$ an integer), Lorch and Russell consider the expansion

$$\left(\frac{\sin z}{z}\right)^m = \sum_{k=0}^{\infty} a_{mk} \sin^{2k} z \quad \left(m = 0, 1, \ldots, |z| < \frac{\pi}{2}\right).$$

Explicit representations of a_{mk} in different forms are given, and trigonometric inequalities involving $(\sin z/z)^m$ (e.g.) are obtained.

In his paper *On polynomial inequalities and extremal problems* Th.M. Rassias gives an account on some fundamental problems about zeros, critical points (i.e. zeros of the derivative) and extremal properties of polynomials. If S is the set of zeros and C the set of critical points, then the Gauß-Lucas theorem states $C \subset \text{conv } S$. A conjecture by Sendov is discussed. It states that if $S \subset B_1(0)$ (closed ball of radius 1 centered at 0) then for any $z \in S$ the ball $B_1(z)$ contains a point of C. The best result (1985) is $B_r(z) \cap C \neq \emptyset$, where $r = 1.083\ldots$. Theorems of Markov and Bernstein type regarding inequalities of the form $||p^{(k)}|| \leq A||p||$ for p with $\deg p \leq n$ and a corrected version of a theorem by M. Marden on zeros in a ring $r_1 \leq |z| \leq r_2$ are given.

R.P. Agarwal's paper *Sharp inequalities in polynomial interpolation* gives a survey of recent progress in polynomial interpolation theory. Given a function $x(t) \in C^n[a, b]$, points $a = a_1 < a_2 < \cdots < a_r = b$ and subsets α_i of $\{0, 1, \ldots, n-1\}$ with $\sum_{i=1}^{r} \text{card } \alpha_i = n$, the problem is to find a polynomial P of degree $\leq n - 1$ such that

$$x^{(j)}(a_i) = P^{(j)}(a_i) \quad \text{for } j \in \alpha_i \quad (i = 1, \ldots, r)$$

and to estimate the difference $e = x - P$. Special consideration is given to Lidstone interpolation ($r = 2$, $\alpha_1 = \alpha_2 = \{0, 2, 4, \ldots, 2m-2\}$, $n = 2m$), (n, p)-interpolation ($r = 2$, $\alpha_1 = \{0, 1, \ldots, n-2\}$, $\alpha_2 = \{p\}$), Abel-Gontscharoff interpolation ($r = n$, $a_i \leq a_{i+1}$, $\alpha_i = \{i-1\}$) and Hermite interpolation. The connection with boundary value problems for ordinary differential equations is stressed.

In his article *Discrete variational inequalities* C.D. Ahlbrandt treats inequalities for discrete quadratic functionals motivated by the theory of second variation for discrete variational problems. These inequalities can be considered as discrete versions of classical integral inequalities associated with quadratic functionals. In discrete Sturmian theory a quadratic functional with the Euler equation

$$-K_n u_{n+1} + B_n u_n - K_{n-1}^T u_{n-1} = 0$$

is given, where B_n and K_n are $m \times m$ matrices, $u_k \in \mathbb{R}^m$. Among others, discrete versions of the Legendre-Clebsch transformation, the Reid roundabout theorem and a Sturm comparison theorem are proved. A discrete variational theory is outlined.

In their paper *Differential inequalities for Minkowski functionals of level sets* G. Chiti and M. Longinetti study the Minkowski functionals w_k of n-dimensional oriented hypersurfaces embedded in $\mathbb{R}^{n+1}$, where the surfaces are level sets $\{u(x) = t\}$ of a smooth function $u : \Omega \subset \mathbb{R}^{n+1} \to \mathbb{R}$. The main results concern solutions u of boundary value problems for the (nonlinear) elliptic equation

$$\sum_{i=1}^{n+1} (a(|\nabla u|^2) u_{x_i})_{x_i} = 0.$$

Working with inequalities for the derivative of Minkowski functionals with respect to the level parameter t, the authors obtain new results, e.g., on the convexity of w_k if Ω is a convex region or convex annulus, and they prove an isoperimetric inequality. The results have important physical applications, e.g., w_2 allows the estimation of quantities in capacity, membrane and torsion problems. For $a \equiv 1$, i.e., $\Delta u = 0$, and $n = 2$ or 3 some of the results go back to Pólya and Szegő.

The chapter on INEQUALITIES OF FUNCTIONAL ANALYSIS starts with the article *Inequalities for summability factors* by W. Beekmann and K. Zeller. The authors consider factors (= infinite triangular matrices) which transform (e.g.) C_α summable series into C_β summable series. The results have a long history with names such as Hardy, Bohr, Bosanquet. The results belong both to summability theory and functional analysis, giving necessary and sufficient conditions on summability factors and description of the continuous linear functionals in certain sequence spaces.

Let $\mathcal{A}$ be an algebra and S a seminorm on $\mathcal{A}$ ($S(x) \geq 0$, $S(\lambda x) = |\lambda| S(x)$, $S(x+y) \leq S(x) + S(y)$). In an expository paper *Seminorms and multiplicativity factors* M. Goldberg investigates multiplicativity factors, i.e., constants $\mu > 0$ for which $S(xy) \leq \mu S(x) S(y)$ in $\mathcal{A}$, in terms of the kernel of S. Special consideration is given operator algebras, the numerical radius $r(A)$, seminorms generated by the supremum norm and quasimonotone seminorms in $\mathbb{C}^n$ (S is quasimonotone if $x, y \in \mathbb{R}^n$, $0 \leq x \leq y$ componentwise implies $S(x) \leq S(y)$).

In 1956 Grothendieck proved the following theorem:

For any $n \in \mathbb{N}$ there is $c(n) > 1$ such that for any finite matrix $(a_{jk})_{j,k=1}^m$ with

$$\left| \sum_{j,k=1}^{m} a_{jk} t_j s_k \right| \leq \sup_{1 \leq j,k \leq m} |t_j| \, |s_k|$$

and any sequence of vectors $(x_j)_{j=1}^m$, $(y_k)_{k=1}^m$ in an n-dimensional Hilbert space H_n one has

$$\left| \sum_{j,k=1}^{m} a_{jk} \langle x_j, y_k \rangle \right| \leq c(n) \sup_{1 \leq j,k \leq m} ||x_j|| \, ||y_k||.$$

In his paper *Some remarks on the Grothendieck inequality* H. König sketches a method of proof for obtaining the best known value of $c(n)$. The inequality has important applications in the geometry of Banach spaces. The value of the best constant is still unknown.

The chapter on Functional Equations and Inequalities starts with an article *Solutions of a functional equation, convex of higher order* by J. Aczél and M. Kuczma, which deals with the functional equation

$$(\star) \qquad f(kx) = k^{\gamma} f(x) \quad \text{in} \quad (0, \infty),$$

where $k > 0$ (the case $k = 1$ is trivial) and $\gamma \in \mathbb{R}$. Starting from the observation that the restriction $y = x$ in the Cauchy equation $f(x+y) = f(x) + f(y)$, which leads to $f(2x) = 2f(x)$, behaves quite differently from the unrestricted Cauchy equation, the authors derive the general solution of $(\star)$ and give various necessary and sufficient conditions such that $f(x) = cx^{\gamma}$.

Let $J = [0,1]$. A strict t-norm is a continuous function $T : J \times J \to J$ which is associative, commutative, 1 is a unit, and T is strictly increasing (a simple example is Prod $(x, y) := xy$). C. Alsina and M.S. Tomas present a study *On positive semidefinite strict t-norms* with an emphasis on t-norms which are positive semidefinite of order 2 and 3. t-norms are a fundamental tool in probabilistic metric spaces and in the study of joint distribution functions with given marginals. Positive semidefiniteness is closely related to (and weaker than) total positivity.

In the paper *On functional inequalities stemming from stability questions* R. Ger considers mappings F from a semigroup, group or Banach space into a normed linear space, which are controlled by a subadditive or subquadratic functional f, i.e., for which

$$||F(x+y) - F(x) - F(y)|| \leq f(x) + f(x) - f(x+y)$$

or

$$||F(x+y) + F(x-y) - 2F(x) - 2F(y)|| \leq 2f(x) + 2f(y) - f(x+y) - f(x-y)$$

holds. Conditions are given such that F is close to an additive function A or quadratic function Q, i.e., $||F(x) - A(x)|| \leq cf(x)$ or $||F(x) - Q(x)|| \leq cf(x)$. Results on the asymptotic behavior of F are also given.

We come to the largest chapter in this volume, INEQUALITIES FOR DIFFERENTIAL OPERATORS. C. Bennewitz and E.J.M. Veling obtain *Optimal bounds for the spectrum of a one-dimensional Schrödinger operator*, i.e., they give sharp estimates on the lowest point s_0 of the spectrum of the eigenvalue problem

$$-u'' + qu = \lambda u \ \text{ in } [0,\infty), \quad u(0)\cos\beta + u'(0)\sin\beta = 0$$

in terms of the $L^p(0,\infty)$ norm $||\cdot||_p$ of q_+ and q_-. The proof is based on the determination of the largest constant λ such that

$$\lambda||u||_2^2 \leq ||u'||_2^2 - \cot\beta|u(0)|^2 + a||u||_r^2$$

for $u \in H^1(0,\infty)$.

In the paper *Interpolation inequalities and nonoscillatory differential equations* M. Ashbaugh, R. Brown and D. Hinton obtain nonoscillation criteria for symmetric differential operators of the form $L(y) = \sum_{i=1}^n (-1)^i (p_i y^{(i)})^{(i)}$ (or generalizations thereof). Weighted interpolation inequalities of the form

$$\left(\int_I |x^\beta u^{(j)}|^p \, dx\right)^{1/p} \leq K \left(\int_I |x^\gamma u|^q \, dx\right)^{(1-\lambda)/q} \left(\int_I |x^\alpha u^{(n)}|^r \, dx\right)^{\lambda/r}$$

(recently found by Brown and Hinton) are used as a basic tool. The results yield information on the spectrum, in particular on disconjugacy and on the number of negative eigenvalues of a Schrödinger operator in $\mathbb{R}^3$ with a spherically symmetric potential.

B.M. Brown, W.D. Evans and W.N. Everitt present an extensive survey paper on *HELP integral and series inequalities.* The starting point is the classical Hardy-Littlewood inequality of 1932

$$\left(\int_0^\infty y'^2\right)^2 < 4\int_0^\infty y^2\,dx \int_0^\infty y''^2\,dx$$

(equality for $y(x) = aY(bx)$, where Y is known) and its discrete analogue for real sequences $(a_n)_0^\infty$

$$\left(\sum_{n=0}^{\infty}(\Delta a_n)^2\right)^2 < 4\sum_{n=0}^{\infty} a_n^2 \sum_{n=0}^{\infty}(\Delta^2 a_n)^2,$$

where $\Delta a_n = a_{n+1} - a_n$ and $\Delta^2 a_n = \Delta(\Delta a_n)$, which goes back to Copson (1979). The HELP (Hardy-Everitt-Littlewood-Pólya) extension, started by Everitt in 1979, deals in its general form with the inequality

$$\left(\int_a^{\to b}(pf'^2 + qf^2)\,dx\right)^2$$
$$\leq K\int_a^b wf^2\,dx \int_a^b w(w^{-1}(-(pf')' + qf))^2\,dx$$

$(-\infty < a < b \leq \infty)$. It is closely connected with the Sturm-Liouville equation

$$-(py')' + qy = \lambda wy \quad \text{on } [a,b),$$

where the endpoint b is a singular point in the limit-point case, in particular with its spectral theory in the Hilbert space $L^2(a,b;w)$ with weight function w. The paper, written by leading experts, gives an up to date account of the theory. It includes the discrete case and describes in detail numerical techniques for obtaining the best constant K, together with numerical results.

In their joint paper *Gewöhnliche Differentialgleichungen mit quasimonton wachsenden rechten Seiten in geordneten Banachräumen* A. Chaljub-Simon, R. Lemmert, S. Schmidt and P. Volkmann prove an existence theorem for the initial value problem

$$u'(t) = f(t, u(t)) \text{ in } J = [0,T], \quad u(0) = a, \quad \text{where } f : J \times E \to E$$

(E Banach space) is continuous and bounded. It is well known that this condition alone guarantees existence if and only if $\dim E < \infty$. In an ordered Banach space, quasiomonotonicity is, roughly speaking, the condition for the validity of a monotonicity theorem (comparison principle) of the form

$$u(a) \leq v(a),\ u' - f(t,u) \leq v' - f(t,v) \text{ in } J \text{ implies } u \leq v \text{ in } J.$$

The authors obtain new existence results under the main assumption that f is quasimonotone and a theorem on nonexistence.

In their contribution *Comparison theorem and its applications* G.S. Ladde, M. Sambandham and S. Sathananthan consider a stochastic integro-differential equation

$$y' = f(t, y, Ty(\omega), \omega), \quad y(t_0, \omega) = y_0(\omega)$$

and a corresponding deterministic initial value problem

$$m' = \hat{f}(t, m, \hat{T}m), \quad m(t_0) = m_0 = E[y_0(\omega)],$$

which is obtained by ignoring the random disturbance. Using corresponding comparison principles for such equations, problems of stability and error estimation are investigated. In particular, a partial solution to the problem of "Stochastic versus Deterministic" in the modelling of dynamical systems in biological, engineering and physical sciences is given.

Parabolic differential inequalities of the form

$$(\star) \qquad u_t \le f(t, x, u, u_x, u_{xx}) \text{ in } G = (0, \infty) \times B$$

are the subject of R.M. Redheffer's articel *Nagumo theory without the Nagumo point.* Here, B is an open set in $\mathbb{R}^n$, u_x is the gradient and u_{xx} the Hessian of $u = u(t, x)$ and $\Gamma = \overline{G} \setminus G$ the parabolic boundary of G. The author develops a new one-step method for proving maximum principles of the form "$u(\Gamma) \le 0$ and $(\star)$ implies $u(G) \le 0$", which bypasses the two-step proof initiated by Nagumo. Another novelty is the use of discontinuous auxiliary functions, which leads to very general theorems. It is pointed out that the method may also be used for monotonicity and comparison theorems, parabolic systems,...

S. Saitoh proves *Inequalities for the solutions of the heat equation* corresponding to the Cauchy problem

$$u_t = \Delta u \text{ in } (0, \infty) \times \mathbb{R}^n, \quad u(0, x) = F(x) \text{ in } \mathbb{R}^n,$$

where $F \in L_2(\mathbb{R}^n)$. Using methods for integral transforms based on the theory of reproducing kernels, the author investigates the time dependence of any derivative $D^\alpha u(t, x)$ for fixed x. It is shown, among others, that $D^\alpha u(\cdot, x)$ can be extended analytically into the right half plane R^+ : Re $\tau = t > 0$ of the complex $\tau = t + i\hat{t}$ plane, and integral inequalities of the form

$$C \cdot \int_{R^+} |D^\alpha u(\tau, x)|^2 t^\beta \, dt \, d\hat{t} \le \int_{\mathbb{R}^n} |F(x)|^2 \, dx$$

are proved.

Dealing with a classical problem of mathematical physics, R.P. Sperb derives *An isoperimetric inequality for the first eigenvalue of the Laplacian under Robin boundary conditions.* Let $\lambda_1(\alpha)$ be the first eigenvalue of the eigenvalue problem

$$\Delta u + \lambda u = 0 \text{ in } \Omega \subset \mathbb{R}^n, \quad \frac{\partial u}{\partial n} + \alpha u = 0 \text{ on } \partial\Omega$$

$(0 \leq \alpha \leq \infty)$; for $\alpha = \infty$ we have the fixed membrane problem, $u = 0$ on $\partial\Omega$. Recently, Payne obtained the lower bound $\lambda_1(\infty) \geq \pi^2/8t_m$, where t_m is the maximum of the "torsion function" $t(x)$, which is the solution of the boundary value problem

$$\Delta t + 1 = 0 \text{ in } \Omega, \quad t = 0 \text{ on } \partial\Omega.$$

Here, an extension of this inequality to arbitrary values $\alpha \geq 0$ is given. It reads $\lambda_1(\alpha) \geq \nu_1(\alpha)$, where $\nu_1(\alpha)$ is the first positive root of $\tan\sqrt{2t_m\nu} = \alpha/\sqrt{\nu}$. Equality holds if Ω degenerates into an infinite slab.

In his paper *Minimum principles for weak solutions of second order ordinary differential equations* W. Walter considers three types of theorems:

$$(p(x)u')' + c(x)u \leq 0 \text{ in } J^\circ, \quad u(a), u(b) \geq 0 \quad (J = [a, b])$$

implies (MP) or (SMP) or (FEC), which means "$u \geq 0$ in J°" (minimum principle) or "$u \equiv 0$ or $u > 0$ in J°" (strong minimum principle) or "$u \equiv 0$ or $u > 0$ in J° or $u =$ first eigenfunction of $(pu')' + cu + \lambda u = 0$ and $\lambda_1 = 0$" (first eigenvalue case). It is only assumed that u and pu' are locally absolutely continuous and $p > 0$ a.e. in J, and a new method of proof is employed.

The notion of viscosity solutions, which was developed in the eighties for Hamilton-Jacobi equations, has been extended to nonlinear second order elliptic equations. A function u is a viscosity solution if it is a viscosity subsolution and supersolution; these notions are defined by differential inequalities via subdifferentials and superdifferentials. In the simplest case $\Delta u = 0$ these notions coincide with those of subharmonic and superharmonic functions. Recently very general theorems on existence, uniqueness and continuous dependence have been obtained for fully nonlinear degenerate elliptic equations of the form

$$F(x, u, Du, D^2u) = 0,$$

where "degenerate" refers to the fact that only weak ellipticity in the form "$F(\ldots, r) - F(\ldots, s) \geq 0$, if the symmetric matrix $s - r$ is positively semidefinite" is assumed. In his article *Uniqueness and existence for viscosity solutions of quasimonotone systems* V. Weckesser extends this theory to quasimonotone elliptic systems. In this approach, where maximum and comparison principles are essential, the restriction to quasimonotone systems is natural.

The chapter INEQUALITIES IN OPTIMIZATION AND APPLICATIONS starts with an article *On some problems concerning best constants for the midpoint and trapezoidal rule* by B. Büttgenbach, G. Lüttgens and R.J. Nessel. Consider the integral $If = \int_0^1 f(u)\,du$ and a sequence of corresponding quadrature formulas $Q_n f := \sum_{k=1}^n a_{kn} f(x_{kn})$ with knots $0 \le x_{1n} < \cdots < x_{nn} \le 1$. The aim is to find upper bounds for the remainder $R_n f = If - Q_n f$ in terms of measures of smoothness of f, namely the classical r-modulus of continuity $\omega(\delta, f)$ if f is continuous and the r-th τ-modulus $\tau_r(\delta, f)$ if f is Riemann integrable. For the compound midpoint rule a Jackson type inequality

$$|R_n^{Mi} f| \le \frac{1}{2}\tau_2\left(\frac{1}{2n}, f\right) \le \frac{1}{2}\omega_2\left(\frac{1}{2n}, f\right)$$

holds. Similar estimates are true for the compound trapezoidal and Simpson rule. The problem of best constants is also discussed.

The article *The uncertainty inequality in quantum holography, Bohr's indeterminacy principle, and synchronized neural networks* by Chen Han-lin and W. Schempp deals with the theoretical foundation of quantum holography and other aspects of quantum physics. It is explained that a rigorous proof of the quantum parallelism cannot be based on the Heisenberg inequality; more generally, N. Bohr's indeterminacy principle cannot be based on any of the known uncertainty principles. After explaining the basics of holographic imaging, it is shown how the holographic transform allows to circumvent the difficulties with the standard deviation by using a group theoretical implementation of the canonical commutation relation of quantum electrodynamics. The paper contains many interesting quotations.

In their paper *Optimization and inequalities on product spaces* Don-lin Mon and Chung-lie Wang support by new examples the general philosophy that an inequality can be used to establish optimization problems while an optimization may be obtained by inequalities. Using this interplay between inequalities and optimization theory, Beckenbach proved in 1966 a generalization of Hölder's inequality by an optimization technique. The authors consider inequalities of the Hölder, Minkowski or weighted mean type on product spaces (products of Banach spaces). In this setting inequalities with a mixture of discrete and continuous data are obtained, pointing to the applicability in statistics and optimization theory.

The last chapter NOTES, PROBLEMS AND REMARKS is of a very special kind — as it has been in the earlier GI volumes. Here inequalities are presented as a colorful, stimulating subject. J.L. Brenner and H. Alzer extend Henrici's inequalities in several directions. R. Lemmert and H. Alzer generalize an inequality of Talenti, provide a simple proof and establish a converse inequality. J. Rätz treats a minimum problem for polygons with corners on the unit circle. J.E. Pečarić gives new results about Jensen's inequality and Hadamard's inequality (which is in fact due to Hermite). B. Saffari proves the inequality $||f'||_\infty^2 \leq \omega(f) H(\sup f'', \inf f'')$ for C^2-functions $f : \mathbb{R} \to \mathbb{R}$, which is sharper than Landau's classical inequality ($\omega(f)$ is the oscillation of f and $H(a,b)$ the harmonic mean). A simple proof of a quadratic inequality due to Saitoh is given by L. Losonczi and Zs. Páles, and the second author generalizes Tarski's plank problem: If a closed set K in $\mathbb{R}^n$ is covered by a finite number of planks (a plank is a closed set bounded by two parallel hyperplanes), then the sum of their widths is greater than or equal to the minimal width of K. H. König (Saarbrücken) shows a simple direct proof of Minkowski's inequality. M. Kuczma has observed that for a convex function f on an interval $I \subset (0,\infty)$ the function $x \mapsto f(x)/x$ is either monotone or first decreasing and then increasing (when x increases). Two different easy roads to this result are described by H. König and by J. Aczél and C.T. Ng. B.M. Brown, W.D. Evans and W.N. Everitt discuss a special HELP inequality $(||f'||^2 - ||f||^2)^2 \leq K||f||^2||f''+f||^2$ in different subspaces of $L^2[0,\pi]$ (L^2-norm).

Finally there are a dozen problems by different authors ranging from the fundamental theorem of algebra in several variables to the parameter estimation for the hypergamma probability distribution.

Mrs. Joy Russell has again provided some beautiful drawings which convey a glance into the Oberwolfach atmosphere. The editor expresses his sincere gratitude to Mrs. Russell for this enrichment of the volume.

The editor expresses his thanks to those who have contributed to this volume and to all who have supported him in the editorial process. Several colleagues have reviewed the articles, sometimes with helpful comments for improvement. Most of the proofreading was done by Dipl.-Math. Volkmar Weckesser who also assisted in technical details. The subject index was assembled by Priv.-Doz. Dr. Raymond Mortini, the name index by Dipl.-Math. Markus Burger, and both helped with proofreading, as well as Dipl.-Math. Gerd Herzog. The editorial secretary, Ms. Irene Jendrasik, has convinced the editor that the volume must be prepared in a uniform TEX style. She wrote most of the articles and carried the correspondence with authors and reviewers. If the technical preparation of the volume deserves recognition, it is due to her work. The editor is especially grateful to her. He thanks Birkhäuser Verlag for a smooth and friendly collaboration.

Karlsruhe, January 1992

Wolfgang Walter
Editor

Contents

Inequalities for Differential Operators

Inequalities in Optimization and Applications

NOTES, PROBLEMS AND REMARKS

Sketches

by

Joy Russell

International Series of Numerical Mathematics, Vol. 103, © 1992 Birkhäuser Verlag Basel

István Fenyő (1917–1987) in memoriam

János Aczél

Professor István (Stefan, Stephen, Étienne) Fenyő, a frequent and active participant of these General Inequalities meetings, passed away since our last conference, less than five months past his 70th birthday after a painful but mercifully short illness.

First undecided whether he wants to be a chemist or a mathematician, he made the decision in favour of the latter not long before writing his thesis under the supervision of L. Fejér on mean values. This thesis became the starting point of much research on functional equations and inequalities, even though it has never been published.

He shared the suffering and tragedy of so many in the first half of the 1940's: his first wife was killed. Several years later, as supervisor of a high school final examination, he met his second wife, Erzsi Fonó, a student of music and conductor in her own right, who survived him.

He was Secretary of the postwar Scientific Council in Hungary. Otherwise, apart from shorter (Rome, Waterloo, Ont.) and longer (Brescia and Milan, Italy, Rostock, Germany) teaching and research appointments abroad, his entire professional career evolved at the Technical University of Budapest, from Assistant to Associate to full Professor till his retirement in 1984. He spent also much time with thankless nonmathematical chores.

He wrote widely used textbooks (with G. Alexits and with T. Frey) and a highly successful series of monographs on integral equations (with H.W. Stolle). He was editor of *Aequationes Mathematicae* and of *Zeitschrift für Analysis und ihre Anwendungen.*

He pursued his early interests in functional and integral equations throughout his life but made also considerable contributions to the theories of generalized functions, elliptic functions, inequalities, iterative processes, computers and applications to medicine and engineering. In different phases of his life the philosophy and history of mathematics and the mathematical aspects of Leonardo da Vinci's work fascinated him.

He was a polyglot, a conversationist and a man of many scientific and human interests. He was creative in all his endeavours, including telling of jokes and anecdotes. He will be sorely missed.

J. Aczél, Department of Pure Mathematics, University of Waterloo, Waterloo, Ont., Canada N2L 3G1

Inequalities for Sums, Series and Integrals

Oberwolfach-Walke under snow

International Series of Numerical Mathematics, Vol. 103, © 1992 Birkhäuser Verlag Basel

Inequalities for pseudo arithmetic and geometric means

Horst Alzer

To the memory of Horst-Siegfried Paul Grässer

Abstract

We prove inequalities for the pseudo arithmetic and geometric means a_n and g_n defined by

$$a_n = \frac{P_n}{p_1}x_1 - \frac{1}{p_1}\sum_{i=2}^{n} p_i x_i \quad \text{and} \quad g_n = x_1^{P_n/p_1} / \prod_{i=2}^{n} x_i^{p_i/p_1},$$

where x_i and p_i $(i = 1, \ldots, n)$ are positive real numbers and $P_n = \sum_{i=1}^{n} p_i$.
Further we prove a Ky Fan-type inequality involving the ratios a_n/a_n' and g_n/g_n' as well as an additive analogue involving $a_n - a_n'$ and $g_n - g_n'$, where a_n' and g_n' will be obtained from a_n and g_n by replacing x_i by $1 - x_i$ with $x_i \in (0, 1/2]$ $(i = 1, \ldots, n)$.

1. Introduction

The classical inequality between the weighted arithmetic and geometric means

$$G_n = G_n(\underline{y}, \underline{q}) = \prod_{i=1}^{n} y_i^{q_i/Q_n} \leq \frac{1}{Q_n}\sum_{i=1}^{n} q_i y_i = A_n(\underline{y}, \underline{q}) = A_n \tag{1.1}$$

is valid for all positive real numbers y_i and q_i $(i = 1, \ldots, n)$ with $Q_n = \sum_{i=1}^{n} q_i$.

This paper is in final form and no version of it will be submitted for publication elsewhere.

Equality holds in (1.1) if and only if $y_1 = \cdots = y_n$. For this inequality which is "probably the most important inequality and certainly a keystone in the theory of inequalities" [4, p.3] many proofs as well as noteworthy extensions, refinements and variants are known; see [4], [7], [8], [10].

A remarkable counterpart of (1.1) is the celebrated Ky Fan inequality which states:

If A'_n and G'_n designate the weighted arithmetic and geometric means of $1 - y_i$ with $y_i \in (0, 1/2]$ $(i = 1, \ldots, n)$, then

$$\frac{G_n}{G'_n} \leq \frac{A_n}{A'_n} \tag{1.2}$$

with equality holding if and only if $y_1 = \cdots = y_n$.

Inequality (1.2) (with $q_1 = \cdots = q_n$) was published for the first time in 1961 in Beckenbach-Bellman's book *Inequalities* [4, p. 5] and since then different proofs, generalizations, sharpenings and variants of Fan's theorem were published; see [3] and the references therein. We mention two results: A remarkable extension of (1.2) was published in 1980 by C.-L. Wang [11] who proved the following Radó-type inequality:

If $y_i \in (0, 1/2]$ $(i = 1, \ldots, n)$, then

$$Q_{n-1}[A_{n-1}G'_{n-1} - A'_{n-1}G_{n-1}] \leq Q_n[A_nG'_n - A'_nG_n]. \tag{1.3}$$

Recently, an additive analogue of (1.2) has been established for the case of equal weights $q_1 = \cdots = q_n$:

If $y_i \in (0, 1/2]$ $(i = 1, \ldots, n)$, then

$$G_n - G'_n \leq A_n - A'_n. \tag{1.4}$$

Equality holds in (1.4) if and only if $y_1 = \cdots = y_n$; see [2].

In this paper we denote by a_n, a'_n, g_n and g'_n the following expressions which are closely connected to A_n, A'_n, G_n and G'_n: For positive real numbers x_i and p_i $(i = 1, \ldots, n)$ with $P_n = \sum_{i=1}^{n} p_i$ we define

$$a_n = a_n(\underline{x}, \underline{p}) = \frac{P_n}{p_1} x_1 - \frac{1}{p_1} \sum_{i=2}^{n} p_i x_i$$

and

$$g_n = g_n(\underline{x}, \underline{p}) = \frac{x_1^{P_n/p_1}}{\prod_{i=2}^{n} x_i^{p_i/p_1}},$$

respectively

$$a'_n = a'_n(\underline{x}, \underline{p}) = \frac{P_n}{p_1}(1 - x_1) - \frac{1}{p_1}\sum_{i=2}^{n} p_i(1 - x_i)$$

and

$$g'_n = g'_n(\underline{x}, \underline{p}) = \frac{(1 - x_1)^{P_n/p_1}}{\prod_{i=2}^{n}(1 - x_i)^{p_i/p_1}}$$

(where the last expression is defined for $x_i < 1$, $i = 1, \ldots, n$).

Although there is no general agreement in literature what constitutes a mean value, most authors consider the intermediate property as the main feature. Since a_n and g_n do not satisfy this condition, this means the double-inequalities

$$\min_{1 \le i \le n} x_i \le a_n \le \max_{1 \le i \le n} x_i \quad \text{and} \quad \min_{1 \le i \le n} x_i \le g_n \le \max_{1 \le i \le n} x_i$$

are not true for all positive x_i's, we call a_n and g_n pseudo arithmetic and geometric means.

In 1986 S. Iwamoto, R.J. Tomkins and C.-L. Wang [3] published for the special case $p_1 = \cdots = p_n$ the following companion of inequality (1.1):

$$a_n \le g_n \tag{1.5}$$

with equality holding if and only if $x_1 = \cdots = x_n$.

The aim of this paper is two-fold: First we shall prove some refinements of inequality (1.5). Thereafter we show that under certain restrictions concerning the x_i's there exist counterparts of (1.2), (1.3) and (1.4) for pseudo arithmetic and geometric means.

We note that inequality (1.5) is an example of a so-called reverse inequality. As far as we could trace it one of the first reverse inequalities was published in 1956 by J. Aczél [1] who proved the following intriguing variant of the Cauchy-Schwarz inequality:

If x_i and y_i $(i = 1, \ldots, n)$ are real numbers with $x_1^2 > \sum_{i=2}^{n} x_i^2$ and $y_1^2 > \sum_{i=2}^{n} y_i^2$, then

$$\left(x_1 y_1 - \sum_{i=2}^{n} x_i y_i\right)^2 \ge \left(x_1^2 - \sum_{i=2}^{n} x_i^2\right)\left(y_1^2 - \sum_{i=2}^{n} y_i^2\right). \tag{1.6}$$

Equality holds in (1.6) if and only if $x_i = cy_i$ $(i = 1, \ldots, n)$, where c is a constant.

It is worth mentioning that Aczél's inequality plays a role in the proof of the famous van der Waerden conjecture; see [12]. Further interesting reverse inequalities were given in [5], [9], [10], [12].

2. Inequalities involving a_n and g_n

"One idea in the theory of inequalities is that every inequality is a consequence of an equality" [6]. Following this statement of R. Bellman we first provide a simple identity from which inequality (1.5) can be deduced immediately.

Theorem 1. *We have*

$$g_n - a_n = \frac{x_1}{p_1} \sum_{i=1}^{n} p_i I\left(\frac{y_i}{x_1}\right), \tag{2.1}$$

where

$$I(a) = (a-1)^2 \int_0^1 \frac{t\,dt}{(a-1)t+1}, \quad y_1 = g_n \quad \textit{and} \quad y_i = x_i \quad (i = 2, \ldots, n).$$

Proof. Because of

$$I(a) = a - \log a - 1$$

we obtain

$$I\left(\frac{y_i}{x_1}\right) = \frac{y_i}{x_1} - \log \frac{y_i}{x_1} - 1 \quad (i = 1, \ldots, n).$$

Multiplication by p_i and summing from $i = 1$ to $i = n$ yields (2.1) after simple manipulations. □

Remarks. 1) Since $I(a) = 0$ if and only if $a = 1$ we conclude from (2.1) that $g_n = a_n$ is valid only if $x_1 = \cdots = x_n$.

2) An application of (1.1) leads to another simple proof of inequality (1.5). Indeed, setting $y_1 = g_n$, $y_i = x_i$ $(i = 2, \ldots, n)$ and $q_i = p_i/P_n$ $(i = 1, \ldots, n)$ in (1.1) we get (1.5).

Two well-known extensions of the arithmetic mean-geometric mean inequality are the following inequalities of Radó and Popoviciu:

$$Q_n(A_n - G_n) \geq Q_{n-1}(A_{n-1} - G_{n-1}) \tag{2.2}$$

and

$$\left(\frac{A_n}{G_n}\right)^{Q_n} \geq \left(\frac{A_{n-1}}{G_{n-1}}\right)^{Q_{n-1}}. \tag{2.3}$$

Equality holds in (2.2) if and only if $y_n = G_{n-1}$ and in (2.3) if and only if $y_n = A_{n-1}$; see [7], [10].

The next propositions provide analogues of (2.2) and (2.3) for pseudo arithmetic and geometric means.

Theorem 2. *For all positive real numbers x_i $(i = 1, \ldots, n;\ n \geq 2)$ we have*

$$g_n - a_n \geq g_{n-1} - a_{n-1} \tag{2.4}$$

with equality holding if and only if $x_1 = x_n$.

Proof. If we put in (2.2): $y_1 = g_n$, $y_i = x_i$ $(i = 2, \ldots, n)$ and $q_i = p_i$ $(i = 1, \ldots, n)$, then we obtain

$$Q_n(A_n - G_n) = p_1(g_n - a_n),$$

which leads to inequality (2.4) with equality holding if and only if $y_n = G_{n-1}$ which is equivalent to $x_1 = x_n$. □

Remarks. 1) Inequality (2.4) is sharper than the corresponding Radó-type inequality

$$P_n(g_n - a_n) \geq P_{n-1}(g_{n-1} - a_{n-1}).$$

2) Repeated application of (2.4) leads to a chain of inequalities interpolating (1.5) written as $g_n - a_n \geq 0$:

$$g_n - a_n \geq g_{n-1} - a_{n-1} \geq \cdots \geq g_2 - a_2 \geq g_1 - a_1 = 0.$$

The special case $p_1 = \cdots = p_n$ yields a particularly simple positive lower bound for $g_n - a_n$:

$$g_n - a_n \geq \max_{2 \leq i \leq n} \frac{(x_1 - x_i)^2}{x_i}.$$

Theorem 3. *Let x_i $(i = 1, \ldots, n;\ n \geq 2)$ be positive real numbers such that $a_n > 0$ and $a_{n-1} > 0$. Then we have*

$$\frac{g_n}{a_n} \geq \frac{g_{n-1}}{a_{n-1}} \tag{2.5}$$

with equality holding if and only if $x_1 = x_n$.

Proof. We set in (2.3): $y_1 = a_n$, $y_i = x_i$ $(i = 2, \ldots, n)$ and $q_i = p_i$ $(i = 1, \ldots, n)$, then we have

$$\left(\frac{A_n}{G_n}\right)^{Q_n} = \left(\frac{g_n}{a_n}\right)^{p_1}$$

which leads to (2.5). Equality is valid in (2.5) if and only if $y_n = A_{n-1}$ which is equivalent to $x_1 = x_n$. □

Remarks. 1) Inequality (2.6) refines the Popoviciu-type inequality

$$\left(\frac{g_n}{a_n}\right)^{P_n} \geq \left(\frac{g_{n-1}}{a_{n-1}}\right)^{P_{n-1}}.$$

2) If we assume $x_1 = \max_{1 \leq i \leq n} x_i$, then we have $a_k > 0$ for $k = 1, \ldots, n$, and we obtain

$$\frac{g_n}{a_n} \geq \frac{g_{n-1}}{a_{n-1}} \geq \cdots \geq \frac{g_2}{a_2} \geq \frac{g_1}{a_1} = 1$$

which leads in the case of equal weights to

$$\frac{g_n}{a_n} \geq \max_{2 \leq i \leq n} \left(1 - \left(\frac{x_i}{x_1} - 1\right)^2\right)^{-1}.$$

3. A Ky Fan-type inequality and related results

In this section we present an inequality involving the ratios a_n/a'_n and g_n/g'_n where we assume that $x_i \in (0, 1/2]$ $(i = 1, \ldots, n)$. Since for every $n \geq 2$ there exist positive numbers x_i and p_i $(i = 1, \ldots, n)$ such that $a'_n = 0$ we need an additional assumption which guarantees that a_n/a'_n is defined. Such an assumption is given by $x_1 = \min_{1 \leq i \leq n} x_i$. (The condition $x_1 = \max_{1 \leq i \leq n} x_i$ does not exclude the case $a'_n = 0$.)

Indeed, if $x_1 = \min_{1 \leq i \leq n} x_i$, then we have $1 - x_1 \geq \frac{1}{P_n} \sum_{i=1}^{n} p_i(1 - x_i)$ which implies

$$a'_n = \frac{P_n}{p_1}(1 - x_1) - \frac{1}{p_1} \sum_{i=2}^{n} p_i(1 - x_i) \geq 1 - x_1 \geq \frac{1}{2}. \tag{3.1}$$

We note that, contrary to the positive term g_n/g'_n, the ratio a_n/a'_n attains (under the assumption $x_1 = \min_{1 \leq i \leq n} x_i$) negative as well as non-negative values.

First we prove a counterpart of Wang's inequality (1.3).

Theorem 4. *If* $x_1 = \min_{1\le i\le n} x_i$*, then we have for all* $x_i \in (0, 1/2]$ $(i = 1, \ldots, n;$ $n \ge 2)$*:*

$$a_n g_n' - a_n' g_n \le a_{n-1} g_{n-1}' - a_{n-1}' g_{n-1}. \tag{3.2}$$

Equality holds in (3.2) *if and only if* $x_1 = x_n$.

Proof. We denote by f the function

$$f : [x_1, 1/2] \to \mathbb{R},$$
$$f(x_n) = a_n g_n' - a_n' g_n.$$

Differentiation with respect to x_n yields

$$\begin{aligned} \frac{p_1}{p_n} f'(x_n) &= \left[-a_{n-1}' + \frac{p_n}{p_1} x_1 + \left(1 - \frac{p_n}{p_1}\right) x_n\right] \cdot \\ &\quad \cdot \left[g_{n-1}' \frac{(1-x_1)^{p_n/p_1}}{(1-x_n)^{1+p_n/p_1}} - g_{n-1} \frac{x_1^{p_n/p_1}}{x_n^{1+p_n/p_1}}\right]. \end{aligned} \tag{3.3}$$

Next we show that the two factors on the right-hand side of (3.3) have opposite signs. From (3.1) we conclude

$$-a_{n-1}' + \frac{p_n}{p_1} x_1 + \left(1 - \frac{p_n}{p_1}\right) x_n \le -\frac{1}{2} + \frac{p_n}{p_1} x_1 + \left(1 - \frac{p_n}{p_1}\right) x_n \le -\frac{1}{2} + x_n \le 0$$

and

$$-a_{n-1}' + \frac{p_n}{p_1} x_1 + \left(1 - \frac{p_n}{p_1}\right) x_n = 0 \iff x_1 = \cdots = x_n = \frac{1}{2}.$$

Further, since $x_1 = \min_{1\le i\le n} x_i$ we have

$$\left(\frac{x_n}{1-x_n}\right)^{p_1/P_n} \prod_{i=2}^{n} \left(\frac{x_i}{1-x_i}\right)^{p_i/P_n} \ge \frac{x_1}{1-x_1}$$

which is equivalent to

$$g_{n-1}' \frac{(1-x_1)^{p_n/p_1}}{(1-x_n)^{1+p_n/p_1}} - g_{n-1} \frac{x_1^{p_n/p_1}}{x_n^{1+p_n/p_1}} \ge 0$$

with equality holding if and only if $x_1 = \cdots = x_n$. Thus, we obtain

$$f'(x_n) \le 0$$

which implies

$$f(x_n) \le f(x_1) = a_{n-1}g'_{n-1} - a'_{n-1}g_{n-1}. \tag{3.4}$$

If $x_1 \neq x_n$, then both terms on the right-hand side of (3.3) are non-zero. Therefore we have

$$f'(x_n) < 0$$

which implies that inequality (3.4) is strict. Hence the sign of equality holds in (3.2) if and only if $x_1 = x_n$. □

Repeated application of (3.2) provides a striking companion of Ky Fan's inequality (1.2) for pseudo arithmetic and geometric means.

Theorem 5. *If* $x_1 = \min_{1\le i\le n} x_i$, *then we have for all* $x_i \in (0, 1/2]$ $(i = 1, \dots, n)$:

$$\frac{a_n}{a'_n} \le \frac{g_n}{g'_n} \tag{3.5}$$

with equality holding if and only if $x_1 = \cdots = x_n$.

It is well known that Ky Fan's theorem can be established by applying Jensen's inequality. We show that an application of Jensen's inequality yields also a short and simple *proof* of (3.5):

Because of $x_1 = \min_{1\le i\le n} x_i$ and $a_n + a'_n = 1$ we obtain from (3.1): $a_n \le 1/2$. If $a_n \le 0$, then inequality (3.5) with "$<$" instead of "$\le$" is obviously true. Next we suppose $0 < a_n \le 1/2$. Then (3.5) follows from

$$f\left(\frac{1}{P_n}\sum_{i=1}^{n} p_i y_i\right) \le \frac{1}{P_n}\sum_{i=1}^{n} p_i f(y_i), \tag{3.6}$$

if we put $y_1 = a_n$, $y_i = x_i$ $(i = 2, \dots, n)$ and $f(x) = \log \frac{1-x}{x}$. Since f is strictly convex on $(0, 1/2]$, equality holds in (3.6) if and only if all the y_i's are equal which is equivalent to $x_1 = \cdots = x_n$.

Remark. If we assume that $a'_n > 0$, then (3.5) is in general not true, as the following example reveals: Setting $x_1 = 1/2$, $x_2 = \cdots = x_n = 1/4$, $p_1 = 1$, $p_2 = \cdots = p_n = \frac{3}{2(n-1)}$, we obtain $a'_n = 1/8$ and $a_n/a'_n = 7 > \sqrt{27} = g_n/g'_n$.

It is natural to ask whether there exist a sharp lower bound for a_n/a'_n and a sharp upper bound for g_n/g'_n which both do not depend on the variables x_i $(i = 1, \dots, n)$.

Theorem 6. *If $x_1 = \min_{1 \le i \le n} x_i$, then we have for all $x_i \in (0, 1/2]$ $(i = 1, \dots, n;$ $n \ge 2)$:*

$$\frac{p_1 - P_n}{p_1 + P_n} < \frac{a_n}{a'_n} \quad \textit{and} \quad \frac{g_n}{g'_n} \le 1. \tag{3.7}$$

Both bounds are best possible. Equality holds in the second inequality if and only if $x_1 = \dots = x_n = 1/2$.

Proof. We have

$$p_1 a_n = P_n x_1 - \sum_{i=2}^{n} p_i x_i > -\frac{1}{2} \sum_{i=2}^{n} p_i$$

which implies

$$a_n \left(2p_1 + \sum_{i=2}^{n} p_i \right) > -(1 - a_n) \sum_{i=2}^{n} p_i = -a'_n \sum_{i=2}^{n} p_i,$$

and since $a'_n > 0$ we obtain the first inequality of (3.7). Setting $x_2 = \dots = x_n = 1/2$ and letting x_1 tend to 0 we conclude that the lower bound for a_n/a'_n is best possible.

Because of $x_1 = \min_{1 \le i \le n} x_i$ and $0 < x_1 \le 1/2$ we obtain

$$\prod_{i=2}^{n} \left(\frac{x_i}{1 - x_i} \right)^{p_i} \ge \left(\frac{x_1}{1 - x_1} \right)^{P_n - p_1} \ge \left(\frac{x_1}{1 - x_1} \right)^{P_n} \tag{3.8}$$

which yields

$$\frac{g_n}{g'_n} \le 1.$$

If equality is valid in the last inequality, then equality holds in both inequalities of (3.8), too, which implies $x_1 = \dots = x_n = 1/2$. □

Finally we derive an inequality involving the differences $a_n - a'_n$ and $g_n - g'_n$. Simple examples reveal that in general these two expressions cannot be compared such that we need (as in Theorem 5) an additional assumption concerning the x_i's. Since $x_1 = \min_{1 \le i \le n} x_i$ is a sufficient condition for a comparison of the ratios a_n/a'_n and g_n/g'_n it is natural to ask whether this assumption is also suitable to compare the differences $a_n - a'_n$ and $g_n - g'_n$. However this conjecture is false! Indeed, putting $p_1 = 1$, $p_2 = \cdots = p_n = 2/(n-1)$, $x_2 = \cdots = x_n = 1/2$ and letting x_1 tend to 0 we obtain

$$a_n - a'_n = -3 > -4 = g_n - g'_n;$$

and setting $p_1 = \cdots = p_n = 1$, $x_1 = \cdots = x_{n-1} = 1/4$, $x_n = 1/3$ we get

$$a_n - a'_n = -\frac{2}{3} < -\frac{21}{32} = g_n - g'_n.$$

The following theorem presents a counterpart of inequality (1.4), respectively an additive analogue of (3.5) under the assumption $x_1 = \max_{1 \le i \le n} x_i$.

Theorem 7. *If $x_1 = \max_{1\le i\le n} x_i$, then we have for all $x_i \in (0, 1/2]$ $(i = 1, \ldots, n)$:*

$$a_n - a'_n \le g_n - g'_n \tag{3.9}$$

with equality holding if and only if $x_1 = \cdots = x_n$.

Proof. If $x_1 = \cdots = x_n$, then the sign of equality holds obviously in (3.9). Now we suppose that the numbers $x_1, \ldots, x_n$ $(n \ge 2)$ are not all equal. Without loss of generality we may assume

$$0 < x_n \le x_{n-1} \le \cdots \le x_1 \le \frac{1}{2}, \quad x_n < x_1.$$

We define

$$\varphi_q : (0, 1/2] \to \mathbb{R},$$
$$\varphi_q(x) = \varphi(x, \ldots, x, x_{q+1}, \ldots, x_n) \quad (1 \le q \le n-1),$$

where

$$\varphi(x_1, \ldots, x_n) = g_n - g'_n - a_n + a'_n.$$

First we prove that φ_q is increasing on $[x_{q+1}, 1/2]$. Differentiation yields

$$\frac{p_1}{2(P_n - P_q + p_1)} \varphi'_q(x) = \frac{1}{2}\left[\left(\frac{x^{P_n - P_q}}{\prod_{i=q+1}^{n} x_i^{p_i}}\right)^{1/p_1} + \left(\frac{(1-x)^{P_n - P_q}}{\prod_{i=q+1}^{n} (1-x_i)^{p_i}}\right)^{1/p_1}\right] - 1,$$

and from the arithmetic mean-geometric mean inequality we get

$$\frac{p_1}{2(P_n - P_q + p_1)} \varphi'_q(x) = \left(\frac{[x(1-x)]^{P_n - P_q}}{\prod_{i=q+1}^{n} [x_i(1-x_i)]^{p_i}}\right)^{1/(2p_1)} - 1. \tag{3.10}$$

Because of

$$0 < x_n \le x_{n-1} \le \cdots \le x_{q+1} \le x \le \frac{1}{2}$$

we obtain

$$\prod_{i=q+1}^{n} [x_i(1-x_i)]^{p_i} \le [x(1-x)]^{P_n - P_q}, \tag{3.11}$$

and from (3.10) and (3.11) we conclude

$$\varphi_q'(x) \ge 0 \quad \text{for} \quad x \in [x_{q+1}, 1/2]$$

with equality holding if and only if $x = x_{q+1} = \cdots = x_n$. From the monotonicity of φ_q and the identity

$$\varphi_{r-1}(x_r) = \varphi_r(x_r) \quad (2 \le r \le n-1)$$

we obtain

$$\varphi(x_1, \ldots, x_n) =$$
$$\varphi_1(x_1) \ge \varphi_1(x_2) = \varphi_2(x_2) \ge \cdots \ge \varphi_{n-2}(x_{n-1}) = \varphi_{n-1}(x_{n-1}) \ge \varphi_{n-1}(x_n) = 0.$$

Since $x_1 > x_n$ we conclude that at least one of these inequalities is strict. Hence we have $\varphi(x_1, \ldots, x_n) > 0$. □

Remark. As simple examples show both expressions $a_n - a_n'$ and $g_n - g_n'$ attain negative as well as non-negative values. Obviously there does not exist an upper bound for $g_n - g_n'$ which is independent of the x_i's. But it is easy to prove that (under the assumption $x_1 = \max\limits_{1 \le i \le n} x_i$) the difference $a_n - a_n'$ is greater than -1. Indeed, from

$$x_1 \ge \frac{1}{P_n} \sum_{i=1}^{n} p_i x_i$$

we conclude

$$a_n = \frac{P_n}{p_1} x_1 - \frac{1}{p_1} \sum_{i=2}^{n} p_i x_i \ge x_1 > 0,$$

which implies

$$a_n - a_n' = 2a_n - 1 > -1.$$

If we let all the x_i's tend to 0, then $a_n - a_n'$ tends to -1; hence -1 is the best possible lower bound for $a_n - a_n'$.

References

1. J. Aczél, *Some general methods in the theory of functional equations in one variable. New applications of functional equations* (Russian). Uspehi Mat. Nauk (N.S.) **11**, No. 3 (69) (1956), 3–58.

2. H. Alzer, *Ungleichungen für geometrische und arithmetische Mittelwerte.* Proc. Kon. Nederl. Akad. Wetensch. **91** (1988), 365–374.

3. H. Alzer, T. Ando and Y. Nakamura, *The inequalities of W. Sierpinski and Ky Fan.* J. Math. Anal. Appl. **149** (1990), 497–512.

4. E.F. Beckenbach and R. Bellman, *Inequalities.* Springer Verlag, Berlin, 1983.

5. R. Bellman, *On an inequality concerning an indefinite form.* Amer. Math. Monthly **63** (1956), 108–109.

6. R. Bellman, *A note on inequalities.* In: General Inequalities I (E.F. Beckenbach, ed.) pp. 3–4, Birkhäuser Verlag, Basel, 1978.

7. P.S. Bullen, D.S. Mitrinović and P.M. Vasić, *Means and Their Inequalities.* Reidel Publ. Co., Dordrecht, 1988.

8. G.H. Hardy, J.E. Littlewood and G. Pólya, *Inequalities.* Cambridge Univ. Press, Cambridge, 1952.

9. S. Iwamoto, R.J. Tomkins and C.-L. Wang, *Some theorems on reverse inequalities.* J. Math. Anal. Appl. **119** (1986), 282–299.

10. D.S. Mitrinović, *Analytic Inequalities.* Springer Verlag, New York, 1970.

11. C.-L. Wang, *On a Ky Fan inequality of the complementary A-G type and its variants.* J. Math. Anal. Appl. **73** (1980), 501–505.

12. C.-L. Wang, *Functional equation approach to inequalities, VI.* J. Math. Anal. Appl. **104** (1984), 95–102.

Horst Alzer, Department of Mathematics, Applied Mathematics and Astronomy, University of South Africa, P.O. Box 392, 0001 Pretoria, South Africa

International Series of Numerical Mathematics, Vol. 103, © 1992 Birkhäuser Verlag Basel

A review of Shapiro's cyclic inequality

Achim Clausing

Abstract

We give a short review of the history of the inequality $\sum_{i=1}^{n} x_i/(x_{i+1}+x_{i+2}) \geq n/2$, conjectured by H.S. Shapiro in 1954, whose exact domain of validity was determined in 1989 by B.A. Troesch, concluding an eventful, sometimes seemingly hopeless search.

1. The problem

In 1954, H.S. Shapiro [3] proposed in the American Mathematical Monthly the following

Problem. *Let* $x_i \geq 0,\ x_i + x_{i+1} > 0\ (i = 1, \ldots, n),\ x_{n+i} = x_i$. *Establish*

$$S_n(x_1, \ldots, x_n) \equiv \sum_{i=1}^{n} \frac{x_i}{(x_{i+1} + x_{i+2})} \geq \frac{n}{2}, \tag{1}$$

equality occurring only if all denominators are equal.

To quote L.J. Mordell [6]: "The inequality attracted considerable attention both from its simplicity and from the difficulty of finding a proof. This is not surprising since the inequality is not true for all n." A counterexample to (1) for $n = 20$ was given by M.J. Lighthill [4], who thus opened the search for the numbers n satisfying Shapiro's inequality.

It added to the interest in the inequality when, more than a decade later, it was proved by V.G. Drinfel'd [30] that inequality (1) missed the truth by only a hairsbreadth. He showed that the following inequality holds and is sharp:

$$\sum_{i=1}^{n} \frac{x_i}{(x_{i+1} + x_{i+2})} \geq 0.989133\ldots \cdot \frac{n}{2}. \tag{2}$$

The search for the domain of validity of Shapiro's inequality (1) itself lasted almost another twenty years. It came to an end recently when B.A. Troesch [51], by resolving the case $n = 23$, completed the proof of the following

This paper is in final form and no version of it will be submitted for publication elsewhere.

Theorem 1. *Inequality* (1) *holds for even* $n \leq 12$ *and for odd* $n \leq 23$. *For all other values of* n, *it is false.*

The 1990 Oberwolfach conference on *General Inequalities* was an appropriate opportunity to recall the long history of Theorem 1. This article is a somewhat enlarged version of the review given at the conference. For the earlier developments, one should also consider the chapter on cyclic inequalities in the book [29] by D.S. Mitrinović.

2. The simple cases

The case $n = 3$ of Shapiro's inequality seems to have been known for a long time; Mitrinović [29] cites a reference from 1903 ([1]).

A proof in that case is quite simple: Let $y_1 = x_2 + x_3$, $y_2 = x_3 + x_1$, $y_3 = x_1 + x_2$. By the inequality between the arithmetic and harmonic means,

$$(y_1 + y_2 + y_3)\left(\frac{1}{y_1} + \frac{1}{y_2} + \frac{1}{y_3}\right) \geq 9$$

which gives

$$\frac{y_2 + y_3}{y_1} + \frac{y_3 + y_1}{y_2} + \frac{y_1 + y_2}{y_3} \geq 6$$

and hence

$$\frac{2x_1}{x_2 + x_3} + \frac{2x_2}{x_3 + x_1} + \frac{2x_3}{x_1 + x_2} \geq 3.$$

Note that the second one of these inequalities can also be obtained from the arithmetic-geometric mean inequality.

It might be interesting to observe that for $n = 3$, an equivalent form of (1) is

$$(x_1 + x_2 + x_3)(x_1^2 + x_2^2 + x_3^2) \leq 3(x_1^3 + x_2^3 + x_3^3). \tag{3}$$

Inequality (3) can be proved by using the monotonicity (in t) of the power means $M_t = \left(\frac{1}{n}\sum_{i=1}^n x_i^t\right)^{1/t}$.

For $n > 3$, one does not find a proof as easy as the above one.

L.J. Mordell [6] gave a simple but ingenious verification of (1) for the cases $n \leq 6$ which is worth reproducing:

Let $y_i = x_{i+1} + x_{i+2}$, indices taken cyclically. Since

$$\left(\sum_{i=1}^n \frac{x_i}{y_i}\right)\left(\sum_{i=1}^n x_i y_i\right) = \left(\sum_{i=1}^n x_i\right)^2 + \sum_{i,j=1}^n x_i x_j \frac{(y_i - y_j)^2}{y_i y_j},$$

the inequality

$$\left(\sum_{i=1}^{n} x_i\right)^2 \geq \frac{n}{2}\sum_{i=1}^{n} x_i y_i \tag{4}$$

clearly would imply (1).

If $n = 4$, then (4) becomes

$$(x_1 + x_2 + x_3 + x_4)^2 \geq 2(x_1(x_2 + x_3) + x_2(x_3 + x_4) + x_3(x_4 + x_1) + x_4(x_1 + x_2)).$$

A simple calculation shows that this is just

$$(x_1 - x_3)^2 + (x_2 - x_4)^2 \geq 0.$$

Hence (4) is true for $n = 4$.

The case $n = 5$ is based on

$$(n-1)\left(\sum_{i=1}^{n} x_i\right)^2 \geq 2n\sum_{i<j} x_i x_j \tag{5}$$

which is true for all n and all real x_i. For $n = 5$, it happens that

$$\sum_{i<j} x_i x_j = \sum_{i=1}^{n} x_i y_i$$

and that $2n/(n-1) = n/2$ so that (5) is equivalent to (4).

Finally, $n = 6$ of (4) rests on the inequality

$$\sum_{i=1}^{m} x_i^2 \geq \frac{1}{m}\left(\sum_{i=1}^{m} x_i\right)^2. \tag{6}$$

Let $s_n = \sum_{i=1}^{n} x_i$. One notes that

$$\begin{aligned} &3(x_1(x_2 + x_3) + \cdots + x_6(x_1 + x_2)) \\ &\quad = 3\left(\sum_{i<j} x_i x_j - x_1 x_4 - x_2 x_5 - x_3 x_6\right) \\ &\quad = \frac{3}{2}\left(s_6^2 - \sum_{i=1}^{n} x_i^2 - 2x_1 x_4 - 2x_2 x_5 - 2x_3 x_4\right) \\ &\quad = \frac{3}{2}(s_6^2 - ((x_1 + x_4)^2 + (x_2 + x_5)^2 + (x_3 + x_6)^2). \end{aligned}$$

Hence (4) is equivalent to

$$(x_1 + x_4)^2 + (x_2 + x_5)^2 + (x_3 + x_6)^2 \geq \frac{1}{3}((x_1 + x_4) + (x_2 + x_5) + (x_3 + x_6))^2,$$

and this is implied by (6), $m = 3$.

Anyone who has tried to prove (1) by himself will appreciate the elegance of these arguments of Mordell.

Unfortunately, they cannot be carried further since for larger n inequality (4) does not hold. Mordell has proved that for $n \geq 7$

$$\left(\sum_{i=1}^{n} x_i\right)^2 \geq 3 \sum_{i=1}^{n} x_i y_i,$$

and that the constant 3 here cannot be replaced by a larger one. Mordell therefore conjectured that Shapiro's inequality should be false for all $n \geq 7$.

In 1963, however, D.Ž. Djoković [24] succeeded in proving (1) for $n = 8$. His proof is elementary in the sense that it uses no analytic methods, but as for the complexity and sheer length of the calculations it is in stark contrast to Mordell's proofs. After that, it was clear that a rather different approach was needed for larger n. This new approach came in 1968 with P. Nowosad's paper [28].

Before discussing this important paper, we first go back to 1956.

3. Counterexamples

The first counterexample to Shapiro's inequality was given by Lighthill [4] for $n = 20$. Mordell's paper [6] was accompanied by a note by A. Zulauf [7], where a counterexample for $n = 14$ was presented. (Apparently, Lighthill had found a similar example. Cf. the footnotes in [8] and [6].)

Zulauf also observed that the following functional relation holds:

$$S_{n+2}(x_1, \ldots, x_n, x_{n-1}, x_n) = S_n(x_1, \ldots, x_n) + 1. \tag{7}$$

Hence, if there is a counterexample to (1) for some n, then there are others for $n + 2, n + 4, \ldots$

In another note ([9]), Zulauf gave a counterexample for the odd case $n = 53$. Therefore, (1) is false for all but finitely many numbers n.

P.H. Diananda [25] noted another useful functional relation between cyclic sums for different n:

$$S_{n+2}(x_1, \ldots, x_n, x_n) = S_n(x_1, \ldots, x_n) + \frac{1}{2} - \frac{(x_n - x_{n+1})(x_n - x_1)}{2x_n(x_n + x_1)}. \tag{8}$$

Thus, if for some $(x_1,\ldots,x_n)$ one has $S_n(x_1,\ldots,x_n) < n/2$ and there is some r, $1 \le r \le n$, such that x_{r-1}, x_r, x_{r+1} (indices cyclically) is monotonic (we may assume that $r = n$), then we obtain that $S_{n+1}(x_1,\ldots,x_n,x_n) < (n+1)/2$ holds too. For odd n, it is clear that there is always such an r since a cyclic sequence of odd length cannot switch between being increasing and decreasing at every position. Using this fact and the functional relations (7) and (8) one readily obtains

Theorem 2 (Diananda [25]). *If* (1) *is false for some odd* $n = n_0$, *then it is false for all* $n \ge n_0$. *If* (1) *is true for some even* $n = n_0$, *then it is true for all* $n \le n_0$.

In an addendum to [25], Diananda gave a counterexample to (1) for $n = 27$. Thus, at this time eleven cases were left undecided, namely $n = 10, 12$ for even n and $n = 9, 11, \ldots, 25$ for odd n.

The same argument as used for Diananda's theorem also implies that equality in (1) can, for even n, hold at all points $x^\alpha \equiv \frac{1}{2}(1+\alpha, 1-\alpha, \ldots, 1+\alpha, 1-\alpha)$ $(0 \le \alpha \le 1)$, whereas for odd n it holds only at $x^0 = (1,\ldots,1)$.

The counterexamples for even n given by Lighthill [4], Zulauf [7], Hershorn and Peck [12], are all perturbations of these points of equality: Let, for even n,

$$x_i = x_i^\alpha + a_i \quad (i = 1,\ldots,n). \tag{9}$$

It is easy to check that

$$S_n(x_1,\ldots,x_n) = \frac{n}{2} + Q(a_1,\ldots,a_n) + O(\varepsilon^3),$$

where

$$Q(a_1,\ldots,a_n) = \sum_{i=1}^n a_i^2 + \alpha \sum_{i=1}^n (-1)^i a_i a_{i+1} - \sum_{i=1}^n a_i a_{i+2} \tag{10}$$

and $C \cdot |a_i| < \varepsilon$ for some $C > 0$. It is clearly a necessary condition for (1) to hold that the quadratic form Q be positive. For $n = 8, 10, 12$ this was proved to be the case by D.G.S. Thomas [15].

The counterexample of Zulauf for $n = 14$ was $\alpha = 1$, $(a_1,\ldots,a_{14}) = (7,7,4,6,1,5,0,2,1,0,4,1,6,4)\cdot\varepsilon$ with $Q(a_1,\ldots,a_{14} = (250 + (-36) - 216)\cdot\varepsilon^2 = -2\varepsilon^2 < 0$.

A better understanding of such counterexamples can be obtained by minimization of the form Q. This was done by J.L. Searcy and B.A. Troesch [46]:

Write (10) as $Q(a_1,\ldots,a_n)=\frac{1}{2}a^tAa$, where A is the symmetric matrix

$$A=\begin{pmatrix} 2 & -\alpha & -1 & & \cdots & & & -1 & \alpha \\ -\alpha & 2 & \alpha & -1 & & & & & -1 \\ -1 & \alpha & 2 & -\alpha & -1 & & & & \\ & & & & \ddots & & & & \\ & & & & -1 & -\alpha & 2 & \alpha & -1 \\ -1 & & & & & -1 & \alpha & 2 & -\alpha \\ \alpha & -1 & & & \cdots & & -1 & -\alpha & 2 \end{pmatrix}.$$

To minimize $Q(a_1,\ldots,a_n)$ one has to find the eigenvectors of A. They are given by

$$a_k^{(j)}=\begin{cases}\delta\sin t_k^{(j)} & \text{for } k \text{ odd},\\ -\delta\cos t_k^{(j)} & \text{for } k \text{ even},\end{cases}$$

where $\delta>0$, $t_k^{(j)}=t_0+(k-1)/h^{(j)}$ ($t_0\in\mathbb{R}$), and $h^{(j)}=2\pi j/n$ ($j=1,\ldots,n$). The corresponding eigenvalues are

$$\lambda^{(j)}=2\sin h^{(j)}(2\sin h^{(j)}-\alpha).$$

We may choose $t_0=0$ and then have the eigenvectors

$$a^{(j)}=\delta(0,-\cos h^{(j)},\sin 2h^{(j)},-\cos 3h^{(j)},\ldots,\sin(n-2)h^{(j)},-\cos(n-1)h^{(j)}).$$

A negative eigenvalue of A now exsists if $0<2\sin h^{(j)}<\alpha$, that is, since $\alpha\le 1$ is arbitrary, if $0<\sin 2\pi j/n<1/2$ or $2\pi j/n<\pi/6$ or $12j<n$. This can be achieved for $n>12$. The corresponding eigenvector yields, by way of (9), the required counterexample.

Similarly, one can construct counterexamples for odd $n\ge 25$ (see [46]). The first counterexample for the critical case $n=25$ was obtained by D.E. Daykin [31] who had found it by a computer search. For other counterexamples with n odd, see also [41], [33], and [25].

4. Lower bounds

Since inequality (1) is not true in general, it is natural to ask how small S_n can be. It was shown by R.A. Rankin [8] that the numbers

$$\gamma_n\equiv\inf_{x_i>0}\frac{2}{n}S_n(x_1,\ldots,x_n)$$

converge to a limit $\gamma\equiv\lim_n\gamma_n=\inf_n\gamma_n>0$. (A similar fact holds for a large class of cyclic sums, cf. [11].)

Rankin stated $\gamma \geq 0.6094\ldots$ [8] and later proved $\gamma \geq 0.66\ldots$ [13]. Diananda [21], [23] improved this to, first, $\gamma \geq 0.9142\ldots$ and then to $\gamma \geq 0.9224\ldots$.

The exact constant (see inequality (2)) was derived by V.G. Drinfel'd in a remarkable paper [30].[1] His proof is short, elementary, and highly ingenious. It is based on the rearrangement inequality [2, p. 261]: If $a_1 \leq \cdots \leq a_n$, and $b_1 \geq \cdots \geq b_n$, then for any permutation $\{k_1, \ldots, k_n\}$ of the indices:

$$\sum_i a_i b_{k_i} \geq \sum_i a_i b_i. \tag{11}$$

Theorem 3 (Drinfel'd [30]). *Let $g(x)$ be the lower convex hull of the two convex, decreasing functions $f(x) = 2(e^x + e^{x/2})^{-1}$ and $h(x) = e^{-x}$. Then $\gamma_n \geq g(0)$ holds for all n.*

Proof. Since $f(0) = h(0) = 1$ and the slope of $f(x)$ at $x = 0$ is slightly less than that of $h(x)$, there are $x^- < 0$ and $x^+ > 0$ such that $g(x) = f(x)$ for $x \leq x^-$, $g(x) = h(x)$ for $x \geq x^+$, and g is linear in $x^- \leq x \leq x^+$. Clearly, $0 < g(0) < 1$. Let $x_i > 0$ and $q_i = x_{i+1}/x_i$ $(i < n)$, $q_n = x_1/x_n$. Then, $\prod_{i=1}^n q_i = 1$ and $x_i/(x_{i+1} + x_{i+2}) = 1/(q_i(1 + q_{i+1}))$. Denote the increasing rearrangement of $q_1, \ldots, q_n$ by $m_1, \ldots, m_n$. Then the sequences $\frac{1}{m_1}, \ldots, \frac{1}{m_n}$ and $\frac{1}{1+m_n}, \ldots, \frac{1}{1+m_1}$ are decreasing, resp. increasing, and therefore, by (11),

$$S_n(x_1, \ldots, x_n) = \sum_{i=1}^n \frac{1}{q_i(1 + q_{i+1})} \geq \sum_{i=1}^n \frac{1}{m_i(1 + m_{n+1-i})}. \tag{12}$$

Let $c_i = m_i m_{n+1-i}$. Then

$$t_i \equiv \frac{1}{m_i(m_{n+1-i})} + \frac{1}{m_{n+1-i}(1 + m_i)} = \frac{1}{c_i}\left(1 + \frac{c_i - 1}{1 + m_i)(1 + m_{n+1-i})}\right).$$

Since

$$(1 + m_i)(1 + m_{n+1-i}) \geq \left(1 + \sqrt{m_i m_{n+1-i}}\right)^2 = \left(1 + \sqrt{c_i}\right)^2, \tag{13}$$

we have

$$t_i \geq \begin{cases} \dfrac{1}{c_i} & \text{for } c_i \geq 1, \\[2ex] \dfrac{2}{c_i + \sqrt{c_i}} & \text{for } c_i < 1, \end{cases}$$

[1] Probably, this is the first published paper of the Fields medal winner of 1990.

as is easily verified. Summing these inequalities, we obtain

$$\begin{aligned} 2S_n(x_1, \ldots, x_n) &= \sum_{i=1}^{n} t_i \geq \sum_{c_i \geq 1} \frac{1}{c_i} + \frac{2}{c_i + \sqrt{c_i}} \\ &= \sum_{z_i \geq 0} h(z_i) + \sum_{z_i < 0} f(z_i) \geq \sum_{i=1}^{n} g(z_i), \end{aligned}$$

where $z_i = \ln c_i$ and g is any function bounding f and h from below. If g is convex, then

$$\frac{2}{n} S_n(x_1, \ldots, x_n) \geq \frac{1}{n} \sum_{i=1}^{n} g(z_i) \geq g\left(\frac{1}{n} \sum_{i=1}^{n} z_i\right) = g(0)$$

since $\sum_{i=1}^{n} z_i = \ln\left(\prod_{i=1}^{n} c_i\right) = \ln\left(\prod_{i=1}^{n} m_i^2\right) = \ln 1 = 0$. This proves $\gamma_n \geq g(0)$.

The actual value of $g(0)$ is quite close to 1: $g(0) = 0.989133\ldots$ (it is a non-algebraic number).

The converse of Theorem 3 is proved by making a careful choice of the ratios q_i, in such a way that approximate equality holds in (12) and (13). For details, we refer to [30].

5. The analytic approach

Mordell [6, p. 230] was the first to remark that finding the minimum of S_n by analytic methods would be a very difficult task since this implies discussing a large number of separate cases.

Let $D_n = \{(x_i, \ldots, x_n) : x_i \geq 0,\ x_{i+1} + x_{i+2} > 0,\ x_{i+n} \equiv x_i\}$ be the domain of definition of $S_n(x_1, \ldots, x_n)$. If the minimum is not attained at a stationary point of S_n, then one has to consider the boundaries of D_n (that is, those parts of D_n where some of the x_i vanish) separately for all choices of vanishing components. Cyclicity helps a little in that one can assume $x_n = 0$ on all boundaries, but still the number of boundaries to discuss increases exponentially with n.

In spite of these considerable difficulties it was this approach that eventually led to a solution of the problem. The main tool was provided by P. Nowosad [28] (who used it to prove (1) for $n = 10$):

Theorem 4 (Nowosad [28]). *The minimum of $S_n(x_1, \ldots, x_n)$ over D_n is either $n/2$ or it is attained at some regular boundary of D_n (a boundary of D_n is called regular if S_n has a continuous extension to it).*

The proof of this theorem needs some machinery from Banach algebras and uses classical eigenvalue theory for matrices. Nowosad also proved a uniqueness result:

Theorem 5 [28, Sec. 5]. *On every regular boundary of D_n there is, up to normalization, at most one point of relative minimum of S_n.*

It is easy to verify that the regular boundaries of D_n are those for which no two cyclically adjacent x_i vanish.

Take, for example, $n = 4$. Of the seven boundaries (we assume $x_4 = 0$), only two are regular: $\{x_2 = x_4 = 0\}$ and $\{x_4 = 0\}$. On these boundaries, S_n becomes $\frac{x_1}{x_3} + \frac{x_3}{x_1}$ and $\frac{x_1}{x_2+x_3} + \frac{x_2}{x_3} + \frac{x_3}{x_1}$. The first form is trivially ≥ 2, for the second one one has

$$\frac{x_1}{x_2+x_3}+\frac{x_2}{x_3}+\frac{x_3}{x_1} = \frac{x_1}{x_2+x_3}+\frac{x_2+x_3}{x_3}+\frac{x_3}{x_1} \geq 3\left(\frac{x_1}{x_2+x_3}\cdot\frac{x_2+x_3}{x_3}\cdot\frac{x_3}{x_1}\right)^{1/3} -1 = 2.$$

For $n = 5$, there are three regular boundaries: $\{x_2 = x_5 = 0\}$, $\{x_3 = x_5 = 0\}$, $\{x_5 = 0\}$. The first one of these is equivalent to the second one (add 3 to the indices), and there one finds

$$S_5(x_1, \ldots, x_5) = \frac{x_1}{x_2} + \frac{x_2}{x_4} + \frac{x_4}{x_1} \geq 3\left(\frac{x_1}{x_2}\cdot\frac{x_2}{x_4}\cdot\frac{x_4}{x_1}\right)^{1/3} = 3 > \frac{5}{2}.$$

In the third case we may assume $x_4 = 1$ and obtain

$$\begin{aligned} S_5(x_1, \ldots, x_5) &= \frac{x_1}{x_2+x_3} + \frac{x_2}{x_3+1} + x_3 + \frac{1}{x_1} \\ &\geq \frac{2}{(x_2+x_3)^{1/2}} + \frac{x_2}{x_3+1} + x_3 \geq 2.787079\ldots > \frac{5}{2} \end{aligned}$$

by a direct computation of the stationary point.

These few examples do not really demonstrate the usefulness of Nowosad's result which becomes much more apparent for larger n. There are 14 regular boundaries for $n = 10$ and 25 for $n = 12$, the largest even n for which the question was open. Nowosad succeeded in solving the former case, mainly by carefully discussing the linear transformation $y_i = x_{i+1} + x_{i+2}$ on the various boundaries.

E.K. Godunova and V.I. Levin [43] used essentially the same methods to prove (1) for $n = 12$, thus finally settling the problem for even n. On four of the 25 boundaries they used, instead of Nowosad's method, a direct evaluation of the ratios $r_i = y_i/y_{i+1}$. The amount of computation involved in Nowosad's and Godunova-Levin's proofs is rather heavy and much of it has become unnecessary by the development described in the next section.

6. The final solution

In 1976, the number of open cases was down to six: $n = 13, 15, \ldots, 23$. The solution for these cases is due to B.A. Troesch, who achieved it in two steps [50], [51]. These papers contain two major contributions: The first one contains results which, in addition to Theorem 4, further reduce the number of boundaries to be inspected for the minimum of S_n (cf. Theorem 6 below). In the second one, a variety of recursive equations and inequalities for the ratios $r_i = y_i/y_{i+1}$ was developed through which the search on a single boundary gets reduced to a two-dimensional problem.

The final solution for $n = 23$ then consisted in solving, by a computer search, 100 two-dimensional minimum problems. It was found that on all boundaries, $S_{23}(x_1, \ldots, x_{23}) \geq 11.512\ldots > 23/2$. Although not all of the details of the algorithm used were described in [51], and there is the theoretical possibility of a numerical error, one can now consider Theorem 1 as proved.

We can only give a sketch of Troesch's arguments here. Let a *segment* of length m in $(x_1, \ldots, x_n)$ be a sequence of (cyclically) maximal length of consecutive nonzero x_i's. The boundaries can be described by the lengths of their segments. For example, the regular boundaries for $n = 5$ mentioned above are (1,2), (2,1), and (4). One then has

Theorem 6 (Troesch [50]). *The minimum values of S_n on two boundaries having the same segment lengths in possibly different orders are equal.*

If all segments of some boundary have length less than 6, Then $S_n(x_1, \ldots, x_n) \geq n/2$ on this boundary.

Finally, the minimum of S_n is never attained on a boundary containing a segment of length 2.

This theorem truly saves work. For example, of the 25 boundaries for $n = 12$ considered in [43], 11 contain a segment of length 2 and a further 9 contain only segments of length < 6. This leaves just 5 boundaries (parts I, II, IV, V, and VII in the notations of [43]) to consider.

The proof of the third statement in Theorem 6 is easy: Assume that a relative minimum of S_n is attained at some boundary containing a 2-segment, say, $(1, 0, x_3, x_4, 0, x_6, x_7, \ldots, x_n)$. Then

$$S_n(1, 0, x_3, \ldots, x_n) = \frac{1}{x_3} + \frac{x_3}{x_4} + \frac{x_4}{x_6} + \frac{x_6}{x_7 + x_8} + \cdots$$

The conditions for a stable point include $\frac{\partial S_n}{\partial x_i} = 0$ $(i = 3, 4, 6)$. Hence $\frac{1}{x_3} = \frac{x_3}{x_4} = \frac{x_4}{x_6} \equiv \frac{1}{\rho} > 0$ at the minimum, that is $(1, 0, x_3, \ldots, x_n) = (1, 0, \rho, \rho^2, 0, \rho^3, x_7, \ldots, x_n)$.

Consider

$$s(\varepsilon) \equiv S_n(1, 0, \rho, \rho^2, \varepsilon, \rho^3, x_7, \ldots, x_n) = \frac{1}{\rho} + \frac{\rho}{\rho^2 + \varepsilon} + \frac{\rho^2}{\varepsilon + \rho^3} + \frac{\varepsilon}{\rho^3 + x_7} + \cdots$$

Since

$$s'(0) = -\frac{1}{\rho^3} - \frac{1}{\rho^4} + \frac{1}{\rho^3 + x_7} < -\frac{1}{\rho^4} < 0,$$

there is some $\varepsilon > 0$ such that $s(\varepsilon) < s(0)$. Hence the minimum on this boundary part is not an absolute minimum of S_n.

Let us now sketch Troesch's final reduction step. He considers segments of length m with the indexing $(x_1, 0, x_2, \ldots, x_{m+1}, 0, x_{m+2}, \ldots)$. Let the denominators of S_n again be y_i, that is, $y_1 = x_2$, $y_i = x_{i+1} + x_{i+2}$ $(i = 2, \ldots, m-1)$, $y_m = x_{m+1}$, $y_{m+1} = x_{m+2}$. Put $c_i = x_i/y_i$ for the terms of S_n and $r_i = y_i/y_{i+1}$ for the ratios of the denominators.

Now S_n is minimized as a function of the variables $y_1, \ldots, y_{m+1}$. To this purpose, the equations for the local minimum are used to establish the following relations:

(14) $$r_1 = r_2 = r_m = r_{m+1}$$

(15) $$r_3 = r_{m-1},\ r_4 = r_{m-2}, \ldots \qquad c_3 = c_m,\ c_4 = c_{m-1}, \ldots$$

(16) $$c_3 = r_3 - 1 \qquad r_{k+1} = c_{k+1} + \frac{c_k}{r_k} \quad (k = 3, \ldots, m).$$

The relation $r_1 = r_{m+1}$ is a crucial one; it tells that this 'pivotal ratio' of $(x_1, \ldots, x_n)$ is the same for all segments. Put $u \equiv r_1 = r_2$ and $v \equiv r_3$. From (16) one sees that the values of u and v determine all other ratios r_i for the segment, and hence the contribution $c_2 + \cdots + c_{m+1}$ of the segment to the sum $S_n(x_1, \ldots, x_n)$. Furthermore, the symmetry condition (15) implies a dependency between u and v. For example, if $m = 5$, on obtains $r_1 = r_2 = r_5 \equiv u$, $r_3 = r_4 \equiv v$, $c_3 = v - 1$, $c_4 = v(u - v + 1)$, $r_4 = c_4 + (v-1)/v$ and hence $u = v + v^{-2} + v^{-1}$.

Since the pivotal ratios for all segments are the same, the value of u for one segment determines the sum $S_n(x_1, \ldots, x_n)$ completely. This reduces finding the relative minimum of S_n on a single boundary to a one-dimensional search. Actually, for larger m, the dependency between v and u is also solved by a one-dimensional search (as the example $m = 5$ indicates, it is easier to find u in terms of v), and there are also other conditions reducing the range of possible values of u to $u < 2.2$.

The essence of Troesch's method, however, is contained in the equations (14)–(16) which provide the necessary reduction of the dimension of the search space.

Troesch reports that the minimum of S_{23} on the boundary of D_{23} is attained on the component having segment lengths (18,1,1) and that $S_{23} \geq 11.512\ldots > 23/2$ on this component. Thus (1) holds for $n = 23$ and hence for all $n \leq 23$.

7. Open problems

There remain many open questions related to cyclic sums similar to Shapiro's. I suggest the inequality $\sum_{i=1}^{n} x_i/(x_{i+2}+x_{i+3}) \geq [n/3]$ for further study. J.C. Boarder and D.E. Daykin [37] is a particularly rich source for open problems concerning cyclic sums.

References

1. A.M. Nesbitt, *Problem 15114.* Educational Times **3 (2)** (1903), 37–38.
2. G.H. Hardy, J.E. Littlewood and G. Pólya, *Inequalities.* Cambridge Univ. Press, pp. 324, 1936.
3. H.S. Shapiro, *Problem 403.* Amer. Math. Monthly **61** (1954), 571.
4. M.J. Lighthill, *An invalid inequality.* Amer. Math. Monthly **63** (1956), 191–192.
5. C.V. Durell, *Query.* Math. Gaz. **40** (1956), 266.
6. L.J. Mordell, *On the inequality $\sum_{r=1}^{n} x_r/(x_{r+1}+x_{r+2}) \geq n/2$ and some others.* Abh. Math. Sem. Hamburg **22** (1958), 229–240.
7. A. Zulauf, *A note on a conjecture of L.J. Mordell.* Abh. Math. Sem. Hamburg **22** (1958), 240–241.
8. R.A. Rankin, *An inequality.* Math. Gaz. **42** (1958), 39–40.
9. A. Zulauf, *A note on a conjecture of L.J. Mordell II.* Math. Gaz. **43** (1959), 42, 182–184.
10. P.H. Diananda, *Extensions of an inequality of H.S. Shapiro.* Amer. Math. Monthly **66** (1959), 489–491.

11. K. Goldberg, *The minima of cyclic sums.* J. London Math. Soc. **35** (1960), 262–264.

12. M. Hershorn and J.E.L. Peck, *An invalid inequality.* Amer. Math. Monthly **67** (1960), 87–88.

13. R.A. Rankin, *A cyclic inequality.* Proc. Edinburgh Math. Soc. **(2) 12** (1960/61), 139–147.

14. P.H. Diananda, *On a conjecture of L.J. Mordell regarding an inequality involving quadradic forms.* J. London Math. Soc. **36** (1961), 185–192.

15. D.G.S. Thomas, *On the definiteness of certain quadratic forms arising in a conjecture of L.J. Mordell.* Amer. Math. Monthly **68** (1961), 472–473.

16. L.J. Mordell, *Note on the inequality* $\sum_{r=1}^{n} x_r/(x_{r+1}+x_{r+2}) \geq n/2$. J. London Math. Soc. **37** (1962), 176–178.

17. B. Bajšanski, *A remark concerning the lower bound of* $x_1/(x_1 + x_3) + \cdots$. Univ. Beograd. Publ. El.-tehn. Fak., Ser. Mat. Fiz. **70** (1962), 19–20.

18. A. Zulauf, *Note on an inequality.* Math. Gaz. **46** (1962), 41–42.

19. P.H. Diananda, *Inequalities for a class of cyclic and other sums.* J. London Math. Soc. **37** (1962), 424–431.

20. P.H. Diananda, *Some cyclic and other inequalities.* Proc. Cambridge Phil. Soc. **58** (1962), 425–427.

21. P.H. Diananda, *A cyclic inequality and an extension of it. I.* Proc. Edinburgh Math. Soc. **(2) 13** (1962), 79–84.

22. P.H. Diananda, *Some cyclic and other inequalities. II.* Proc. Cambridge Phil. Soc. **58** (1962), 703–705.

23. P.H. Diananda, *A cyclic inequality and an extension of it. II.* Proc. Edinburgh Math. Soc. **(2) 13** (1962), 143–152.

24. P.Ž. Djoković, *Sur une inégalité.* Proc. Glasgow Math. Assoc. **6** (1963), 1–10.

25. P.H. Diananda, *On a cyclic sum.* Proc. Glasgow Math. Assoc. **6** (1963), 11–13.

26. V.J.D. Baston, *On a cyclic sum of Mordell.* Acta Arithmetica **11** (1965), 133–146.

27. D.S. Mitrinović, *Problem 75.* Math. Vesnik **4** **(19)** (1967), 103.

28. P. Nowosad, *Isoperimetric eigenvalue problems in algebras.* Comm. Pure Appl. Math. **21** (1968), 401–465.

29. D.S. Mitrinović, *Analytic inequalities.* Springer Verlag, Berlin – Heidelberg – New York, pp. 400, 1970.

30. V.G. Drinfel'd, *A cyclic inequality.* Math. Notes USSR **9** (1971), 68–71.

31. D.E. Daykin, *Inequalities for functions of a cyclic nature.* J. London Math. Soc. **(2)** **3** (1971), 453–462.

32. D.E. Daykin, *Inequalities for certain cyclic sums.* Proc. Edinburgh Math. Soc. **17** (1971), 257–262.

33. M.A. Malcolm, *A note on a conjecture of L.J. Mordell.* Math. Comp. **25** (1971), 375–377.

34. A. Elbert, *On a cyclic inequality.* Per. Math. Hungarica **4** (1973), 163–168.

35. P.H. Diananda, *Some cyclic and other inequalities. III.* Proc. Cambridge Phil. Soc. **73** (1973), 69–71.

36. V.J. Baston, *Some cyclic inequalities.* Proc. Edinburgh Math. Soc. **18** (1973), 115–118.

37. J.C. Boarder and D.E. Daykin, *Inequalities for certain cyclic sums II.* Proc. Edinburgh Math. Soc. **18** (1973), 209–218.

38. E.K. Godunova and V.I. Levin, *Some cyclic inequalities* (Russian). Differential Equations and Inequalities (1973), 33–39.

39. E.K. Godunova and V.I. Levin, *Sharpening certain cyclic inequalities.* Math. Notes USSR **14** (1973), 735–741.

40. P.H. Diananda, *Some cyclic and other inequalities. IV.* Proc. Cambridge Phil. Soc. **76** (1974), 183–186.

41. P.J. Bushell and A.H. Craven, *On Shapiro's cyclic inequality.* Proc. Roy. Soc. Edinburgh 75A **26** (1975/76), 333–338.

42. E.K. Godunova and V.I. Levin, *An estimate of the cyclic sum with three summands* (Russian). Num. Anal. Math. Phys. **1** (1975), 36–57.

43. E.K. Godunova and V.I. Levin, *A cyclic sum with 12 terms.* Math. Notes USSR **19** (1976), 510–517.

44. E.K. Godunova and V.I. Levin, *Exactness of a nontrivial estimate in a cyclic inequality.* Math. Notes USSR **20** (1976), 673–675.

45. P.H. Diananda, *Inequalities for some cyclic sums.* Proc. Singapore Mat. Coll., Part 2. Math. Medley **5** (1977), 171–177.

46. J.L. Searcy and B.A. Troesch, *A cyclic inequality and a related eigenvalue problem.* Pac. J. Math. **81** (1979), 217–226.

47. J.L. Searcy and B.A. Troesch, *Correction to: A cyclic inequality and a related eigenvalue problem.* Pac. J. Math. **85** (1979), 501.

48. B.A. Troesch, *The shooting method applied to a cyclic inequality.* Math. Comp. **34** (1980), 175–184.

49. E.K. Godunova and V.I. Levin, *Lower bound for a cyclic sum.* Math. Notes USSR **32** (1982), 481–485.

50. B.A. Troesch, *On Shapiro's cyclic inequality for $N = 13$.* Math. Comp. **45** (1985), 199–207.

51. B.A. Troesch, *The validity of Shapiro's cyclic inequality.* Math. Comp. **53** (1989), 657–664.

52. D.S. Mitrinović and J.E. Pečarić, *Cyclic inequalities and cyclic functional equations* (Serbo-Croatian). Beograd, 1991. To appear.

Achim Clausing, Institut für Numerische und Instrumentelle Mathematik, Einsteinstr. 62, D-4400 Münster, Germany

International Series of Numerical Mathematics, Vol. 103, © 1992 Birkhäuser Verlag Basel

Some remarks on the Hardy inequality for higher order derivatives

Alois Kufner and Andreas Wannebo

Abstract

The paper deals with necessary and sufficient conditions on p, q and non-negative weights w_0, w_k for the validity of the inequality

$$\left(\int_0^1 |u(x)|^q w_0(x)dx\right)^{1/q} \leq C\left(\int_0^1 |u^{(k)}(x)|^p w_k(x)dx\right)^{1/p}$$

for some classes of functions u satisfying certain conditions at the endpoints 0, 1.

1. Introduction

To begin with we fix some standard notation. For the exponents p and q let $p' = p/(p-1)$ etc. Let LHS $\sim$ RHS denote that there is a positive constant C such that $\frac{1}{C}\cdot$ RHS $\leq$ LHS $\leq C\cdot$ RHS .

The "Hardy inequality of order k",

$$\left(\int_a^b |u(x)|^q w_0(x)dx\right)^{1/q} \leq C\left(\int_a^b |u^{(k)}(x)|^p w_k(x)dx\right)^{1/p}, \tag{1}$$

where C is a positive constant independent of u, is investigated in [3] and [1] and necessary and sufficient conditions on the weights (which only are supposed to be measurable and nonnegative) are derived when $(a,b) = (0,\infty)$ with some of the possible boundary conditions. Note that we always interpret the statement of the existence of an inequality of this type in the following way: If the RHS is finite then the LHS is finite and the inequality holds. It is shown that for $1 < p \leq q < \infty$ (1) holds for all functions u with $u^{(k-1)}$ locally absolutely continuous on $[0,\infty)$ and such that

$$u(0) = u'(0) = \cdots = u^{(k-1)}(0) = 0 \tag{2}$$

This paper is in final form and no version of it will be submitted for publication elsewhere.

if and only if the functions

$$\bar{B}_1(x) = \left(\int_x^\infty w_0(t)dt\right)^{1/q} \left(\int_0^x w_k^{1-p'}(t)(x-t)^{(k-1)p'} dt\right)^{1/p'} \tag{3}$$

$$\bar{B}_2(x) = \left(\int_x^\infty w_0(t)(t-x)^{(k-1)q} dt\right)^{1/q} \left(\int_0^x w_k^{1-p'}(t)dt\right)^{1/p'} \tag{4}$$

are defined and bounded on $(0,\infty)$, see [3] (in (3), (4) it is understood that $0\cdot\infty = 0$ and $0^{1-p'} = \infty$). It is also shown that (1) holds (again for $1 < p \le q < \infty$) for all functions u with $u^{(k-1)}$ locally absolutely continuous and such that

$$\begin{gathered} u(0) = u'(0) = \cdots = u^{(m-1)}(0) = 0 \\ u^{(m)}(\infty) = u^{(m+1)}(\infty) = \cdots = u^{(m+n-1)}(\infty) = 0 \end{gathered} \tag{5}$$

with

$$k = m + n, \quad m, n \in \mathbb{N} \tag{6}$$

if and only if the functions

$$\bar{B}_3(x) = \left(\int_x^\infty w_0(t)t^{(m-1)q} dt\right)^{1/q} \left(\int_0^x w_k^{1-p'}(t)t^{np'} dt\right)^{1/p'} \tag{7}$$

$$\bar{B}_4(x) = \left(\int_0^x w_0(t)t^{mq} dt\right)^{1/q} \left(\int_x^\infty w_k^{1-p'}(t)t^{(n-1)p'} dt\right)^{1/p'} \tag{8}$$

are defined and bounded on $(0,\infty)$, see [1]. Moreover in [1] it is shown that:

(i) That $\bar{B}_3(x)$, with $m = k$, $n = 0$, is defined and bounded is sufficient for the validity of (1) for functions u with $u^{(k-1)}$ locally absolutely continuous on $[0,\infty)$ and satisfying (5) where $m = k$, $n = 0$ (i.e. for functions satisfying (2));

(ii) that $\bar{B}_4(x)$, with $m = 0$, $n = k$, is defined and bounded is sufficient for the validity of (1) for functions u with $u^{(k-1)}$ locally absolutely continuous on $[0,\infty)$ satisfying (5) where $m = 0$, $n = k$.

The exact formulation of the boundary conditions in (5) is important. For example in the case $k = 3$ we can consider functions u satisfying

$$u(0) = u'(0) = u''(0) = 0 \quad (m = 3,\ n = 0),$$

$$u(0) = u'(0) = u''(\infty) = 0 \quad (m = 2,\ n = 1),$$

$$u(0) = u'(\infty) = u''(\infty) = 0 \quad (m = 1,\ n = 2)$$

or

$$u(\infty) = u'(\infty) = u''(\infty) = 0 \quad (m = 0,\ n = 3),$$

and derive necessary and sufficient conditions for the validity of the Hardy inequality (of order three) for these classes. However for the respective boundary conditions

$$u(0) = u'(\infty) = u''(0) = 0$$

or

$$u(\infty) = u'(0) = u''(0) = 0$$

etc. there is a problem that arises from the method used. Indeed, even in the case $k = 2$ with

$$u(0) = u'(\infty) = 0 \quad (m = 1,\ n = 1)$$

we have the "representation"

$$u(x) = \int_0^x \left(\int_t^\infty f(s)ds \right) dt, \quad f \geq 0;$$

for the non-admissible case

$$u(\infty) = u'(0) = 0$$

the corresponding "representation"

$$u(x) = \int_x^\infty \left(\int_0^t f(s)ds \right) dt$$

fails for $f \neq 0$. This problem seems to be substantial (cf. the counterexample in [2, Example 10.16]).

The situation changes if we consider *bounded* intervals (a, b). Without loss of generality we can assume that $(a, b) = (0, 1)$ and consider the inequality

$$\left(\int_0^1 |u(x)|^q w_0(x)dx \right)^{1/q} \leq C \left(\int_0^1 |u^{(k)}(x)|^p w_k(x)dx \right)^{1/p} \tag{9}$$

Let $u \in AC^{k-1}_{m,n}(0,1)$ denote that $u^{(k-1)}$ is absolutely continuous on [0,1] with the boundary conditions

$$\begin{aligned} &u(0) = u'(0) = \cdots = u^{(m-1)}(0) = 0 \\ &u^{(m)}(1) = u^{(m+1)}(1) = \cdots = u^{(m+n-1)}(1) = 0, \quad k = m + n, \end{aligned} \tag{10}$$

with m, n nonnegative integers [note that also the values $m = 0$ and $n = 0$ are admissible]. Similarly $u \in AC^{k-1}_{m,0}(0,1)$ denotes that

$$u(0) = u'(0) = \ldots = u^{(k-1)}(0) = 0, \quad (k = m)$$

while $u \in AC^{k-1}_{0,n}(0,1)$ denotes that

$$u(1) = u'(1) = \ldots = u^{(k-1)}(1) = 0, \quad (k = n)].$$

The constant C in the inequality (9) will not be estimated here.

2. Results

Using methods analogous to those of [3] and [1], we can prove the following assertions.

Theorem 1. *Let $1 < p \le q < \infty$. Then the Hardy inequality*

$$\left(\int_0^1 |u(x)|^q w_0(x)dx\right)^{1/q} \le C\left(\int_0^1 |u^{(k)}(x)|^p w_k(x)dx\right)^{1/p}, \tag{11}$$

with w_0, w_k measurable nonnegative functions on $(0,1)$, holds (with a constant C independent of u)

(i) *for every $u \in AC^{k-1}_{m,0}(0,1)$ if and only if the functions*

$$B_1(x) = \left(\int_x^1 w_0(t)dt\right)^{1/q}\left(\int_0^x w_k^{1-p'}(t)(x-t)^{(k-1)p'}dt\right)^{1/p'} \tag{12}$$

$$B_2(x) = \left(\int_x^1 w_0(t)(t-x)^{(k-1)q}dt\right)^{1/q}\left(\int_0^x w_k^{1-p'}(t)dt\right)^{1/p'} \tag{13}$$

are defined and bounded on $(0,1)$;

(ii) *for every $u \in AC^{k-1}_{0,n}(0,1)$ if and only if the functions*

$$B_3(x) = \left(\int_0^x w_0(t)dt\right)^{1/q}\left(\int_x^1 w_k^{1-p'}(t)(t-x)^{(k-1)p'}dt\right)^{1/p'} \tag{14}$$

$$B_4(x) = \left(\int_0^x w_0(t)(x-t)^{(k-1)q}dt\right)^{1/q}\left(\int_x^1 w_k^{1-p'}(t)dt\right)^{1/p'} \tag{15}$$

are defined and bounded on $(0,1)$;

(iii) *for every* $u \in AC^{k-1}_{m,n}(0,1)$ *with* $k = m+n$, $m \geq 1$, $n \geq 1$, *if and only if the functions*

$$B_5(x) = \left(\int_x^1 w_0(t)t^{(m-1)q}dt\right)^{1/q}\left(\int_0^x w_k^{1-p'}(t)t^{np'}dt\right)^{1/p'}, \tag{16}$$

$$B_6(x) = \left(\int_0^x w_0(t)t^{mq}dt\right)^{1/q}\left(\int_x^1 w_k^{1-p'}(t)t^{(n-1)p'}dt\right)^{1/p'} \tag{17}$$

are defined and bounded on $(0,1)$.

Moreover, the inequality (11) *holds for* $u \in AC^{k-1}_{m,0}(0,1)$ *[for* $u \in AC^{k-1}_{0,n}(0,1)$*] if the function* $B_5(x)$ *with* $m = k$, $n = 0$ *[the function* $B_6(x)$ *with* $m = 0$, $n = k$*] is bounded on* $(0,1)$.

Furthermore, we can easily derive conditions for the validity of (11) for functions u satisfying the conditions

$$\begin{gathered} u(1) = u'(1) = \cdots = u^{(m-1)}(1) = 0 \\ u^{(m)}(0) = u^{(m+1)}(0) = \cdots = u^{(m+n-1)}(0) = 0, \quad k = m+n. \end{gathered} \tag{18}$$

Theorem 2. *Let* $1 < p \leq q < \infty$. *Then the Hardy inequality*

$$\left(\int_0^1 |u(x)|^q w_0(x)dx\right)^{1/q} \leq C\left(\int_0^1 |u^{(k)}(x)|^p w_k(x)dx\right)^{1/p}, \tag{19}$$

with w_0, w_k *measurable nonnegative functions on* $(0,1)$, *holds (with a constant* C *independent of* u*) for every* u *with* $u^{(k-1)}$ *absolutely continuous on* $[0,1]$ *and satisfying*

$$\begin{gathered} u(1) = u'(1) = \cdots = u^{(m-1)}(1) = 0 \\ u^{(m)}(0) = u^{(m+1)}(0) = \cdots = u^{(m+n-1)}(0) = 0, \quad k = m+n, \end{gathered} \tag{20}$$

with $m \geq 1$, $n \geq 1$ *if and only if the functions*

$$B_7(x) = \left(\int_x^1 w_0(t)(1-t)^{mq}dt\right)^{1/q}\left(\int_0^x w_k^{1-p'}(t)(1-t)^{(n-1)p'}dt\right)^{1/p'}, \tag{21}$$

$$(22) \qquad B_8(x) = \left(\int_0^x w_0(t)(1-t)^{(m-1)q} dt\right)^{1/q} \left(\int_x^1 w_k^{1-p'}(t)(1-t)^{np'} dt\right)^{1/p'}$$

are defined and bounded on $(0,1)$.

Proof. If u satisfies (20), then the function

$$U(x) = u(1-x)$$

belongs to $AC_{m,n}^{k-1}(0,1)$ and the Hardy inequality (19) for u with weight functions w_0, w_k can be rewritten into the Hardy inequality (9) for U with new weight functions $W_0(x) = w_0(1-x)$, $W_k(x) = w_k(1-x)$. Consequently we can use Theorem 1, (iii), and obtain that the corresponding functions B_5 and B_6 (calculated for W_0, W_k) have to be bounded. But for $B_5(x)$ (with respect to W_0, W_k) we obtain

$$\begin{aligned}
&\left(\int_x^1 W_0(t)t^{(m-1)q} dt\right)^{1/q} \left(\int_0^x W_k^{1-p'}(t)t^{np'} dt\right)^{1/p'} \\
&= \left(\int_x^1 w_0(1-t)t^{(m-1)q} dt\right)^{1/q} \left(\int_0^x w_k^{1-p'}(1-t)t^{np'} dt\right)^{1/p'} \\
&= \left(\int_0^{1-x} w(s)(1-s)^{(m-1)q} ds\right)^{1/q} \left(\int_{1-x}^1 w_k^{1-p'}(s)(1-s)^{np'} ds\right)^{1/p'} \\
&= B_8(1-x)
\end{aligned}$$

and similarly, $B_6(x)$ (with respect to W_0, W_k) equals to $B_7(1-x)$ (with respect to w_0, w_k). Thus by replacing $1-x$ by x, we obtain the assertion. This completes the proof.

Now we turn to some further possibilities as far as concerns the boundary conditions. Since a general investigation seems to be technically difficult, let us consider some special (small) values of k.

The case $k = 2$

The Hardy inequality

$$(23) \qquad \left(\int_0^1 |u(x)|^q w_0(x) dx\right)^{1/q} \le C \left(\int_0^1 |u''(x)|^p w_2(x) dx\right)^{1/p}$$

is fully described by the foregoing theorems if we consider functions u such that either

$$u(0) = u'(0) = 0 \quad (m = 2,\ n = 0),$$

$$u(1) = u'(1) = 0 \quad (m = 0,\ n = 2),$$
$$u(0) = u'(1) = 0 \quad (m = 1,\ n = 1)$$

(cf. Theorem 1) or

$$u(1) = u'(0) = 0 \quad (m = 1,\ n = 1)$$

(cf. Theorem 2).

Another possibility for boundary conditions not described by Theorems 1 and 2 the following

$$u(0) = u(1) = 0. \tag{24}$$

In this case we have

Theorem 3. *Let $1 < p \le q < \infty$. Then the Hardy inequality*

$$\left(\int_0^1 |u(x)|^q w_0(x)dx\right)^{1/q} \le C\left(\int_0^1 |u''(x)|^p w_2(x)dx\right)^{1/p}, \tag{25}$$

with w_0, w_2 measurable nonnegative functions on $(0,1)$, holds for every function u with u' absolutely continuous on $[0,1]$ and satisfying

$$u(0) = u(1) = 0. \tag{26}$$

if and only if the functions

$$B_9(x) = \left(\int_x^1 w_0(t)(1-t)^q dt\right)^{1/q}\left(\int_0^x w_2^{1-p'}(t)t^{p'}dt\right)^{1/p'}, \tag{27}$$

$$B_{10}(x) = \left(\int_0^x w_0(t)t^q dt\right)^{1/q}\left(\int_x^1 w_2^{1-p'}(t)(1-t)^{p'}dt\right)^{1/p'} \tag{28}$$

are defined and bounded on $(0,1)$.

Proof. Analogously to the treatment in [3] and [1] instead of (25) we will consider the inequality

$$\left(\int_0^1 |(Tf)(x)|^q w_0(x)dx\right)^{1/q} \le C\left(\int_0^1 |f(x)|^p w_2(x)dx\right)^{1/p} \tag{29}$$

for $f \ge 0$ with

$$(Tf)(x) = \int_0^x (1-x)tf(t)dt + \int_x^1 x(1-t)f(t)dt. \tag{30}$$

Obviously the function $u(x) = (Tf)(x)$ satisfies (26) and differentiating we get $u'' = -f$. Furthermore u' is absolutely continuous on $[0,1]$.

We denote

$$Tf = I_1 f + I_2 f, \tag{31}$$

where

$$\begin{aligned} (I_1 f)(x) &= \int_0^x (1-x)tf(t)dt, \\ (I_2 f)(x) &= \int_x^1 x(1-t)f(t)dt. \end{aligned} \tag{32}$$

If we write $F(t) = tf(t)$, $W_0(t) = (1-t)^q w_0(t)$ and $W_1(t) = t^{-p}w_2(t)$, then

$$\begin{aligned} \int_0^1 |(I_1 f)(x)|^q w_0(x)dt &= \int_0^1 \left(\int_0^x tf(t)dt\right)^q (1-x)^q w_0(x)dx \\ &= \int_0^1 \left(\int_0^x F(t)dt\right)^q W_0(x)dx \end{aligned}$$

and

$$\begin{aligned} \int_0^1 f^p(x)w_2 dx &= \int_0^1 |xf(x)|^p x^{-p} w_2(x)dx \\ &= \int_0^1 F^p(x)W_1(x)dx. \end{aligned}$$

The inequality

$$\left(\int_0^1 |(I_1 f)(x)|^q w_0(x)dx\right)^{1/q} \le C \left(\int_0^1 f^p(x)w_2(x)dx\right)^{1/p}, \tag{33}$$

i.e. the inequality

$$\left(\int_0^1 \left(\int_0^x F(t)dt\right)^q W_0(x)dx\right)^{1/q} \le C \left(\int_0^1 F^p(x)W_1(x)dx\right)^{1/p} \tag{34}$$

holds if and only if the function

$$B_1^*(x) = \left(\int_x^1 W_0(t)dt\right)^{1/q} \left(\int_0^x W_1^{1-p'}(t)dt\right)^{1/p'} \tag{35}$$

is defined and bounded on $(0,1)$ (see, e.g., [2]). But, in view of our notation, we have

$$B_1^*(x) = B_9(x).$$

If we write $\tilde{F}(t) = (1-t)f(t)$, $\tilde{W}_0(t) = t^q w_0(t)$ and $\tilde{W}_1(t) = (1-t)^{-p} w_2(t)$, then we have analogously

$$\begin{aligned}\int_0^1 |(I_2 f)(x)|^q w_0(x)dx &= \int_0^1 \left(\int_x^1 (1-t)f(t)dt\right)^q x^q w_0(x)dx \\ &= \int_0^1 \left(\int_x^1 \tilde{F}(t)dt\right)^q \tilde{W}_0(x)dx,\end{aligned}$$

$$\begin{aligned}\int_0^1 f^p(x)w_2(x)dx &= \int_0^1 |(1-x)f(x)|^p (1-x)^{-p} w_2(x)dx \\ &= \int_0^1 \tilde{F}^p(x)\tilde{W}_2(x)dx\end{aligned}$$

and the inequality

$$\left(\int_0^1 |(I_2 f)(x)|^q w_0(x)dx\right)^{1/q} \le C\left(\int_0^1 f^p(x)w_2(x)dx\right)^{1/p}. \tag{36}$$

Thus the inequality

$$\left(\int_0^1 \left(\int_x^1 \tilde{F}(t)dt\right)^q \tilde{W}_0(x)dx\right)^{1/q} \le C\left(\int_0^1 \tilde{F}^p(x)\tilde{W}_1(x)dx\right)^{1/p} \tag{37}$$

holds if and only if the function

$$B_2^*(x) = \left(\int_0^x \tilde{W}_0(t)\right)^{1/q} \left(\int_x^1 \tilde{W}_1^{1-p'}(t)dt\right)^{1/p'} \tag{38}$$

is defined and bounded on $(0,1)$ (see again [2]). But, in view of our notation, we have

$$B_2^*(x) = B_{10}(x).$$

Consequently, if the functions B_9 and B_{10} are defined and bounded, then the inequality (29) holds in view of (31), (33) and (36).

Conversely, suppose that (29) holds. Since the operators T, I_1 and I_2 are positive, we have that $(I_i f)(x) \le (Tf)(x)$, $i = 1,2$, and the inequalities (33) and (36) hold, too. But this implies that B_1^*, B_2^* – i.e. B_9, B_{10} – are defined and bounded. This completes the proof.

The case $k = 3$

The Hardy inequality

$$\left(\int_0^1 |u(x)|^q w_0(x)dx\right)^{1/q} \le C\left(\int_0^1 |u'''(x)|^p w_3(x)dx\right)^{1/p} \tag{39}$$

is fully described by Theorems 1 and 2 if we consider functions u such that either

$$u(0) = u'(0) = u''(0) = 0 \quad (m = 3,\ n = 0),$$

$$u(1) = u'(1) = u''(1) = 0 \quad (m = 0,\ n = 3),$$

$$u(0) = u'(0) = u''(1) = 0 \quad (m = 2,\ n = 1),$$

$$u(0) = u'(1) = u''(1) = 0 \quad (m = 1,\ n = 2),$$

(cf. Theorem 1)

$$u(1) = u'(1) = u''(0) \quad (m = 2,\ n = 1)$$

or

$$u(1) = u'(0) = u''(0) = 0 \quad (m = 1,\ n = 2)$$

(cf. Theorem 2).

Other possibilities for the boundary conditions not described by Theorems 1 and 2 read as follows

$$u(0) = u'(1) = u''(0) = 0 \tag{40}$$

or

$$u(1) = u'(0) = u''(1) = 0. \tag{41}$$

In these cases we have

Theorem 4. *Let $1 < p \le q < \infty$. Then the Hardy inequality*

$$\left(\int_0^1 |u(x)|^q w_0(x)dx\right)^{1/q} \le C\left(\int_0^1 |u'''(x)|^p w_3(x)dx\right)^{1/p}, \tag{42}$$

with w_0, w_3 measurable nonnegative functions on $(0,1)$ holds for every function u with u'' absolutely continuous on $[0,1]$ and a constant C independent of u and satisfying in case I

$$u(0) = u'(1) = u''(0) = 0 \tag{43}$$

and in case II

$$u(1) = u'(0) = u''(1) = 0 \tag{44}$$

if and only if in case I *the functions*

$$B_{11}(x) = \left(\int_x^1 w_0(t)t^q dt\right)^{1/q} \left(\int_0^x w_3^{1-p'}(t)(1-t)^{p'} dt\right)^{1/p'}, \tag{45}$$

$$B_{12}(x) = \left(\int_0^x w_0(t)t^q dt\right)^{1/q} \left(\int_x^1 w_3^{1-p'}(t)(1-t)^{p'} dt\right)^{1/p'}, \tag{46}$$

and in case II *the functions*

$$B_{13}(x) = \left(\int_x^1 w_0(t)(1-t)^q dt\right)^{1/q} \left(\int_0^x w_3^{1-p'}(t)t^{p'} dt\right)^{1/p'}, \tag{47}$$

$$B_{14}(x) = \left(\int_0^x w_0(t)(1-t)^q dt\right)^{1/q} \left(\int_x^1 w_3^{1-p'}(t)t^{p'} dt\right)^{1/p'} \tag{48}$$

are defined and bounded on $(0,1)$.

Proof. (i) Let us consider functions u satisfying (43). Then we can put

$$u(x) = (Tf)(x) = \int_0^x \left(\int_t^1 \left(\int_0^s f(r)dr\right) ds\right) dt \tag{49}$$

and instead of (42), investigate the inequality

$$\left(\int_0^1 (Tf)^q(x)w_0(x)dx\right)^{1/q} \le C \left(\int_0^1 f^p(x)w_3(x)dx\right)^{1/p} \tag{50}$$

for $f \ge 0$.

Differentiating successively three times we can see that for $x \in [0,1]$

$$(Tf)(x) = (I_1 f)(x) + (I_2 f)(x), \tag{51}$$

where

$$(I_1 f)(x) = \frac{1}{2}\int_0^x [2x - x^2 - t^2] f(t)dt, \tag{52}$$

$$(53)\qquad (I_2 f)(x) = \int_x^1 x(1-t)f(t)dt.$$

holds. Furthermore we have that u'' is absolutely continuous on $[0,1]$ and that (43) holds.

Since we have $0 < t < x < 1$ in $(I_1 f)(x)$, the trivial inequality $x^2 + t^2 > 2xt$ implies that $2x - x^2 - t^2 < 2x - 2xt = 2x(1-t)$. Furthermore, we have $2x - x^2 - t^2 > x(1-t)$ since $0 < t < x < 1$ implies that $x^2 < x$ and $t^2 < xt$. Consequently,

$$(54)\qquad \frac{1}{2}\int_0^x x(1-t)f(t)dt \le (I_1 f)(x) \le \int_0^x x(1-t)f(t)dt$$

and the inequality

$$(55)\qquad \left(\int_0^1 (I_1 f)^q(x)w_0(x)dx\right)^{1/p} \le C\left(\int_0^1 f^p(x)w_3(x)dx\right)^{1/p}$$

holds if and only if the inequality

$$(56)\qquad \left(\int_0^1\left(\int_0^x x(1-t)f(t)dt\right)^q w_0(x)dx\right)^{1/q} \le C_1\left(\int_0^1 f^p(x)w_3(x)dx\right)^{1/p}$$

is valid. But in virtue of the formulas

$$\int_0^1\left(\int_0^x x(1-t)f(t)dt\right)^q w_0(x)dx = \int_0^1\left(\int_0^x (1-t)f(t)dt\right)^q x^q w_0(x)dx,$$

$$\int_0^1 f^p(x)w_3(x)dx = \int_0^1 [(1-x)f(x)]^p(1-x)^{-p}w_3(x)dx$$

we obtain – similar to the proof of Theorem 3 – that the necessary and sufficient condition for the validity of (56) [i.e. for the validity of (55)] is that the B_{11} from (45) is defined and bounded.

Similarly, the inequality

$$(57)\qquad \begin{aligned} \left(\int_0^1 (I_2 f)f^q(x)w_0(x)dx\right)^{1/q} &= \left(\int_0^1\left(\int_x^1 (1-t)f(t)dt\right)^q x^q w_0(x)dx\right)^{1/q} \\ \le C\left(\int_0^1 [(1-x)f(x)]^p(1-x)^{-p}w_3(x)dx\right)^{1/p} &= C\left(\int_0^1 f^p(x)w_3(x)dx\right)^{1/p} \end{aligned}$$

holds if and only if the function B_{12} from (46) is defined and bounded.

In view of (51), (55) and (57) we have shown that the inequality (50) - and consequently, the Hardy inequality (42) for functions u satisfying (43) - holds if B_{11}, B_{12} are defined and bounded. Moreover that B_{11} and B_{12} are defined and bounded is also necessary since T, I_1 and I_2 are positive operators (see the end of the proof of Theorem 3).

(ii) Let us consider functions u satisfying (44). Then we can use the arguments from Theorem 2, since the function $U(x) = u(1-x)$ satisfies (43) and $B_{13}(x)$ and $B_{14}(x)$ are in fact the functions $B_{12}(1-x)$ and $B_{11}(1-x)$, respectively, calculated for the weight functions $w_0(1-x)$, $w_3(1-x)$. The proof is complete.

Further possible boundary conditions for the case $k = 3$ are

(58) $$u(0) = u(1) = u'(0) = 0,$$

(59) $$u(0) = u(1) = u''(0) = 0$$

and

(60) $$u(0) = u'(0) = u'(1) = 0$$

respectively. In these cases, we use the fact that the function u can be expressed in the form

(61) $$u(x) = (Tf)(x) = (I_1f)(x) + (I_2f)(x)$$

where

(62) $$(I_1f)(x) = \frac{1}{2}\int_0^x (1-x)t(2x - tx - t)f(t)dt \sim \int_0^x x(1-x)t(1-t)f(t)dt$$
$$(I_2f)(x) = \frac{1}{2}\int_x^1 x^2(1-t)^2 f(t)dt$$

for u satisfying (58),

(63) $$(I_1f)(x) = \frac{1}{2}\int_0^x (1-x)(x-t^2)f(t)dt \sim \int_0^x x(1-x)(1-t)f(t)dt,$$
$$(I_2)f(x) = \frac{1}{2}\int_x^1 x(1-t)^2 f(t)dt$$

for u satisfying (59), and

$$(64)\qquad \begin{aligned}(I_1 f)(x) &= \frac{1}{2}\int_0^x t(2x - x^2 - t)f(t)dt \sim \int_0^x xt(1-t)f(t)dt,\\ (I_2 f)(x) &= \frac{1}{2}\int_x^1 x^2(1-t)f(t)dt\end{aligned}$$

for u satisfying (60), and investigate – instead of (42) – the inequality (50).

The equivalence ($\sim$) in (62) for instance is obtained by using the fact that for $0 < t < x < 1$, we have $2x - tx - t > 2x - tx - x = x - tx = x(1-t)$ and $tx < t$ implies $2x - tx - t < 2x - 2tx = 2x(1-t)$.

Proceeding analogously as in the proof of Theorem 4, we obtain in view of (61) and of the corresponding estimates for I_1 the following result:

Theorem 5. *Let $1 < p \le q < \infty$. Then the Hardy inequality*

$$(65)\qquad \left(\int_0^1 |u(x)|^q w_0(x)dx\right)^{1/q} \le C\left(\int_0^1 |u'''(x)|^p w_3(x)dx\right)^{1/p},$$

with w_0, w_3 measurable nonnegative functions on $(0,1)$, holds for every function u with u'' absolutely continuous on $[0,1]$ with a constant C independent of u and satisfying in case I

$$(66)\qquad u(0) = u(1) = u'(0) = 0,$$

in case II

$$(67)\qquad u(0) = u(1) = u''(0) = 0$$

and in case III

$$(68)\qquad u(0) = u'(0) = u'(1) = 0,$$

if and only if the following functions are defined and bounded on $(0,1)$: In case I,

$$B_{15}(x) = \left(\int_x^1 w_0(t)t^q(1-t)^q dt\right)^{1/q}\left(\int_0^x w_3^{1-p'}(t)t^{p'}(1-t)^{p'}dt\right)^{1/p'},$$

$$B_{16}(x) = \left(\int_0^x w_0(t)t^{2q}dt\right)^{1/q}\left(\int_x^1 w_3^{1-p'}(t)(1-t)^{2p'}dt\right)^{1/p'};$$

in case II,

$$B_{17}(x) = \left(\int_x^1 w_0(t) t^q (1-t)^q dt \right)^{1/q} \left(\int_0^x w_3^{1-p'}(t)(1-t)^{p'} dt \right)^{1/p'},$$

$$B_{18}(x) = \left(\int_0^x w_0(t) t^q dt \right)^{1/q} \left(\int_x^1 w_3^{1-p'}(t)(1-t)^{2p'} dt \right)^{1/p'};$$

and in case III,

$$B_{19}(x) = \left(\int_x^1 w_0(t) t^q dt \right)^{1/q} \left(\int_0^x w_3^{1-p'}(t) t^{p'} (1-t)^{p'} dt \right)^{1/p'},$$

$$B_{20}(x) = \left(\int_0^x w_0(t) t^{2q} dt \right)^{1/q} \left(\int_x^1 w_3^{1-p'}(t)(1-t)^{p'} dt \right)^{1/p'}.$$

Necessary and sufficient conditions for the validity of the Hardy inequality (65) for the functions u satisfying

(69) $$u(0) = u(1) = u'(1) = 0,$$

(70) $$u(0) = u(1) = u''(1) = 0$$

or

(71) $$u(1) = u'(0) = u'(1) = 0$$

can again be obtained from the conditions in Theorem 5 using the method of the proof of Theorem 2. Indeed: if u satisfies (69), (70) or (71), then $U(x) = u(1-x)$ satisfies (66), (67) or (68), respectively.

An investigation of the Hardy inequality (65) for u satisfying some of the remaining boundary conditions is meaningless for trivial reasons. For instance, if we consider functions u satisfying

$$u'(0) = u'(1) = u''(0) = 0,$$

which are fulfilled for the constant function $u(x) = 1$, then the right hand side of (39) vanishes and (39) holds only with $w_0(x) = 0$ almost everywhere in $(0,1)$.

Remark. In the foregoing theorems, we considered inequality (1) only for the case $1 < p \le q < \infty$ and our main tool have been criteria for the validity of the inequality

(72) $$\left(\int_0^1 (SF)^q(x) w_0(x) dx \right)^{1/q} \le C \left(\int_0^1 F^p(x) w_k(x) dx \right)^{1/p}$$

where S is one of the integral operators

$$\int_0^x F(t)dt, \quad \int_x^1 F(t)dt.$$

Since necessary and sufficient conditions for the validity of (72) are known also for the case

$$p > q, \quad 1 < p < \infty, \quad 0 < q < \infty, \tag{73}$$

(see, eg. [2, Theorems 5.9, 6.3, 9.3]), we can investigate the inequality (1) also for $p > q$.

For instance, the Hardy inequality (65) holds for p, q satisfying (73) for all functions u with u'' absolutely continuous and satisfying (60) if and only if

$$\int_0^1 \left(\int_x^1 w_0(t)t^q dt\right)^{r/q} \left(\int_0^x w_3^{1-p'}(t)t^{p'}(1-t)^{p'} dt\right)^{r/q'} w_3^{1-p'}(x)x^{p'}(1-x)^{p'} dx < \infty,$$

$$\int_0^1 \left(\int_0^x w_0(t)t^{2q} dt\right)^{r/q} \left(\int_x^1 w_3^{1-p'}(t)(1-t)^{p'} dt\right)^{r/q'} w_3^{1-p'}(x)(1-x)^{p'} dx < \infty,$$

where $\frac{1}{r} = \frac{1}{q} - \frac{1}{p}$. [Compare the last two expressions with the functions B_{19}, B_{20} in Theorem 5.]

References

1. A. Kufner and H.P. Heinig, *Hardy's inequality for higher order derivatives.* Trudy Mat. Inst. Steklov **192** (1990), 105–113 (Russian).

2. B. Opic and A. Kufner, *Hardy-type inequalities.* Longman Scientific & Technical (Pitman Research Notes in Mathematics Series 219), 1990.

3. V.D. Stepanov, *Two-weighted estimates for Riemann-Liouville integrals.* Preprint no. 39, Math. Inst. Czech. Acad. Sci., Prague, 1988.

Alois Kufner, Math. Inst., Acad. Sc., Žitná 25, 11567 Praha 1, Czechoslovakia

Andreas Wannebo, Royal Institute of Technology, Department of Mathematics, S-100 44 Stockholm, Sweden

International Series of Numerical Mathematics, Vol. 103, © 1992 Birkhäuser Verlag Basel

Copson's inequalities for series and their variations

R.N. Mohapatra, D.C. Russell and K. Vajravelu

Abstract

Two inequalities related to Copson's and Hardy's inequalities for series are established, together with their interpretations in terms of inclusion between sequence spaces.

1. Introduction

Following a method of Elliott [6] to provide a simple proof of Hardy's inequality for a series of positive terms, Copson [3] proved the following:

Theorem A [3, Theorems A and B]. *Suppose that*

$$p > 1, \quad q_n > 0, \quad a_n \geq 0 \ (n = 1, 2, \ldots), \quad Q_n := q_1 + \cdots + q_n, \quad \sum_{n=1}^{\infty} q_n a_n^p < \infty.$$

Then

$$\sum_{n=1}^{\infty} q_n \left(Q_n^{-1} \sum_{k=1}^{n} q_k a_k \right)^p \leq \left(\frac{p}{p-1} \right)^p \sum_{n=1}^{\infty} q_n a_n^p; \tag{1.1}$$

$$\sum_{n=1}^{\infty} q_n \left(\sum_{k=n}^{\infty} \frac{q_k a_k}{Q_k} \right)^p \leq p^p \sum_{n=1}^{\infty} q_n a_n^p. \tag{1.2}$$

Copson also showed that the constants on the right of (1.1) and (1.2) are best possible. Hardy's inequality [7, # 326] is the case $q_n = 1$ of (1.1), namely

$$\sum_{n=1}^{\infty} \left(\frac{1}{n} \sum_{k=1}^{n} a_k \right)^p \leq \left(\frac{p}{p-1} \right)^p \sum_{n=1}^{\infty} a_n^p \quad (p > 1, \ a_n \geq 0). \tag{1.3}$$

This paper is in final form and no version of it will be submitted for publication elsewhere.

Inequalities involving lower triangular matrices (in place of the q, Q kernels) have been studied by Johnson and Mohapatra [8].

Love [11] generalized the Hardy and Copson inequalities by replacing the kernels by general infinite matrices. He has proved:

Theorem B [11, Theorem 1]. *Let $p > 1$ and $\alpha(\cdot)$ be a non-negative and decreasing function in $(0,1]$ with $A := \int_0^1 \alpha(t)t^{-1/p}\,dt < \infty$. Let $\lambda_n > 0$, $\Lambda_m := \sum_{n=1}^{m} \lambda_n$ and $|a_{mn}| \le (\lambda_n/\Lambda_n)\alpha(\Lambda_n/\Lambda_m)$, $1 \le n \le m$. Then*

$$\left\{\sum_{m=1}^{\infty} \lambda_m \left|\sum_{n=1}^{m} a_{mn}x_n\right|^p\right\}^{1/p} \le A\left\{\sum_{m=1}^{\infty} \lambda_m |x_m|^p\right\}^{1/p}. \tag{1.4}$$

Copson [4] also considered inequalities analogous to Theorem A for an index satisfying $0 < p < 1$ (e.g. see [7, p. 255]). Bennett [1], [2] has obtained a number of generalizations and variants of these inequalities of Hardy and Copson, among them Theorems C, D, E below.

Theorem C [2, Theorem 3]. *If $0 < p < 1$ and $w_k \ge 0$ then*

$$\sum_{n=1}^{\infty}\left(\frac{1}{n}\sum_{k=n}^{\infty} w_k\right)^p < \frac{\pi p}{\sin \pi p}\sum_{n=1}^{\infty} \sup_{k \ge n} w_k^p \tag{1.5}$$

unless $w_k \equiv 0$. The constant on the right of (1.5) is best possible.

Theorem D [1; 2, Theorem 3]. *Let $r > s \ge 1$ and let* **u**,**v**,**w** *be N-tuples with non-negative entries. If, for $m = 1, 2, \ldots, N$,*

$$\sum_{n=1}^{m} u_n \left(\sum_{k=1}^{n} v_k\right)^r \le \left(\sum_{k=1}^{m} v_k\right)^s, \tag{1.6}$$

then

$$\sum_{n=1}^{N} u_n \left(\sum_{k=1}^{n} v_k w_k\right)^r \le \left(\frac{r}{r-s}\right)^r \left(\sum_{k=1}^{N} v_k w_k^{r/s}\right)^s. \tag{1.7}$$

Theorem E [2, Theorem 1]. *Let $0 < r < s \le 1$ and let* **u**,**v**,**w** *be N-tuples with non-negative entries. If, for $m = 1, 2, \ldots, N$,*

$$\sum_{k=1}^{m} v_k \left(\sum_{n=1}^{k} u_n\right)^{r/(r-s)} \le \left(\sum_{n=1}^{m} u_n\right)^{(1-r)/(s-r)}, \tag{1.8}$$

then

$$\sum_{n=1}^{N} u_n \left(\sum_{k=n}^{N} v_k w_k \right)^r \geq r^r \left(\sum_{k=1}^{N} v_k w_k^{r/s} \right)^s . \tag{1.9}$$

For example, if in Theorem E we set $s = 1$, $v_k = 1$, $u_n = n^{-r}$ we obtain [7, # 345]; while setting $s = 1$, $u_n = v_n / \left(\sum_{k=1}^{n} v_k\right)^r$ in Theorems D and E gives us (1.1) and its analogue for $0 < p < 1$.

2. Statement of results

Our results will consider an index $p > 1$, with $(1/p) + (1/p') = 1$, and $K(p)$ will denote a positive constant (which may be different at different occurrences) depending on p only. We shall assume throughout that $q_n > 0$ (though $q_1 > 0$, $q_n \geq 0$ is enough), $Q_n := q_1 + \cdots + q_n$, $\Delta q_n := q_n - q_{n+1}$, $n = 1, 2, \ldots$. We shall prove the following theorems.

Theorem 1. *Let* $1 < p < \infty$, $p^{-1} + (p')^{-1} = 1$ *and let* (q_n) *satisfy* $q_n > 0$ *and*

$$\text{(a)}\quad nq_n \leq AQ_n, \qquad \text{(b)}\quad n\left|\Delta q_n^{1/p'}\right| \leq B_p q_n^{1/p'}, \qquad n = 1, 2, \ldots. \tag{2.1}$$

Then for $a_n \geq 0$ *and for each* $N \geq 1$ *we have*

$$\sum_{n=1}^{N} q_n \left(Q_n^{-1} \sum_{k=1}^{n} q_k a_k \right)^p \leq K(p) \sum_{n=1}^{N} \left(n^{-1} \sum_{k=1}^{n} q_k^{1/p} a_k \right)^p , \tag{2.2}$$

where $K(p) \leq [A + p^2 B_p/(p-1)]^p$.

Theorem 2. *Let* $1 < p < \infty$, $p^{-1} + (p')^{-1} = 1$ *and let* (q_n) *satisfy* $q_n > 0$ *and*

$$\text{(a)}\quad nq_n \leq AQ_n, \qquad \text{(b)}\quad \left|\Delta\left(\frac{nq_n^{1/p'}}{Q_n}\right)\right| \leq \frac{C_p q_n^{1/p'}}{Q_n}, \qquad n = 1, 2, \ldots. \tag{2.3}$$

Then for $a_n \geq 0$ *and for each* $N \geq 1$ *we have*

$$\sum_{n=1}^{N} q_n \left(\sum_{k=n}^{N} Q_k^{-1} q_k a_k \right)^p \leq K(p) \sum_{n=1}^{N} \left(\sum_{k=n}^{N} k^{-1} q_k^{1/p} a_k \right)^p , \tag{2.4}$$

where $K(p) \leq [A + pC_p]^p$.

Remark 1. Hypothesis (2.1) implies (2.3). For:

$$\Delta\left(nq_n^{1/p'}Q_n^{-1}\right) = (n+1)Q_{n+1}^{-1}\Delta q_n^{1/p'} + q_n^{1/p'}\Delta\left(nQ_n^{-1}\right).$$

Now

$$(n+1)Q_{n+1}^{-1}\left|\Delta q_n^{1/p'}\right| = \left(1+n^{-1}\right)Q_{n+1}^{-1}n\left|\Delta q_n^{1/p'}\right| \le 2Q_n^{-1}B_p q_n^{1/p'}$$

by (2.1) (b), while

$$\left|\Delta\left(nQ_n^{-1}\right)\right| = Q_n^{-1}\left|(n+1)q_{n+1}Q_{n+1}^{-1} - 1\right| \le Q_n^{-1}\max(1, A-1)$$

by (2.1) (a). Hence if (2.1) holds, we have

$$\left|\Delta\left(nq_n^{1/p'}Q_n^{-1}\right)\right| \le C_p q_n^{1/p'}Q_n^{-1}, \quad \text{where } C_p \le 2B_p + \max(1, A-1). \tag{2.5}$$

If we put $q_n := 1/n$ in Theorems 1 and 2, then $Q_n := 1 + (1/2) + \cdots + (1/n)$; thus $Q_n \ge \log(n+1)$ and (2.1), (2.2) hold with $A = 1$, $B_p = 1/p'$, $C_p = 1$. Letting $N \to \infty$ we obtain the following corollaries.

Corollary 1. *Let* $1 < p < \infty$ *and* $a_n \ge 0$ $(n = 1, 2, \ldots)$. *Then*

$$\sum_{n=1}^{\infty}\frac{1}{n}\left(\frac{1}{\log(n+1)}\sum_{k=1}^{n}\frac{a_k}{k}\right)^p \le (p+1)^p\sum_{n=1}^{\infty}\left(\frac{1}{n}\sum_{k=1}^{n}\frac{a_k}{k^{1/p}}\right)^p. \tag{2.6.}$$

Corollary 2. *Let* $1 < p < \infty$ *and* $a_n \ge 0$ $(n = 1, 2, \ldots)$. *Then*

$$\sum_{n=1}^{\infty}\frac{1}{n}\left(\sum_{k=n}^{\infty}\frac{a_k}{k\log(k+1)}\right)^p \le (p+1)^p\sum_{n=1}^{\infty}\left(\sum_{k=n}^{\infty}\frac{a_k}{k^{1+1/p}}\right)^p. \tag{2.7.}$$

Remark 2. If we apply (1.3) to the right side of (2.2) $(N \to \infty)$ we get

$$\sum_{n=1}^{\infty}\left(\frac{1}{n}\sum_{k=1}^{n}q_k^{1/p}a_k\right)^p \le \left(\frac{p}{p-1}\right)^p\sum_{n=1}^{\infty}q_n a_n^p. \tag{2.8}$$

It is a plausible conjecture that $0 \le K(p) \le 1$ in (2.2), though the use of partial summation and Minkowski's inequality at the beginning of our proof (see § 3 below) does not permit this conclusion, and a different method would have to be found. However, if this were the case, we would be able, by combining (2.2) and (2.8), to obtain an improvement on Theorem A, in which the expression on the right of (2.2) is inserted between the two sides of (1.1). For the relevance of (2.2) and (2.8) to sequence spaces see § 4 below.

3. Proof of the theorems

We first require the following elementary result.

Lemma 1. *Let $1 < p < \infty$ and $c_n \geq 0$ $(n = 1, 2, \ldots)$. Then, for $n = 1, 2, \ldots$,*

$$\left(\sum_{k=1}^{n} c_k\right)^p \leq p \sum_{k=1}^{n} c_k \left(\sum_{j=1}^{k} c_j\right)^{p-1}; \tag{3.1}$$

$$\left(\sum_{k=n}^{N} c_k\right)^p \leq p \sum_{k=n}^{N} c_k \left(\sum_{j=k}^{N} c_j\right)^{p-1}, \quad N \geq n. \tag{3.2}$$

The result in (3.1) is due to Davies and Petersen [5, Lemma 1]; see also the discussion and related result in Johnson and Mohapatra [8, Remark (d) following Lemma 5.1]. (3.2) is deducible from (3.1), though it is stated explicitly by Németh [13, Lemma 4]. An integral analogue is used by Mohapatra and Russell [12, Lemma 1].

Proof of Theorem 1

Let $p > 1$, $\dfrac{1}{p} + \dfrac{1}{p'} = 1$, and $\sigma_n := \dfrac{1}{n} \displaystyle\sum_{k=1}^{n} q_k^{1/p} a_k$. Thus

$$\sum_{k=1}^{n} q_k a_k = \sum_{k=1}^{n} q_k^{1/p'} \left(q_k^{1/p} a_k\right) = \sum_{k=1}^{n-1} \left(\Delta q_k^{1/p'}\right) k\sigma_k + q_n^{1/p'} n\sigma_n. \tag{3.3}$$

Then, by (3.3) and Minkowski's inequality,

$$\begin{aligned} &\left[\sum_{n=1}^{N} q_n \left(Q_n^{-1} \sum_{k=1}^{n} q_k a_k\right)^p\right]^{1/p} \\ &\quad \leq \left[\sum_{n=1}^{N} q_n \left(Q_n^{-1} \sum_{k=1}^{n} \left|\Delta q_k^{1/p'}\right| k\sigma_k\right)^p\right]^{1/p} + \left[\sum_{n=1}^{N} \left(Q_n^{-1} q_n n\sigma_n\right)^p\right]^{1/p} \\ &\quad =: \Sigma_1^{1/p} + \Sigma_2^{1/p}. \end{aligned}$$

By (2.1) (a),

$$\Sigma_2 \leq A^p \sum_{n=1}^{N} \sigma_n^p. \tag{3.4}$$

We show that a similar inequality holds also for Σ_1.

Writing $c_k := k\sigma_k \left|\Delta q_k^{1/p'}\right|$ we have by Lemma 1 (3.1)

$$\Sigma_1 := \sum_{n=1}^{N} q_n \left(Q_n^{-1}\sum_{k=1}^{n} c_k\right)^p \leq \sum_{n=1}^{N} q_n Q_n^{-1} p \sum_{k=1}^{n} c_k \left(\sum_{j=1}^{k} c_j\right)^{p-1}$$

(3.5)

$$= p\sum_{k=1}^{N} c_k \left(\sum_{j=1}^{k} c_j\right)^{p-1} \sum_{n=k}^{N} q_n Q_n^{-p}.$$

Now

$$\sum_{n=k}^{N} q_n Q_n^{-p} \leq q_k Q_k^{-p} + \sum_{n=k+1}^{N} \int_{Q_{n-1}}^{Q_n} x^{-p}\, dx \leq Q_k Q_k^{-p} + \frac{Q_k^{1-p}}{p-1} = \frac{pQ_k^{1-p}}{p-1};$$

substituting this in (3.5) and using Hölder's inequality we obtain

$$\begin{aligned} \Sigma_1 &\leq \frac{p^2}{p-1}\sum_{k=1}^{N} c_k \left(Q_k^{-1}\sum_{j=1}^{k} c_j\right)^{p-1} = \frac{p^2}{p-1}\sum_{k=1}^{N} c_k q_k^{(1-p)/p} \left(q_k^{1/p} Q_k^{-1}\sum_{j=1}^{k} c_j\right)^{p-1} \\ &\leq \frac{p^2}{p-1}\left[\sum_{k=1}^{N} c_k^p q_k^{1-p}\right]^{1/p} \left[\sum_{k=1}^{N} q_k \left(Q_k^{-1}\sum_{j=1}^{k} c_j\right)^p\right]^{1/p'} \\ &= \frac{p^2}{p-1}\left[\sum_{k=1}^{N} c_k^p q_k^{1-p}\right]^{1/p} \Sigma_1^{1/p'}. \end{aligned}$$

Divide through by $\Sigma_1^{1/p'}$ and raise both sides to the power p, and we get

$$\Sigma_1 \leq \left(\frac{p^2}{p-1}\right)^p \sum_{k=1}^{N} \left(c_k q_k^{-1/p'}\right)^p,$$

since $1-p = -p/p'$.

Now, by (2.1) (b), $c_k q_k^{-1/p'} = \sigma_k k \left|\Delta q_k^{1/p'}\right| q_k^{-1/p'} \leq \sigma_k B_p$, and hence

(3.6)
$$\Sigma_1 \leq \left(\frac{p^2 B_p}{p-1}\right)^p \sum_{k=1}^{N} \sigma_k^p.$$

Finally, combining (3.4) and (3.6), we get (2.2), namely

$$\sum_{n=1}^{N} q_n \left(Q_n^{-1}\sum_{k=1}^{n} q_k a_k\right)^p \leq \left(A + \frac{p^2 B_p}{p-1}\right)^p \sum_{n=1}^{N} \sigma_n^p. \qquad \square$$

Proof of Theorem 2

Let $p > 1$, $\dfrac{1}{p} + \dfrac{1}{p'} = 1$, and $t_n := \sum_{k=n}^{N} \dfrac{q_k^{1/p} a_k}{k}$ $(t_{N+1} := 0)$. Then

$$\sum_{k=n}^{N} \frac{q_k a_k}{Q_k} = \sum_{k=n}^{N} \frac{k q_k^{1/p'}}{Q_k} \Delta t_k = -\sum_{k=n}^{N-1} t_{k+1} \Delta \left(\frac{k q_k^{1/p'}}{Q_k} \right) + n q_n^{1/p'} \frac{t_n}{Q_n},$$

and applying Minkowski's inequality we get

$$\begin{aligned} \left[\sum_{n=1}^{N} q_n \left(\sum_{k=n}^{N} \frac{q_k a_k}{Q_k} \right)^p \right]^{1/p} &\leq \left[\sum_{n=1}^{N} q_n \left(\sum_{k=n}^{N} d_k \right)^p \right]^{1/p} \\ &\quad + \left[\sum_{n=1}^{N} \left(\frac{n q_n t_n}{Q_n} \right)^p \right]^{1/p} \\ &=: \; S_1^{1/p} + S_2^{1/p}, \end{aligned} \tag{3.7}$$

where $d_k := \left| \Delta \left(k q_k^{1/p'} / Q_k \right) \right| t_{k+1}$. Then, by (2.3) (a),

$$S_2 \leq A^p \sum_{n=1}^{N} t_n^p. \tag{3.8}$$

We now apply Lemma 1 (3.2) to S_1 to give

$$\begin{aligned} S_1 &\leq \sum_{n=1}^{N} q_n p \sum_{k=n}^{N} d_k \left(\sum_{j=k}^{N} d_j \right)^{p-1} = p \sum_{k=1}^{N} d_k \left(\sum_{j=k}^{N} d_j \right)^{p-1} \sum_{n=1}^{k} q_n \\ &= p \sum_{k=1}^{N} d_k Q_k q_k^{(1-p)/p} \left(q_k^{1/p} \sum_{j=k}^{N} d_j \right)^{p-1} \\ &\leq p \left[\sum_{k=1}^{N} d_k^p Q_k^p q_k^{1-p} \right]^{1/p} S_1^{1/p'} \end{aligned}$$

by Hölder's inequality.

Dividing by $S_1^{1/p'}$ and raising both sides to the power p, we get

$$S_1 \leq p^p \sum_{k=1}^{N} \left(d_k Q_k q_k^{-1/p'} \right)^p.$$

However, by (2.3) (b) and the definition of d_k,

$$d_k Q_k q_k^{-1/p'} \le C_p t_{k+1}$$

and hence

$$S_1 \le (pC_p)^p \sum_{k=2}^{N} t_k^p. \tag{3.9}$$

The combination of (3.7), (3.8), (3.9) now yields (2.4). □

4. Inclusion among sequence spaces

If ω is the vector space of all real- or complex-valued sequences and φ is the subspace of all sequences with finitely many non-zero entries, we regard a *sequence space* as any subspace of ω containing φ.

Given $q = (q_n)$ $(n \ge 1)$, $q_n > 0$, $Q_n := q_1 + \cdots + q_n$, $p > 0$, examples are

$$q^{-1/p} l_p := \left\{ x \in \omega | \sum_{n=1}^{\infty} q_n |x_n|^p < \infty \right\},$$

$$\text{ces}_p := \left\{ x \in \omega | \sum_{n=1}^{\infty} \left(\frac{1}{n} \sum_{k=1}^{n} |x_k| \right)^p < \infty \right\},$$

$$q^{-1/p}\text{ces}_p := \left\{ x \in \omega | \sum_{n=1}^{\infty} \left(\frac{1}{n} \sum_{k=1}^{n} q_k^{1/p} |x_k| \right)^p < \infty \right\}.$$

Given an infinite matrix of real numbers $A = (a_{nk})$ $(n, k \ge 1)$, the transform of a sequence x is $Ax = ((Ax)_n)$ defined by

$$(Ax)_n := \sum_{k=1}^{\infty} a_{nk} x_k \quad (n = 1, 2, \dots)$$

and we write $|x| := (|x_k|)$. If $a_{nk} := q_k/Q_n$ $(1 \le k \le n)$, $a_{nk} := 0$ $(k \ge n)$, then $A = M_q$, the matrix of weighted means. Johnson and Mohapatra [10] have studied, for a sequence space λ, the topological properties of

$$\text{nor}-A^{-1}(\lambda) := \{x \in \omega | \; A|x| \in \lambda\},$$

and in [8] they study its application to inequalities involving lower triangular matrices A. Thus, for instance,

$$\text{nor}-M_q^{-1}\left(q^{-1/p} l_p\right) := \left\{ x \in \omega | \sum_{n=1}^{\infty} q_n \left(Q_n^{-1} \sum_{k=1}^{n} q_k |x_x| \right)^p < \infty \right\}.$$

Many of the inequalities we have referred to in the present paper may be regarded as inclusion relations between sequence spaces. Thus Hardy's inequality (1.3) asserts that

$$l_p \subseteq \text{ces}_p \quad (p > 1)$$

and that the norm of the operator mapping l_p to ces_p is $p/(p-1)$. The replacement of x_k by $q_k^{1/p} x_k$ (as in (2.8)) gives of course

$$q^{-1/p} l_p \subseteq q^{-1/p} \text{ces}_p \quad (p > 1). \tag{4.1}$$

Likewise Copson's inequality in Theorem A (1.1) gives

$$q^{-1/p} l_p \subseteq \text{nor}-M_q^{-1}\left(q^{-1/p} l_p\right) \quad (p > 1). \tag{4.2}$$

What our Theorem 1 shows is that

$$q^{-1/p} \text{ces}_p \subseteq \text{nor}-M_q^{-1}\left(q^{-1/p} l_p\right) \quad (p > 1); \tag{4.3}$$

in other words, we can insert $q^{-1/p}\text{ces}_p$ between the two sides of (4.2). The question of the exact operator norm between these spaces has not been considered here. Similar inclusion relations can be stated for Theorem 2, and related results on Copson's inequality and inclusion between sequence spaces have been considered by Johnson and Mohapatra [9].

References

1. G. Bennett, *Some elementary inequalities*. Quart. J. Math. (Oxford) (2) **38** (1987), 401–425.

2. G. Bennett, *Some elementary inequalities, II*. Quart. J. Math. (Oxford) (2) **39** (1988), 385–400.

3. E.T. Copson, *Note on series of positive terms*. J. London Math. Soc. **2** (1927), 9–12.

4. E.T. Copson, *Note on series of positive terms*. J. London Math. Soc. **3** (1928), 49–51.

5. G.S. Davies and G.M. Petersen, *On an inequality of Hardy's (II)*. Quart. J. Math. (Oxford) (2) **15** (1964), 35–40.

6. E.B. Elliott, *A simple exposition of some recently proved facts as to convergence.* J. London Math. Soc. **1** (1926), 93–96.

7. G.H. Hardy, J.E. Littlewood and G. Pólya, *Inequalities.* Univ. Press, Cambridge, 2nd Ed., 1952.

8. P.D. Johnson Jr. and R.N. Mohapatra, *Inequalities involving lower triangular matrices.* Proc. London Math. Soc. (3) **41** (1980), 83–137.

9. P.D. Johnson Jr. and R.N. Mohapatra, *On inequalities related to sequence spaces* ces $[p,q]$. In: General Inequalities 4, 191–201 (ed. W. Walter); Birkhäuser Verlag, Basel, 1984.

10. P.D. Johnson Jr. and R.N. Mohapatra, *The maximal normal subspace of the inverse image of a normal space of sequences by a non-negative matrix transformation.* Ann. Polon. Math. **45** (1985), 104–120.

11. E.R. Love, *Generalizations of Hardy's and Copson's inequalities.* J. London Math. Soc. **30** (1984), 431–440.

12. R.N. Mohapatra and D.C. Russell, *Integral inequalities related to Hardy's inequality.* Aequat. Math. **28** (1985), 199–207.

13. J. Németh, *Generalizations of the Hardy-Littlewood inequality.* Acta Sci. Math. (Szeged) **32** (1971), 295–299.

R.N. Mohapatra and K. Vajravelu, Department of Mathematics, University of Central Florida, Orlando, Florida 32816-6990, USA

D.C. Russell, Department of Mathematics & Statistics, York University, Toronto-Downsview, Ontario M3J 1P3, Canada

International Series of Numerical Mathematics, Vol. 103, © 1992 Birkhäuser Verlag Basel

Comparison of two variable homogeneous means

Zsolt Páles

Abstract

In this paper we give necessary and sufficient conditions for the inequality

(*) $$M(x,y) \le N(x,y), \qquad x,y \in [\alpha,\beta],$$

where $0 < \alpha < \beta < \infty$ are fixed values and $M : \mathbb{R}_+ \times \mathbb{R}_+ \to \mathbb{R}_+$ and $N : \mathbb{R}_+ \times \mathbb{R}_+ \to \mathbb{R}_+$ belong to one of the following classes of means:

$$D_{a,b}(x,y) = \left(\frac{x^a - y^a}{a}\frac{b}{x^b - y^b}\right)^{\frac{1}{a-b}}, \qquad S_{a,b}(x,y) = \left(\frac{x^a + y^a}{x^b + y^b}\right)^{\frac{1}{a-b}}.$$

The following two inequalities are simple necessary conditions for (*) to hold:

$$-\partial_1\partial_2 M(1,1) \le -\partial_1\partial_2 N(1,1), \qquad M(\alpha,\beta) \le N(\alpha,\beta).$$

In the class of power means, already the first inequality is a sufficient condition. The aim of this paper is to show that these two inequalities together are sufficient conditions for the comparison of the above D and S means as well.

1. Introduction

In the present paper we deal with inequalities of the form

(1) $$M(x,y) \le N(x,y), \qquad x,y \in [\alpha,\beta],$$

where $M : \mathbb{R}_+ \times \mathbb{R}_+ \to \mathbb{R}_+$ and $N : \mathbb{R}_+ \times \mathbb{R}_+ \to \mathbb{R}_+$ are homogeneous two variable symmetric means in a given class of means $\mathcal{M}$ and $0 < \alpha < \beta < \infty$ are fixed values. Our aim is to find necessary and sufficient conditions for (1) to hold.

This paper is in final form and no version of it will be submitted for publication elsewhere.

If M and N are C^2 functions then a simple necessary condition can be obtained in the following way: Let $\alpha < x_0 < \beta$ be fixed. Then tx_0 is also betveen α and β if t is close to 1 enough. Substituting $x = x_0$ and $y = tx_0$ into (1) and using homogeneity we obtain

$$0 \le N(1,t) - M(1,t) =: \delta(t)$$

for values of t from a neighbourhood of 1. Since $\delta(1) = 0$, hence $\delta''(1) \ge 0$. Differentiating the identities $M(t,t) = t$, $N(t,t) = t$ with respect to t two times, the symmetry of M and N yields

$$\partial_2^2 M(t,t) = -\partial_1\partial_2 M(t,t), \qquad \partial_2^2 N(t,t) = -\partial_1\partial_2 N(t,t).$$

Therefore the condition $\delta''(1) \ge 0$ can be expressed in the form

$$-\partial_1\partial_2 M(1,1) \le -\partial_1\partial_2 N(1,1). \tag{2}$$

Sometimes this condition is also sufficient. For instance, if $\mathcal{M}$ is the class of power means, then (2) is equivalent to the comparison of the corresponding powers.

Another obvious necessary condition is

$$M(\alpha,\beta) \le N(\alpha,\beta). \tag{3}$$

In what follows, we define two classes of means in which (2) and (3) together turn out to be sufficient conditions.

The class of *difference means* $\mathcal{D}$ consists of the means

$$D_{a,b}(x,y) = D(a,b;x,y) = \left(\frac{x^a - y^a}{a}\frac{b}{x^b - y^b}\right)^{\frac{1}{a-b}}$$

if $a, b \in \mathbb{R}$, $x, y > 0$ and $ab(a-b)(x-y) \ne 0$. It can be shown that D can be extended continuously to the domain

$$\{(a,b;x,y) : a, b \in \mathbb{R}, x, y > 0\}. \tag{4}$$

(See Stolarsky [13] who was the first to define this class of means.) If $a = 2p$, $b = p$ then it is easy to see that $D_{a,b}$ reduces to the pth power mean.

The second class to be considered here consists of the two variable Gini means, called *sum means* here:

$$S_{a,b}(x,y) = S(a,b;x,y) = \left(\frac{x^a + y^a}{x^b + y^b}\right)^{\frac{1}{a-b}}$$

where $a, b \in \mathbb{R}$, $x, y > 0$ and $a \neq b$. It can be shown that S can be extended to the same domain (4) as D. (See Gini [5]). The power means are also contained in this class, namely if $a = p$ and $b = 0$ then $S_{a,b}$ is obviously the pth power mean.

There is a number of papers concerning the comparison of difference and sum means. (See, for instance [1], [2], [3], [4], [7], [11].) The problem of comparison of the difference means on the infinite interval $(0, \infty)$ was completely solved first by Leach and Sholander [7]. The same problem for the sum means was treated and solved by the author in [10], Using the same method, a new proof for the result of Leach and Sholander was presented in [9].

The subject of the present paper is to offer a unified treatment for these comparison problems on bounded intervals of the positive reals. The main results of the paper are the following two theorems.

Theorem D. *Let a, b, c, d be real parameters and $0 < \alpha < \beta < \infty$. Then*

$$D_{a,b}(x, y) \leq D_{c,d}(x, y)$$

holds for all $x, y \in [\alpha, \beta]$ if and only if

$$a + b \leq c + d \quad \text{and} \quad D_{a,b}(\alpha, \beta) \leq D_{c,d}(\alpha, \beta).$$

Theorem S. *Let a, b, c, d be real parameters and $0 < \alpha < \beta < \infty$. Then*

$$S_{a,b}(x, y) \leq S_{c,d}(x, y)$$

holds for all $x, y \in [\alpha, \beta]$ if and only if

$$a + b \leq c + d \quad \text{and} \quad S_{a,b}(\alpha, \beta) \leq S_{c,d}(\alpha, \beta).$$

The comparison of means on bounded intervals is clearly more difficult than on $(0, \infty)$. Inspite of this, the formulation of the results are more complicated for the $(0, \infty)$ interval (cf.[9], [10]).

2. Basic Tools

The following lemma plays an important role in the sequel. It is a sharper form of an intermediate value theorem.

Lemma. *Let $f : \mathbb{R} \to \mathbb{R}$ be a three times differentiable even function and assume that there exist values $a_1 < a_2 < a_3 < a_4$ with $a_1 + a_4 > 0$, $a_2 + a_3 > 0$ and a linear function $l : \mathbb{R} \to \mathbb{R}$ such that*

$$f(a_i) = l(a_i) \qquad \text{for} \qquad i = 1, 2, 3, 4.$$

Then f'' has at least two distinct positive zeros and consequently $f'''(x)$ vanishes for some positive x.

Proof. We prove by contradiction. Assume that $f''(x)$ vanishes for at most one positive x.

The function $f - l$ vanishes at four points, therefore its second derivative, f'' must vanish at two distinct points, $\xi > \xi'$. Then only ξ can be positive and thus $\xi' = -\xi$ since f'' is even. We may assume that $f''(x)$ is positive for $0 < |x| < \xi$, otherwise we would work with $-f$ instead of f. Then $f''(x)$ must be negative for $|x| > \xi$, since otherwise f would be strictly convex, and a strictly convex function cannot intersect a linear function at four points. In other words, f is convex on $[-\xi, \xi]$ and is concave on the complementer set.

Now we show that f' vanishes at a point $\eta > \xi$. As we have proved, f' is strictly increasing on $[-\xi, \xi]$ and is strictly decreasing on $(-\infty, -\xi]$ and on $[\xi, \infty)$. The function $f' - l'$ vanishes at three distinct points, that is f' assumes the value $c = l'(1)$ at three different places. If $c \geq 0$, then $f'(x)$ can assume c only for two nonnegative values of x. Therefore $f'(x_0) = c \geq 0$ for some negative x_0. Since $f'(x) < f'(0) = 0$ for $-\xi < x < 0$, thus $f'(x)$ must vanish for some value $x = -\eta < -\xi$. Then $f'(\eta) = f'(-\eta) = 0$. For $c \leq 0$ the proof is similar.

Thus f' is an odd function which is positive on $(-\infty, -\eta)$ and on $(0, \eta)$ and negative on $(-\eta, 0)$ and on (η, ∞). Therefore f is an even function which is strictly increasing on $(-\infty, -\eta]$ and on $[0, \eta]$ and strictly decreasing on $[-\eta, 0]$ and on $[\eta, \infty)$ and has global maximum at $x = -\eta$ and $x = \eta$.

Now we state that $-\eta < a_2 < a_3 < \eta$. If a_2 were $\leq -\eta$, then by $l(a_1) = f(a_1) < f(a_2) = l(a_2)$, l would be strictly increasing, and since f is concave on $(-\infty, -\eta]$, $l(-\eta) \geq f(-\eta)$. Because f has a global maximum at $-\eta$, the function l could not intersect f at a_3 and a_4. Similarly we see that $a_3 < \eta$.

The function f is increasing on $[0, \eta]$, therefore the assumption $a_2 + a_3 > 0$ implies $f(a_2) = f(-a_2) < f(a_3)$. Thus l must be an increasing function. For a_1 we have $a_1 < -\xi$. Otherwise $f - l$ would have at least four zeros for $x \geq -\xi$ and hence $f'(x) - l'(1)$ at least three zeros for $x \geq -\xi$. This contradicts that f' is strictly increasing on $[-\xi, \xi]$ and strictly decreasing on $[\xi, \infty)$. In fact we have $a_1 < -\eta$. If a_1 were in $[-\eta, -\xi)$, we would have $a_2 > \xi$ since l is increasing, and $f(x) < f(\xi) = f(-\xi) \leq f(a_1)$ for $|x| < \xi$ owing to the monotonicity of f. But l intersects f at most twice in $[\xi, \infty)$ because f is strictly concave on $[\xi, \infty)$.

Now $a_1 + a_4 > 0$ implies $a_4 > \eta$. Because l is increasing we get $f(a_1) = l(a_1) < l(a_4) = f(a_4) = f(-a_4)$. Since f is strictly increasing on $(-\infty, -\eta]$ we get $a_1 < -a_4$, which contradicts $a_4 + a_1 > 0$ and completes the proof.

Theorem 1. *Let $g : \mathbb{R} \to \mathbb{R}$ be a four times differentiable even function such that $g''(0) > 0$ and*

$$g''(x) > 0, \qquad g'''(x) < 0, \qquad \text{and} \qquad (xg'''(x)/g''(x))' < 0$$

holds for all $x > 0$. Let further $a_1 < a_2 < a_3 < a_4$ and $\alpha_1, \ldots, \alpha_4$ be arbitrary values with

$$(5) \quad (a_1 + a_4)(a_2 + a_3) > 0, \qquad \text{and} \qquad \alpha_1 + \cdots + \alpha_4 = 0 = \alpha_1 a_1 + \cdots + \alpha_4 a_4.$$

Let $s > 0$ be a fixed value. Then the inequality

$$(6) \qquad 0 \leq \alpha_1 g(a_1 t) + \cdots + \alpha_4 g(a_4 t)$$

is valid for $-s \leq t \leq s$ if and only if

$$(7) \qquad 0 \leq \alpha_1 a_1^2 + \cdots + \alpha_4 a_4^2 \qquad \text{and} \qquad 0 \leq \alpha_1 g(a_1 s) + \cdots + \alpha_4 g(a_4 s).$$

Proof. First we show that (6) implies (7). By (6) the function $t \mapsto \alpha_1 g(a_1 t) + \cdots + \alpha_4 g(a_4 t)$ has a minimum at the point $t = 0$ therefore its second derivative is nonnegative at $t = 0$. Thus we get that the first inequality in (7) is necessary. The necessity of the second inequality is obvious.

Now we prove the sufficiency of (7). Without loss of generality we can assume that $a_1 + a_4 > 0$ and $a_2 + a_3 > 0$, otherwise we would work with $-a_4 < -a_3 < -a_2 < -a_1$ instead of $a_1 < a_2 < a_3 < a_4$, respectively.

Let $t \in (0, s)$ be fixed arbitrarily. We are going to show first that there exist real values u, v and nonnegative real values w, z such that

$$(8) \qquad g(a_i t) = u + v a_i + w a_i^2 + z g(a_i s), \qquad i = 1, 2, 3, 4.$$

This is a linear system of equations for the unknown values u, v, w, z. First we show that the homogeneous system corresponding to (8) has only trivial solution for u, v, w, z. Let

$$f_1(x) := wx^2 + zg(xs), \qquad l_1(x) := -u - vx, \qquad x \in \mathbb{R}.$$

Then this homogeneous system can be written as

$$f_1(a_i) = l_1(a_i), \qquad i = 1, 2, 3, 4.$$

By the lemma there exists a value $\xi > 0$ for which $f_1'''(\xi) = 0$. This yields $zs^3g'''(\xi s) = 0$, whence $z = 0$. With this value of z the homogeneous system has the form

$$wa_i^2 + va_i + u = 0, \qquad i = 1,2,3,4,$$

whence $u = v = w = 0$ follows. Therefore the homogeneous system admits only the trivial solution. Then (8) can be solved uniquely for u, v, w, z. We want to show now that w and z are nonnegative. Define

$$f_2(x) := wx^2 + zg(xs) - g(xt), \qquad l_2(x) := -u - vx, \qquad x \in \mathbb{R}.$$

Then $f_2(a_i) = l_2(a_i)$ holds for $i = 1,2,3,4$ by (8). Applying the lemma again, we find three values $0 < \xi_1 < \xi < \xi_2$ such that $f_2''(\xi_1) = f_2''(\xi_2) = 0$ and $f_2'''(\xi) = 0$. The latter relation yields

$$z = \frac{t^3 g'''(\xi t)}{s^3 g'''(\xi s)} > 0.$$

We also have

$$2w + zs^2g''(\xi_1 s) = t^2g''(\xi_1 t), \qquad \text{and} \qquad 2w + zs^2g''(\xi_2 s) = t^2g''(\xi_2 t).$$

Eliminating z, we get the following representation for w:

$$w = \frac{t^2[g''(\xi_1 t)g''(\xi_2 s) - g''(\xi_2 t)g''(\xi_1 s)]}{2[g''(\xi_2 s) - g''(\xi_1 s)]} = \frac{t^2}{2} \frac{\dfrac{g''(\xi_1 t)}{g''(\xi_1 s)} - \dfrac{g''(\xi_2 t)}{g''(\xi_2 s)}}{\dfrac{1}{g''(\xi_1 s)} - \dfrac{1}{g''(\xi_2 s)}}.$$

Applying the Cauchy mean value theorem, we can find a value $\eta \in (\xi_1, \xi_2)$ such that

$$\begin{aligned} w &= \frac{t^2}{2} \frac{tg'''(\eta t)g''(\eta s) - sg'''(\eta s)g''(\eta t)}{-sg'''(\eta s)} \\ &= \frac{t^2 g''(\eta t)g''(\eta s)}{-2\eta s g'''(\eta s)} \left[\frac{\eta t g'''(\eta t)}{g''(\eta t)} - \frac{\eta s g'''(\eta s)}{g''(\eta s)} \right] > 0, \end{aligned}$$

where the last inequality follows from $t < s$ and the properties of g listed in the theorem.

Now we are able to prove (6) using (5), (7) and (8). Multiply (8) by α_i and add the inequalities obtained. Then we get

$$\sum \alpha_i g(a_i t) = u \sum \alpha_i + v \sum \alpha_i a_i + w \sum \alpha_i a_i^2 + z \sum \alpha_i g(a_i s) \geq 0.$$

The proof is completed.

Theorem 2. *Let $g : \mathbb{R} \to \mathbb{R}$ be a four times differentiable even function such that $g''(0) > 0$ and*

$$g''(x) > 0, \qquad g'''(x) < 0, \qquad \text{and} \qquad (xg'''(x)/g''(x))' < 0$$

holds for all $x > 0$. Define $G_{a,b}(t)$ by

$$G_{a,b}(t) := \begin{cases} \dfrac{g(at) - g(bt)}{a - b}, & \text{if } a \neq b, \\ \\ tg'(at), & \text{if } a = b. \end{cases}$$

Let a, b, c, d and $s > 0$ be fixed real values. Then the inequality

$$G_{a,b}(t) \leq G_{c,d}(t) \tag{9}$$

holds true for all $t \in [-s, s]$ if and only if

$$a + b \leq c + d \qquad \text{and} \qquad G_{a,b}(s) \leq G_{c,d}(s). \tag{10}$$

Proof. First we prove the necessity of the conditions. The necessity of the second inequality in (10) is obvious. On the other hand, the function

$$t \mapsto G_{c,d}(t) - G_{a,b}(t) = \gamma(t), \qquad t \in \mathbb{R},$$

has a minimum at $t = 0$. Thus $\gamma''(0) \geq 0$. Since $G''_{a,b}(0) = (a + b)g''(0)$, therefore $\gamma(0) \geq 0$ reduces to the first inequality in (10).

In the proof of the sufficiency *we assume first that the absolute values of a, b, c, d are pairwise distinct values.*

The inequality (9) can be written in the following form:

$$\frac{1}{b - a}g(at) + \frac{1}{a - b}g(bt) + \frac{1}{c - d}g(ct) + \frac{1}{d - c}g(dt) \geq 0. \tag{11}$$

Let $((a_1, \alpha_1), (a_2, \alpha_2), (a_3, \alpha_3), (a_4, \alpha_4))$ be a permutation of

$$\left(\left(a, \frac{1}{b - a}\right), \left(b, \frac{1}{a - b}\right), \left(c, \frac{1}{c - d}\right), \left(d, \frac{1}{d - c}\right)\right)$$

such that $a_1 < a_2 < a_3 < a_4$ holds true.

Then (11) can be rewritten as

$$\alpha_1 g(a_1 t) + \cdots + \alpha_4 g(a_4 t) \geq 0. \tag{12}$$

Moreover the inequality $a+b \le c+d$ is equivalent to

$$0 \le \alpha_1 a_1^2 + \cdots + \alpha_4 a_4^2. \tag{13}$$

Because of symmetry we may assume that $a < b$ and $c < d$. On the other hand we have $a+b \le c+d$ by (10). Thus we obtain the following possibilities for $a_1 < \ldots < a_4$:

$$a<b<c<d, \qquad a<c<b<d, \qquad a<c<d<b, \qquad c<a<b<d.$$

If $0 < a+b \le c+d$ then one can check that $a_1+a_4 > 0$ and $a_2+a_3 > 0$ holds in each of the above cases. Therefore Theorem 1 can be applied. Thus (12) with $s=t$ and (13) are imply (12) for $t \in (-s, s)$.

The case $a+b \le c+d < 0$ can be treated similarly. Then $a_1+a_4 < 0$ and $a_2+a_3 < 0$, and Theorem 1 can also be applied.

The last case we have to deal with is $a+b < 0 < c+d$. Since $g''(x)$ is nonegative for all $x \in \mathbb{R}$, hence g' is an increasing function; $g'(0)=0$ thus $g(x)$ decreases for $x<0$ and increases for $x>0$. The inequalities $a<b$ and $a<-b$ yield $a<-|b|$, whence we get $g(at) > g(bt)$. Thus $G_{a,b}(t) < 0$. Similarly, $G_{c,d}(t) > 0$. Therefore (9) is valid in this case as well.

Now we eliminate the assumption that the absolute values of a, b, c, d are pairwise distinct.

Let a, b, c, d be arbitrary parameters satisfying (10). Let $\{a_n\}$, $\{b_n\}$ be increasing sequences and $\{c_n\}$, $\{d_n\}$ be decreasing sequences such that

$$\lim_{n\to\infty} a_n = a, \qquad \lim_{n\to\infty} b_n = b, \qquad \lim_{n\to\infty} c_n = c, \qquad \lim_{n\to\infty} d_n = d,$$

and the absolute values of a_n, b_n, c_n, d_n are pairwise distinct values.

Since g is a convex function, therefore

$$G_{a_n,b_n}(t) \le G_{a,b}(t) \qquad \text{and} \qquad G_{c_n,d_n}(t) \ge G_{c,d}(t).$$

Thus it follows from (10) that

$$a_n + b_n \le c_n + d_n \qquad \text{and} \qquad G_{a_n,b_n}(s) \le G_{c_n,d_n}(s).$$

Applying what we have already proved, these conditions imply

$$G_{a_n,b_n}(t) \le G_{c_n,d_n}(t)$$

for all $t \in (-s, s)$. Taking now the limit $n \to \infty$, we obtain (9) and complete the proof of the theorem.

3. The comparison of difference means

Define the function g by the formula:

$$(14) \qquad g(x) := \begin{cases} \ln\left(\dfrac{\sinh x}{x}\right), & \text{if} \quad x \neq 0, \\ 0, & \text{if} \quad x = 0. \end{cases}$$

Then we have the following representation for difference means:

$$D_{a,b}(x,y) = \sqrt{xy}\exp\left(\frac{g(a\ln\sqrt{x/y}) - g(b\ln\sqrt{x/y})}{a-b}\right).$$

for $a \neq b$. More exactly we have

$$D_{a,b}(x,y) = \sqrt{xy}\exp\big(G_{a,b}(\ln\sqrt{x/y})\big)$$

with the notation introduced in Theorem 2.

Therefore the comparison inequality

$$D_{a,b}(x,y) \leq D_{c,d}(x,y), \qquad x, y \in [\alpha, \beta]$$

is equivalent to

$$G_{a,b}(t) \leq G_{c,d}(t), \qquad t \in \big[\ln\sqrt{\alpha/\beta}, \ln\sqrt{\beta/\alpha}\big].$$

To complete the proof of Theorem D, we have to show the following

Lemma D. *The function g defined by (14) is an even analytic function, $g''(0) > 0$ and*

$$g''(x) > 0, \qquad g'''(x) < 0, \qquad \textit{and} \qquad (xg'''(x)/g''(x))' < 0$$

holds for all $x > 0$.

Proof. A simple calculation yields that $g''(0) = 1/3$ and

$$(15) \qquad g''(x) = \frac{\sinh^2 x - x^2}{(x\sinh x)^2}, \qquad g'''(x) = 2\frac{x^3\cosh x - \sinh^3 x}{(x\sinh x)^3}$$

for $x > 0$. Applying the inequalities

$$\sinh x > x, \qquad x^3\cosh x < \sinh^3 x \qquad (x > 0)$$

proved in [9] we get $g''(x) > 0$ and $g'''(x) < 0$ for $x > 0$.

One of the fundamental lemmas of Leach and Sholander [7, Lemma 1, formula (12)] states that the function $\gamma(t) = g''(t)$ satisfies

$$\frac{\partial}{\partial x}[\gamma(xy)]\frac{\partial}{\partial y}[\gamma(xy)] - \gamma(xy)\frac{\partial}{\partial x}\frac{\partial}{\partial y}[\gamma(xy)] > 0 \qquad x, y > 0.$$

In other words, this means that

$$t(g'''(t))^2 - tg''(t)g''''(t) - g''(t)g'''(t) > 0$$

for $t > 0$. An easy calculation shows that this inequality is enough to obtain

$$\left(\frac{xg'''(x)}{g''(x)}\right)' < 0, \qquad x > 0.$$

Thus the proof is complete.

4. The comparison of sum means

The sum means have the following representation:

$$S_{a,b}(x, y) = \sqrt{xy}\exp\big(G_{a,b}(\ln\sqrt{x/y})\big)$$

where the function G is introduced in Theorem 2 with the help of g defined by:

$$g(t) := \ln(\cosh x), \qquad x \in \mathbb{R}. \tag{16}$$

Thus the comparison inequality

$$S_{a,b}(x, y) \le S_{c,d}(x, y), \qquad x, y \in [\alpha, \beta]$$

is equivalent to

$$G_{a,b}(t) \le G_{c,d}(t), \qquad t \in \left[\ln\sqrt{\alpha/\beta}, \ln\sqrt{\beta/\alpha}\right].$$

Theorem S is an easy consequence of the following

Lemma S. *The function g defined by (16) is an even analytic function, $g''(0) > 0$ and*

$$g''(x) > 0, \qquad g'''(x) < 0, \qquad \textit{and} \qquad (xg'''(x)/g''(x))' < 0 \tag{17}$$

holds for all $x > 0$.

Proof. A simple calculation yields

$$g''(0) = 1, \qquad g''(x) = \frac{1}{\cosh^2 x}, \qquad g'''(x) = -2\frac{\sinh x}{\cosh^3 x}.$$

Thus the first two inequalities in (17) are clear. To prove the last one, observe that

$$\frac{xg'''(x)}{g''(x)} = -2x \cdot \tanh x.$$

The product of positive increasing functions is increasing, therefore the right hand side of the latter equality is a decreasing function.

The proof is complete.

5. Open problem

The definition of difference and sum means can be extended for conjugate complex parameters a, b. Then the function $D_{a,b}$ and $S_{a,b}$ still have real values, but their real-function form contains trigonometric functions instead of hyperbolic functions. To obtain a two variable mean, their domain should also be restricted. The problem is to find necessary and sufficient conditions for the comparison in this more general case. A possible conjecture is that Theorems D and S remain valid in this extended case as well (however the argument of their proofs clearly cannot be followed then). Some results in this direction for sum means (without making restriction for the number of variables) can be found in [11].

Acknowledgements. This research was supported by Hungarian National Science Foundation Operating Grant number OTKA 1652.

The author also wishes to thank the referee whose suggestions and criticism greatly improved the clarity of the proofs. The second part of the proof of the lemma is due to Mr. Weckesser and the editor.

References

1. J. L. Brenner, *A unified treatment and extension of some means of classical analysis I. Comparison theorems.* J. Combin. Inform. System Sci. **3** (1978), 175–199.

2. F. Burk, *By all means.* Amer. Math. Monthly **92** (1985), 50.

3. B. C. Carlson, *The logarithmic mean.* Amer. Math. Monthly **79** (1972), 615–618.

4. E. L. Dodd, *Some generalizations of the logarithmic mean and of similar means of two variates which become indeterminate when the two variates are equal.* Ann. Math. Statist. **12** (1971), 422–428.

5. C. Gini, *Di una formula comprensiva delle medie.* Metron, **13** (1938), 3–22.

6. E. Leach and M. Sholander, *Extended mean values.* Amer. Math. Monthly **85** (1978), 84–90.

7. E. Leach and M. Sholander, *Extended mean values II.* J. Math. Anal. Appl. **92** (1983), 207–223.

8. T. P. Lin, *The power mean and the logarithmic mean.* Amer. Math. Monthly **81** (1974), 879–883.

9. Zs. Páles, *Inequalities for differences of powers.* J. Math. Anal. Appl., **131** (1988), 271–281.

10. Zs. Páles, *Inequalities for sums of powers.* J. Math. Anal. Appl. **131** (1988), 265–270.

11. Zs. Páles, *On comparison of homogeneous means.* Annales Univ. Sci. **32** (1989), 261–266.

12. A. O. Pittinger, *Inequalities between arithmetic and logarithmic means.* Univ. Beograd Publ. Elektrotechn. Fak. Ser. Mat. Fiz. **680** (1980), 15–18.

13. K. B. Stolarsky, *Generalization of the logarithmic mean.* Math. Mag. **48** (1975), 87–92.

Zsolt Páles, Institute of Mathematics, L. Kossuth University, H-4010 Debrecen, Pf. 12, Hungary

Inequalities in Analysis and Approximation

Black forest fir tree

International Series of Numerical Mathematics, Vol. 103, © 1992 Birkhäuser Verlag Basel

Sharp inequalities in polynomial interpolation

Ravi P. Agarwal

Abstract

This paper surveys recently established inequalities in classical polynomial interpolation theory. Besides their applications in numerical computation these inequalities are of immense value in the study of boundary value problems for ordinary differential equations.

1. Introduction

Let $x(t) \in C^{(n)}[a,b]$, $n \geq 2, -\infty < a < b < \infty$ be given and let $P_{n-1}(t)$ be the Hermite interpolating polynomial of degree $(n-1)$ which agrees with $x(t)$ and its first $k_i \geq 0$ derivatives at a_i, i.e.,

$$P^{(j)}_{n-1}(a_i) = x^{(j)}(a_i); \quad 0 \leq j \leq k_i, \quad 1 \leq i \leq r \ (\geq 2) \tag{1.1}$$

where $a = a_1 < a_2 < \ldots < a_r = b$ and $\sum_{i=1}^{r} k_i + r = n$. Let $e(t) = x(t) - P_{n-1}(t)$ be the associated error. In interpolation theory it is known, e.g. Agarwal [4], that

$$e(t) = x(t) - P_{n-1}(t) = \int_a^b g(t,s)x^{(n)}(s)ds, \tag{1.2}$$

where $g(t,s)$ is the Green's function (Peano's kernel) of the boundary value problem

$$\begin{aligned} x^{(n)}(t) &= 0 \\ x^{(j)}(a_i) &= 0; \quad 0 \leq j \leq k_i,\ 1 \leq i \leq r. \end{aligned}$$

Differentiating (1.2), k times immediately yields

$$|e^{(k)}(t)| \leq C_{n,k}(b-a)^{n-k}\|x^{(n)}(t)\|_{\infty}, \quad 0 \leq k \leq n-1 \tag{1.3}$$

where the best possible constants $C_{n,k}$ are

$$C_{n,k} = \max_{a \leq t \leq b} \left(\frac{1}{(b-a)^{n-k}} \int_a^b \left| \frac{\partial^k g(t,s)}{\partial t^k} \right| ds \right), \quad 0 \leq k \leq n-1. \tag{1.4}$$

This paper is in final form and no version of it will be submitted for publication elsewhere.

However, from (1.4) the explicit computation of these constants, except for the case $n = 2$, is unwieldy. For example, Birkhoff and Priver [14] could get these constants $C_{n,k}$ for $n = 4$, $r = 2$, $k_1 = k_2 = 1$ and $n = 6$, $r = 2$, $k_1 = k_2 = 2$ only by using computer. The difficulty lies in the fact that the derivative $\frac{\partial^k g(t,s)}{\partial t^k}$ $(1 \le k \le n-1$, $n > 2)$ is messy, further as a function of s it changes sign in the interval $[a, b]$. Similar difficulties arise in other known representations of $e(t)$, e.g., Agarwal [1].

In (1.3) let $x(t) = x_1(t) = \frac{1}{n!} \prod_{i=1}^{r} (t - a_i)^{k_i+1}$, for which $P_{n-1}(t) \equiv 0$, to obtain

$$|x_1^{(k)}(t)| \le C_{n,k}(b-a)^{n-k}, \quad 0 \le k \le n-1$$

and hence it is clear that in (1.3)

$$C_{n,k} \ge \max_{a \le t \le b} \left(\frac{1}{(b-a)^{n-k}} \, |x_1^{(k)}(t)| \right), \quad 0 \le k \le n-1. \tag{1.5}$$

We conjecture that the inequalities (1.3) hold with

$$C_{n,k} = \max_{a \le t \le b} \left(\frac{1}{(b-a)^{n-k}} \, |x_1^{(k)}(t)| \right), \quad 0 \le k \le n-1. \tag{1.6}$$

If our claim is true, then from (1.5) it is clear that the inequalities (1.3) with these constants $C_{n,k}$ are best possible, as in (1.3) equality holds for $x(t) = x_1(t)$ and only for this function up to a multiplicative constant.

The purpose of this paper is to survey recently established results which support our conjecture. Similar inequalities involving $\|e(t)\|_p$ and $\|x^{(n)}(t)\|_\nu$ are also included. Although, these inequalities are motivated by the polynomial interpolation theory (Wong and Agarwal [29]) these are known to play fundamental role in the study of boundary value problems for ordinary differential equations (Agarwal [1,4]).

2. Lidstone Interpolation

For a given function $x(t) \in C^{(2m)}[0,1]$ the Lidstone interpolating polynomial $P_{2m-1}(t)$ of degree $(2m-1)$ matches $x(t)$ and its $(m-1)$ even derivatives $x^{(2i)}(t)$, $0 \le i \le m-1$ at 0 and 1, i.e.,

$$\begin{aligned} P_{2m-1}^{(2i)}(0) &= x^{(2i)}(0) \\ P_{2m-1}^{(2i)}(1) &= x^{(2i)}(1), \quad 0 \le i \le m-1. \end{aligned}$$

Definition 2.1. The unique polynomial $\Lambda_n(t)$ of degree $(2n+1)$ defined by the relations

$$\begin{aligned} \Lambda_0(t) &= t \\ \Lambda_n''(t) &= \Lambda_{n-1}(t) \\ \Lambda_n(0) &= \Lambda_n(1) = 0, \quad n \geq 1 \end{aligned}$$

is called *Lidstone Polynomial.*

Lemma 2.1. *The Lidstone polynomial $\Lambda_n(t)$ can be expressed*

i) *in terms of Green's function (Widder [28]) as*

$$\Lambda_n(t) = \int_0^1 G_n(t,s)s\,ds,$$

where

$$G_1(t,s) = \begin{cases} (t-1)s, & s \leq t \\ (s-1)t, & t \leq s \end{cases}$$

$$G_n(t,s) = \int_0^1 G_1(t,t_1)G_{n-1}(t_1,s)dt_1, \; n \geq 2;$$

(ii) *in terms of Fourier series expansion (Widder [28]) as*

$$\Lambda_n(t) = (-1)^n \frac{2}{\pi^{2n+1}} \sum_{k=1}^{\infty} \frac{(-1)^{k+1}}{k^{2n+1}} \sin k\pi t;$$

(iii) *in terms of Bernoulli polynomials (Whittaker [27]) as*

$$\Lambda_n(t) = \frac{2^{2n+1}}{(2n+1)!} B_{2n+1}\left(\tfrac{1+t}{2}\right);$$

(iv) *in terms of Bernoulli numbers (Agarwal and Wong [6]) as*

$$\Lambda_n(t) = \frac{1}{6}\left[\frac{6t^{2n+1}}{(2n+1)!} - \frac{t^{2n-1}}{(2n-1)!}\right] - \sum_{k=0}^{n-2} \frac{2(2^{2k+3}-1)}{(2k+4)!} B_{2k+4} \frac{t^{2n-2k-3}}{(2n-2k-3)!}.$$

Lemma 2.2 (Widder [28]). *The Lidstone interpolating polynomial $P_{2m-1}(t)$ can be expressed as*

$$P_{2m-1}(t) = \sum_{k=0}^{m-1} \left[x^{(2k)}(0)\Lambda_k(1-t) + x^{(2k)}(1)\Lambda_k(t)\right].$$

Further,

$$P_{2m-1}^{(2i)}(t) = \sum_{k=0}^{m-i-1} \left[x^{(2k+2i)}(0)\Lambda_k(1-t) + x^{(2k+2i)}(1)\Lambda_k(t)\right], \quad 0 \le i \le m-1.$$

Theorem 2.3. *If $x(t) \in C^{(2m)}[0,1]$, then the following holds:*

(i) *(Widder [28])*

$$e(t) = x(t) - P_{2m-1}(t) = \int_0^1 G_m(t,s)x^{(2m)}(s)ds;$$

(ii) *(Agarwal and Wong [6])*

$$\begin{aligned} |e^{(2i)}(t)| &\le (-1)^{m-i}E_{2m-2i}(t)M_{2m} \\ &\le \frac{(-1)^{m-i}E_{2m-2i}}{2^{2m-2i}(2m-2i)!}M_{2m}, \quad 0 \le i \le m-1 \end{aligned} \tag{2.1}$$

where $E_{2m}(t)$ is the Euler polynomial of degree $2m$ and E_{2m} is the $2m$-th Euler number, and $M_{2m} = \max_{0\le t\le 1} |x^{(2m)}(t)|$.

(iii) *(Agarwal and Wong [6])*

$$\begin{aligned} |e^{(2i+1)}(t)| &\le (-1)^{m-i}\left[2E_{2m-2i}(t) + (1-2t)E_{2m-2i-1}(t)\right]M_{2m} \\ &\le (-1)^{m-i+1}\frac{2(2^{2m-2i}-1)}{(2m-2i)!}B_{2m-2i}M_{2m}, \end{aligned} \tag{2.2}$$

$0 \le i \le m-1$.

Remark 2.1. Inequalities (2.1) and (2.2) are the best possible as they are exact for the function $x_1(t) = (-1)^m E_{2m}(t)$ for which $P_{2m-1}(t) \equiv 0$, and only for this function up to a multiplicative constant. It is interesting to note that $x_1(t)$ is the unique solution of the boundary value problem $x^{(2m)} = 1$, $x^{(2i)}(0) = x^{(2i)}(1) = 0$, $0 \le i \le m-1$.

3. (n,p) Interpolation

For a given function $x(t) \in C^{(n)}[a,b]$ the (n,p) interpolating polynomial $P_{n-1}(t)$ of degree $(n-1)$ satisfies the conditions

$$\begin{aligned} P_{n-1}^{(i)}(a) &= x^{(i)}(a) \ \text{ for } \ 0 \le i \le n-2, \\ P_{n-1}^{(p)}(b) &= x^{(p)}(b) \ \text{ for a fixed } p \text{ with } \ 0 \le p \le n-1. \end{aligned}$$

For a given n-th order differential equation the (n,p) boundary conditions occur frequently in determining the intervals of nonoscillation, see, e.g., Agarwal [3], and Levin [21, 22].

Lemma 3.1 (Agarwal [3]). *The (n,p) interpolating polynomial $P_{n-1}(t)$ can be expressed as*

$$P_{n-1}(t) = \sum_{i=0}^{n-2} \frac{(t-a)^i}{i!} x^{(i)}(a) + \left[x^{(p)}(b) - \sum_{i=0}^{n-p-2} \frac{(b-a)^i}{i!} x^{(p+i)}(a) \right] \cdot \frac{(n-p-1)!}{(n-1)!} \frac{(t-a)^{n-1}}{(b-a)^{n-p-1}}.$$

Theorem 3.2 (Agarwal [3]). *If $x(t) \in C^{(n)}[a,b]$, then the following holds:*

$$|e^{(k)}(t)| = |x^{(k)}(t) - P_{n-1}^{(k)}(t)| \le A_{n,k}(b-a)^{n-k} \max_{a \le t \le b} |x^{(n)}(t)|, \quad 0 \le k \le n-1, \tag{3.1}$$

where

$$A_{n,k} = \begin{cases} \dfrac{(n-k-1)^{n-k-1}}{(n-k)!(n-p)^{n-k}}, & k = p \le n-1 \\ \dfrac{(p-k)}{(n-p)(n-k)!}, & k+1 \le p \le n-1 \\ \dfrac{(k-p)}{(n-p)(n-k)!}, & p+1 \le k \le n-1. \end{cases} \tag{3.2}$$

Remark 3.1. For each $0 \le k \le n-1$ the inequality (3.1) is best possible as it is exact for the function

$$x_1(t) = \frac{(t-a)^{n-1}}{(n-1)!} \left[\frac{t-a}{n} - \frac{b-a}{n-p} \right]$$

and only for this function up to a multiplicative constant. This function $x_1(t)$ is the unique solution of the boundary value problem $x^{(n)} = 1$, $x^{(i)}(a) = 0$ for $0 \le i \le n-2$, $x^{(p)}(b) = 0$ for a fixed p with $0 \le p \le n-1$.

4. (p, n) Interpolation

For a given function $x(t) \in C^{(n)}[a, b]$ the (p, n) interpolating polynomial $\bar{P}_{n-1}(t)$ of degree $(n-1)$ satisfies the conditions

$$\begin{aligned} \bar{P}_{n-1}^{(p)}(a) &= x^{(p)}(a) \text{ for a fixed } p \text{ with } 0 \le p \le n-1, \\ \bar{P}_{n-1}^{(i)}(b) &= x^{(i)}(b) \text{ for } 0 \le i \le n-2. \end{aligned}$$

Lemma 4.1 (Agarwal [3]). *The (p, n) interpolating polynomial $\bar{P}_{n-1}(t)$ can be expressed as*

$$\bar{P}_{n-1}(t) = \sum_{i=0}^{n-2} (-1)^i \frac{(b-t)^i}{i!} x^{(i)}(b) + \left[x^{(p)}(a) - \sum_{i=0}^{n-p-2} (-1)^i \frac{(b-a)^i}{i!} x^{(p+i)}(b) \right] \cdot (-1)^p \frac{(n-p-1)!}{(n-1)!} \frac{(b-t)^{n-1}}{(b-a)^{n-p-1}}.$$

Theorem 4.2 (Agarwal [3]). *If $x(t) \in C^{(n)}[a, b]$, then the following holds:*

$$(4.1) \quad |e^{(k)}(t)| = |x^{(k)}(t) - \bar{P}_{n-1}^{(k)}(t)| \le A_{n,k}(b-a)^{n-k} \max_{a \le t \le b} |x^{(n)}(t)|, \quad 0 \le k \le n-1,$$

where the constants $A_{n,k}$ are the same as in (3.2).

Remark 4.1. A remark similar to Remark 3.1 holds for the function

$$x_1(t) = \frac{(t-b)^{n-1}}{(n-1)!} \left[\frac{b-t}{n} - \frac{b-a}{n-p} \right],$$

which is the unique solution of the boundary value problem $x^{(n)} = 1$, $x^{(p)}(a) = 0$ for a fixed p with $0 \le p \le n-1$ and $x^{(i)}(b) = 0$ for $0 \le i \le n-2$.

5. Abel-Gontscharoff interpolation

For a given function $x(t) \in C^{(n)}[a, b]$ the Abel-Gontscharoff interpolating polynomial $P_{n-1}(t)$ of degree $(n-1)$ satisfies the conditions

$$P_{n-1}^{(i)}(a_{i+1}) = x^{(i)}(a_{i+1}), \quad 0 \le i \le n-1,$$

where $a \le a_1 \le a_2 \le \ldots \le a_n \le b$.

Definition 5.1. The unique polynomial $Q_i(t)$ of degree i defined in terms of i-fold iterated integral

$$Q_i(t) = \int_{a_1}^{t} dt_1 \int_{a_2}^{t_1} dt_2 \ldots \int_{a_i}^{t_{i-1}} dt_i, \quad i \geq 1 \quad (Q_0(t) = 1)$$

is called *Abel-Gontscharoff polynomial.*

Lemma 5.1. *The Abel-Gontscharoff interpolating polynomial $P_{n-1}(t)$ can be expressed as*

$$P_{n-1}(t) = \sum_{i=0}^{n-1} Q_i(t) x^{(i)}(a_{i+1}).$$

Theorem 5.2 (Levin [22]). *If $x(t) \in C^{(n)}[a,b]$, then the following holds:*
(5.1)

$$\begin{aligned} |e^{(k)}(t)| = |x^{(k)}(t) - P_{n-1}^{(k)}(t)| &\leq M_n \frac{(b-a)^{n-k}}{(n-k)!} \binom{n-k-1}{\left[\frac{n-k-1}{2}\right]} \\ &= M_n \frac{(b-a)^{n-k}}{(n-k)\left[\frac{n-k-1}{2}\right]!\,\left[\frac{n-k}{2}\right]!}, \quad 0 \leq k \leq n-1, \end{aligned}$$

where $M_n = \max_{a \leq t \leq b} |x^{(n)}(t)|$.

Remark 5.1. For each k, $0 \leq k \leq n-1$ the inequality (5.1) is best possible in the sense that equality holds if and only if $x^{(n)}(t) = M$. However, if $a_1 = a$ and $a_n = b$, then (5.1) can be improved.

Theorem 5.3 (Wong and Agarwal [30]). *If $a = a_1 = \cdots = a_{\alpha+1} < a_{\alpha+2} \leq \cdots \leq a_{n-\beta} \leq a_{n-\beta+1} = \cdots = a_n = b$, then for any function $x(t) \in C^{(n)}[a,b]$ the estimate*

$$|e^{(k)}(t)| \leq M_n \frac{(b-a)^{n-k}}{(n-k)!} \begin{cases} \binom{n-k-1}{r^*}, & 0 \leq k \leq n-\beta-1 \\ 1, & n-\beta \leq k \leq n-1 \end{cases} \tag{5.2}$$

holds, where $r^* = \max\left\{\alpha - k, \beta, \left[\frac{n-k-1}{2}\right]\right\}$.

Remark 5.2. If $\alpha + 1 + \beta = n$, i.e., $\beta = n - \alpha - 1$ then since $\binom{n-k-1}{\alpha-k} = \binom{n-k-1}{n-\alpha-1}$, $r^* = \max\{\alpha - k, \beta\}$. Thus, in this case (5.2) reduce to recently obtained (Agarwal and Usmani [5]) inequalities

$$|e^{(k)}(t)| \le M_n \frac{(b-a)^{n-k}}{(n-k)!} \begin{cases} \binom{n-k-1}{\alpha-k}, & 0 \le k \le \alpha \\ 1 , & \alpha+1 \le k \le n-1. \end{cases} \tag{5.3}$$

Obviously, for each $0 \le k \le n-1$ the inequality (5.3) is best possible as it is exact for the function

$$x_1(t) = \frac{1}{n!} \sum_{i=\alpha+1}^{n} \binom{n}{i} (a-b)^{n-i}(t-a)^i$$

and only for this function up to a multiplicative constant. This function $x_1(t)$ is the unique solution of the boundary value problem $x^{(n)} = 1$, $x^{(i)}(a) = 0$, $0 \le i \le \alpha$, $x^{(i)}(b) = 0$, $\alpha+1 \le i \le n-1$.

Remark 5.3. For each $0 \le k \le n-1$, (5.2) is sharper than the corresponding (5.1). For this, it suffices to note that

$$\binom{n-k-1}{\left[\frac{n-k-1}{2}\right]} \ge \begin{cases} \binom{n-k-1}{r^*}, & 0 \le k \le n-\beta-1 \quad (5.4) \\ 1 , & n-\beta \le k \le n-1. \quad (5.5) \end{cases}$$

Obviously, in (5.4) the strict inequality holds if

$$\left[\frac{n-k-1}{2}\right] < \begin{cases} \max\{\alpha-k, \beta\}, & (n-k-1) \text{ even}, \\ \max\{\alpha-k, \beta\}-1, & (n-k-1) \text{ odd} \end{cases}$$

and in (5.5) provided $n-\beta \le k \le n-3$.

Remark 5.4. Inequalities (5.2) are best possible. For this, we define an integer $n(k), 1 \le n(k) \le n-1$ for a fixed k, $0 \le k \le n-1$ as follows

$$n(k) = \begin{cases} \left[\frac{n+k+1}{2}\right], & r^* = \left[\frac{n-k-1}{2}\right] \\ n-\beta, & r^* = \beta \\ \alpha+1, & r^* = \alpha-k. \end{cases}$$

The n-th degree polynomial

$$x_{n(k)}(t) = \frac{1}{n!} \sum_{j=n(k)}^{n} (-1)^j \, t^j \binom{n}{j}, \quad 0 \le t \le 1,$$

obviously satisfies $|x^{(n)}_{n(k)}(t)| = 1$, and

$$x^{(i)}_{n(k)}(0) = 0, \quad 0 \le i \le n(k)-1,$$

also

$$x_{n(k)}^{(i)}(1) = \frac{i!}{n!}\sum_{j=i}^{n}(-1)^j\binom{n}{j}\binom{j}{i}$$
$$= \frac{i!}{n!}(-1)^i\delta_{ni} = 0$$

for $n(k) \leq i \leq n-1$.

For the function $x_{n(k)}(t)$ for which $e(t) = x_{n(k)}(t)$, it is easy to obtain (5.6)

$$\max_{0\leq t\leq 1}|x_{n(k)}^{(k)}(t)| = \begin{cases} |x_{n(k)}^{(k)}(1)| = \frac{1}{(n-k)!}\binom{n-k-1}{n(k)-k-1}, & 0\leq k\leq n(k)-1 \\ |x_{n(k)}^{(k)}(0)| = \frac{1}{(n-k)!}, & n(k)\leq k\leq n-1. \end{cases}$$

Further, we note that if $0 \leq k \leq n-\beta-1$ then $k \leq n(k)-1$, and if $n-\beta \leq k \leq n-1$ then $n(k) = n-\beta$. Therefore, the relation $\binom{n-k-1}{n(k)-k-1} = \binom{n-k-1}{r^*}$ in (5.6) leads to

$$\max_{0\leq t\leq 1}|x_{n(k)}^{(k)}(t)| = \frac{1}{(n-k)!}\begin{cases}\binom{n-k-1}{r^*}, & 0\leq k\leq n-\beta-1 \\ 1 \quad , & n-\beta\leq k\leq n-1. \end{cases}$$

Thus, for this function $x_{n(k)}(t)$ in (5.2) equality holds.

6. Hermite interpolation

Lemma 6.1. *The Hermite interpolating polynomial $P_{n-1}(k)$ can be written as*

$$P_{n-1}(t) = \sum_{i=1}^{r}\sum_{j=0}^{k_i}\sum_{\ell=0}^{k_i-j}\frac{1}{j!\ell!}\left[\frac{(t-a_i)^{k_i+1}}{\omega(t)}\right]_{t=a_i}^{(\ell)}\frac{\omega(t)}{(t-a_i)^{k_i+1-j-\ell}}x^{(j)}(a_i),$$

where $\omega(t) = \prod_{i=1}^{r}(t-a_i)^{k_i+1}$.

Let μ and ν be conjugate exponents and $1 \leq \mu \leq \infty$, then (1.2) gives, using Hölder's inequality,

$$|e(t)| \leq \left(\int_a^b |g(t,s)|^{\mu}ds\right)^{1/\mu}\left(\int_a^b |x^{(n)}(s)|^{\nu}ds\right)^{1/\nu},$$

which for $p \geq 1$ is the same as

$$\int_a^b |e(t)|^p dt \leq \left[\int_a^b \left(\int_a^b |g(t,s)|^\mu ds\right)^{p/\mu} dt\right] \left[\int_a^b |x^{(n)}(s)|^\nu ds\right]^{p/\nu}$$

and hence

$$\|e(t)\|_p \leq \left[\int_a^b \left(\int_a^b |g(t,s)|^\mu ds\right)^{p/\mu} dt\right]^{1/p} \|x^{(n)}(t)\|_\nu. \tag{6.1}$$

Lemma 6.2. *The following holds:*

(i) *(Agarwal* [1]*)*

$$\int_a^b |g(t,s)| ds = \frac{1}{n!} |\omega(t)|; \tag{6.2}$$

(ii) *(Beesack* [11], *Nehari* [25]*)*

$$|g(t,s)| \leq \frac{1}{(n-1)!} \frac{|\omega(t)|}{(b-a)}. \tag{6.3}$$

In view of Lemma 6.2 the inequality (6.1) can be written as

(6.4)

$$\begin{aligned} \|e(t)\|_p &\leq \left[\int_a^b \left[\left\{\frac{1}{(n-1)!(b-a)}|\omega(t)|\right\}^{\mu-1} \left\{\frac{1}{n!}|\omega(t)|\right\}\right]^{p/\mu} dt\right]^{1/p} \|x^{(n)}(t)\|_\nu \\ &= \frac{1}{((n-1)!(b-a))^{1/\nu}(n!)^{1/\mu}} \left(\int_a^b |\omega(t)|^p dt\right)^{1/p} \|x^{(n)}(t)\|_\nu. \end{aligned}$$

Inequality (6.4) suffers from the facts that: (i) inequality (6.3) is not always best possible; (ii) (6.1) requires the splitting of $\int_a^b |g(t,s)|^\mu ds$ so that both (6.2) and (6.3) can be used; and (iii) it contains the term $\int_a^b |\omega(t)|^p dt$ which is unwieldy. Nevertheless, it covers several cases of frequent occurrence.

Case 1 *Equidistant Points*: $r = n$, $a_{i+1} - a_i = h = \frac{b-a}{n-1}$, $1 \leq i \leq n-1$, $p = \infty$.

The following holds (Agarwal [7])

(6.5)

$$\begin{aligned} \|e(t)\|_\infty \leq &\frac{(b-a)^{n-1/\nu}}{(n-1)^n((n-1)!)^{1/\nu}(n!)^{1/\mu}} \left(\frac{3(n-1) - \sqrt{(n-1)(5n-9)}}{2n}\right) \cdot \\ &\left(1 - \frac{3(n-1) - \sqrt{(n-1)(5n-9)}}{2n}\right) \cdot \\ &\prod_{i=2}^{n-1} \left(i - \frac{(n+3) - \sqrt{n^2 - 2n + 9}}{2n}\right) \|x^{(n)}(t)\|_\nu \end{aligned}$$

$$\text{(6.6)} \qquad \le \frac{(b-a)^{n-1/\nu}}{(n-1)^n((n-1)!)^{1/\nu}(n!)^{1/\mu}} \, \frac{1}{4} \prod_{i=2}^{n-1} \left(i - \frac{1}{n}\right) \|x^{(n)}(t)\|_\nu .$$

Remark 6.1. For $\nu = \infty$ the estimate (6.5) is best possible for $n = 2$ and 3 and for $n \ge 4$ it compares favourably with the best possible result obtained over the computer.

Remark 6.2. For $\nu = \infty$, estimate (6.6) is the same as

$$\|e(t)\|_\infty \le \frac{(b-a)^n}{(n-1)^n n!} \, \frac{1}{4} \prod_{i=2}^{n-1} \left(i - \frac{1}{n}\right) \|x^{(n)}(t)\|_\infty ,$$

which is obviously better than the known error bound

$$\text{(6.7)} \qquad \|e(t)\|_\infty \le \frac{(b-a)^n}{4(n-1)^n n} \|x^{(n)}(t)\|_\infty .$$

Further, (6.7) is sharper than

$$\|e(t)\|_\infty \le \frac{(b-a)^n}{2^n n!} \|x^{(n)}(t)\|_\infty ,$$

which is easily obtained by using the arithmetic-geometric mean inequality.

Case 2 *Simple Points*: $r = n$, but no condition on the distribution of $a_i's$, $p = \infty$.
The following holds (Agarwal [7])

$$\text{(6.8)} \qquad \|e(t)\|_\infty \le \frac{1}{((n-1)!(b-a))^{1/\nu}(n!)^{1/\mu}} \, \frac{1}{4} \prod_{i=2}^{n-1} \left(i - \frac{1}{n}\right) \delta^n \|x^{(n)}(t)\|_\nu ,$$

where $\delta = \max_{1 \le i \le n-1} (a_{i+1} - a_i)$.

Remark 6.3. If $\delta = \frac{b-a}{n-1}$, i.e., the case of equidistant points, then (6.8) reduces to (6.6).

Remark 6.4. For the function $x(t) = \prod_{i=1}^{n} (t - a_i)$ we have $P_{n-1}(t) \equiv 0$, and hence (6.8) with $\nu = \infty$ reduces to

$$\prod_{i=1}^{n} |t - a_i| \le \frac{1}{4} \prod_{i=2}^{n-1} \left(i - \frac{1}{n}\right) \delta^n ,$$

which is sharper than (Das and Vatsala [18])

$$\prod_{i=1}^{n} |t - a_i| \le (n-1)^{n-1}\left(\frac{\delta}{2}\right)^n.$$

Case 3 *Multiple Points*: $p = \infty$.

Define $\alpha = \min\{k_1, k_r\}$, $\beta = \min_{2\le i\le r-1} k_i$, $\gamma = \min\{\alpha, \beta\}$, $\delta = \max_{1\le i\le r-1}(a_{i+1} - a_i)$, and

$$C(n,r,\alpha,\beta,\gamma) = \frac{(n-1-\alpha-(\beta+1)(r-1))^{(n-1-\alpha-(\beta+1)(r-1))}(\alpha+1)^{(\alpha+1)}}{(n-(\beta+1)(r-1))^{(n-(\beta+1)(r-1))}} \cdot \prod_{\ell=1}^{r-1}\left(\ell - \frac{\gamma+1}{n-\alpha+\gamma}\right)^{(\beta+1)}(b-a)^{(n-(\beta+1)(r-1))}\delta^{(\beta+1)(r-1)}.$$

The following inequalities hold (Agarwal [7]):

$$\|e(t)\|_\infty \le \frac{1}{((n-1)!(b-a))^{1/\nu}(n!)^{1/\mu}}\, C(n,r,\alpha,\beta,\gamma)\, \|x^{(n)}(t)\|_\nu \tag{6.9}$$

and

$$\|e(t)\|_\infty \le \frac{(n-1-\alpha)^{(n-1-\alpha)}(\alpha+1)^{(\alpha+1)}}{((n-1)!)^{1/\nu}(n!)^{1/\mu}n^n}\,(b-a)^{n-1/\nu}\|x^{(n)}(t)\|_\nu. \tag{6.10}$$

Remark 6.5. Inequalities (6.9) and (6.10) cannot be compared. Further, from these inequalities (6.6) and (6.8) cannot be deduced or compared.

Case 4 $p \ge 1$.

The following holds (Agarwal [7])

$$\begin{aligned}\|e(t)\|_p \;\le\;& \frac{1}{((n-1)!)^{1/\nu}(n!)^{1/\mu}}\,(b-a)^{n+1/p-1/\nu}. \\ & (2B_{1/2}(p(\alpha+1)+1,\; p(n-1-\alpha)+1))^{1/p}\|x^{(n)}(t)\|_\nu,\end{aligned} \tag{6.11}$$

where $B_{1/2}$ is the incomplete Beta function.

Remark 6.6. If $r = 2$, then (6.11) can be improved to

$$\begin{aligned}\|e(t)\|_p \;\le\;& \frac{1}{((n-1)!)^{1/\nu}(n!)^{1/\mu}}\,(b-a)^{n+1/p-1/\nu}. \\ & (B(p(\alpha+1)+1,\; p(n-\alpha-1)+1))^{1/p}\|x^{(n)}(t)\|_\nu.\end{aligned} \tag{6.12}$$

For $\nu = \infty$, (6.12) is best possible. Further, for $p = 1$, $\nu = \infty$ it reduces to

$$\|e(t)\|_1 \le \frac{(n-1-\alpha)!(\alpha+1)!}{n!(n+1)!}\ (b-a)^{n+1}\|x^{(n)}(t)\|_\infty, \tag{6.13}$$

which is obtained in Agarwal [2].

Remark 6.7. For $n = 2m$, $\alpha = m-1$, $p = 1$, $\nu = \infty$ both inequalities (6.11) and (6.12) reduce to

$$\|e(t)\|_1 \le \frac{(m!)^2}{(2m)!(2m+1)!}\ (b-a)^{2m+1}\|x^{(2m)}(t)\|_\infty, \tag{6.14}$$

which is sharper than Anon's problem [10], and includes Zaidman's problem [32], as well as Iyengar and Mahajani's well known inequalities, e.g., Mitrinović [24].

A new representation of the error function $e(t)$ is embodied in the following:

Theorem 6.3 (Wong and Agarwal [31]). *Let the functions $F_k(t)$, $k = n, n-1, \dots, 1$, be recursively defined as follows*

$$F_n(t) = \int_{a_r}^{t} (a_r - t_n)^{n-1} x^{(n)}(t_n)dt_n,$$

$$F_k(t) = \begin{cases} \displaystyle\int_{a_{s+1}}^{t} \frac{(a_{s+1}-t_k)^{k-1}}{(a_{s+2}-t_k)^{k+1}} F_{k+1}(t_k)dt_k, & k = p(s+1)+s+1 \\ \displaystyle\int_{a_{s+1}}^{t} (a_{s+1}-t_k)^{-2} F_{k+1}(t_k)dt_k, & p(s)+s+1 \le k \le p(s+1)+s \end{cases}$$

$(0 \le s \le r-1)$, where $p(0) = 0$, and for $1 \le s \le r$, $p(s) = \sum_{j=1}^{s} k_j$.

Then, the error function $e(t)$ can be explicitly expressed as

$$e(t) = (a_1 - t)^{k_1} \prod_{i=2}^{r} (a_i - t)^{k_i+1} F_1(t). \tag{6.15}$$

This representation of $e(t)$ is used to obtain the following results.

Theorem 6.4 (Wong and Agarwal [31]). *If $(t-a) \ge (b-t)$, then the following pointwise error estimates hold*

$$|e^{(k)}(t)| \le \frac{k!}{n!}\ g_k(t)\ \ \|x^{(n)}(t)\|_\infty,\ 0 \le k \le n-1, \tag{6.16}$$

where

$$g_k(t) = \begin{cases} (t-a)^{n-\alpha-2}(b-t)^{\alpha+1-k}\left\{\binom{n-\alpha-1}{k}(t-a)+\right. \\ \qquad \left. +\left[\binom{n}{k}-\binom{n-\alpha-1}{k}\right](b-t)\right\}, \quad 0 \le k \le \alpha, \\ (t-a)^{n-k-1}\left\{\binom{n-\alpha-1}{k-\alpha-1}(t-a)+\left[\binom{n}{k}-\binom{n-\alpha-1}{k-\alpha-1}\right](b-t)\right\}, \\ \qquad \alpha+1 \le k \le n-1. \end{cases}$$

Theorem 6.5 (Wong and Agarwal [31]). *If $(t-a) \le (b-t)$, then the following pointwise error estimates hold*

$$|e^{(k)}(t)| \le \frac{k!}{n!}\, g_k(a+b-t)\, \|x^{(n)}(t)\|_\infty, \quad 0 \le k \le n-1. \tag{6.17}$$

In our next result we provide uniform error estimates for $|e^{(k)}(t)|$, $0 \le k \le n-1$. For this, we define the following: If $0 \le k \le \alpha$ then

$$c_1(k) = \binom{n-\alpha-1}{k}, \quad c_2(k) = \binom{n}{k} - c_1(k),$$

$$\begin{aligned} c_3(k) = {} & [(n-\alpha-1)(c_2(k)-c_1(k)) - c_2(k)(n-\alpha-1)]^2 \\ & +4(\alpha+1-k)c_2(k)(c_2(k)-c_1(k)), \end{aligned}$$

and if $\alpha+1 \le k \le n-1$, then

$$d_1(k) = \binom{n-\alpha-1}{k-\alpha-1}, \quad d_2(k) = \binom{n}{k} - d_1(k).$$

Theorem 6.6 (Wong and Agarwal [31]). *The following uniform estimates hold*

$$|e^{(k)}(t)| \le \frac{k!}{n!}\, g_k(t_k^*)\, \|x^{(n)}(t)\|_\infty, \tag{6.18}$$

where for $0 \le k \le \alpha$

$$t_k^* = \begin{cases} a + (b-a)\,\dfrac{(n-\alpha-1)(c_2(k)-c_1(k)) + c_2(k)(n-k-1) - \sqrt{c_3(k)}}{2(n-k)(c_2(k)-c_1(k))}, & \textit{if } c_1(k) \ne c_2(k) \\ a + (b-a)\dfrac{(n-\alpha-2)}{(n-k-1)}, & \textit{if } c_1(k) = c_2(k) \end{cases}$$

and for $\alpha+1 \le k \le n-1$

$$t_k^* = \begin{cases} a + (b-a)\,\frac{(n-k-1)d_2(k)}{(n-k)(d_2(k)-d_1(k))}, & \text{if it is in the interval } ((a+b)/2, b], \\ b, & \text{otherwise.} \end{cases}$$

Remark 6.8. If $\alpha = 0$ then for $k = 0$, $c_1(0) = 1$, $c_2(0) = 0$, $c_3(0) = (n-1)^2$, $t_0^* = a + \frac{n-1}{n}(b-a)$; and for $1 \le k \le n-1$, $d_1(k) = \binom{n-1}{k-1}$, $d_2(k) = \binom{n}{k} - \binom{n-1}{k-1}$, $t_k^* = b$. Therefore, inequalities (6.18) reduce to

$$|e^{(k)}(t)| \le B_{n,k}(b-a)^{n-k}\|x^{(n)}(t)\|_\infty, \quad 0 \le k \le n-1, \tag{6.19}$$

where

$$B_{n,0} = \frac{1}{n!}\,\frac{(n-1)^{n-1}}{n^n}, \qquad B_{n,k} = \frac{k}{(n-k)!n}, \; 1 \le k \le n-1.$$

In (6.19) the constants $B_{n,k}$ are best possible as equality holds for the functions $x_1(t) = \frac{1}{n!}(t-a)^{n-1}(t-b)$ and $x_2(t) = \frac{1}{n!}(t-a)(t-b)^{n-1}$ and only for these functions up to a constant factor. Obviously, $x_1(t)$ and $x_2(t)$ are the solutions of the respective boundary value problems $x^{(n)} = 1$, $x^{(i)}(a) = 0$, $0 \le i \le n-1$, $x(b) = 0$; and $x^{(n)} = 1$, $x(a) = 0$, $x^{(i)}(b) = 0$, $0 \le i \le n-1$. Hukuhara [19] indicates that Tumura [26] obtained the inequalities (6.19). These inequalities have also appeared in the work of Agarwal [1], Bessmertnyh and Levin [12], Bessmertnyh [13], Brink [15], Coppel [17], Kobelkov and Kobyakov [20], Levin [23] and several others.

Remark 6.9. For $k = 0$, $c_1(0) = 1$, $c_2(0) = 0$, $c_3(0) = (n-\alpha-1)^2$, $t_0^* = a + \frac{(n-\alpha-1)}{n}(b-a)$ and for $k = n-1$, $d_1(n-1) = n-\alpha-1$, $d_2(n-1) = \alpha+1$, $t_{n-1}^* = b$. Therefore, inequalities (6.18) for $k = 0$ and $n-1$ reduce to

$$|e(t)| \le \frac{1}{n!}\,\frac{(n-\alpha-1)^{n-\alpha-1}(\alpha+1)^{\alpha+1}}{n^n}\,(b-a)^n\,\|x^{(n)}(t)\|_\infty \tag{6.20}$$

and

$$|e^{(n-1)}(t)| \le \frac{n-\alpha-1}{n}\,(b-a)\,\|x^{(n)}(t)\|_\infty. \tag{6.21}$$

Both of these inequalities are best possible as equality holds for the functions $x_1(t) = (t-a)^{n-\alpha-1}(b-t)^{\alpha+1}$ and $x_2(t) = (t-a)^{\alpha+1}(b-t)^{n-\alpha-1}$, and only for these functions up to a constant factor.

Remark 6.10. A simple calculation shows that $g_k(t_k^*) \le C_{n,k}^{\alpha}(b-a)^{n-k}$, where

$$C_{n,k}^{\alpha} = \begin{cases} \dfrac{(n-\alpha-1^{n-\alpha-1}(\alpha+1-k)^{\alpha+1-k}}{(n-k)!\,(n-k)^{n-k}}, & 0 \le k \le \alpha, \\ \dfrac{(k-\alpha)}{(n-k)!(n-\alpha)}, & \alpha+1 \le k \le n-1. \end{cases}$$

Therefore, (6.18) can be replaced by

$$|e^{(k)}(t)| \le C_{n,k}^{\alpha}(b-a)^{n-k}\|x^{(n)}(t)\|_{\infty}. \tag{6.22}$$

It is clear that $C_{n,k}^{\alpha} \le B_{n,k}$, $0 \le k \le n-1$ with equality only for $\alpha = 0$.

Theorem 6.7 (Agarwal and Wong [9]). *For the two point Hermite interpolation, i.e.,*

$$\begin{aligned} P_{n-1}^{(i)}(a) &= x^{(i)}(a), \quad 0 \le i \le \alpha \\ P_{n-1}^{(i)}(b) &= x^{(i)}(b), \quad 0 \le i \le n-\alpha-2 \end{aligned}$$

with $0 \le \alpha \le n-2$ *and* $2 \le n \le 6$ *the following uniform estimates hold*

$$|e^{(k)}(t)| \le D_{n,k}^{\alpha}(b-a)^{n-k}\|x^{(n)}(t)\|_{\infty}, \quad 0 \le k \le n-1,$$

where $D_{n,k}^{\alpha} = D_{n,k}^{n-\alpha}$, *and*

$$D_{n,k}^{\alpha} = \max_{0 \le t \le 1} \left| \frac{d^k}{dt^k} \frac{1}{n!} t^{\alpha+1}(1-t)^{n-\alpha-1} \right|.$$

Remark 6.11. In the following table we provide the constants $D_{n,k}^{\alpha}$. For the comparison purpose the constants $E_{n,k}^{\alpha} = \frac{k!}{n!} g_k(t_k^*)/(b-a)^{n-k}$ which are different from $D_{n,k}^{\alpha}$ have also been included in parentheses.

		k					
n	α	0	1	2	3	4	5
2	0	$\frac{1}{8}$	$\frac{1}{2}$				
3	0	$\frac{2}{81}$	$\frac{1}{6}$	$\frac{2}{3}$			
4	0	$\frac{9}{2048}$	$\frac{1}{24}$	$\frac{1}{4}$	$\frac{3}{4}$		
4	1	$\frac{1}{348}$	$\frac{7}{72\sqrt{3}}\left(\frac{1}{48}\right)$	$\frac{1}{12}\left(\frac{29}{192}\right)$	$\frac{1}{2}$		
5	0	$\frac{32}{46875}$	$\frac{1}{120}$	$\frac{1}{15}$	$\frac{3}{10}$	$\frac{4}{5}$	
5	1	$\frac{9}{31250}$	$\frac{3+8\sqrt{6}}{20000}\left(\frac{827-73\sqrt{73}}{61440}\right)$	$\frac{1}{60}\left(\frac{9}{320}\right)$	$\frac{3}{20}\left(\frac{49}{320}\right)$	$\frac{3}{5}$	
6	0	$\frac{625}{6718464}$	$\frac{1}{720}$	$\frac{1}{72}$	$\frac{1}{12}$	$\frac{1}{3}$	$\frac{5}{6}$
6	1	$\frac{1}{32805}$	$\frac{25+34\sqrt{10}}{911250}\left(\frac{\sqrt{15}}{7500}\right)$	$\frac{1}{360}\left(\frac{7203}{1406080}\right)$	$\frac{1}{30}\left(\frac{128}{3645}\right)$	$\frac{1}{5}$	$\frac{2}{3}$
6	2	$\frac{1}{46080}$	$\frac{\sqrt{5}}{30000}\left(\frac{1}{3840}\right)$	$\frac{1}{1920}\left(\frac{471+133\sqrt{57}}{552960}\right)$	$\frac{1}{120}\left(\frac{6859}{262440}\right)$	$\frac{1}{10}\left(\frac{2}{15}\right)$	$\frac{1}{2}$

Remark 6.12 (Agarwal [8]). For the two point Hermite interpolation, i.e.,

$$(6.23)\qquad \begin{aligned} P_{2m-1}^{(i)}(a) &= x^{(i)}(a) \\ P_{2m-1}^{(i)}(b) &= x^{(i)}(b), \quad 0 \le i \le m-1, \end{aligned}$$

the functions $F_k(t)$ defined in Theorem 6.4 can be written as

$$F_{2m-k+1}(t) = \frac{(b-t)^{1-k}}{(k-1)!}\int_b^t (t-s)^{k-1}(b-s)^{2m-k}x^{(2m)}(s)ds, \quad 1 \le k \le m,$$

$$F_{m-k+1}(t) = \frac{(a-t)^{1-k}}{(k-1)!}\int_a^t \frac{(t-s)^{k-1}}{(b-s)^{m+1}}(a-s)^{m-k}\, F_{m+1}(s)ds, \quad 1 \le k \le m,$$

and hence the error function $e(t)$ can be written as

$$e(t) = \frac{(t-b)^m}{((m-1)!)^2}\int_a^t \left[\int_a^{\min(s,t)} \frac{[(t-r)(s-r)]^{m-1}}{(b-r)^{2m}}\, dr\right](b-s)^m x^{(2m)}(s)ds.$$

Remark 6.13. For (6.23) obviously $\alpha = m - 1$, and thus from Theorems 6.5 and 6.6 it follows that for all $a \le t \le b$

(6.24)

$$|e^{(k)}(t)| \le \begin{cases} \frac{1}{(2m)!}\,[(t-a)(b-t)]^m \|x^{(n)}(t)\|_\infty \\ \frac{1}{2(2m-k)!}\,[(t-a)(b-t)]^{m-k}(b-a)^k \|x^{(n)}(t)\|_\infty, \; 1 \le k \le m-1. \end{cases}$$

For the same case Ciarlet, Schultz and Varga [16] obtained the bounds

$$(6.25)\quad |e^{(k)}(t)| \le \frac{1}{k!(2m-2k)!}\,[(t-a)(b-t)]^{m-k}(b-a)^k \|x^{(n)}(t)\|_\infty, \; 0 \le k \le m-1.$$

A comparison of (6.24) with (6.25) shows that (6.24) is sharper than (6.25) by a factor $1/2\binom{2m-k}{k}$, $1 \le k \le m-1$.

References

1. R.P. Agarwal, *Some inequalities for a function having n zeros.* General Inequalities 3, eds. E. F. Beckenbach and W. Walter, ISNM **64** (1983), 371–378.

2. R.P. Agarwal, *Boundary value problems for higher order integro-differential equations.* Nonlinear Analysis **7** (1983), 259–270.

3. R.P. Agarwal, *Some new results on two-point problems for higher order differential equations.* Funkcialaj Ekvacioj **29** (1986), 197–212.

4. R.P. Agarwal, *Boundary Value Problems for Higher Order Differential Equations.* World Scientific, Singapore - Philadelphia, 1986.

5. R.P. Agarwal and R.A. Usmani, *Iterative methods for solving right focal point boundary value problems.* J. Comput. Appl. Math. **14** (1986), 371–390.

6. R.P. Agarwal and P.J.Y. Wong, *Lidstone polynomials and boundary value problems.* Computers Math. Applic. **17** (1989), 1397-1421.

7. R.P. Agarwal, *Better error estimates in polynomial interpolation.* J. Math. Anal. Appl., to appear.

8. R.P. Agarwal, *Sharp Hermite interpolation error bounds for derivatives.* Nonlinear Analysis, to appear.

9. R.P. Agarwal and P.J.Y. Wong, *Optimal error bounds for the derivatives of two point Hermite interpolation.* Computers Math. Applic., to appear.

10. S. Anon, *Problem E 2155.* Amer. Math. Monthly **76** (1969), 188.

11. P.R. Beesack, *On the Green's function of an N-point boundary value problem.* Pacific J. Math. **12** (1962), 801–812.

12. G.A. Bessmertnyh and A.Ju. Levin, *Some inequalities satisfied by differentiable functions of one variable.* Soviet Math. Dokl. **3** (1962), 737–740.

13. G.A. Bessmertnyh, *The existence and uniqueness of the solutions of a multipoint de la Vallée Poussin problem for nonlinear differential equations* (Russian). Differential Equations **6** (1970), 298–310.

14. G. Birkoff and A. Priver, *Hermite interpolation errors for derivatives.* J. Math. and Phys. **46** (1967), 440–447.

15. J. Brink, *Inequalities involving* $\|f\|_p$ *and* $\|f^{(n)}\|_q$ *for* f *with* n *zeros.* Pacific J. Math. **42** (1972), 289–311.

16. P.G. Ciarlet, M.H. Schultz and R.S. Varga, *Numerical methods of high-order accuracy for nonlinear boundary value problems, I. One dimensional problem.* Numerische Mathematik **9** (1967), 394–430.

17. W.A. Coppel, *Disconjugacy.* Lecture Notes in Mathematics 220, Springer-Verlag, New York, 1971.

18. K.M. Das and A.S. Vatsala, *On Green's function of an n-point boundary value problem.* Trans. Amer. Math. Soc. **182** (1973), 469–480.

19. M. Hukuhara, *On the zeros of solutions of linear ordinary differential equations.* Súgaku **15** (1963), 108–109.

20. Y.P. Kobelkov and I.I. Kobyakov, *The existence and uniqueness of solutions of boundary value problems for differential equations with deviating arguments.* Differential Equations **10** (1974), 610–614.

21. A.Ju. Levin, *Some problems bearing on the oscillation of solutions of linear differential equations.* Dokl. Akad. Nauk SSSR, **148** (1963), 512–515.

22. A.Ju. Levin, *Distribution of the zeros of solutions of a linear differential equation.* Dokl. Akad. Nauk SSSR, **156** (1964), 1281–1284.

23. A.Ju. Levin, *A bound for a function with monotonely distributed zeros of successive derivatives.* Math. Sb. **64**(106) (1964), 396–409.

24. D.S. Mitrinović , *Analytic Inequalities.* Springer-Verlag, Berlin and New York, 1970.

25. Z. Nehari, *On an inequality of P.R. Beesack.* Pacific J. Math. **14** (1964), 261–263.

26. M. Tumura, *Kôkai zyôbibunhôteisiki ni tuite.* Kansû Hôteisiki **30** (1941), 20–35.

27. J.M. Whittaker, *On Lidstone's series and two-point expansions of analytic functions.* Proc. Lond. Math. Soc. **36** (1933-34), 451–469.

28. D.V. Widder, *Completely convex functions and Lidstone series.* Trans. Amer. Math. Soc. **51** (1942), 387–398.

29. P.J.Y. Wong and R.P. Agarwal, *Explicit error estimates for quintic and biquintic spline interpolation.* Computers Math. Appl. **18** (1989), 701–722.

30. P.J.Y. Wong and R.P. Agarwal, *Abel-Gontscharoff interpolation error bounds for derivatives.* Proc. R. Soc. Edinburgh, to appear.

31. P.J.Y. Wong and R.P. Agarwal, *A new representation for the error function for the Hermite interpolation and sharper pointwise and uniform error bounds for the derivatives.* Submitted for publication.

32. S. Zaidman, *Problem E 2622.* Amer. Math. Monthly **83** (1976), 740–741.

Ravi P. Agarwal, Department of Mathematics, National University of Singapore, Kent Ridge, Singapore 0511.

International Series of Numerical Mathematics, Vol. 103, © 1992 Birkhäuser Verlag Basel

Discrete variational inequalities

Calvin D. Ahlbrandt

Abstract

The inequalities considered are for discrete quadratic functionals motivated by the theory of the second variation for discrete variational problems. They are discrete versions of classical integral inequalties associated with quadratic functionals. Connections are made between positive definiteness of discrete quadratic functionals and discrete analogues of Jacobi conditions and Sturmian theorems for associated three term recurrence relations

$$-K_n y_{n+1} + B_n y_n - K_{n-1}^T y_{n-1} = 0.$$

These equations arise in the study of symplectic continued fractions.

1. Discrete Sturmian theory

We will be concerned with inequalities associated with discrete quadratic functionals. For $M < N - 1$ and $\eta_n \in \mathbb{R}^m$ the quadratic functionals considered have the form

$$\mathcal{J}[\eta] = \sum_{n=M}^{N} \{\eta_n^T B_n \eta_n - \eta_{n-1}^T K_{n-1} \eta_n - \eta_n^T K_{n-1}^T \eta_{n-1}\}. \tag{1}$$

Closely related is the matrix-vector three term recurrence relation

$$-K_n u_{n+1} + B_n u_n - K_{n-1}^T u_{n-1} = 0. \tag{2}$$

This equation is the Euler equation for $\mathcal{J}[\eta]$. The second variation in discrete variational theory is a special case of (1). The variational motivation of these forms will be developed in Section 2.

This paper is in final form and no version of it will be submitted for publication elsewhere.

The primary results presented here are discrete versions of the classical Legendre-Clebsch transformation and Reid Roundabout Theorem [10, page 398]. These make possible discrete versions of the Sturm separation and comparison theorems for symmetric three term recurrences of the generality of (2). However, the Sturm separation theorem is an improvement of the previous result of [5] and is apparently not included in known results even in the scalar case. In contrast to conjugate point theory for differential equations, the corresponding discrete theory of conjugate intervals involves inequalities. Peterson and Ridenhour [9] use the terminology of *generalized zeros* in this context.

Assume that B_n and K_n are $m \times m$ real matrices with B_n symmetric and K_n nonsingular. We seek criteria for $\mathcal{J}[\eta]$ to be positive definite on the class of *admissible* η; namely, those sequences $\eta_n \in \mathbb{R}^m$ with $\eta_{M-1} = 0 = \eta_N$.

The matrix version of (2) is

$$L[X]_n = -K_n X_{n+1} + B_n X_n - K_{n-1}^T X_{n-1} = 0. \tag{3}$$

Proposition 1. *If U_n and V_n are $m \times r$ solutions of (3) with real entries, then the bilinear form*

$$\{U_n, V_n\} \equiv U_{n-1}^T K_{n-1} V_n - U_n^T K_{n-1}^T V_{n-1}$$

is constant. Furthermore, $\{U_n, V_n\}^T = -\{V_n, U_n\}$.

Proof.

$$(L[U]_n)^T V_n - U_n^T L[V]_n = 0 \quad \Longrightarrow \quad \{U_n, V_n\} = \{U_{n+1}, V_{n+1}\}. \qquad \square$$

An $m \times m$ solution X_n of (3) is called *prepared* if $\{X_n, X_n\} \equiv 0$; that is, the expression $X_{n-1}^T K_{n-1} X_n$ is symmetric for each n. Note that if X_n is the zero matrix at one point n, then the solution is prepared.

Proposition 2 (Discrete Legendre-Clebsch Transformation). *Suppose that (3) has an $m \times m$ prepared matrix solution X_n which is nonsingular for $n \in [M, N-1]$. For admissible variations η and ψ related by $\eta_n = X_n \psi_n$, we have*

$$\mathcal{J}[\eta] = \sum_{n=M}^{N} (\Delta\psi_{n-1})^T X_{n-1}^T K_{n-1} X_n (\Delta\psi_{n-1}). \tag{4}$$

Proof. Use (3) to obtain $X_n^T L[X]_n = 0$ and the relation

$$X_n^T B_n X_n = X_n^T K_n X_{n+1} + X_n^T K_{n-1}^T X_{n-1} \quad \text{for} \quad n = M, \ldots, N-1.$$

Because η is admissible, the upper limit on the sum in $\mathcal{J}[\eta]$ can be replaced by $N-1$. Substitute $\eta = X\psi$ and replace the B term as above in $\mathcal{J}[\eta]$ to obtain

$$\mathcal{J}[\eta] = \sum_{n=M}^{N-1} \{\psi_n^T X_n^T K_n X_{n+1}\psi_n + \psi_n^T X_n^T K_{n-1}^T X_{n-1}\psi_n$$
$$-\psi_{n-1}^T X_{n-1}^T K_{n-1} X_n \psi_n - \psi_n^T X_n^T K_{n-1}^T X_{n-1}\psi_{n-1}\}$$

Because of the zero end conditions on ψ_n the first term can be rewritten as $\psi_{n-1}^T X_{n-1}^T K_{n-1} X_n \psi_{n-1}$ and because X_n is prepared, we have (4). □

Proposition 3. *Suppose that $M-1 \le p < q < N$. Let u_n be a real vector solution of (2). The admissible variation*

$$\eta_n = \begin{cases} u_n, & p+1 \le n \le q, \\ 0, & \textit{otherwise}, \end{cases}$$

gives

$$\mathcal{J}[\eta] = u_{p+1}^T K_p^T u_p + u_q^T K_q u_{q+1}.$$

Proof. If $p+1=q$, then

$$\mathcal{J}[\eta] = u_{p+1}^T B_{p+1} u_{p+1} = u_{p+1}^T (K_p^T u_p + K_{p+1} u_{p+2}).$$

For the case where $p+1<q$ break off the $n=p+1$ term in $\mathcal{J}[\eta]$, i.e.,

$$\mathcal{J}[\eta] = u_{p+1}^T B_{p+1} u_{p+1} + \sum_{n=p+2}^{q} \{u_n^T B_n u_n - u_{n-1}^T K_{n-1} u_n - u_n^T K_{n-1}^T u_{n-1}\}.$$

Equation (2) gives

$$\mathcal{J}[\eta] = u_{p+1}^T K_p^T u_p + u_{p+1}^T K_{p+1} u_{p+2} + \sum_{p+2}^{q} \{u_n^T K_n u_{n+1} - u_{n-1}^T K_{n-1} u_n\}$$

which telescopes to the stated conclusion. □

This result motivates the following definition of *conjugate intervals* [1, 4]. For integers p and q with $p<q$, the intervals $[p, p+1]$ and $[q, q+1]$ are *conjugate* if there exists a real vector solution u_n of (2) such that

$$u_p^T K_p u_{p+1} \le 0 \quad \text{and} \quad u_q^T K_q u_{q+1} \le 0$$

with $u_{p+1} \ne 0$ and $u_q \ne 0$. Equation (2) is called *disconjugate* on $[M-1, N]$ if the interval $[M-1, N]$ contains no pair of conjugate intervals.

Theorem 4 (Discrete Reid Roundabout Theorem). *The following conditions are equivalent:*

(i) *$\mathcal{J}[\eta]$ is positive definite on the class of admissible η.*

(ii) *Equation (2) is disconjugate on $[M-1, N]$.*

(iii) *If U_n is the solution of (3) with $U_{M-1} = 0$ and $U_M = I$, then*

$$U_{n-1}^T K_{n-1} U_n > 0, \quad \textit{for} \quad n \in [M+1, N].$$

(iv) *If V_n is the solution of (3) with $V_{N-1} = I$ and $V_N = 0$, then*

$$V_{n-1}^T K_{n-1} V_n > 0, \quad \textit{for} \quad n \in [M, N-1].$$

(v) *There exists a prepared solution X_n of (3) with*

$$X_{n-1}^T K_{n-1} X_n > 0, \quad \textit{for} \quad n \in [M, N].$$

Outline of Proof.

$$\text{(i)} \Rightarrow \text{(ii)} \Rightarrow \begin{cases} \text{(iii)} \Rightarrow \text{(i)} \\ \text{(iv)} \Rightarrow \text{(i)} \end{cases} \quad \text{and} \quad \text{(i)} \Rightarrow \text{(v)} \Rightarrow \text{(i)}.$$

(i) $\Rightarrow$ (ii): Proposition 3 implies that no intervals $[p, p+1]$ and $[q, q+1]$ for $M - 1 \le p < q \le N - 1$ can be conjugate.

Conditions (iii) and (iv) follow from (ii). For example, for each unit vector c, the vector solution $u_n = U_n c$, has $u_{n-1}^T K_{n-1} u_n > 0$ for $n = M+1, \ldots, N$ and (iii) follows.

Condition (i) follows from each of the conditions (iii), (iv), and (v) because of the Legendre-Clebsch transformation of Proposition 2.

It remains to show that (i) $\Rightarrow$ (v). Assume (i). Then (iii) and (iv) also hold and we may use U and V in order to construct X of (v). Modify V_n of (iv) to be a solution with $V_N = 0$ and $V_{N-1} = \Gamma$, for some nonsingular matrix Γ. The solution X_n of (v) is constructed by setting $X_n = U_n + V_n$. Select Γ so that $\{U_n, V_n\} = -I$. Use $n = N$ to obtain the choice $\Gamma = (U_N^T K_{N-1}^T)^{-1}$. Bilinearity and skew symmetry give

$$\{X_n, X_n\} = \{U_n, U_n\} + \{U_n, V_n\} + \{V_n, U_n\} + \{V_n, V_n\} = 0.$$

Thus X_n is prepared. Next show that $X_{n-1}^T K_{n-1} X_n > 0$ for $n = M, \ldots, N$. To see this for $n = M$ note that $\{U_M, V_M\} = -I$ implies that $V_{M-1} = (K_{M-1}^T)^{-1}$ and

$$X_{M-1}^T K_{M-1} X_M = I + V_{M-1}^T K_{M-1} V_M > 0.$$

At N, $X_{N-1}^T K_{N-1} X_N = U_{N-1}^T K_{N-1} U_N + I > 0$. It remains to be seen that for $M < k < N$, we have $X_{k-1}^T K_{k-1} X_k > 0$. Let c be an arbitrary unit vector in $\mathbb{R}^m$. We show that $c^T X_{k-1}^T K_{k-1} X_k c > 0$. Define an admissible variation by

$$\eta_n = \begin{cases} U_n c, & M-1 \le n \le k-1, \\ -V_n c, & k \le n \le N. \end{cases}$$

Then $\eta_M = c$ and $\eta_{N-1} = -V_{N-1}c = -(U_N^T K_{N-1}^T)^{-1} c$. Break the sum in $\mathcal{J}[\eta]$ up in three parts; the sum from $n = M$ to $k-1$, the term for $n = k$ and the sum from $k+1$ to N. Replace the B terms as before to obtain

$$0 < \mathcal{J}[\eta] = c^T X_{k-1}^T K_{k-1} X_{k-1} c + c^T \{U, V\} c.$$

As the last term is -1 and c is arbitrary we have $X_{k-1}^T K_{k-1} X_k > 0$. □

Note that this statement of the Reid Roundabout Theorem does not involve the usual equivalence in terms of special symmetric solutions of matrix Riccati equations. (See for example, Ahlbrandt and Hooker [4, Theorem 3.1, pages 1185–1186].) Thus the earlier proofs [4, 1] of the Reid Roundabout Theorem for K_n symmetric can be simplified by using the Legendre-Clebsch transformation. However, the existence of an appropriate Riccati equation associated with equation (2) is unknown to the author except in the case where K_n is symmetric.

Define a symmetric block tridiagonal matrix S and a vector z by

$$S = \begin{bmatrix} B_M & -K_M & 0 & \\ -K_M^T & B_{M+1} & -K_{M+1} & \\ & \ddots & \ddots & -K_{N-2} \\ & & -K_{N-2}^T & B_{N-1} \end{bmatrix}, \qquad z = \begin{bmatrix} \eta_M \\ \vdots \\ \eta_{N-1} \end{bmatrix}. \tag{5}$$

Then

$$\mathcal{J}[\eta] = z^T S z \tag{6}$$

Theorem 5 (Legendre). *If $\mathcal{J}[\eta]$ is nonnegative for admissible η, then B_n is positive semidefinite ($B_n \ge 0$) for $n \in [M, N-1]$ (Legendre condition). If $\mathcal{J}[\eta]$ is positive definite on the class of admissible η, then B_n is positive definite ($B_n > 0$) for $n \in [M, N-1]$ (strengthened Legendre condition).*

Proof. Let z be a column vector whose components are in $\mathbb{R}^m$ with kth entry chosen as x, a nonzero vector in $\mathbb{R}^m$, and zero in other entries. Then $z^T S z = x^T B_{M+k-1} x$. Hence, the diagonal blocks are positive semidefinite (positive definite) if S is positive semidefinite (positive definite). □

We now present the Sturm theorems: The following comparison theorem is an immediate consequence of the Reid Roundabout Theorem.

Theorem 6 (Sturm Comparison Theorem). *Suppose that* (2) *is disconjugate on* $[M-1, N]$, S *is defined by* (5) *and* $\hat{S}$ *corresponds to an equation*

$$-\hat{K}_n u_{n+1} + \hat{B}_n u_n - \hat{K}_{n-1}^T u_{n-1} = 0 \tag{7}$$

with $\hat{K}_n$ *nonsingular and* $\hat{B}_n$ *symmetric. If* $\hat{S} \geq S$, *then* (7) *is disconjugate on* $[M-1, N]$.

The separation theorem comes much harder. Introduce the matrices

$$\mathcal{U}_n = \begin{bmatrix} U_{n-1} \\ U_n \end{bmatrix} \quad \text{and} \quad \mathcal{K}_{n-1} = \begin{bmatrix} 0 & K_{n-1} \\ -K_{n-1}^T & 0 \end{bmatrix}.$$

Proposition 7. *Suppose that* U_n *is an* $m \times r$ *solution of* (3). *Then* $\mathcal{U}_n$ *is of constant rank. If* $\{U_n, U_n\} = 0$ *and* $\mathcal{U}_n$ *has rank* r, *then* $r \leq m$.

Proof. Assume $\mathcal{U}_n c = 0$. Use $U_{n-1}c = 0 = U_n c$ in (3) for $U_{n+1}c = 0$. Thus $\mathcal{U}_{n+1}c = 0$. Conversely, $\mathcal{U}_{n+1}c = 0$ implies $\mathcal{U}_n c = 0$. Thus $\mathcal{U}_n$ and $\mathcal{U}_{n+1}$ have the same nullity and hence the same rank. From the identity $\{U_n, U_n\} = \mathcal{U}_n^T \mathcal{K}_{n-1} \mathcal{U}_n$ we have $\mathcal{U}_n^T \mathcal{K}_{n-1} \mathcal{U}_n = 0$. Then the rank r matrix $\mathcal{U}_n^T$ has nullity at least r since $\mathcal{K}_{n-1}\mathcal{U}_n$ has full rank. For $\mathcal{U}_n^T$, *rank* + *nullity* $= 2m$ implies $r + r \leq r +$ *nullity* $= 2m$ and $r \leq m$. □

Vector solutions u_n and v_n of (2) are called *mutually conjoined* if $\{u_n, v_n\} \equiv 0$. If U_n is an $m \times m$ prepared solution of (3), then any pair of columns of U_n are mutually conjoined. Indeed, that follows from the more general observation that if c and d are constant vectors, then solutions $u_n = U_n c$ and $v_n = U_n d$ have

$$\{u_n, v_n\} = c^T U_{n-1}^T K_{n-1} U_n d - c^T U_n^T K_{n-1}^T U_{n-1} d = c^T \{U_n, U_n\} d = 0.$$

An $m \times m$ prepared solution U_n of (3) with $\mathcal{U}_n$ of rank m will be called a *prepared basis* or a *conjoined basis* [10, p. 307].

Proposition 8. *Suppose that* U_n *and* V_n *are* $m \times m$ *prepared solutions of* (3) *with* $\{U_n, V_n\}$ *nonsingular. Then* $\mathcal{U}_n$ *is a prepared basis and* $\mathcal{V}_n$ *is a prepared basis.*

Proof. From $\{U_n, V_n\} = \mathcal{U}_n^T \mathcal{K}_{n-1} \mathcal{V}_n$, we note that $\mathcal{V}_n c = 0$ implies $\{U_n, V_n\}c = 0$ and hence $c = 0$. Thus $\mathcal{V}_n$ has full rank. Similary, $c^T \mathcal{U}_n^T = 0$ implies $c = 0$. □

Suppose that U_n is an $m \times m$ solution of (3) which is a prepared basis. Define an index set of *focal points* $\mathcal{F}_P = \mathcal{F}_P[M,N]$ of U by

$$\mathcal{F}_P = \{k \mid k \in [M,N] \text{ and } U_k \text{ is singular}\}.$$

Define an index set of *focal intervals* $\mathcal{F}_I = \mathcal{F}_I[M,N]$ of U by (n.s. = nonsingular)

$$\mathcal{F}_I = \{k \mid k \in [M, N-1] \text{ with } U_k^T K_k U_{k+1} \text{ n.s. and not positive definite}\}.$$

It follows that for $k \in \mathcal{F}_I$, the symmetric matrix $D_k = U_k^T K_k U_{k+1}$ has a negative eigenvalue, but zero is not an eigenvalue. Note that $\mathcal{F}_P$ and $\mathcal{F}_I$ are disjoint and the set $\mathcal{F} = \mathcal{F}_P \cup \mathcal{F}_I$ has cardinality at most $N - M + 1$. If f denotes the cardinality of $\mathcal{F}[M,N]$ then we will say that f is the *focality of* U *on* $[M,N]$.

Theorem 9 (Discrete Sturm Separation Theorem). *If* (2) *is disconjugate on* $[M-1,N]$ *and* U_n *is any* $m \times m$ *prepared basis for* (3), *then the focality of* U *on* $[M,N]$ *is at most* m.

Before establishing this result, we list three easy corollaries.

Corollary 10. *Suppose* (2) *is disconjugate on* $[M,\infty)$ *and* U_n *is any* $m \times m$ *prepared basis for* (3). *Then there exists an integer* P *in* $[M,\infty)$ *such that*

$$D_n \equiv U_n^T K_n U_{n+1} > 0 \quad for \quad n \geq P;$$

i.e., D_n *is positive definite on* $[P,\infty)$. *In particular, every prepared basis* U_n *is eventually nonsingular.*

Corollary 11. *If one prepared basis has bounded focality on* $[M,\infty)$, *then every prepared basis has bounded focality on* $[M,\infty)$. *Also, if one prepared basis has unbounded focality on* $[M,\infty)$ *then every prepared basis has unbounded focality on* $[M,\infty)$.

Let **BF** denote the condition that every prepared basis for (3) has bounded focality on $[M,\infty)$.

Let **UF** denote the negation of condition **BF**. Thus **UF** is the condition that every prepared basis has unbounded focality on $[M,\infty)$. That these conditions are disjunctive alternatives is in contrast with the concept of nonoscillation in the scalar case. In particular, the Fibonacci recurrence $y_{n+1} = y_n + y_{n-1}$ has both oscillatory and nonoscillatory real solutions. However, when put in self-adjoint form

$$-(-1)^n y_{n+1} + (-1)^n y_n - (-1)^{n-1} y_{n-1} = 0$$

every real solution has unbounded focality on $[1, \infty)$ because the strengthened Legendre condition is violated on every $[M, \infty)$. Thus this equation (as well as the negative of this equation) satisfies condition **UF**. It has long been thought that there was no Sturm separation theorem for the Fibonacci equation. (See [6, page 176] and [8, page 21].)

Corollary 12. *If W_n is positive definite, then there exists an extended real number μ such that the equation*

$$-K_n x_{n+1} + B_n x_n - K_{n-1}^T x_{n-1} = \lambda W_n x_n$$

satisfies **BF** *for $\lambda < \mu$ and satisfies* **UF** *for $\lambda > \mu$.*

In the scalar case with $K_n > 0$, Smith [11] has shown that μ is the infimum of the essential spectrum of an associated self-adjoint operator.

In order to establish the theorem, we define a bilinear form $\mathcal{B}[\phi, \theta]$ by

$$\mathcal{B}[\phi, \theta] = \sum_{M}^{N} (B_n \phi_n, \theta_n) - (K_{n-1}\phi_n, \theta_{n-1}) - (K_{n-1}^T \phi_{n-1}, \theta_n).$$

Lemma 13. *Suppose that u and v are mutually conjoined solutions of* (2). *For $p, q \in [M, N]$ define sequences ϕ, θ by*

$$\phi_n = \begin{cases} u_n, & M-1 \le n \le p, \\ 0, & \text{otherwise}, \end{cases} \quad \text{and} \quad \theta_n = \begin{cases} v_n, & M-1 \le n \le q, \\ 0, & \text{otherwise}. \end{cases}$$

Then

$$\mathcal{B}[\phi, \theta] = \begin{cases} -(u_{M-1}, K_{M-1} v_M), & \text{if } p \ne q, \\ -(u_{M-1}, K_{M-1} v_M) + (u_q, K_q v_{q+1}), & \text{if } p = q. \end{cases}$$

Proof of the Lemma. Suppose $M \le p \le q$. Use (2) for v_n to obtain

$$(\phi_n, B_n v_n) = (\phi_n, K_n v_{n+1}) + (\phi_n, K_{n-1}^T v_{n-1}),$$

i.e.,

$$(B_n \phi_n, v_n) = (\phi_n, K_n v_{n+1}) + (K_{n-1}\phi_n, v_{n-1}).$$

Then replace this term in

$$\mathcal{B}[\phi, \theta] = \sum_{M}^{q} (B_n \phi_n, v_n) - (K_{n-1}\phi_n, v_{n-1}) - (K_{n-1}^T \phi_{n-1}, v_n)$$

to obtain

$$\mathcal{B}[\phi, \theta] = \sum_{M}^{q} (\phi_n, K_n v_{n+1}) - (\phi_{n-1}, K_{n-1} v_n)$$

which telecopes to $(\phi_q, K_q v_{q+1}) - (u_{M-1}, K_{M-1} v_M)$. If $p < q$, then the first term is zero. If $p = q$, then $\phi_q = u_q$ and the result holds. The case $p > q$ follows from $\mathcal{B}[\theta, \phi] = (\mathcal{B}[\phi, \theta])^T$ and the assumption that u and v are mutually conjoined. □

Proof of the Discrete Sturm Separation Theorem. Suppose that there exists an $m \times m$ prepared basis U_n with focality $f = \mathcal{F}[M, N] > m$. For each $k \in \mathcal{F}_P$, there exists a constant unit m-vector $c(k)$ such that $U_k c(k) = 0$. For each $k \in \mathcal{F}_I$, there exists a constant unit m-vector $c(k)$ such that

$$c(k)^T U_k^T K_k U_{k+1} c(k) < 0.$$

For $k \in \mathcal{F}$, define mutually conjoined solutions $u_n(k)$ by $u_n(k) = U_n c(k)$ and define sequences $\eta_n(k)$ by

$$\eta_n(k) = \begin{cases} u_n(k), & M - 1 \le n \le k, \\ 0, & \text{otherwise.} \end{cases}$$

Note that $\eta_N(k) = 0$. Since the set of f vectors, $\{\eta_{M-1}(k) \mid k \in \mathcal{F}\}$ in m-space, is linearly dependent, there exist scalars γ_k, not all zero, such that

$$\sum_{k \in \mathcal{F}} \gamma_k \eta_{M-1}(k) = 0.$$

Define an admissible variation η_n by

$$\eta_n = \sum_{k \in \mathcal{F}} \gamma_k \eta_n(k).$$

We will show that $\mathcal{J}[\eta] = 0$. From

$$\mathcal{J}[\eta] = \mathcal{B}[\sum_{p \in \mathcal{F}} \gamma_p \eta_n(p), \sum_{q \in \mathcal{F}} \gamma_q \eta_n(q)] = \sum_{p \in \mathcal{F}} \sum_{q \in \mathcal{F}} \gamma_p \gamma_q \mathcal{B}[\eta_n(p), \eta_n(q)]$$

and the Lemma, we have

$$\mathcal{J}[\eta] = -\sum_{p \in \mathcal{F}} \sum_{q \in \mathcal{F}} \gamma_p \gamma_q (u_{M-1}(p), K_{M-1} u_M(q)) + \sum_{q \in \mathcal{F}} \gamma_q^2 (u_q(q), K_q u_{q+1}(q)).$$

The last term is nonpositive and

$$\mathcal{J}[\eta] \le -\sum_{q \in \mathcal{F}} \gamma_q (\underbrace{\sum_{p \in \mathcal{F}} \gamma_p u_{M-1}(p)}_{0}, K_{M-1} u_M(q)) = 0$$

Since $\mathcal{J}$ is positive definite, we conclude that $\mathcal{J}[\eta] = 0$ and $\eta_n \equiv 0$ on $[M-1, N]$. Without loss of generality, omit any indices k from $\mathcal{F}$ for which $\gamma_k = 0$. Let j be the maximum index in $\mathcal{F}$. First, we will show that j is not in the set $\mathcal{F}_I$. Assume $j \in \mathcal{F}_I$. Then $j < N$ and U_j and U_{j+1} are nonsingular. Thus $0 = \eta_j = \gamma_j u_j(j)$ contrary to $\gamma_j \neq 0$ and $u_j(j) = U_j c(j) \neq 0$. Thus $j \in \mathcal{F}_P$. It might be that no other γ_k is nonzero. If that is the case, then for $n \leq j, 0 = \eta_n = \gamma_j \eta_n(j)$ and $u_n(j) \equiv 0$ for $n \leq j$. Since $j \geq M$, $u_n(j)$ vanishes at $M-1$ and M contrary to U_n being a prepared basis. If there are other nonzero γ_k, then let the predessor of j in $\mathcal{F}$ be labeled as i. If $i < j-1$, then $u_n(j)$ vanishes at j and $j-1$, contrary to U_n being a prepared basis. Thus $i = j-1$. Two cases remain:

Case I: $j \in \mathcal{F}_P$, $i = j-1$, and i is in $\mathcal{F}_I$. Then U_i and $U_{i+1} = U_j$ are nonsingular. But this is contrary to $j \in \mathcal{F}_P$.

Case II: $j \in \mathcal{F}_P$, $i = j-1$, and i is in $\mathcal{F}_P$. Then $u_j(j) = 0$ and

$$0 = \eta_i = \gamma_i \underbrace{u_i(i)}_{0} + \gamma_j u_i(j).$$

Thus $u_n(j)$ vanishes at j and $j-1$ contrary to U_n being a prepared basis. □

We close this section with a result which connects this conjugate interval discussion with the classical Jacobi conditions for discrete variational problems. The proof of this result depends upon the fact that the three term recurrence (2) can be regarded as the Euler and Jacobi equations for $\mathcal{J}[\eta]$ and the fact that a minimizing arc must satisfy the Euler equation.

Theorem 14 (Jacobi). *Let u_n be a real vector valued solution of (2) with*

$$u_{M-1}^T K_{M-1} u_M \leq 0 \qquad \text{and} \qquad u_M \neq 0.$$

Set $d_n = u_n^T K_n u_{n+1}$. If $\mathcal{J}[\eta]$ is nonnegative for admissible η, then (Jacobi condition)

$$d_n > 0, \quad \text{for} \quad n \in [M, N-2] \quad \text{and} \quad d_{N-1} \geq 0.$$

If $\mathcal{J}$ is positive definite, then (strengthened Jacobi condition)

$$d_n > 0, \quad \text{for} \quad n \in [M, N-1].$$

Proof. Apply Proposition 3 with $p = M-1 < q < N$. Then

$$d_{M-1} = u_{M-1}^T K_{M-1} u_M = d_{M-1}^T = u_M^T K_{M-1}^T u_{M-1} \leq 0$$

gives

$$d_q \geq d_{M-1}^T + d_q = u_M^T K_{M-1}^T u_{M-1} + u_q^T K_q u_{q+1} = \mathcal{J}[\eta] \geq 0$$

for $q = M, \ldots, N-1$.

Suppose for some $q \in [M, N-2]$ that $d_q = 0$. Then $\hat{\eta}_n$ defined by

$$\hat{\eta}_n = \begin{cases} 0, & n = M-1, \\ u_n, & n = M, \ldots, q, \\ 0, & n = q+1, \ldots, N \end{cases}$$

is admissible and $0 = d_q \geq \mathcal{J}[\hat{\eta}]$. But $\mathcal{J}$ was assumed positive semidefinite. Thus, $\hat{\eta}_n$ provides a minimum for $\mathcal{J}$ and satisfies the Euler equation for $\mathcal{J}$, namely, the Jacobi equation

$$-K_n\hat{\eta}_{n+1} + B_n\hat{\eta}_n - K_{n-1}^T\hat{\eta}_{n-1} = 0,$$

for $n = M, \ldots, N-1$. But $\hat{\eta}_N = 0 = \hat{\eta}_{N-1}$ and uniqueness of solutions of the three term recurrence relation gives $\hat{\eta}_n \equiv 0$, which is contrary to $\hat{\eta}_M = u_M \neq 0$. Furthermore, if $\mathcal{J}$ is positive definite, then apply Proposition 3 with $p = M-1$ and $q = N-1$ for

$$d_{N-1} = u_{N-1}^T K_{N-1} u_N \geq u_M^T K_{M-1}^T u_{M-1} + u_{N-1}^T K_{N-1} u_N = \mathcal{J}[\eta] > 0,$$

since $\eta_M = u_M \neq 0$. □

2. Discrete Variational Theory

As motivation for the quadratic functional and three term recurrence of the previous section, we present an outline of discrete variational theory. Variational theory is useful in determining appropriate standard forms for differential equations and difference equations. Assume $M \leq N-1$ and $y(n) \in \mathbb{R}^m$ is a finite sequence of vectors for $n \in [M-1, N]$.

Let Δ denote the forward difference operator $\Delta y(n-1) = y(n) - y(n-1)$. Consider a functional

$$J[y] = \sum_{n=M}^{N} f(n, y(n), \Delta y(n-1)) \tag{8}$$

on the class of vectors $y(n)$ which satisfy fixed end conditions

$$y(M-1) = c \in \mathbb{R}^m, \qquad y(N) = d \in \mathbb{R}^m.$$

An *admissible variation* is a sequence of vectors $\eta(n) \in \mathbb{R}^m$ for $n \in [M-1, N]$ which satisfy zero end conditions $\eta(M-1) = 0$ and $\eta(N) = 0$. The basic idea is to assume that $\hat{y}$ provides a relative minimum for this fixed endpoint variational problem. Then for any admissible variation $\eta(n)$, the real valued function

$$\phi(t) = J[\hat{y} + t\eta]$$

has a relative minimum at the interior point $t = 0$. The first and second variations of $J[y]$ defined by $J_1[\eta] = \phi'(0)$ and $J_2[\eta] = \phi''(0)$ must satisfy $J_1[\eta] = 0$ and $J_2[\eta] \geq 0$ for all admissible variations η. Application of the chain rule gives

$$\phi'(t) = \sum_{n=M}^{N} \langle f_y, \eta(n)\rangle + \langle f_r, \Delta\eta(n-1)\rangle = \sum_{n=M}^{N} \{\eta^T(n) f_y + (\Delta\eta(n-1))^T f_r\}$$

where

$$f_y(n,y,r) = \begin{bmatrix} f_{y_1}(n,y,r) \\ \vdots \\ f_{y_m}(n,y,r) \end{bmatrix} \quad \text{and} \quad f_r(n,y,r) = \begin{bmatrix} f_{r_1}(n,y,r) \\ \vdots \\ f_{r_m}(n,y,r) \end{bmatrix}$$

are evaluated at $(n, \hat{y}(n) + t\eta(n), \Delta\hat{y}(n-1) + t\Delta\eta(n-1))$. The second variation is

$$\begin{aligned} \phi''(0) \quad = \quad & \sum_{n=M}^{N} \left(\sum_{i=1}^{m} \eta_i(n) \left[\sum_{j=1}^{m} f_{y_i y_j} \eta_j(n) + f_{y_i r_j} \Delta\eta_j(n-1) \right] \right. \\ & \left. + \sum_{i=1}^{m} \Delta\eta_i(n-1) \left[\sum_{j=1}^{m} f_{r_i y_j} \eta_j(n) + f_{r_i r_j} \Delta\eta_j(n-1) \right] \right) \end{aligned}$$

Define matrices

$$P_n = [\, f_{y_i y_j} \,], \qquad Q_n = [\, f_{y_i r_j} \,], \qquad R_n = [\, f_{r_i r_j} \,]$$

where the partials are evaluated at $(n, \hat{y}(n), \Delta\hat{y}(n-1))$. Then

(9) $$J_2[\eta] = \sum_{n=M}^{N} 2\omega(n, \eta(n), \Delta\eta(n-1))$$

where

(10) $$2\omega(n, \eta, \zeta) = \eta^T[P_n\eta + Q_n\zeta] + \zeta^T[Q_n^T\eta + R_n\zeta].$$

Theorem 15 (Euler Necessary Condition). *If $\hat{y}$ provides a local minimum for $J[y]$, then the Euler equation*

(11) $$f_y(n, \hat{y}(n), \Delta\hat{y}(n-1)) = \Delta f_r(n, \hat{y}(n), \Delta\hat{y}(n-1))$$

must be satisfied for $n \in [M, N-1]$.

Proof.

$$J_1[\eta] = \sum_{n=M}^{N} \langle f_y(n), \eta(n) \rangle + \langle f_r(n), \Delta\eta(n-1) \rangle = 0$$

where $f_y(n)$ and $f_r(n)$ denote f_y and f_r evaluated at $(n, \hat{y}(n), \Delta\hat{y}(n-1))$. From

$$\begin{aligned} J_1[\eta] &= \langle f_y(N), \eta(N) \rangle + \langle f_r(N), \Delta\eta(N-1) \rangle \\ &\quad + \sum_{n=M}^{N-1} \{ \langle f_y(n), \eta(n) \rangle + \langle f_r(n), \Delta\eta(n-1) \rangle \} \end{aligned}$$

and use of the summation by parts formula

$$\sum_{n=M}^{N-1} \langle a_n, \Delta b_{n-1} \rangle = \langle a_n, b_{n-1} \rangle |_M^N - \sum_{n=M}^{N-1} \langle \Delta a_n, b_n \rangle$$

on the last term gives

$$\sum_{n=M}^{N-1} \langle f_r(n), \Delta\eta(n-1) \rangle = \langle f_r(n), \eta(n-1) \rangle |_M^N - \sum_{n=M}^{N-1} \langle \Delta f_r(n), \eta(n) \rangle.$$

Thus

$$\begin{aligned} J_1[\eta] &= \langle f_y(N) + f_r(N), \eta(N) \rangle - \langle f_r(M), \eta(M-1) \rangle \\ &\quad + \sum_{n=M}^{N-1} \langle f_y(n) - \Delta f_r(n), \eta(n) \rangle = 0 \end{aligned}$$

and, because $\eta(n)$ vanishes at $M-1$ and N, we have

$$J_1[\eta] = \sum_{n=M}^{N-1} \langle f_y(n) - \Delta f_r(n), \eta(n) \rangle = 0.$$

Since $\eta(M), \dots, \eta(N-1)$ are arbitrary, the condition that $J_1[\eta] = 0$ for all admissible variations η is equivalent to the condition that the Euler equation $f_y(n) = \Delta f_r(n)$ holds for the values $n = M, \dots, N-1$. □

The Jacobi equation is the Euler equation for $J_2[\eta]$; namely,

$$\omega_\eta(n) = \Delta\omega_\zeta(n), \tag{12}$$

which reduces to

$$P_n\eta(n) + Q_n\Delta\eta(n-1) = \Delta(Q_n^T\eta(n) + R_n\Delta\eta(n-1)). \tag{13}$$

This can be written as

$$-\Delta(R_n\Delta\eta_{n-1} + Q_n^T\eta_n) + Q_n\Delta\eta_{n-1} + P_n\eta_n = 0 \tag{14}$$

if $\eta(n)$ is denoted by η_n. Also, the Jacobi equation can be expanded as a three term recurrence

$$-K_n\eta_{n+1} + B_n\eta_n - K_{n-1}^T\eta_{n-1} = 0 \tag{15}$$

for

$$K_n = R_{n+1} + Q_{n+1}^T \quad \text{and} \quad B_n = R_{n+1} + R_n + Q_n^T + Q_n + P_n. \tag{16}$$

In this notation the quadratic form $J_2[\eta]$ can be written as

$$J_2[\eta] = \sum_{n=M}^{N} \{\eta_n^T B_n\eta_n - \eta_{n-1}^T K_{n-1}\eta_n - \eta_n^T K_{n-1}^T\eta_{n-1}\}. \tag{17}$$

These are special cases of equations (1) and (2).

This study of discrete vector variational problems improves and extends the presentation in [3] (see [2, pp. 13-15]). Related applications are given in [7]. If f of (8) were $f(n, y(n), \Delta y(n-1), \Delta^2 y(n-2))$ for y scalar, then the Euler equation would be a fourth order scalar difference equation; some cases could be written in the form of (2) with m=2. That may help explain Copson's "whole line" inequality which was discussed in [2, § 6].

References

1. C.D. Ahlbrandt, *Continued fraction representations of maximal and minimal solutions of a discrete matrix Riccati equation.* SIAM J. Math. Analysis. To appear.

2. C.D. Ahlbrandt, *Variational Inequalities.* In: Inequalities — Fifty years on from Hardy, Littlewood and Pólya, W.N. Everitt, ed., Lecture Notes in Pure and Applied Mathematics **129**, Marcel Dekker, New York, 1991, pp. 1–19.

3. C.D. Ahlbrandt and J.W. Hooker, *A variational view of nonoscillation theory for linear difference equations.* In: Proc. Thirteenth Midwest Differential Equations Conf., J.L. Henderson, ed., Institute of Applied Mathematics, University of Missouri-Rolla, Rolla, MO, 1985, pp. 1–21.

4. C.D. Ahlbrandt and J.W. Hooker, *Riccati matrix difference equations and disconjugacy of discrete linear systems.* SIAM J. Math. Anal. **19** (1988), 1183–1197.

5. C.D. Ahlbrandt and J.W. Hooker, *Recessive solutions of symmetric three term recurrence relations.* In: Canadian Mathematical Society, Conference Proceedings **8**, 1987, 3–42.

6. M. Bocher, *Boundary problems in one dimension.* In: Proceedings of the Fifth International Congress of Mathematcicans, August 1912, Cambridge Univ. Press **1**, 1913, 163–196.

7. S.S. Cheng and R.F. Lu, *A generalization of the discrete Hardy's inequality.* Author preprint, October 1990.

8. A.B. Mingarelli, *Volterra-Stieltjes integral equations and generalized ordinary differential expressions.* Lecture Notes in Math. **989**, Springer-Verlag, New York, 1983.

9. A. Peterson and J. Ridenhour, *A disconjugacy criterion of W.T. Reid for difference equations.* Author preprint, December 1990.

10. W.T. Reid, *Ordinary Differential Equations.* John Wiley, New York, 1971.

11. D.T. Smith, *On the spectral analysis of self adjoint operators generated by second order difference equations.* Proc. Royal Society Edinburgh. To appear.

Calvin D. Ahlbrandt, Department of Mathematics, University of Missouri, Columbia, MO 65211, USA

International Series of Numerical Mathematics, Vol. 103, © 1992 Birkhäuser Verlag Basel

Differential inequalities for Minkowski functionals of level sets

Giuseppe Chiti and Marco Longinetti

Abstract

The Minkowski functionals of the level surfaces of a function u are differentiated with respect to the level parameter. When u is a solution to some classical boundary value problems the Minkowski functionals satisfy sharp differential inequalities.

1. Introduction

In this paper we consider the Minkowski functionals associated to a one parameter family of n-dimensional oriented hypersurfaces M embedded in $\mathbb{R}^{n+1}$.

If M is of class C^2 the k-Minkowski functional for $k = 1, \dots, n+1$ is defined by:

$$w_k = (n+1)^{-1} \int_M \sigma_{k-1} dM. \tag{1}$$

In (1) σ_k is the normalized symmetric elementary function of degree k of the principal curvatures K_i of M, i.e.

$$\binom{n}{k} \sigma_k = S_k, \tag{2}$$

where

$$S_0 = 1, \quad S_1 = \sum_{1 \le i \le n} K_i, \quad S_2 = \sum_{1 \le i < j \le n} K_i K_j, \dots, \quad S_n = \prod_{i=1}^{n} K_i.$$

We extend the definition of S_k by setting $S_k = 0$ for $k > n$; this turns out to unify the formulas in the sequel. When $M = \partial\Omega$ among the Minkowski functionals of M we include also the volume of Ω given by:

$$w_0 = (n+1)^{-1} \int_M h dM\ , \tag{3}$$

This paper is in final form and no version of it will be submitted for publication elsewhere.

where h is the support function of Ω. If Ω is strictly convex, that is to say the curvatures K_i are positives, we can consider the normalized elementary functions $\widehat{\sigma}_k$ of the radii of principal curvatures $R_i = K_i^{-1}$ and it turns out, see [3], that

$$w_k = (n+1)^{-1} \int_B \widehat{\sigma}_{n-k+1} dB, \qquad 1 \le k \le n+1,$$

and

$$w_k = (n+1)^{-1} \int_B h\widehat{\sigma}_{n-k} dB, \qquad 0 \le k \le n. \tag{4}$$

B is the surface of the unit ball in $\mathbb{R}^{n+1}$ and $\widehat{\sigma}$ and h are considered as functions on B by the Gauss map between B and M. The Minkowski functionals can be introduced in different ways and under different regularity assumptions on M, see [4]. The Minkowski functionals are geometric invariants of Ω: w_1 is proportional to the measure of the hypersurface $\partial\Omega$, w_{n+1} has the same value for all simply-connected sets Ω and w_n is related to the mean value of the width of Ω.

Pólya and Szegő have considered in [15] for a surface $\Omega \subset \mathbb{R}^3$ the Minkowski functional w_2 which is important in extimating relevant physical quantities in capacity, membrane and torsion problems. We refer the reader to [12] and [13] for applications of isoperimetric inequalities involving w_2 in these problems.

Here we would like to show that the solutions to classical boundary value problems of mathematical physics have level sets with Minkowski functionals satisfying sharp differential inequalities. This implies sharp estimates on the shape of the level sets of u and of related physical quantities. In some cases we are able to show that the differential inequalities are isoperimetric in the sense that we have equality in a certain geometry. We deal with solutions of certain elliptic partial differential equations of the type

$$\operatorname{div}(a(|\nabla u^2|)\nabla u) = 0. \tag{5}$$

In particular we consider solutions to the p-Laplacian and the minimal surface equation. Results of this type were previously proven in $\mathbb{R}^2$ by M. Longinetti [9], [10] for solutions with convex level lines; P. Laurence [7] and G. Alessandrini [1] generalized to solutions whose level lines are simple closed curves and whose gradient does not vanish. To the knowledge of the authors the only results of this type in n dimensions, $n > 2$, are given by L. Payne and G. Philippin in [14].

The paper is organized as follows. In Section 2 we give the most important formulas in order to compute the derivatives of w_k with respect to a deformation field of M. In Section 3 we compute the first and the second fundamental form of a strictly convex hypersurface M in terms of the support function h and its partial derivatives with respect to local coordinates. When M is a regular level hypersurface $\{u = t\}$ of a given real smooth function u we compute first and second derivatives of w_k with respect to the level parameter t. In Section 4 we apply the previous computation to solutions of (5). We obtain convexity properties for the Minkowsi functionals related to some classical capacity problems. In Section 5 we get a sharp estimate involving the mean value of the width of the level sets of the Green's function for the Laplacian.

We conclude this section proving some inequalities for Minkowski functionals. The classical isoperimetric inequality for a bounded domain Ω in $\mathbb{R}^{n+1}$ is, in our notation:

$$w_1^{n+1} \geq C w_0^n \tag{6}$$

where C is the volume of the unit ball in $\mathbb{R}^{n+1}$. This is extended to convex bodies for all Minkowski functionals w_k (see [3]):

$$w_k^{n+2-k} \geq C w_{k-1}^{n+1-k} \qquad (1 \leq k \leq n) . \tag{7}$$

As in (6) equality holds in (7) if and only if Ω is a ball.

Lemma 1.1. *Suppose Ω is a regular convex body in $\mathbb{R}^{n+1}$, $n > 1$ and let $M = \partial\Omega$. Then we have:*

$$\int_M S_k S_1^{-1} dM \leq c_k w_k \qquad (1 \leq k \leq n) , \tag{8}$$

with $c_k = \frac{n+1}{n}\binom{n}{k}$, and

$$\int_M S_k S_1 dM \geq d_k w_{k+2} \qquad (0 \leq k \leq n-1) , \tag{9}$$

with $d_k = n(n+1)\binom{n}{k}$. Moreover,

$$\int_M S_n S_1 dM \geq C^2 n(n+1) w_n^{-1}. \tag{10}$$

Equality holds in (8), (9), (10) if and only if Ω is a ball.

Proof. In [5], p. 52, Theorem 51, it is proved that:

$$\sigma_{r-1}\sigma_{r+1} < \sigma_r^2 \quad (1 \leq r < n) ,$$

unless all the positive numbers K_i are equal. From this it follows that:

$$\sigma_1 \sigma_{k-1} > \sigma_k \quad (1 \leq k \leq n) ,$$

unless all the positive numbers K_i are equal. Therefore by (2) we have

$$\frac{n+1-k}{n\ k} S_{k-1} \geq S_k S_1^{-1} \quad (1 \leq k \leq n)\,, \tag{11}$$

and by integration on M we obtain:

$$\int_M S_k S_1^{-1} dM \leq \frac{n+1-k}{nk} \binom{n}{k-1} (n+1) w_k = c_k w_k \ ,$$

and (8) is proved. By changing $k-1$ with k in (11) we get after integration for $0 \leq k \leq n-1$:

$$\int_M S_1 S_k dM \geq \frac{n(k+1)}{n-k} \binom{n}{k+1} (n+1) w_{k+2},$$

and (9) is proved. To prove (10) we observe that changing coordinates with the Gauss map:

$$\int_M S_1 S_n dM = \int_B S_1 dB.$$

By Schwarz's inequality

$$\int_B S_1 dB \geq C^2 (n+1)^2 \left(\int_B S_1^{-1} dB \right)^{-1}. \tag{12}$$

From (11), with $k = n$, we have

$$S_1^{-1} \leq \frac{1}{n^2} S_{n-1} S_n^{-1}\ ,$$

and by integration

$$\begin{aligned} \int_B S_1^{-1} dB &\leq \frac{1}{n^2} \int_B S_{n-1} S_n^{-1} dB \\ &= \frac{1}{n^2} \int_M S_{n-1} dM \\ &= \frac{n+1}{n} w_n. \end{aligned} \tag{13}$$

Our result follows from (12) and (13). Moreover in (8), (9), (10) equality holds if and only if it holds in (11), i.e., if and only if all the principal curvatures K_i are equal; this means that all points of M are umbilic, and therefore (see [19, p. 128]) M is a sphere. □

2. Derivatives of Minkowski functionals

First we briefly recall the principal results of [16], which give us the most important formulas and a technique to compute the derivatives of Minkowski functionals. Let us consider a one parameter family:

$$X \equiv X(t) : M \hookrightarrow \mathbb{R}^{n+1}, \tag{14}$$

of immersions of the n-manifold M into $\mathbb{R}^{n+1}$. In the sequel in order to simplify the notations M stands for the immersed manifold $X(t)(M)$.

We suppose that M is a smooth compact hypersurface with empty boundary. The letters i, j, l, m, p, q ,when indices (with or without subscripts) are used in the sense of classical tensor analysis. In particular, the summation convention is in effect for these indices, all calculations will be in terms of local coordinates $x^1, \ldots, x^n$ on M. The normal component of the deformation field $\partial X/\partial t$ is given by:

$$\lambda = < \frac{\partial X}{\partial t}, N > \ , \tag{15}$$

where N denotes the unit inner normal in $\mathbb{R}^{n+1}$ to M. Using the notation μ^j for the tangential components of $\partial X/\partial t$ we have

$$\frac{\partial X}{\partial t} = \lambda N + \mu^j \frac{\partial X}{\partial x^j}. \tag{16}$$

Let g_{ij} represent the metric tensor

$$g_{ij} = < \frac{\partial X}{\partial x^i}, \frac{\partial X}{\partial x^j} >,$$

and g^{ij} the inverse of the metric tensor. The coefficients of the second fundamental form on M are

$$b_{ij} = - < \frac{\partial X}{\partial x^i}, \frac{\partial N}{\partial x^j} > .$$

We also represent the shape operator E on M by the tensor:

$$b_i^j = g^{jl} b_{li}. \tag{17}$$

Let us recall that the principal curvatures K_i of M are the eigenvalues of the operator E.

The r-th *Newton tensor* associated to the shape operator E is the tensor:

$$\Gamma_r = S_r I - S_{r-1} E + \cdots + (-1)^r S_0 E^r, \quad r = 0, \ldots, n. \tag{18}$$

It is obvious that the Newton tensors can be defined inductively by

$$(19) \qquad \Gamma_0 = I, \quad \Gamma_{r+1} = S_{r+1}I - E\Gamma_r .$$

The following property of the Newton tensors are proved in [17]:

Theorem A. *The Newton tensors are free-divergence, i.e.:*

$$(20) \qquad \Gamma_r^{ij}{}_{,j} = 0, \qquad r = 0, \ldots, n.$$

In the previous formula and in the sequel the commas indicate covariant differentiation relative to the metric induced by X. Another property of Γ_r is the following

Proposition 2.1. *If the shape operator E is nonnegative the same holds for each Γ_r.*

Proof. The principal curvatures K_i on M are the eigenvalues of E. Therefore by assumption $K_i \geq 0$. From (19) it is easy to see by induction on r that the eigenvalues Λ_j of Γ_r are given by

$$\Lambda_j = \sum_{i_1 < i_2 < \cdots < i_r \, , \, i_p \neq j} K_{i_1} K_{i_2} \cdots K_{i_r}$$

which are obviously non negative. □

In next theorem we summarize the results of [16] we need in the present paper.

Theorem B. *Let $P = < X, N >$ and $Q = \frac{1}{2}|x|^2$. Then for any smooth function f of $n+2$ variables we have:*

$$\begin{aligned}
&\frac{d}{dt}\int_M f(S_1, \ldots, S_n, P, Q)dM = \\
&\int_M \lambda\{-S_1 f + \sum_{r=1}^{n}(S_r S_1 - (r+1)S_{r+1})D_r f \\
&+ \sum_{r=1}^{n}((D_r f)\ {}_{,ij}\Gamma_{r-1}^{ij}) + (PS_1 + (n+1))D_{n+1} f \\
&+ (D_{n+1} f)\ {}_{,j} Q^j{}_{,} + PD_{n+2} f\} dM.
\end{aligned}$$

Here $D_r f$ denotes the usual partial derivative with respect to the $r-th$ variable and the commas indicate covariant differentiation relative to the metric induced by X. The proof gived by Reilly is based on standard differential geometry and computations in terms of local coordinates $x^1, \dots, x^n$ on M. The principal part of Reilly's proof turns out to be the evaluation of the following first variations:

$$\frac{\partial(dM)}{\partial t} = (-S_1\lambda + \mu^j_{,\,j})dM, \tag{21}$$

$$\frac{\partial N}{\partial t} = -g^{ij}(\lambda_{,\,j} + \mu^l b_{li})\frac{\partial X}{\partial x^j}, \tag{22}$$

and for $0 \le r \le n-1$:

$$\frac{\partial S_{r+1}}{\partial t} = \lambda(S_1 S_{r+1} - (r+2)S_{r+2}) + \Gamma^{ij}_r\ \lambda_{,\,ij} + S_{r+1\ ,\ j}\ \mu^j. \tag{23}$$

Now we apply Theorem B to get first and second variation of the k-Minkowski functionals defined in (1).

Theorem 2.1. *Let w_k be the Minkowski functionals related to the one parameter family* (14). *Then, for $k = 0, \dots, n$ we have*

$$\frac{d}{dt}w_k = -\frac{n+1-k}{n+1}\int_M \lambda\sigma_k dM. \tag{24}$$

Moreover for $k = 1, \dots, n$ we have

$$\begin{aligned}\binom{n+1}{k}\frac{d^2}{dt^2}w_k &= -\int_M \frac{\partial\lambda}{\partial t}S_k dM + (k+1)\int_M \lambda^2 S_{k+1}dM \\ &\quad + \int_M \Gamma^{ij}_{k-1}\ \lambda_{,\,i}\lambda_{,\,j}dM + \int_M S_k\mu^j\lambda_{,\,j}dM,\end{aligned} \tag{25}$$

and

$$\frac{d^2}{dt^2}w_0 = \int_M (-\frac{\partial\lambda}{\partial t} + \lambda^2 S_1 + \mu^j\lambda_{,\,j})dM. \tag{26}$$

Proof. By applying Theorem B with $f = S_{k-1}$ we easily obtain (24) for $1 \le k \le n$. Starting from (3) and choosing $f = -P = h$ we derive also the case $k = 0$. Now by differentiating (24) we get for $k = 1, \dots, n$

$$\frac{d^2}{dt^2}w_k = -\frac{n+1-k}{n+1}\binom{n}{k}^{-1}\int_M \left(\frac{\partial\lambda}{\partial t}S_k + \lambda\frac{\partial S_k}{\partial t}\right)dM + \lambda S_k\frac{\partial(dM)}{\partial t}. \tag{27}$$

From (23) and (21) we obtain

$$\int_M \lambda\left(\left(\frac{\partial S_k}{\partial t}\right) dM + S_k \frac{\partial(dM)}{\partial t}\right) =$$

$$\int_M \lambda\left((\lambda(S_1 S_k - (k+1)S_{k+1}) + \Gamma^{ij}_{k-1}\,\lambda_{,\,ij} + S_{k,\,j}\mu^j)\right) dM \tag{28}$$

$$+\int_M \lambda S_k(-S_1\lambda + \mu^j_{,\,j})dM.$$

By Theorem A and "integration by parts" we have:

$$\int_M \lambda\lambda_{,\,ij}\Gamma^{ij}_{k-1}\, dM = -\int_M \lambda_{,\,i}\lambda_{,\,j}\Gamma^{ij}_{k-1}\, dM,$$

and

$$\int_M \lambda(S_k\mu^j_{,\,j} + S_{k,\,j}\mu^j)dM = \int_M S_k\mu^j\lambda_{,\,j}dM.$$

Substituting in (28) we obtain

$$\int_M \lambda\left(\left(\frac{\partial S_k}{\partial t}\right) dM + S_k \frac{\partial(dM)}{\partial t}\right) =$$

$$\int_M (\Gamma^{ij}_{k-1}\lambda_{,\,i}\lambda_{,\,j} + S_k\mu^j\lambda_{,\,j} + \lambda^2(k+1)S_{k+1})dM.$$

Formula (25) now follows from (27). The proof of (26) is similar. □

3. Using the support function

Let G be the Gauss map between the unit sphere $B \subset \mathbb{R}^{n+1}$ and the strictly convex hypersurface M, i.e., for $Y \in B$ let $G(Y)$ be the point $X \in M$ such that Y is the unit exterior normal to M at X. As usual we define the support function of M by

$$h = \langle X, Y \rangle\,. \tag{29}$$

If we choose local parameters $\theta^1, \cdots, \theta^n$ on B, then the support function h is a function of the variables θ^t. In the sequel h_i and h_{ij} denote first and second partial derivatives of h with respect to θ^i. Since Y is orthogonal to the hypersurface M differentiating (29) we obtain

$$h_j = \langle X, \frac{\partial Y}{\partial \theta^j} \rangle\,. \tag{30}$$

We choose on B a system of geodesics (great circles) C_i and let θ^i be the circular coordinate on C_i. Then

(31) $$\frac{\partial}{\partial\theta^i}Y = T_i \,,$$

where T_i is the tangent field to C_i. From the previous formulas we derive

(32) $$X = hY + h_i T^i \,,$$

where $\{T^i\}$ is the dual system of the set $\{T_i\}$ on the tangent plane at X to M, i.e.

(33) $$< T^i,\, T_j >= \delta^i_j.$$

Let now O be a fixed point on M. We can suppose that the parametric curves C_i are orthogonal at O; therefore we have at O:

(34) $$T^i = T_i \,,$$

and

(35) $$\frac{\partial}{\partial\theta^j}T_i = \frac{\partial}{\partial\theta^j}T^i = -\delta^i_j Y.$$

Differentiating (32) with respect to θ^j we get, using (33):

(36) $$\frac{\partial X}{\partial\theta^j} = h_j Y + hT_j + h_{ij}T^i + h_i\frac{\partial T^i}{\partial\theta^j}.$$

From (34) and (35) we have at O:

(37) $$\frac{\partial X}{\partial\theta^j}(O) = h_j Y + hT_j + h_{ij}T^i - h_j Y = hT_j + h_{ij}T_i$$

The coefficients of the second fundamental form on M at O from (31), (35) and (37) are given by:

(38) $$b_{ij} =< \frac{\partial X}{\partial\theta^i}, \frac{\partial Y}{\partial\theta^j} >= h\delta^i_j + h_{ij}.$$

Now let us consider the parameter family (14). For each t we consider the support function $h(\theta^1,\dots,\theta^n,t)$ related to M and we introduce in the previous computations the dependence on t. Let us observe that Y, T_i and T^i does not depend on t. Our aim is to obtain the first and the second variation of Minkowski functionals, computed in Section 2, in terms of the support function h and its derivatives. From (32) we get

(39) $$X_t = h_t Y + h_{ti}T^i,$$

where the subscript t is for the partial derivatives with respect to t; therefore, since $Y = -N$, the normal component of X_t is

$$\lambda = -h_t. \tag{40}$$

Since N does not depend on t we get from (22) that

$$\lambda_{,\,i} + \mu^l b_{li} = 0.$$

Denoting with P^{li} the inverse tensor of the second fundamental form b_{li} we derive from the two previous equations:

$$\mu^l = P^{li} h_{ti}. \tag{41}$$

Theorem 3.1. *If M is a one-parameter family of strictly convex hypersurface, with empty boundary, then for $k = 0, \ldots, n$*

$$\frac{d}{dt} w_k = \binom{n+1}{k}^{-1} \int_M (h_t S_k) dM, \tag{42}$$

and for $k = 1, \ldots, n$

$$\binom{n+1}{k} \frac{d^2}{dt^2} w_k = \int_M (h_{tt} S_k + (k+1) h_t^2 S_{k+1} + (\Gamma^{ij}_{k-1} - S_k P^{ij}) h_{ti} h_{tj}) dM; \tag{43}$$

moreover

$$\frac{d^2}{dt^2} w_0 = \int_M (h_{tt} + h_t^2 S_1 + P^{ij} h_{ti} h_{tj}) dM. \tag{44}$$

Proof. From (24) and (40) we immediately obtain

$$\frac{d}{dt} w_k = \frac{n+1-k}{n+1} \int_M h_t \sigma_k dM, \tag{45}$$

and using (2) we derive (42). Similarly, using (40) and (41) in (25) and (26) we obtain (43) and (44), respectively. □

Remark. Theorem 3.1 can be proved without the assumption on strict convexity of M. One can prove that (42), (43) and (44) hold for regular manifolds M with shape operator almost everywhere not singular. In fact what we essentially need is to locally parametrise M by the Gauss map. This is possible in a neighbourhood of all points where the shape operator is not singular.

We conclude this section by considering the family M of the level surfaces of a given real function u. In this case our assumptions are the convexity of each level surface and the condition $\nabla u \neq 0$ on M. The map $G : (\theta^1, \dots, \theta^n, t) \to X$ has Jacobian given by

$$\left(\frac{\partial X}{\partial \theta^1}, \dots, \frac{\partial X}{\partial \theta^n}, \frac{\partial X}{\partial t}\right).$$

These components are computed in (37) and (39). The inverse G^{-1} has Jacobian given by the following expressions:

$$\nabla u \equiv \left(\frac{\partial t}{\partial x^1}, \dots, \frac{\partial t}{\partial x^{n+1}}\right) = h_t^{-1} Y \ , \tag{46}$$

$$\nabla \theta^i \equiv \left(\frac{\partial \theta^i}{\partial x^1}, \dots, \frac{\partial \theta^i}{\partial x^{n+1}}\right) = P^{ij}(T_j - h_t^{-1} h_{tj} Y). \tag{47}$$

We observe that formula (46) holds in M, while (47) is obtained only for the points where the parametric curves are orthogonal. By differentiating (46) with respect to θ^i and to t we obtain:

$$\frac{\partial \nabla u}{\partial \theta^i} = -h_t^{-2} h_{ti} Y + h_t^{-1} T_i \ ,$$

$$\frac{\partial \nabla u}{\partial t} = -h_t^{-2} h_{tt} Y.$$

From the chain rule it follows that:

$$u_{lm} = \frac{\partial u_l}{\partial \theta^i} \cdot \frac{\partial \theta^i}{\partial x^m} + \frac{\partial u_l}{\partial t} \cdot \frac{\partial t}{\partial x^m}$$

and from the previous equalities we get:

$$u_{lm} = (-h_t^{-2} h_{ti} Y + h_t^{-1} T_i)_l P^{ij}(T_j - h_t^{-1} h_{tj} Y)_m - h_t^{-3} h_{tt} (Y)_l (Y)_m, \tag{48}$$

where $(\ )_l$ and $(\ )_m$ denotes the l and m components of the vectors in the brackets.

Theorem 3.2. *If u is a smooth function on a domain $\Omega \subset \mathbb{R}^{n+1}$, with regular level surfaces, then at each point where the second fundamental form is not singular we have that*

$$|\nabla u| = h_t^{-1} \quad , \tag{49}$$

and the second normal derivative of u is

$$u_{NN} = h_t^{-3}(P^{ij}h_{ti}h_{tj} - h_{tt}) \quad . \tag{50}$$

Proof. The formula (49) follows directly from (46). The second normal derivative is $u_{NN} = |\nabla u|^{-2}(u_{lm}u_l u_m)$. We can suppose that the parametric curves are orthogonal at a fixed point. Formula (50) follows now from (48) and (46), since T_i and T_j are orthogonal to the normal field $Y = \nabla u/|\nabla u|$. This concludes the proof. □

From Theorem 3.1 and 3.2 we obtain:

Theorem 3.3. *If u is any smooth function on a domain $\Omega \subset \mathbb{R}^{n+1}$, with regular t-level surfaces M, without boundary, then for $k = 0, \ldots, n$:*

$$\frac{d}{dt}w_k = \binom{n+1}{k}^{-1} \int_M |\nabla u|^{-1} S_k dM, \tag{51}$$

moreover if the t-level surfaces are convex:

$$\frac{d^2}{dt^2}w_k \geq \binom{n+1}{k}^{-1} \int_M (-u_{NN}|\nabla u|^{-3}S_k + (k+1)|\nabla u|^{-2}S_{k+1})dM. \tag{52}$$

For $k \neq 0$ equality holds if and only if $|\nabla u|$ is constant on each M.

Proof. First we suppose M strictly convex. Then from (49) and (42) formula (51) follows immediately. To prove (52) we use Proposition 2.1. For $k \geq 1$ we have $\Gamma^{ij}_{k-1}\, h_{ti}h_{tj} \geq 0$, inequality (52) follows from (43) and (50). For $k = 0$ (52) is obtained similarly using (43). This concludes the proof of Theorem 3.3 under the additional strict convexity assumption on M; by an approximation argument we get the proof for convex M. For $k \geq 1$ let us observe that equality holds in (52) if and only if $\Gamma^{ij}_{k-1}\, h_{ti}h_{tj} \equiv 0$ which implies $h_{ti} \equiv 0$. This means by (49) that the tangential derivatives of $|\nabla u|$ vanish, therefore $|\nabla u|$ is constant on M. □

4. Differential inequalities

In this section we present differential inequalities for level Minkowski functionals w_k of solutions in the open set $\Omega \subset \mathbb{R}^{n+1}$ to the following equation

$$(a(q^2)u_i)_i = 0. \tag{53}$$

The function a is C^1 and q^2 stands for $|\nabla u|^2$. The equation (53) will usually be assumed to be elliptic. This is the case for a satisfying the conditions:

$$a \geq 0 \ , \ a(s) + 2sa'(s) > 0 \ . \tag{54}$$

We look for differential inequalities involving only w_k and its derivatives. These inequalities will be sharp in the sense that equality holds for special geometry, that is to say in the case that all the level surfaces M are radially symmetric and concentric, so that u is radial.

In the following theorem d_k is the constant in (9) of Lemma 1 and

$$e_k = \binom{n+1}{k} \frac{n}{n+1-k} \quad , \quad f_k = \binom{n+1}{k+1} \frac{1}{n-k} .$$

Theorem 4.1. *Let u be a smooth solution to (53) with closed regular and convex level surfaces. If there exist non negative constants α, β such that*

$$a(a + 2a'q^2)^{-1} \geq \alpha q^2 + \beta \ , \tag{55}$$

then for $k = 1, \ldots, n-1$:

$$\binom{n+1}{k} \frac{d^2}{dt^2} w_k \geq \alpha d_k w_{k+2} + \beta e_k w_k^{-1} (\frac{d}{dt} w_k)^2 + f_k w_{k+2}^{-1} (\frac{d}{dt} w_{k+1})^2 \ , \tag{56}$$

and

$$\frac{d^2}{dt^2} w_0 \geq \alpha d_0 w_2 + (\beta + 1) \frac{n+1}{n} w_2^{-1} (\frac{d}{dt} w_1)^2 \ , \tag{57}$$

$$\frac{d^2}{dt^2} w_n \geq C^2 n w_n^{-1} + \beta n w_n^{-1} (\frac{d}{dt} w_n)^2 \ . \tag{58}$$

Equality holds in (56), (57) *or* (58) *if and only if it holds in* (55) *and u is radial.*

Proof. Since u is a solution to (53) we have

$$2a' u_{ij} u_i u_j + a u_{ii} = 0 \ . \tag{59}$$

Therefore by recalling that $u_{NN} = u_{ij} u_i u_j q^{-2}$ and that $u_{ii} = u_{NN} + S_1 q$ we deduce from (59) that

$$(2a'q^2 + a) u_{NN} + a S_1 q = 0.$$

Substituting u_{NN} in (52) we have

$$\binom{n+1}{k}\frac{d^2}{dt^2}w_k \geq \int_M (a(a+2a'q^2)^{-1}q^{-2}S_1S_k + (k+1)q^{-2}S_{k+1})dM. \tag{60}$$

By assumption (55) the first term on the right is greater than

$$\int_M (\alpha+\beta q^{-2})S_1S_k dM = \alpha\int_M S_1S_k dM + \beta\int_M q^{-2}S_1S_k dM.$$

By applying Lemma 1, formula (9), to the first integral on the right in the previous equality we obtain the first term on the right in (56). Then from Schwarz's inequality

$$\int_M q^{-2}S_1S_k dM \geq \left(\int_M q^{-1}S_k dM\right)^2\left(\int_M S_kS_1^{-1}dM\right)^{-1}. \tag{61}$$

From Lemma 1, formula (8), and from (51) we obtain that

$$\int_M q^{-2}S_1S_k dM \geq \binom{n+1}{k}^2\left(\frac{d}{dt}w_k\right)^2 c_k^{-1}\, w_k^{-1},$$

which is the second term on the right in (56). Similarly

$$\int_M q^{-2}S_{k+1}dM \geq \left(\int_M q^{-1}S_{k+1}dM\right)^2\left(\int_M S_{k+1}dM\right)^{-1},$$

and therefore:

$$\int_M q^{-2}S_{k+1}dM \geq \frac{1}{n-k}\binom{n+1}{k+1}\left(\frac{d}{dt}w_{k+1}\right)^2 w_{k+2}^{-1}\ ;$$

this is the last term on the right in (56). Finally, if equality holds in (56) then we have also equality in (61), (52) and (9): the functions $q=|\nabla u|$ and S_1 must be proportional, $|\nabla u|$ has to be constant on M, at each point on M all the principal curvatures must be equal. This implies that u is radially symmetric.

The proof of (57) is similar. First we obtain the inequalities:

$$\frac{d^2}{dt^2}w_0 \geq \int_M (a(a+2a'q^2)^{-1}q^{-2}S_1 + q^{-2}S_1)dM \geq \int_M (\alpha+(\beta+1)q^{-2})S_1 dM.$$

This gives, using Lemma 1, the term on the right in (57). The proof of (58) is obtained with the same arguments. □

Now we consider the functionals w_k for solutions of classical capacity problems when the domain is a convex annulus. We obtain in the following theorems convexity properties of w_k which extend to n-dimensions analogous results contained in [1], [7], [10], [14].

Theorem 4.2. *Let u be harmonic in a convex annulus $\Omega \subset \mathbb{R}^{n+1}$ bounded by two convex level surfaces. Then*

i) $(w_k)^\tau$ *is concave for* $k = 2, \cdots, n$ *and* $\tau = \frac{1-k}{n+1-k}$;

ii) w_1 *is* log *convex;*

iii) w_0 *is convex.*

Proof. In [8], it is proved that the level surfaces of u in Ω are also convex. Therefore Theorem 4.1 applies with $\alpha = 0$, $\beta = 1$. We observe that by convexity all the w_k's are positive. From (56) it follows that

$$\frac{d^2}{dt^2} w_k \geq \frac{n}{n+1-k} w_k^{-1} \left(\frac{d}{dt} w_k \right)^2 \qquad (1 \leq k \leq n-1).$$

This inequality implies ii), and it implies i) except for the case $k = n$. This case and the convexity of w_0 can be proved by similar arguments starting from (58) and (57). □

Theorem 4.3. *Let u be a solution to the p-Laplacian equation*

$$(|\nabla u|^{p-2} u_i)_i = 0\ , \quad p > 1,$$

in a convex annulus $\Omega \subset \mathbb{R}^{n+1}$, bounded by two convex level surfaces of u. Then

i) $\frac{1}{\tau}(w_k)^\tau$ *is convex if* $0 \neq \tau = 1 - n(n+1-k)^{-1}(p-1)^{-1}$;

ii) w_k *is* log *convex if* $k = 1 - (p-2)(p-1)^{-1} n$;

iii) w_0 *is convex.*

Proof. In [8] it is proved that u has convex level surfaces, then Theorem 4.1 applies with $a(a + 2a'q^2)^{-1} = (p-1)^{-1}$, $\alpha = 0$ and $\beta = (p-1)^{-1}$. □

Theorem 4.4. *Let u be a solution to the minimal surface equation*

$$((1 + |\nabla u|^2)^{\frac{-1}{2}} u_i)_i = 0$$

in a convex annulus $\Omega \subset \mathbb{R}^{n+1}$ bounded by two convex level surfaces. Then for $k = 1, \ldots, n-1$

$$\binom{n+1}{k} \frac{d^2}{dt^2} w_k \geq d_k w_{k+2} + e_k (w_k)^{-1} \left(\frac{d}{dt} w_k \right)^2 ; \tag{62}$$

and

$$\frac{d^2}{dt^2}w_n \geq C^2 n w_n^{-1} + n w_n^{-1} \left(\frac{d}{dt}w_n\right)^2 . \tag{63}$$

Equality holds in (63) *if and only if* u *is radially symmetric.*

Proof. In [6] it is proved that u has convex level surfaces. In this case we have $a = (1+q^2)^{-1/2}$ and (55) holds with $\alpha = \beta = 1$. Then (62) and (63) follow from (57) and (58), respectively. □

Remark. The results of Theorem 4.4 can be extended also to parametrical minimal surfaces, the case $n = 2$ is considered in [9].

5. An isoperimetric inequality

Let us consider the Green's function of the Laplace operator Δ in a domain $\Omega \subset \mathbb{R}^{n+1}$. We have

$$g(x,y) = \frac{1}{2\pi}\log\frac{R(y)}{|x-y|} + H(x,y) \qquad (n=1)\ ,$$

and

$$g(x,y) = \frac{1}{(n-2)C}(|x-y|^{2-n} - R^{2-n}(y)) + H(x,y) \quad (n>1)\ ,$$

where $H(x,y)$ is a harmonic function with $H(y,y) = 0$, $R(y)$ is called the harmonic radius of Ω with respect to y and $g(x,y)$ vanishes for $x \in \partial\Omega$. The following theorem gives sharp inequalities for the Minkowski functional w_n of the level surfaces of the Green's function in a convex domain $\Omega \subset \mathbb{R}^{n+1}$.

Theorem 5.1. *Let* Ω *be a convex domain in* $\mathbb{R}^{n+1}$ *and let* y *be a fixed point in* Ω. *Then the* n*-Minkowski functional of the level surfaces* $\{g = t\}$ *satisfies the following inequalities:*

$$w_1(t) \leq w_1(0)\exp(-2\pi t) \qquad (n=1)\ , \tag{64}$$

and

$$w_n(t) \leq (C^{2-n}(n^2-1)t + w_n(0)^{1-n})^{\frac{1}{1-n}} \qquad (n>1). \tag{65}$$

Equality holds in (64) *or* (65) *for some* t *if and only if* Ω *is a ball centered at* y.

Proof. The support function h of $\{g = t\}$ with respect to the point y is given by $h = - < x - y, N >$. Then h is positive and $h \leq |x - y|$. From (4) it follows that

$$(n+1)w_n = \int_B h dB \leq \int_B |x - y| dB.$$

For $n > 1$ the level surface $\{g = t\}$ is the set of points x such that

$$|x - y|^{1-n} = R^{1-n}(y) + (n^2 - 1)C(t - H(x, y)) .$$

Hence

$$(n+1)w_n \leq \int_B (R^{1-n}(y) + (n^2 - 1)C(t - H))^{\frac{1}{1-n}} dB.$$

Therefore

$$(n+1)^{1-n} w_n^{1-n} \geq \left(\int_B (R_y^{1-n} + (n^2 - 1)C(t - H))^{\frac{1}{1-n}} dB\right)^{1-n}.$$

When t goes to $+\infty$ the level surfaces of $\{g = t\}$ converge to y and $H(x, y)$ goes to zero. Hence for $\epsilon > 0$ and t large:

$$w_n^{1-n} \geq C^{1-n}(R^{1-n}(y) + (n^2 - 1)Ct) - \epsilon.$$

From the above inequality and by the concavity of w_n^{1-n} proved in Theorem 4.2 we derive that for each $t > 0$:

$$\frac{d}{dt} w_n^{1-n} \geq C^{2-n}(n^2 - 1) \quad .$$

Then by integration on t

$$w_n^{1-n}(t) \geq C^{2-n}(n^2 - 1)t + w_n^{1-n}(0) ,$$

and (65) is proved. Inequality (64) is proved similarly, see also [10], p. 387. □

References

1. G. Alessandrini, *Isoperimetric inequalities for the length of level lines of solutions of quasilinear capacity problems in the plane.* J. Appl. Math. Phys. **40** (1989), 258–283.

2. C. Bandle. *Isoperimetric inequalities and their applications.* Pitman, London 1980.

3. T. Bonnesen and W. Fenchel, *Theorie der konvexen Körper.* Springer, Berlin 1934.

4. H. Busemann, *Convex Surfaces.* Wiley, 1958.

5. G.H. Hardy, J.E. Littlewood and G. Pólya, *Inequalities.* Cambridge University Press, Cambridge 1952.

6. N.J. Korevaar, *Convexity of level sets for solutions to elliptic ring problems.* Preprint.

7. P. Laurence, *On the convexity of geometric functionals of level for solutions of certain elliptic partial differential equations.* J. Appl. Math. Phys. **40** (1989), 258–283.

8. J. Lewis, *Capacitary functions in convex rings.* Arch. Rat. Mech. Anal. **66** (1977), 201–224.

9. M. Longinetti, *On minimal surfaces bounded by two convex curves in parallel planes.* J. Diff. Eq. **67** (1987), 344–358.

10. M. Longinetti, *Some isoperimetric inequalities for the level curves of capacity and Green's functions on convex plane domains.* SIAM J. Math. Anal. **19** (1988), 377–389.

11. M. Longinetti, *A strong convexity principle for convex solutions to elliptic P.D.E.'s.* To appear in Indiana Univ. Math. J.

12. L.E. Payne, *New isoperimetric inequalities for eigenvalues and other physical quantities.* Comm. Pure Appl. Math. **9** (1956), 531–542.

13. L.E. Payne, *Isoperimetric inequalities and their applications.* SIAM Rev. **9** (1967), 453–488.

14. L.E. Payne and G.A. Philippin, *On the conformal capacity problem.* Symposia Mathematica XXX (1989), 119–136.

15. G. Pólya and G. Szegő, *Isoperimetric inequalities in mathematical physics.* Princeton Univ. Press, Princeton 1951.

16. R.C. Reilly, *Variational properties of functions of the mean curvatures for hypersurfaces in space forms.* J. Diff. Geo. **8** (1973), 465–477.

17. R.C. Reilly, *On the Hessian of a function and the curvatures of his graph.* Michigan Math. J. **20** (1973–74), 373–383.

18. K. Tso, *On symmetrization and Hessian equations.* J. Analyse Math. **52** (1989), 94–106.

19. T.J. Willmore, *An Introduction to differential geometry.* Oxford Univ. Press, London and New York, 1959.

Giuseppe Chiti, Politecnico di Torino, Corso Duca degli Abruzzi 24, I-10129 Torino, Italy

Marco Longinetti, Università di Torino, via Carlo Alberto 10, I-10123 Torino, Italy

International Series of Numerical Mathematics, Vol. 103, © 1992 Birkhäuser Verlag Basel

Some best constant inequalities for conjugate functions

Matts Essén

Abstract

Let $F = f + i\tilde{f}$ be analytic in the unit disc U or the upper half-plane D. We discuss inequalities of the type

$$||F||_p^p \le C_p ||f||_p^p, \quad 1 < p < \infty,$$

where $||\cdot||_p$ denotes Hardy norms in U or D. We consider also inequalities of the $L \log L$-type and some related questions. The proofs are via superharmonic functions. In all cases, they give the best constants.

0. Introduction

Let $F = f + i\tilde{f}$ be analytic in the unit disc U with $\tilde{f}(0) = 0$. It is known that there are constants C_p, A and B such that

$$(0.1) \qquad ||F||_p^p \le C_p ||f||_p^p, \quad 1 < p < \infty,$$

$$(0.2) \qquad ||F||_1 \le A \sup_{0<r<1} \frac{1}{2\pi} \int_0^{2\pi} |f(re^{i\theta})| \log^+ |f(re^{i\theta})| \, d\theta + B.$$

The best constants are known (cf. [4]). In [2], we gave a simple proof of (0.1) using superharmonic functions: this proof gave also the best constant. Another superharmonic proof can be found in [5]. Here, our purpose is to give best constant proofs of inequalities of this type in U and in the half-plane $D = \{z \in \mathbb{C} : \operatorname{Im} z > 0\}$ and to discuss some related questions (the results in the $L \log L$-case in D are incomplete). In all cases, we start from simple inequalities involving superharmonic functions (cf. Section 1). Our inequalities are given in Sections 2, 3 and 4. An alternative proof of a result from [3] is a direct consequence of the results in Section 2.

For basic facts on Hardy spaces, we refer to [1].

This paper is in final form and no version of it will be submitted for publication elsewhere.

1. Inequalities for superharmonic functions

If $z = x + iy$ and $x \geq 0$, we define

$$G(z) = -\text{Re}\,\{z \log z\} = y \arg z - x \log |z|, \quad G(0) = 0.$$

We assume that $\arg z$ vanishes on the real axis.

Lemma 1. *If $x > 0$, then*

$$|z| - \frac{2}{\pi} x \log |z| - x \leq \frac{2}{\pi} G(z), \tag{1.1}$$

$$|z| - \frac{2}{\pi} x \log |z| - \frac{2}{\pi} x \geq \frac{2}{\pi} G(z), \tag{1.2}$$

$$G(z) = e^{-1} - (e/2)(x - e^{-1})^2 + O(|z - e^{-1}|^3), \quad z \to e^{-1}. \tag{1.3}$$

The proofs are elementary and are omitted. We note that G is harmonic in $\{\text{Re}\, z > 0\}$.

Lemma 2. *We define $G_1(z) = G(|x| + iy + e^{-1})$. Then G_1 is superharmonic in $\mathbb{C}$.*

Proof. It is clear that G_1 is continuous in $\mathbb{C}$ and harmonic outside the imaginary axis. It is easy to see that if $y \neq 0$, then (if $G_{1x} = \partial G_1/\partial x$),

$$\lim_{x \to 0+} G_{1x}(x + iy) = -1 - \log |y + ie^{-1}| < 0,$$

$$\lim_{x \to 0-} G_{1x}(x + iy) = 1 + \log |y + ie^{-1}| > 0.$$

Thus, if $y_0 \neq 0$, then locally around y_0, we have

$$G_1(z) = \min\,(G(x + iy + e^{-1}), G(-x + iy + e^{-1})).$$

Since G_1 is the minimum of two harmonic functions, it is superharmonic in a neighborhood of iy_0. To prove that G_1 has the superharmonic mean value property at the origin, it suffices to note that

$$G_1(z) = e^{-1} - \frac{1}{2} ex^2 + O(|z|^3), \quad z \to 0, \tag{1.4}$$

(cf. (1.3)). This completes the proof of Lemma 2.

Lemma 3. *The function* $H_0(z) = ((1+|x|)^2 + y^2)^{1/2} - (1+|x|)\log(1+|x|)$ *is superharmonic in* $\mathbb{C}$. *Furthermore, we have*

$$H_0(z) = 1 + \frac{1}{2}(y^2 - x^2) - \frac{1}{2}|x|y^2 + \frac{|x|^3}{6} + O(|z|^4), \quad z \to 0. \tag{1.5}$$

Proof. If $x > -1$, we consider the function $H_1(z) = |z+1| - (1+x)\log(1+x)$. Since $\Delta H_1(z) = |z+1|^{-1} - (1+x)^{-1}$ is non-positive, H_1 is superharmonic if $x > -1$. Thus H_0 is superharmonic outside the imaginary axis. Since

$$H_{1x}(iy) = |1+iy|^{-1} - 1 < 0, \quad y \neq 0,$$

it is clear that locally around iy_0, $y_0 \neq 0$, we have

$$H_0(z) = \min\,(H_1(x+iy), H_1(-x+iy)).$$

Thus H_0 is superharmonic everywhere except possibly at the origin. There, we use (1.5) to prove that H_0 has the superharmonic mean value property. This concludes the proof of Lemma 3.

Lemma 4. *If* $1 < p < 2$ *and* $\alpha = \pi/(2p)$, *we define*

$$G(z) = \begin{cases} |z|^p - (\cos\alpha)^{-p}|x|^p, & \alpha < \arg z < \pi - \alpha, \\ -\tan\alpha\,|z|^p\cos p\theta, & |\theta| < \alpha, \text{ where } \theta = \arg z, \\ -\tan\alpha\,|z|^p\cos p(\pi - |\theta|), & 0 \le \pi - |\theta| \le \alpha, \end{cases}$$

and $G(0) = 0$. *Then* G *is superharmonic in* $\mathbb{C}$ *and*

$$|z|^p - (\cos\alpha)^{-p}|x|^p \le G(z), \quad z \in \mathbb{C}\,. \tag{1.6}$$

For the proof, we refer to [2].

Lemma 5. *Let* H *be superharmonic in* $\mathbb{C}$ *and such that* $H(0) = 0$ *and*

$$|\nabla H(z)| = o(1), \quad z \to 0.$$

Let F *be analytic in the closed upper halfplane* $\overline{D}$ *and assume furthermore that we have the estimates*

$$F(z) = O(|z|^{-1}), \quad z \to \infty, \quad z \in \overline{D}; \qquad F'(z) = O(|z|^{-2}), \quad z \to \infty, \quad z \in \overline{D}.$$

Then we have

$$\int_{-\infty}^{\infty} H(F(x))\,dx \le 0. \tag{1.7}$$

Proof. Let $D_R = \overline{D} \cap \{|z| < R\}$. Since $H \circ F$ is superharmonic in D, we can apply Green's formula to obtain

$$0 \ge \iint_{D_R} y\Delta H \circ F(z)\,dx\,dy = \int_{-R}^{R} H(F(x))\,dx + \int_{C_R} \Big(y\frac{\partial}{\partial r} H \circ F(z) - H \circ F(z)\frac{\partial y}{\partial r}\Big)\,ds,$$

where $C_R = D \cap \{|z| = R\}$. It is easy to estimate the integral over C_R. Letting $R \to \infty$, we obtain Lemma 5.

2. Inequalities in the unit disc

It is known that (0.1) holds with $C_p = (\cos(\pi/2p))^{-p}$, $1 < p \le 2$, and $C_p = (\cos(\pi/2q))^{-p}$, $2 < p < \infty$, where $p^{-1} + q^{-1} = 1$ (cf. [2]). To deduce this estimate in the case $1 < p < 2$, we use (1.6) and conclude that

$$|F(z)|^p - (\cos\alpha)^{-p}|f(z)|^p \le G \circ F(z). \tag{2.1}$$

Integrating over $\{|z| = r\}$, applying the superharmonic mean value inequality to the right hand member and letting $r \to 1$, we obtain

$$\|F\|_p^p - (\cos\alpha)^{-p}\|f\|_p^p \le G \circ F(0) \le 0.$$

The associated extremal case is a mapping of U onto the sector $\{w : |\arg w| < \pi/(2p)\}$. Details are given in [2].

In the $L\log L$-case, we introduce

$$L_U(f) = \sup_{0 \le r < 1} \frac{1}{2\pi}\int_0^{2\pi} |f(re^{i\theta})| \log|f(re^{i\theta})|\,d\theta,$$

$$L_U(f, F) = \sup_{0 \le r < 1} \frac{1}{2\pi}\int_0^{2\pi} |f(re^{i\theta})| \log|F(re^{i\theta})|\,d\theta.$$

Lemma 6. *$L_U(f)$ is finite if and only if $L_U(f, F)$ is finite.*

Proof. From Lemma 3, we see that

$$|F(z)| - (1 + |f(z)|)\log(1 + |f(z)|) \le H_0 \circ F(z).$$

Integrating over $\{|z| = r\}$ and arguing as above, we obtain

$$||F||_1 - L_U(1 + |f|) \leq H_0(F(0)).$$

Thus, if $L_U(f)$ is finite, then $F \in H^1(U)$.

To prove the lemma, we consider the elementary inequality

$$|x| \log |z| \leq |x| \log |x| + |y|,$$

and deduce that

$$L_U(f) \leq L_U(f, F) \leq L_U(f) + ||\tilde{f}||_1.$$

If $L_U(f, F)$ is infinite and $L_U(f)$ is finite, $||\tilde{f}||_1$ must be infinite which contradicts the first part of our argument. We have proved Lemma 6.

Theorem 1. *We have the inequality*

$$||F||_1 \leq \frac{2}{\pi} L_U(f, F) + B_1, \tag{2.2}$$

where the constant $2/\pi$ *is best possible and*

$$B_1 = ||f||_1 + \frac{2}{\pi e} \int_0^{2\pi} \log\left(e^{-1} + |F(e^{i\theta})|\right) \frac{d\theta}{2\pi} + e^{-1}\left(1 + \frac{4}{\pi}\right).$$

Remark. Theorem 1 should be compared with Theorem 3.4 in [4].

Proof. From Lemma 1 with z replaced by $w = u + iv$, we see that

$$|w| - \frac{2}{\pi}(|u| + e^{-1}) \log \left||u| + e^{-1} + iv\right| - (|u| + e^{-1}) \leq G_1(w).$$

To prove (2.2), we put $w = F(z)$, use the elementary inequality

$$(|u| + e^{-1}) \log \left||u| + e^{-1} + iv\right| \leq e^{-1} \log (e^{-1} + |w|) + e^{-1} + |u| \log |w|,$$

and the superharmonic mean value inequality. We note also that

$$G_1(F(0)) \leq G_1(0) = e^{-1}.$$

To prove that $2/\pi$ is the best constant in (2.2), we consider a mapping $F : U \to \{\mathrm{Re}\, z > 0\}$ with $\tilde{f}(0) = 0$. Since G is harmonic in $F(U)$, it follows from (1.1) and (1.2) that

$$\begin{aligned} &\frac{2}{\pi} G(F(0)) \leq ||F||_1 - \frac{2}{\pi} L_U(f, F) - \frac{2}{\pi} ||f||_1, \\ &||F||_1 - \frac{2}{\pi} L_U(f, F) - ||f||_1 \leq \frac{2}{\pi} G(F(0)). \end{aligned} \tag{2.3}$$

Let $E_a = \{u+iv : |w| < au \log |w|\}$ and let $F_a : U \to E_a$ be a conformal mapping onto E_a with $\tilde{f}_a(0) = 0$. It is known that $F_a \in H^1(U)$ if and only if $a < 2/\pi$ (cf. [3]). If $a \to (2/\pi)^-$ and $F_a(0) = ||f_a||_1$ is fixed, we deduce from (2.3) that

$$||F_a||_1 - \frac{2}{\pi} L_U(f_a, F_a) = O(1), \quad ||F_a||_1 \to \infty.$$

It follows that $L_U(f_a, F_a) \to \infty$. It is now clear that

$$\frac{||F_a||_1}{L_U(f_a, F_a)} \to \frac{2}{\pi}, \quad a \to \left(\frac{2}{\pi}\right)^-,$$

and our assertion is proved.

Remark. Our technique gives also an alternative proof of Theorem 1 in [3]. If F is analytic in U, we consider the Nevanlinna counting function

$$N(r,w) = \int_0^r \frac{n(t,w)}{t}\, dt,$$

where $n(r,w)$ is the number of roots of the equation $F(z) = w$ in the disc $\{|z| \le r\}$. It is known that $N(1,w) = \lim_{r \to 1-} N(r,w)$ exists. Let

$$N(w) = \limsup_{\zeta \to w} N(1,\zeta)$$

be the upper regularization of $N(1,w)$. Then $N(w)$ is subharmonic in $\mathbb{C} \setminus \{F(0)\}$ and the function $N(w) + \log |w - F(0)|$ can be extended to $F(0)$ to give us a function which is harmonic in $\mathbb{C}$. We note that $N(w)$ and $N(1,w)$ coincide outside an exceptional set of w-values of at most capacity 0 (cf. Section 2 in [3]).

The following identity of Stanton is given in Theorem 2 in [3]. Let F be analytic in U and let Φ be δ-subharmonic in $\mathbb{C}$ with $\Phi(F(0))$ finite (a δ-subharmonic function is the difference of two subharmonic functions). Then, if μ denotes the signed measure $\Delta\Phi/2\pi$, we have

$$\text{(2.4)} \qquad \frac{1}{2\pi}\int_0^{2\pi} \Phi(F(re^{i\theta}))\, d\theta = \int_{\mathbb{C}} N(r,w)\, d\mu(w) + \Phi(F(0)), \quad 0 \le r < 1.$$

If G is the function defined in Lemma 1, we let $G_2(z) = G(|x| + iy)$; G_2 is harmonic in $\mathbb{C}$ except on the imaginary axis, and we have

$$\Delta G_2(z) = -2(1 + \log |y|)\delta_x.$$

It follows from (2.4) that

$$\frac{1}{2\pi}\int_0^{2\pi} G_2(F(re^{i\theta}))\,d\theta = -\frac{1}{\pi}\int_{-\infty}^{\infty} N(r,iy)\log(1+|y|)\,dy + G_2(F(0)).$$

Letting $r \to 1$ and applying Lemma 1, we obtain

$$\begin{aligned}(2.5)\qquad & G_2(F(0)) + \|f\|_1 \le \\ & \frac{\pi}{2}\|F\|_1 - L_U(f,F) + \frac{1}{\pi}\int_{-\infty}^{\infty} N(iy)\log(1+|y|)\,dy \le G_2(F(0)) + \frac{\pi}{2}\|f\|_1.\end{aligned}$$

The following result (cf. Theorem 1 in [3]) is an immediate consequence of (2.5) and Lemma 6.

Theorem A. *Let $F \in H^1(U)$. The following statements are equivalent:*

a) $\displaystyle\int_{-\infty}^{\infty} N(iy)\log(1+|y|)\,dy < \infty$,

b) $L_U(f) < \infty$,

c) $L_U(f,F) < \infty$.

3. Inequalities for the Hilbert transform

If $p \in (1,\infty)$ and $f \in L^p(\mathbb{R})$ is real-valued, we consider for z in the upper half-plane D

$$(3.1)\qquad \begin{cases} F(z) = \dfrac{1}{\pi i}\displaystyle\int_{-\infty}^{\infty} \frac{f(t)}{(t-z)}\,dt, \\ f(z) = \operatorname{Re} F(z) = \dfrac{1}{\pi}\displaystyle\int_{-\infty}^{\infty} f(t)y|t-z|^{-2}\,dt = P_y * f(x), \\ \tilde f(z) = \operatorname{Im} F(z) = \dfrac{1}{\pi}\displaystyle\int_{-\infty}^{\infty} f(t)(x-t)|t-z|^{-2}\,dt, \end{cases}$$

where $P_y(x) = \pi^{-1}y(x^2+y^2)^{-1}$ is the Poisson kernel. It is known that

$$(3.2)\qquad \|P_y * f - f\|_p = \left(\int_{-\infty}^{\infty} |P_y * f(x) - f(x)|^p\,dx\right)^{1/p} \to 0,\quad y \to 0+,$$

$$(3.3)\qquad \tilde f(x) = \lim_{y\to 0+} \tilde f(x+iy) \quad \text{a.e. on } \mathbb{R},$$

$$(3.4) \qquad \tilde{f}(x) = Hf(x) = \lim_{\varepsilon \to 0+} \frac{1}{\pi} \int_{|t|>\varepsilon} \frac{f(x-t)}{t}\, dt \quad \text{a.e. on } \mathbb{R}.$$

We shall give a new proof of

Theorem 2. *Let $F = f + i\tilde{f}$ be as above. Then, with C_p defined as in Section 2, we have*

$$(3.5) \qquad ||F||_p^p \le C_p ||f||_p^p, \quad 1 < p < \infty,$$

and the constant C_p is best possible for every $p \in (1, \infty)$.

Remark. The proof will also give us the following inequalities, where the constants are also best possible:

$$(3.6) \qquad ||\tilde{f}||_p^p \le C_p' ||f||_p^p, \quad 1 < p < \infty,$$

where

$$C_p' = \left(\tan \frac{\pi}{2p}\right)^p, \quad 1 < p \le 2,$$

$$C_p' = \left(\tan \frac{\pi}{2q}\right)^p, \quad 2 < p < \infty, \quad q = \frac{p}{p-1}.$$

In the case of the unit disc, the analogue of (3.6) was proved by Pichorides in 1972 (cf. [4]). As mentioned by Pichorides, it is possible to prove (3.6) via a procedure given in XVI.3.8. in [6]. An alternative was used in our proof of (0.1) and will be used in the proof of Theorem 2.

Proof of Theorem 2. Let us first assume that supp f is a compact subset of $\mathbb{R}$. If ε is a small fixed positive number and

$$F(z + i\varepsilon) = F_\varepsilon(z) = f_\varepsilon(z) + i\tilde{f}_\varepsilon(z),$$

we shall prove that

$$(3.7) \qquad ||F_\varepsilon||_p^p \le C_p ||f_\varepsilon||_p^p.$$

Let us first assume that (3.7) holds for all small positive ε. It follows from (3.2) that $f_\varepsilon = P_\varepsilon * f$ converges to f in $L^p(\mathbb{R})$ when $\varepsilon \to 0$. It is now clear from (3.7) that $\{\tilde{f}_{\varepsilon_n}\}$ is a Cauchy sequence in $L^p(\mathbb{R})$ for any sequence $\{\varepsilon_n\}$ tending to 0. According to (3.3), the limit must be $\tilde{f}$. From Fatou's lemma, we see that $\tilde{f} \in L^p(\mathbb{R})$ and that (3.5) holds.

It remains to remove our assumption that supp f is compact. For this purpose, we consider $\{f_n = f\chi_n\}$, where χ_n is the characteristic function of the interval $(-n, n)$. By (3.5), we have

$$||F_n||_p^p \leq C_p||f_n||_p^p.$$

Since f_n converges to f in $L^p(\mathbb{R})$, it is easy to see that $\lim_{n\to\infty} \tilde{f}_n$ exists in $L^p(\mathbb{R})$ and that the limit is $Hf(x)$ a.e. on $\mathbb{R}$. It follows that (3.5) holds in general.

We shall now prove that (3.7) holds in the case $1 < p \leq 2$, assuming that supp f is compact. It is clear that

$$\text{(3.8)} \qquad F_\varepsilon \text{ is analytic and bounded in } \{\text{Im } z \geq 0\},$$

$$\text{(3.9)} \qquad |F_\varepsilon(z)| \leq \text{const. } |z|^{-1}, \text{ for all large } z \text{ in } \{\text{Im } z \geq 0\}.$$

It follows from (1.6) that we have

$$\text{(3.10)} \qquad |F_\varepsilon(x)|^p - C_p|f_\varepsilon(x)|^p \leq G(F_\varepsilon(x)), \quad x \in \mathbb{R}.$$

Choosing the functions H and F in Lemma 5 as G and F_ε, we use the superharmonicity of $G \circ F_\varepsilon$ to deduce that we have

$$\int_{-\infty}^{\infty} G(F_\varepsilon(x))\, dx = \iint_D y\Delta G \circ F_\varepsilon(z)\, dx\, dy \leq 0.$$

Integrating (3.10) over $\mathbb{R}$, we obtain (3.7).

So far, we have proved that (3.5) holds in the case $1 < p \leq 2$. Arguing as in Remark 2 in [2], we conclude that (3.6) holds in the case $1 < p \leq 2$. Using duality in a standard way, we shall now prove that (3.6) holds in the case $2 < p < \infty$. Let us start with two real-valued functions f and g with compact support on $\mathbb{R}$ and form analytic functions F and G as in (3.1). Since the complex integral of $F_\varepsilon(z)G_\eta(z)$ over a closed curve in $\{\text{Im } z \geq 0\}$ vanishes when ε and η are positive, we obtain

$$\int_{-\infty}^{\infty} (\tilde{f}_\varepsilon g_\eta)(x)\, dx = -\int_{-\infty}^{\infty} (f_\varepsilon \tilde{g}_\eta)(x)\, dx.$$

If $2 < p < \infty$, we have $1 < q < 2$. Applying Hölder's inequality and (3.6), we see that

$$\left|\int_{-\infty}^{\infty} (\tilde{f}_\varepsilon g_\eta)(x)\, dx\right| \leq ||f_\varepsilon||_p||\tilde{g}_\eta||_q \leq (C'_p)^{1/q}||g_\eta||_q||f_\varepsilon||_p.$$

Letting $\eta \to 0+$ and arguing in a standard way, we obtain

$$||\tilde{f}_\varepsilon||_p = \sup \left|\int_{-\infty}^{\infty} \tilde{f}_\varepsilon g\, dx\right|, \quad ||g||_q = 1.$$

Letting $\varepsilon \to 0+$, we deduce that (3.6) holds in the case $2 < p < \infty$.

Finally, our inequality (3.5) is a direct consequence of (3.6) when $2 < p < \infty$. For details, we refer to Remark 2 in [2].

To prove that our constants are best possible, we consider in the case $1 < p < 2$ the function

$$F(z) = \left(\frac{i}{z+i}\right)^b, \quad z \in D,$$

where $bp > 1$ and $F(i) = 2^{-b}$. Then

$$F(x) = (x^2+1)^{-b/2} \exp(ib \arctan x),$$

$$\lim \frac{||f||_p^p}{||F||_p^p} = \left(\cos \frac{\pi}{2p}\right)^p, \quad b \to (p^{-1})^-.$$

Thus the constant C_p is best possible in the case $1 < p \le 2$.

In the case $2 < p < \infty$, we consider the function $F_1(z) = i(F(z))^{q-1}$ which is such that $|F_1(z)|^p = |F(z)|^q$. Here $F(z)$ is defined as above and $bq > 1$. We leave the details to the reader. This completes the proof of Theorem 2.

Remark. In U, the inequality of M. Riesz is obtained by integrating our inequality (2.1) over a circle and applying the superharmonic mean value inequality. We work with the mean of the boundary values with respect to harmonic measure evaluated at the origin.

Let us use the same kind of argument in D. Let us assume that f is a real-valued function on the real line with compact support and that we have extended f to a harmonic function in D via (3.1). Taking the mean of f with respect to harmonic measure evaluated at ia, we obtain

$$f(ia) = \frac{a}{\pi} \int_{-\infty}^{\infty} \frac{f(t)}{a^2+t^2}\, dt.$$

Letting $a \to \infty$, we see that $af(ia) \to \pi \int_{-\infty}^{\infty} f(t)\, dt$. Thus, the integral of the boundary values of a function harmonic in D is associated with the point at infinity. In Lemma 5, we study such an integral for a function superharmonic in D (cf. (1.7)).

Intuitively, to compute the mean with respect to harmonic measure at ia, multiplying by a and letting $a \to \infty$ means that we go from working at a point in the interior to working with the minimal harmonic function at infinity in D (cf. the proof of Lemma 5). We have gone from a compact situation in U to a non-compact situation in D.

The connection between inequalities for conjugate functions and harmonic measure is also discussed in Section 3 in [2].

4. $L \log L$-Inequalities in a half-plane

If $F : D \to \mathbb{C}$ is a given analytic function, we define

$$L_D(f) = \sup_{y>0} \int_{-\infty}^{\infty} (1 + |f(x+iy)|) \log(1 + |f(x+iy)|)\, dy.$$

Let us first assume that F is defined by (3.1) for an integrable function f with compact support. If H_0 is given by Lemma 3, we apply Lemma 5 with $H = H_0 - 1$ to the function $F_\varepsilon(z) = F(z + i\varepsilon)$ and deduce that

$$\int_{-\infty}^{\infty} ((1 + |F_\varepsilon(x)|^2)^{1/2} - 1)\, dx \le \int_{-\infty}^{\infty} ((1 + |f_\varepsilon(x)| \log(1 + |f_\varepsilon(x)|)\, dx.$$

Letting $\varepsilon \to 0$, we see that

$$\int_{-\infty}^{\infty} |F(x)| \inf(|F(x)|, 1)\, dx \le (1 + \sqrt{2})\, L_D(f).$$

It is now easy to remove our assumption that supp f is compact.

If the right hand member is finite, we can control that part of the $H^1(D)$-norm of F which comes from the set where F is "large" but not the contribution from the set where F is "small". Is there an inequality of the type

$$||F||_{H^1(D)} \le \text{const. } L_D(f)?$$

The following example shows that the answer is no in general. We define

$$F(z, A) = (1 - iz)^{-1}(\log(1 - iz))^{-A}, \quad z \in D,$$

where we assume that the logarithm is real on the imaginary axis. If $|x|$ is large, it is easy to see that, as $|x| \to \infty$, we have

$$\arg F(x, A) = (\arctan x)(1 + O((\log|x|)^{-1}),$$

$$|\cos \arg(F(x, A)| = |x|^{-1}(1 + o(1)),$$

$$|f(x, A)| = |x|^{-2}(\log|x|)^{-A}(1 + o(1)),$$

$$|F(x, A)| = |x|^{-1}(\log|x|)^{-A}(1 + o(1)).$$

It follows that $L_D(f(\cdot, A))$ is uniformly bounded for $A \ge 1$ while $||F(\cdot, A)||_{H^1(D)}$ tends to infinity when $A \to 1-$.

The results of the present section are incomplete. We hope to return to these questions in future work.

References

1. P.L. Duren, *Theory of H^p-spaces.* Academic Press, 1970.

2. M. Essén, *A superharmonic proof of the M. Riesz conjugate function theorem.* Ark. Mat. **22** (1984), 241–249.

3. M. Essén, D.F. Shea and C.S. Stanton, *A value-distribution criterion for the class $L \log L$ and some related questions.* Ann. Inst. Fourier (Grenoble) XXXV (1985), 127–150.

4. S.K. Pichorides, *On the best value of the constants in the theorems of M. Riesz, Zygmund and Kolmogorov.* Studia Math. **44** (1972), 165–179.

5. I.E. Verbitskii, *An estimate of the norm of a function in a Hardy space in terms of the norms of its real and imaginary parts.* Mat. Issled. Vyp. **54** (1980), 16–20; Amer. Math. Soc. Transl. (2) **124** (1984), 11–15.

6. A. Zygmund, *Trigonometric Series.* Volumes 1 and 2. Cambridge Univ. Press, 1977.

Matts Essén, Department of Mathematics, University of Uppsala, Thunbergsvägen 3, S-75238 Uppsala, Sweden

International Series of Numerical Mathematics, Vol. 103, © 1992 Birkhäuser Verlag Basel

Remarks on trigonometric inequalities

L. Lorch and D.C. Russell

Abstract

We consider expansions of $((\sin z)/z)^m$ and $((\sin z)/z)^m/\cos z$ in powers of $\sin z$, which can be used to generate a sequence of two-sided inequalities for $((\sin x)/x)^m$ $(m = 1, 2, 3, \ldots)$.

1. Motivation

In Turán's posthumously published book [6] he is concerned with *generalized power-sums* of the form $g(\nu) := \sum_{j=1}^{n} b_j z_j^\nu$ where $b_j, z_j \in \mathbb{C}$ $(j = 1, \ldots, n)$, $\nu \in \mathbb{Z}_+ := \{0, 1, \ldots\}$, $\min_j |z_j| = 1$. Bounds for sums like $g(\nu)$ occur frequently in analysis; e.g. a key lemma is Harald Bohr's work on the Riemann ζ-function was

$$\max_{\nu=1,\ldots,5^n} \frac{\left|\sum_{j=1}^{n} b_j e^{i\lambda_j \nu}\right|}{\sum_{j=1}^{n} b_j} \geq \cos\frac{2\pi}{5} \quad (b_j > 0,\ \lambda_j \in \mathbb{R}).$$

Halász (who edited Turán's book) gives the inequality [6, p. 73]:

$$\max_{m+1 \leq \nu \leq m+kn} |g(\nu)| \geq \frac{|g(0)|}{2^{m+1}(kn+1)^{1/2}\rho_k^{3n}},$$

where $m \in \mathbb{Z}_+$, $k \in \{2, 3, \ldots\}$, and $\rho_k := \max_\theta |\pi_k(e^{i\theta})|$ with π_k a certain polynomial (and Halász has $\rho_k < e^{2/k}$). In improving this theorem the estimate for ρ_k depended on the integral $\int_T^\infty -\log\left[1 - ((\sin x)/x)^2\right] dx$ (see Lorch and Russell [2]), and hence we examined some inequalities for $((\sin x)/x)^{2n}$. In the end [3] we did the estimation directly, but some of the trigonometric inequalities we obtained in this work have independent interest. In particular if we consider the expansion

$$\left(\frac{\sin z}{z}\right)^m = \sum_{k=0}^{\infty} a_{mk} \sin^{2k} z \quad \left(m = 0, 1, \ldots,\ |z| < \frac{\pi}{2}\right), \tag{1.1}$$

This paper is in final form and no version of it will be submitted for publication elsewhere.

it turns out that the matrix (a_{mk}) has an interesting pattern of signs which can be used to generate a sequence of trigonometric inequalities; see Table A for some of these values. Inequalities in the opposite direction can be obtained by considering the related expansion

$$(1.2)\qquad \left(\frac{\sin z}{z}\right)^m \frac{1}{\cos z} = \sum_{k=0}^{\infty} b_{mk} \sin^{2k} z \quad \left(m = 0, 1, \ldots, |z| < \frac{\pi}{2}\right).$$

2. The trigonometric expansions

We shall need to refer to the Bernoulli numbers B_k which we define (see [7, p. 127] or [1, p. 203 and footnote 32 on p. 183]) by

$$(2.1)\qquad \frac{z}{e^z - 1} = 1 - \frac{1}{2} z + B_1 \frac{z^2}{2!} - B_2 \frac{z^4}{4!} + B_3 \frac{z^6}{6!} - \cdots \quad (|z| < 2\pi);$$

thus $B_1 = \frac{1}{6}$, $B_2 = \frac{1}{30}$, $B_3 = \frac{1}{42}$, $B_4 = \frac{1}{30}$, $B_5 = \frac{5}{66}$, $B_6 = \frac{691}{2730}$, $B_7 = \frac{7}{6}$, $B_8 = \frac{3617}{510}$. Some further properties of the series (2.1) for real z are given in [5, Lemma]. Series readily derivable from (2.1) (e.g. see [1, p. 204]) are

$$(2.2)\qquad z \cot z = 1 - \sum_{k=1}^{\infty} \frac{2^{2k} B_k z^{2k}}{(2k)!} \quad (|z| < \pi),$$

$$(2.3)\qquad \frac{\tan z}{z} = \sum_{k=1}^{\infty} \frac{2^{2k}(2^{2k} - 1) B_k z^{2k-2}}{(2k)!} \quad \left(|z| < \frac{\pi}{2}\right)$$

$$(2.4)\qquad \frac{1}{\sin z} = \frac{1}{z} + \sum_{k=1}^{\infty} \frac{(2^{2k} - 2) B_k z^{2k-1}}{(2k)!} \quad (0 < |z| < \pi),$$

$$(2.5)\qquad \frac{1}{\sin^2 z} = \frac{1}{z^2} + \sum_{k=1}^{\infty} \frac{(2k-1) 2^{2k} B_k z^{2k-2}}{(2k)!} \quad (0 < |z| < \pi).$$

For (2.5), divide (2.2) by z and differentiate. However,

$$(2.6)\qquad \frac{1}{\cos z} = \sum_{k=0}^{\infty} \frac{E_k z^{2k}}{(2k)!} \quad \left(|z| < \frac{\pi}{2}\right),$$

where E_k are Euler numbers (see [1, p. 239 with a slight notation change]); thus $E_0 = 1$, $E_1 = 1$, $E_2 = 5$, $E_3 = 61$, $E_4 = 1385$, $E_5 = 50521$, $E_6 = 2702765$, $E_7 = 199360981$.

We shall prove the following results.

Theorem 1. *Let $m = 1, 2, \ldots$ and*

$$\left(\frac{\sin z}{z}\right)^m = \sum_{k=0}^{\infty} a_{mk} \sin^{2k} z \quad \left(|z| < \frac{\pi}{2}\right). \tag{2.7}$$

Then

$$a_{mk} = \frac{m}{m-2k} \cdot \frac{1}{2\pi i} \int_{|z|=r<\pi/2} z^{-m-1} (\sin z)^{m-2k}\, dz \quad (2k \neq m), \tag{2.8a}$$

$$a_{mk} = -\frac{2^m}{m!} B_{m/2} \quad \left(m \text{ even}, k = \frac{m}{2}\right); \tag{2.8b}$$

$$(-1)^k a_{mk} > 0 \quad \left(0 \leq k < \frac{m}{2}\right), \qquad a_{mk} < 0 \quad \left(k \geq \frac{m}{2}\right); \tag{2.9}$$

$$\sum_{k=0}^{\infty} a_{mk} \quad \text{converges and} \quad \sum_{k=0}^{\infty} a_{mk} = \left(\frac{2}{\pi}\right)^m; \tag{2.10}$$

$$\sum_{0 \leq k < m/2} a_{mk} = \frac{2^m(2^m - 1)}{m!} B_{m/2} \quad (m = 2, 4, 6, \ldots), \tag{2.11a}$$

$$\sum_{0 \leq k < m/2} a_{mk} = \frac{1}{(m-1)!} E_{(m-1)/2} \quad (m = 1, 3, 5, \ldots). \tag{2.11b}$$

Theorem 2. *Let $m = 1, 2, \ldots$ and*

$$\left(\frac{\sin z}{z}\right)^m \frac{1}{\cos z} = \sum_{k=0}^{\infty} b_{mk} \sin^{2k} z \quad \left(|z| < \frac{\pi}{2}\right). \tag{2.12}$$

If $m = 1$ then

$$b_{1k} = \frac{1}{2\pi i} \int_{|z|=r<\pi/2} z^{-1} (\sin z)^{-2k}\, dz > 0 \quad \text{for } k = 0, 1, 2, \ldots. \tag{2.13}$$

If $m \geq 2$ *then*

$$b_{mk} = \frac{m-1-2k}{m-1}\, a_{m-1,k} \text{ for } k = 0, 1, 2, \ldots, \tag{2.14}$$

where (a_{mk}) *is given in Theorem* 1*; moreover, for* $m \geq 2$,

$$\begin{gathered} (-1)^k b_{mk} > 0 \quad \left(0 \leq k < \tfrac{m-1}{2}\right), \qquad b_{mk} = 0 \quad \left(m \text{ odd}, \ k = \tfrac{m-1}{2}\right), \\ b_{mk} > 0 \quad \left(k > \tfrac{m-1}{2}\right). \end{gathered} \tag{2.15}$$

Proof of Theorem 1. Whittaker and Watson [7, pp. 131–132] consider the expansion of one function in powers of another. Under suitable conditions,

$$f(z) = \sum_{p=0}^{\infty} A_p [\theta(z)]^p, \text{ where } A_p = \frac{1}{2\pi i} \int_C \frac{f(z)\theta'(z)}{[\theta(z)]^{p+1}}\, dz.$$

The choice $f(z) := ((\sin z)/z)^m$, $\theta(z) := \sin z$ satisfies the conditions if we take, say, C as the circle $|z| = r < \pi/2$. Thus if $((\sin z)/z)^m = \sum_{p=0}^{\infty} A_{mp} \sin^p z$, $m \in \{1, 2, \ldots\}$, we get

$$A_{mp} = \frac{1}{2\pi i} \int_C \left(\frac{\sin z}{z}\right)^m \frac{\cos z}{(\sin z)^{p+1}}\, dz \tag{2.16}$$

$$= \frac{m}{m-p} \cdot \frac{1}{2\pi i} \int_C \frac{\sin^{m-p} z}{z^{m+1}}\, dz \quad (m \neq p), \quad \text{integrating by parts.} \tag{2.17}$$

If p is odd, the integrand in (2.16) is an even function with residue zero, and then $A_{mp} = 0$. The a_{mk} in (2.7) thus correponds to $a_{mk} := A_{m,2k}$, and then (2.17) gives us (2.8a). In case m is even we get, from (2.16),

$$a_{m,m/2} = A_{mm} = \frac{1}{2\pi i} \int_C z^{-m} \cot z\, dz; \tag{2.18}$$

but the residue of the integrand is the coefficient of z^m in the power series for $z \cot z$ which, from (2.2), is $-(2^m/m!)B_{m/2}$, which gives (2.8b).

To investigate the signs of a_{mk} note that, from (2.4), the power series for $1/(\sin z)$ has all coefficients positive, so the same is true of the power series for $1/(\sin^p z)$ $(p = 1, 2, \ldots)$. Then, from (2.8a),

$$\begin{aligned} a_{mk} &= \frac{m}{m-2k} \ (\text{coefficient of } z^m \text{ in the power series} \\ &\qquad\qquad \text{for } \sin^{m-2k} z) \quad (2k \neq m) \\ &< 0 \qquad \text{if } \ k > \tfrac{m}{2}; \end{aligned} \tag{2.19}$$

clearly also $a_{mk} < 0$ if $k = m/2$, by (2.8b).

If $0 \le k < m/2$ we may proceed by induction. If, for $m = 0, 1, \ldots$, we write $\sin^m z = \sum_{n=0}^{\infty} c_{mn} z^n$, then $c_{mn} = 0$ if $n < m$ or if $n - m$ is odd; moreover $n!\, c_{mn} = (D^n \sin^m z)_{z=0}$, which leads to the recurrence relation

$$(2.20) \qquad (n+1)(n+2)c_{m,n+2} = m(m-1)c_{m-2,n} - m^2 c_{mn};$$

it is now easy to prove from this, by induction, that $(-1)^{(n-m)/2} c_{mn} > 0$ for all $n \ge m$ and $n - m$ even. Since, by (2.19),

$$(2.21) \qquad a_{mk} = \frac{m}{m-2k}\, c_{m-2k,m} \quad (m - 2k > 0)$$

it follows immediately that $(-1)^k a_{mk} > 0$ for $0 \le k < m/2$. This proves (2.9).

On taking z real and $z \to \pi/2 - 0$ in (2.7) (so $y := \sin^2 z \to 1 - 0$) we get

$$(2.22) \qquad \lim_{y \to 1-0} \sum_{k=0}^{\infty} a_{mk} y^k = \left(\frac{2}{\pi}\right)^m.$$

However, since by (2.9), a_{mk} is ultimately of constant sign for m fixed, if $\sum_{k=0}^{\infty} a_{mk}$ were (properly) divergent then the limit in (2.22) could not exist finitely. Hence $\sum_{k=0}^{\infty} a_{mk}$ must converge and (by Abel's theorem) (2.10) follows.

To obtain (2.11) we first let $m = 2n$ $(n = 1, 2, \ldots)$. Then (2.16) $(p = 2k)$ gives

$$\begin{aligned}
\sum_{0 \le k < m/2} a_{mk} &= \frac{1}{2\pi i} \int_C \frac{\cot z}{z^{2n}} \sum_{k=0}^{n-1} (\sin z)^{2n-2k}\, dz = \frac{1}{2\pi i} \int_C \frac{\cot z}{z^{2n}} \sum_{j=1}^{n} (\sin^2 z)^j\, dz \\
&= \frac{1}{2\pi i} \int_C \frac{\cos z \sin z}{z^{2n}} \cdot \frac{1 - \sin^{2n} z}{1 - \sin^2 z}\, dz \\
&= \frac{1}{2\pi i} \int_C \frac{\tan z}{z^{2n}}\, dz,
\end{aligned}$$

since $(\tan z \sin^{2n} z)/z^{2n}$ is holomorphic inside C. The residue of $z^{-2n} \tan z$ is the coefficient of z^{2n-2} in the power series for $(\tan z)/z$, which we obtain from (2.3), and (2.11a) follows. However, if $m = 2n - 1$ $(n = 1, 2, \ldots)$ a similar procedure yields

$$\sum_{0 \le k < m/2} a_{mk} = \frac{1}{2\pi i} \int_C \frac{dz}{z^{2n-1} \cos z} = \frac{E_{n-1}}{(2n-2)!} = \frac{E_{(m-1)/2}}{(m-1)!},$$

by making use of (2.6). Thus (2.11b) follows and the proof is complete. □

Proof of Theorem 2. If $((\sin z)/z)^m/\cos z = \sum_{k=0}^{\infty} b_{mk} \sin^{2k} z$ $(m = 1, 2, \ldots)$ then, following the procedure used in the proof of Theorem 1 to obtain (2.16), we have

$$(2.23)\quad b_{mk} = \frac{1}{2\pi i}\int_C \left(\frac{\sin z}{z}\right)^m \frac{1}{\cos z}\frac{\cos z}{(\sin z)^{2k+1}}\,dz = \frac{1}{2\pi i}\int_C z^{-m}(\sin z)^{m-1-2k}\,dz.$$

If $m = 1$ this gives (2.13); and $b_{1k} > 0$ for all k because the power series for $(\sin z)^{-2k}$ has all positive coefficients.

If $m \geq 2$ and $2k \neq m - 1$ then comparison of (2.23) with (2.8a) shows that

$$(2.24)\qquad b_{mk} = \frac{m-1-2k}{m-1}\,a_{m-1,k};$$

however, if $2k = m - 1$ ($m \geq 2$ and m odd) then $b_{mk} = (2\pi i)^{-1}\int_C z^{-m}\,dz = 0$ and so (2.24) can still be used in this case also, as stated in (2.14). Now (2.15) follows directly from (2.24) and (2.9). □

Remarks. (a) If we wish to compute numerical values of a_{mk} there are several possibilities. For example, we may use (2.19); if $0 \leq k < m/2$ we can expand $\sin^{m-2k} z$ as a finite sum of sines or cosines of multiples of z (depending on whether m is odd or even) and pick out the relevant coefficients from the sine or cosine power series. This method leads to the expression

$$(2.25)\qquad a_{mk} = \frac{(-1)^k}{2^{m-2k-1}(m-1)!\,(m-2k)} \sum_{0\leq j<(m/2)-k} (-1)^j \binom{m-2k}{j} (m-2k-2j)^m$$

for $0 \leq k < m/2$, $m = 1, 2, \ldots$. It follows from (2.9) that the sum in (2.25) must be positive, though this is not immediately obvious from (2.25) itself.

(b) It is also possible, by using suitable recurrence relations, to develop a_{mk}, for fixed k, as a polynomial in m; the first few of these are

$$(2.26)\qquad \begin{aligned} a_{m0} &= 1\\ a_{m1} &= -\tfrac{m}{6}\\ a_{m2} &= \tfrac{m}{360}(5m-22)\\ a_{m3} &= -\tfrac{m}{45360}(35m^2-462m+1528)\\ a_{m4} &= \tfrac{m}{5443200}(175m^3-4620m^2+40724m+119856)\\ a_{m5} &= -\tfrac{m}{359251200}(385m^4-16940m^3+279884m^2-2057968m\\ &\qquad\qquad +5682048). \end{aligned}$$

(c) For $m \geq 2$, values of b_{mk} may be computed by reference to (2.25), (2.26), or Table A, by appealing to the relation (2.24). If $m = 1$ we may compute the residues in (2.13) to give us the series

$$\frac{\tan z}{z} = 1 + \frac{1}{3}\sin^2 z + \frac{11}{45}\sin^4 z + \frac{191}{945}\sin^6 z + \frac{2497}{14175}\sin^8 z + \cdots. \tag{2.27}$$

(d) By a change of variable, (2.7) and (2.12) give us power series for $(\arcsin t)^{-m}$ and $(1-t^2)^{-1/2}(\arcsin t)^{-m}$, $0 < |t| < 1$.

3. Trigonometric inequalities

The inequalities $a_{mk} < 0$ $(k \geq m/2)$ in (2.9) and $b_{mk} > 0$ $(k > (m-1)/2)$ in (2.15) allow us to say that, for real x and $m = 1, 2, 3, \ldots$,

$$\left(\frac{\sin x}{x}\right)^m < \sum_{k=0}^{r} a_{mk}\sin^{2k} x \quad \left(r \geq 0,\ r \geq \frac{m-2}{2},\ \text{all } x \neq 0\right), \tag{3.1}$$

$$\left(\frac{\sin x}{x}\right)^m > \cos x \sum_{k=0}^{s} b_{mk}\sin^{2k} x \quad \left(s \geq 0,\ s > \frac{m-3}{2},\ 0 < |x| \leq \pi\right). \tag{3.2}$$

For example, if $m = 1$ and $r = s = 1$ we get

$$\cos x\left(1 + \frac{1}{3}\sin^2 x\right) < \frac{\sin x}{x} < 1 - \frac{1}{6}\sin^2 x \quad (0 < |x| \leq \pi), \tag{3.3}$$

while $m = 3$, $r = s = 2$ gives
(3.4)

$$\cos x\left(1 + \frac{1}{15}\sin^4 x\right) < \left(\frac{\sin x}{x}\right)^3 < 1 - \frac{1}{2}\sin^2 x - \frac{7}{120}\sin^4 x \quad (0 < |x| \leq \pi);$$

for instance, the elementary inequality $\cos x < ((\sin x)/x)^3$ (a consequence of the left side of (3.4)) appears in [4, p. 238].

Incidentally, some of these inequalities may also be obtained from the power series for $g(x) := (\sin x)^{-1} - x^{-1}$ and $h(x) := (\sin x)^{-2} - x^{-2}$ given in (2.4) and (2.5). For a given $n \in \{0, 1, 2, \ldots\}$ we first calculate $g^{(n)}(x)$ or $h^{(n)}(x)$ directly, and then use the fact that since the series (2.4) and (2.5) have all positive coefficients, we have $g^{(n)}(x) > 0$ and $h^{(n)}(x) > 0$, for $0 < x < \pi$ and $n = 0, 1, 2, \ldots$.

Table A

$$\left(\frac{\sin z}{z}\right)^m = \sum_{k=0}^{\infty} a_{mk} \sin^{2k} z \qquad \left(|z| < \frac{\pi}{2}\right)$$

Matrix of numerical values of a_{mk}

m	a_{m0}	a_{m1}	a_{m2}	a_{m3}	a_{m4}	a_{m5}	a_{m6}
0	+1	0	0	0	0	0	0
1	+1	$-\frac{1}{6}$	$-\frac{17}{360}$	$-\frac{367}{15120}$	$-\frac{27859}{1814400}$	$-\frac{1295803}{119750400}$	$-\frac{5329242827}{653837184000}$
2	+1	$-\frac{1}{3}$	$-\frac{1}{15}$	$-\frac{31}{945}$	$-\frac{289}{14175}$	$-\frac{317}{22275}$	$-\frac{6803477}{638512875}$
3	+1	$-\frac{1}{2}$	$-\frac{7}{120}$	$-\frac{457}{15120}$	$-\frac{11513}{604800}$	$-\frac{106727}{7983360}$	$-\frac{1313487619}{130767436800}$
4	+1	$-\frac{2}{3}$	$-\frac{1}{45}$	$-\frac{4}{189}$	$-\frac{41}{2835}$	$-\frac{982}{93555}$	$-\frac{341749}{42567525}$
5	+1	$-\frac{5}{6}$	$+\frac{1}{24}$	$-\frac{31}{3024}$	$-\frac{3287}{362880}$	$-\frac{18979}{2661120}$	$-\frac{247496773}{43589145600}$
6	+1	-1	$+\frac{2}{15}$	$-\frac{2}{945}$	$-\frac{1}{225}$	$-\frac{31}{7425}$	$-\frac{109028}{30405375}$
7	+1	$-\frac{7}{6}$	$+\frac{91}{360}$	$-\frac{1}{720}$	$-\frac{127}{86400}$	$-\frac{11617}{5702400}$	$-\frac{62451523}{6227020800}$
8	+1	$-\frac{4}{3}$	$+\frac{2}{5}$	$-\frac{4}{315}$	$-\frac{1}{4275}$	$-\frac{8}{10395}$	$-\frac{41324}{42567525}$
9	+1	$-\frac{3}{2}$	$+\frac{23}{40}$	$-\frac{41}{1008}$	$+\frac{1}{40320}$	$-\frac{73}{380160}$	$-\frac{16954277}{43589145600}$
10	+1	$-\frac{5}{3}$	$+\frac{7}{9}$	$-\frac{17}{189}$	$+\frac{2}{2835}$	$-\frac{2}{93555}$	$-\frac{1382}{11609325}$
11	+1	$-\frac{11}{6}$	$+\frac{121}{120}$	$-\frac{2497}{15120}$	$+\frac{7381}{1814400}$	$-\frac{1}{3628800}$	$-\frac{1414477}{59439744000}$
12	+1	-2	$+\frac{19}{15}$	$-\frac{256}{945}$	$+\frac{62}{4725}$	$-\frac{4}{155925}$	$-\frac{1382}{638512875}$

References

1. K. Knopp, *Theory and application of infinite series.* Blackie, London, 2nd English Ed., 1966.

2. L. Lorch and D.C. Russell, *On some contributions of Halász to the Turán power-sum theory.* Proc. Int. Sem. on Approximation and Optimization, Havana 1987; Lecture Notes in Math. No. 1354, pp. 169–177 (ed. A. Gómez et al.). Springer, Berlin, 1988.

3. L. Lorch and D.C. Russell, *On some contributions of Halász to the Turán power-sum theory, II.* Proc. Int. Conf. on Constructive Theory of Functions, Varna 1987, pp. 297–304 (ed. Bl. Sendov et al.). Bulgar. Acad. Sc., Sofia, 1988.

4. D.S. Mitrinović, *Analytic inequalities.* Springer, Berlin, 1970.

5. D.C. Russell, *A note on Mathieu's inequality.* Aequat. Math. **36** (1988), 294–302.

6. P. Turán, *On a new method of analysis and its applications.* Wiley, New York, 1984.

7. E.T. Whittaker and G.N. Watson, *A course of modern analysis.* Univ. Press, Cambridge, 3rd Ed., 1920.

L. Lorch and D.C. Russell, Department of Mathematics & Statistics, York University, Toronto-Downsview, Ontario M3J 1P3, Canada

International Series of Numerical Mathematics, Vol. 103, © 1992 Birkhäuser Verlag Basel

On an extremal property of nonnegative trigonometric polynomials

László Losonczi

Abstract

Extending an inequality of Egerváry and Szász [2] we prove that for the coefficients of any nonnegative trigonometric polynomial $T_n(x) = a_0/2 + \sum_{k=1}^{n}(a_k \cos kx + b_k \sin kx) \geq 0$, $x \in [0, 2\pi[$, the inequality

$$-a_0 \cos \frac{\pi}{2\left[\frac{n-k}{l-k}\right]+3} \leq \operatorname{Re}\left((a_k - ib_k)\gamma + (a_l - ib_l)\delta\right) \leq a_0 \cos \frac{\pi}{2\left[\frac{n-k}{l-k}\right]+3}$$

holds, where $n \geq 2$, k, l are fixed natural numbers with

$$\frac{n+1}{2} \leq k < l \leq n$$

and γ, δ are complex numbers of absolute value 1. In particular we have

$$(a_k^2 + b_k^2)^{1/2} + (a_l^2 + b_l^2)^{1/2} \leq a_0 \cos \frac{\pi}{2\left[\frac{n-k}{l-k}\right]+3}.$$

The cases of equality are discussed too.

Introduction

Any nonnegative trigonometric polynomial

$$(1) \qquad T_n(x) = \frac{a_0}{2} + \sum_{k=1}^{n} (a_k \cos kx + b_k \sin kx) \geq 0, \qquad x \in [0, 2\pi[,$$

can be represented, by a classical theorem of L. Fejér [4] as

$$T_n(x) = |p_n(e^{ix})|^2,$$

This paper is in final form and no version of it will be submitted for publication elsewhere.

where $p_n(z) = \sum_{j=0}^n u_j z^j$ is an algebraic polynomial. If T_n is a pure cosine polynomial (i.e. $b_k = 0$; $k = 1, \dots, n$) then the coefficients u_k $(k = 0, \dots, n)$ are real. By this representation we get

$$T_n(x) = \sum_{r=0}^{n}\sum_{s=0}^{n} u_r \overline{u}_s e^{i(r-s)x} = \sum_{r=0}^{n}\sum_{s=0}^{n} u_r \overline{u}_s \left[\cos(r-s)x + i\sin(r-s)x\right].$$

Hence, comparing the coefficients

$$\text{(2)} \qquad \begin{cases} a_0 &= 2\sum_{j=0}^{n} u_j \overline{u}_j \\ a_k &= \sum_{j=0}^{n-k} (u_j \overline{u}_{j+k} + \overline{u}_j u_{j+k}) \\ b_k &= i\sum_{j=0}^{n-k} (-u_j \overline{u}_{j+k} + \overline{u}_j u_{j+k}) \end{cases}$$

$(k = 1, \dots, n)$ or, with $u_k = x_k + iy_k$ $(x_k, y_k \in \mathbb{R},\ k = 0, \dots, n)$

$$\text{(3)} \qquad \begin{cases} a_0 &= 2\sum_{j=0}^{n} (x_j^2 + y_j^2) \\ a_k &= 2\sum_{j=0}^{n-k} (x_j x_{j+k} + y_j y_{j+k}) \\ b_k &= 2\sum_{j=0}^{n-k} (-x_j y_{j+k} + y_j x_{j+k}) \end{cases}$$

$(k = 1, \dots, n)$ where $y_j = 0$ $(j = 0, \dots, n)$ if $b_k = 0$ $\quad (k = 1, \dots, n)$.

(2) can also be written as

$$\text{(4)} \qquad \begin{cases} a_0 &= 2\sum_{j=0}^{n} u_j \overline{u}_j \\ a_k - ib_k &= 2\sum_{j=0}^{n-k} u_{j+k} \overline{u}_j \quad (k = 1, \dots, n). \end{cases}$$

Keeping a_0 fixed,one can ask about the maximal and minimal values of the coefficients a_k, b_k.

If T_n is a nonnegative cosine polynomial, $a_0/2 = 1$, the extremal values of a_1 and a_k with $k > \left[\frac{n}{2}\right]$ have been determined by L. Fejér [4]. He found that

$$-2\cos\frac{\pi}{n+2} \le \ a_1 \ \le 2\cos\frac{\pi}{n+2},$$
$$-1 \le \ a_k \ \le 1 \qquad \text{if} \qquad k > \left[\frac{n}{2}\right].$$

In the general case the extremal values of the coefficients have been found by G. Szegő [9] proving that

$$\text{(5)} \qquad (a_k^2 + b_k^2)^{1/2} \le a_0 \cos\frac{\pi}{2\left[\frac{n-k}{l-k}\right] + 3} \qquad (k = 1, \dots, n).$$

One can consider a more general extremal problem concerning T_n. Let now a linear function of the coefficients of T_n be given:

$$L(T_n) = \alpha_0 a_0 + \sum_{j=1}^{n} (\alpha_j a_j + \beta_j b_j), \tag{6}$$

where $\alpha_0, \alpha_j, \beta_j$ $(j = 1, \ldots, n)$ are (given) real numbers. With the notations

$$\gamma_0 = \alpha_0, \qquad \gamma_j = \alpha_j + i\beta_j$$

$$c_o = a_0, \qquad c_j = a_j - ib_j$$

(j=1,...,n) we get

$$L(T_n) = \text{Re} \sum_{j=0}^{n} \gamma_j c_j. \tag{7}$$

Let further

$$u = (u_0, \ldots, u_n)^*$$

and

$$A = \begin{pmatrix} \gamma_0 & \gamma_1 & \cdots & \gamma_n \\ \overline{\gamma}_1 & \gamma_0 & \cdots & \gamma_{n-1} \\ \vdots & \vdots & & \vdots \\ \overline{\gamma}_n & \overline{\gamma}_{n-1} & \cdots & \gamma_0 \end{pmatrix} \tag{8}$$

where * denotes the transposition.Using (4) we can rewrite (7) in the form

$$L(T_n) = \langle Au, u \rangle \tag{9}$$

and we also have $a_0/2 = \langle u, u \rangle$ where $\langle ., . \rangle$ denotes the usual inner product in the $n+1$-dimensional Euclidean space.

What are the extremal values of $L(T_n)$ if a_0 is a fixed (necessarily nonnegative) value?

Let $\lambda_0 \geq \lambda_1 \geq \ldots \geq \lambda_n$ be the eigenvalues of the (Hermitian) matrix A and let $u^0, u^1, \ldots, u^n$ be the corresponding linearly independent eigenvectors. It is known that

$$\lambda_n \langle u, u \rangle \leq \langle Au, u \rangle \leq \lambda_0 \langle u, u \rangle \tag{10}$$

holds for every vector $u \in \mathbb{C}^{n+1}$. The equality $\lambda_n \langle u, u \rangle = \langle Au, u \rangle$ holds if and only if $u = 0$ or u is an eigenvector corresponding to λ_n (i.e. u is a solution of

$Au = \lambda_n u$). Similarly $\langle Au, u\rangle = \lambda_0 \langle u, u\rangle$ holds if and only if u is a solution of $\langle Au, u\rangle = \lambda_0 u$ (see e.g. [1]). Concerning inequalities of type (10) see e.g. [3], [7], [8].

By (9), (10) we have for any nonnegative trigonometric polynomial T_n the inequality

$$\frac{a_0}{2}\lambda_n \le L(T_n) \le \frac{a_0}{2}\lambda_0. \tag{11}$$

Thus the minimal (and maximal) value of $L(T_n)$ is $(a_0/2)\lambda_n$ (and $(a_0/2)\lambda_0$) where λ_n (λ_0) is the minimal (maximal) eigenvalue of A. Egerváry and Szász [2] established several inequalities of type (11). They proved (5) and the inequalities

$$-\left(\frac{n+1}{2}\binom{n+2}{3}\right)^{1/2} \le T_n'(x) \le \left(\frac{n+1}{2}\binom{n+2}{3}\right)^{1/2} \tag{12}$$

$$(a_k^2 + b_k^2)^{1/2} + (a_{n+1-k}^2 + b_{n+1-k}^2)^{1/2} \le 2 \qquad (k = 1, \ldots, n) \tag{13}$$

for any nonnegative trigonometric polynomial T_n with constant term $a_0/2 = 1$ and determined the cases of equality.

The aim of this paper is to extend the inequality (13).

The main result

Theorem 1. *Let $n \ge 2, k, l$ be fixed natural numbers with*

$$\frac{n+1}{2} \le k < l \le n \tag{14}$$

and let γ, δ complex numbers of absolute value 1. Then for the coefficients of any nonnegative trigonometric polynomial T_n of the form (1) the inequality

$$-a_0 \cos\frac{\pi}{2\left[\frac{n-k}{l-k}\right]+3} \le \operatorname{Re}\left((a_k - ib_k)\gamma + (a_l - ib_l)\delta\right) \le a_0 \cos\frac{\pi}{2\left[\frac{n-k}{l-k}\right]+3} \tag{15}$$

holds. In particular we have

$$(a_k^2 + b_k^2)^{1/2} + (a_l^2 + b_l^2)^{1/2} \le a_0 \cos\frac{\pi}{2\left[\frac{n-k}{l-k}\right]+3}. \tag{16}$$

Proof. Let

$$\tilde{\gamma}_j = \begin{cases} \gamma & \text{if } j = k, \\ \delta & \text{if } j = l, \\ 0 & \text{otherwise} \end{cases}$$

then by (7), (9)

$$\tilde{L}(T_n) := \operatorname{Re}\sum_{j=0}^{n}\tilde{\gamma}_j c_j = \operatorname{Re}\left((a_k - ib_k)\gamma + (a_l - ib_l)\delta\right) = \left\langle \tilde{A}u, u\right\rangle, \tag{17}$$

and $a_0/2 = \langle u, u\rangle$ where

$$u = (u_0, \ldots, u_n)^*$$

and $\tilde{A} = (a_{pr})$ with

$$a_{pr} = \begin{cases} \gamma & \text{if } r - p = k, \\ \overline{\gamma} & \text{if } p - r = k, \\ \delta & \text{if } r - p = l, \\ \overline{\delta} & \text{if } p - r = l, \\ 0 & \text{otherwise} \end{cases} \qquad (p, r = 1, \ldots, n+1).$$

We shall determine the eigenvalues of $\tilde{A}$. These are the zeros of the determinant

$$D(\lambda) = \det(\tilde{A} - \lambda E)$$

E being the $(n+1)\times(n+1)$ unit matrix. $D(\lambda)$ has $-\lambda$'s in its main diagonal, γ's ($\overline{\gamma}$'s) and δ's ($\overline{\delta}$'s) in its $(k+1)$th and $(l+1)$th upper (lower) side diagonals (the numbering of side diagonals starts with the main diagonal). All other elements are zero.

Suppose that $\lambda \neq 0$. In four steps we reduce $D(\lambda)$ to the product of a known determinant and a diagonal determinant.

Step 1. We multiply the rows $k+1, k+2, \ldots, n+1$ of $D(\lambda)$ by γ/λ and add them to the rows $1, 2, \ldots, n+1-k$. Then the $(k+1)$th upper side diagonal will disappear,the first $n+1-k$ elements of the main diagonal change from $-\lambda$ to $-\lambda + |\gamma|^2/\lambda = -\lambda + 1/\lambda$ and in the $(n+1-k)\times(n+1-k)$ left upper block of the determinant (and only there) the elements of the $(l-k+1)$th lower side diagonal change from zero to $\overline{\delta}\gamma/\lambda$.

Step 2. Multiplying the rows $l+1, l+2, \ldots, n+1$ of the determinant (obtained after Step 1) by δ/λ and adding them to the rows $1, 2, \ldots, n+1-l$ even the $(l+1)$th upper side diagonal disappears. The first $n+1-l$ elements of the main diagonal change from $-\lambda + 1/\lambda$ to $-\lambda + 1/\lambda + |\delta|^2/\lambda = -\lambda + 2/\lambda$ and in the $(n+1-k)\times(n+1-k)$ left upper block of the determinant (and only there the (formerly zero) elements of the $(l-k+1)$th upper side diagonal change to $\delta\overline{\gamma}/\lambda$.

Step 3. We repeat Step 1 with columns instead of rows,i.e. we multiply the columns $k+1, k+2, \ldots, n+1$ of the determinant (obtained after Step 2) by $\overline{\gamma}/\lambda$ and add the to the columns $1, 2, \ldots, n+1-k$. The result of this step will be the disappearance of the $(k+1)$th lower side diagonal.

Step 4. Repeating Step 2 with columns instead of rows, i.e. multiplying the columns $l+1, l+2, \ldots, n+1$ of our determinant by $\overline{\delta}/\lambda$ and add them to the columns $1, 2, \ldots, n+1-l$ the $(l+1)$th lower side diagonal of the determinant will disappear.

Hence

$$D(\lambda) = \det \begin{pmatrix} A(\lambda) & O \\ O^* & -\lambda E' \end{pmatrix} \tag{18}$$

where O is the $(n+1-k) \times k$ zero matrix, O^* is its transpose, E' is the $k \times k$ unit matrix and $A(\lambda)$ can be obtained from the matrix

$$A_0(v) = \left(\begin{array}{ccc|ccc|ccc} a+v & & & t & & & & & \\ & \ddots & & & \ddots & & & & \\ & & \ddots & & & & & & \\ & & a+v & & & & & & \\ \hline \bar{t} & & & v & & & & & \\ & \ddots & & & \ddots & & & \ddots & \\ & & & & & v & & & t \\ \hline & & & & & & b+v & & \\ & & & & & & & \ddots & \\ & & & & \ddots & & & & \ddots \\ & & & & & \bar{t} & & & b+v \end{array} \right) \tag{19}$$

by dividing each of its rows by λ.

$A_0(v)$ is an $(n+1-k) \times (n+1-k)$ matrix, $t = \gamma\overline{\delta}$ (and $\bar{t}$) are located in the $(l-k+1)$th upper (and lower) side diagonals, $a = 0, b = -1, v = 2 - \lambda^2$.

Thus, by (18) we have

$$D(\lambda) = (-1)^k \lambda^{2k-n-1} \det A_0(2 - \lambda^2) \tag{20}$$

which is valid for $\lambda = 0$ too (since both sides are polynomials of λ, these polynomials are equal if $\lambda \neq 0$). The eigenvalues and eigenvectors of $A_0(v)$ have been determined by the author [6], Theorems 1, 4, 6 (see also [5] where $\det A_0(v)$ was evaluated if $t = 1$).

Next, we summarize the results of [6] we need.

Theorem 2. *Let $n'+1=n+1-k, k'=l-k, a=0, b=1, |t|=1$ in (19). Let further q, r' be nonnegative integers defined by $n'+1=k'q+r'(0\le r'<k')$. Then*

$$\det A_0(v)=\prod_{j=0}^{q}\left(v-2\cos\frac{(2j+1)\Pi}{2q+3}\right)^{r'}\prod_{j=0}^{q-1}\left(v-2\cos\frac{(2j+1)\Pi}{2q+1}\right)^{k'-r'} \tag{21}$$

further all solutions $u_0, u_1, \ldots, u_{n'}$ of the linear homogeneous system of equations

$$A_0(v)(u_0, u_1, \ldots, u_{n'})^* = 0$$

are given by

$$u_{s+hk'}=(-\bar{t})^h C_s \sin(h+1)\theta \tag{22}$$

where $s=0,1,\ldots,r'-1$; $h=0,1,\ldots,q$ and $s=r',r'+1,\ldots,k'-1$; $h=0,1,\ldots,q-1$, provided that $v=2\cos\theta$ is a zero of $\det A_0(v)$. $C_0, C_1, \ldots, C_{r'-1}$ are arbitrary constants, $C_{r'}=C_{r'+1}=\ldots=C_{k'-1}=0$ if $\theta=\dfrac{(2j+1)\Pi}{2q+3}$ for some $j=0,1,\ldots,q$ while $C_0=C_1=\ldots=C_{r'-1}=0$ and $C_{r'}, C_{r'+1}, \ldots, C_{k'-1}$ are arbitrary constants if $\theta=\dfrac{(2j+1)\Pi}{2q+1}$ for some $j=0,1,\ldots,q-1$.

Returning to the *proof* of Theorem 1 we have

$$\begin{aligned} v-2\cos\frac{(2j+1)\Pi}{2q+3} &= 2-\lambda^2-2\cos\frac{(2j+1)\Pi}{2q+3}=4\sin^2\frac{(2j+1)\Pi}{2(2q+3)}-\lambda^2 \\ &= \left(2\sin\frac{(2j+1)\Pi}{2(2q+3)}-\lambda\right)\left(2\sin\frac{(2j+1)\Pi}{2(2q+3)}+\lambda\right) \end{aligned}$$

thus, by (20), (21) the zeros λ of $D(\lambda)$ and their multiplicities $m(\lambda)$ are

$$\begin{array}{lll} \lambda=0 & m(\lambda) = 2k-n-1 & \\ \lambda=\pm 2\sin\dfrac{(2j+1)\Pi}{2(2q+3)} & m(\lambda) = r' & (j=0,1,\ldots,q) \\ \lambda=\pm 2\sin\dfrac{(2j+1)\Pi}{2(2q+1)} & m(\lambda) = k'-r' & (j=0,1,\ldots,q-1). \end{array}$$

It is easy to see that the minimal and maximal zeros of $D(\lambda)$ are

$$\begin{aligned} \tilde{\lambda}_n &= -2\sin\frac{(2q-1)\Pi}{2(2q+1)}=-2\cos\frac{\Pi}{2q+1} \\ \tilde{\lambda}_0 &= 2\cos\frac{\Pi}{2q+1} \end{aligned}$$

if $r' = 0$ while

$$\tilde{\lambda}_n = -2\cos\frac{\Pi}{2q+3}, \qquad \tilde{\lambda}_0 = 2\cos\frac{\Pi}{2q+3}$$

if $r' > 0$. Since

$$\left[\frac{n'}{k'}\right] = \left[\frac{n-k}{l-k}\right] = \begin{cases} q-1 & \text{if } r' = 0 \\ q & \text{if } r' > 0 \end{cases}$$

we have

$$\tilde{\lambda}_n = -2\cos\frac{\Pi}{2\left[\frac{n-k}{l-k}\right]+3} = -\tilde{\lambda}_0 \tag{23}$$

in both cases. By (11), (7), (17) the inequality (15) follows. With

$$\gamma = \frac{a_k - ib_k}{|a_k - ib_k|}, \qquad \delta = \frac{a_l - ib_l}{|a_l - ib_l|}$$

we get (16) from (15).

The cases of equality

Equality holds in the right hand side inequality of (15) if and only if

$$T_n(x) = |p_n(e^{ix})|^2 = \left|\sum_{j=0}^{n} u_j e^{ijx}\right|^2 \tag{24}$$

where $u = (u_0, \ldots, u_n)^*$ is a solution of

$$(\tilde{A} - \tilde{\lambda}_0 E)u = 0. \tag{25}$$

To find the solutions of (25) let us first consider an arbitrary homogeneous system

$$Bu = 0 \tag{26}$$

with $b = (b_{pr})$, $(p, r = 1, \ldots, n+1)$ and observe how the system (26) is transformed if we add a constant multiple of a row (or column) of B to another row (or column). Let B_1 (B_2) be the matrix obtained from B by multiplying its t-th row vector (column vector) by c and adding it to the s-th row vector (column vector) of B ($t \neq s$). It is clear that the system $Bu = 0$ is equivalent to $B_1 u = 0$. The equation

$$b_{ps}u_s + b_{pt}u_t = (b_{ps} + cb_{pt})u_s + b_{pt}(u_t - cu_s)$$

shows that the system (26) is equivalent to

$$B_2 u' = 0$$

where $u' = (u'_0, \ldots, u'_n)^*$ with

(27) $$u'_j = \begin{cases} u_j & \text{if } j \neq t \\ u_t - cu_s & \text{if } j = t. \end{cases}$$

If we perform Steps 1-4 on the matrix $\tilde{A} - \lambda E$ (with $\lambda = \tilde{\lambda}_0$) instead of the determinant $D(\lambda)$ we get the linear homogeneous system

(28) $$\frac{1}{\tilde{\lambda}_0^{n+1-k}} A_0(2 - \tilde{\lambda}_0^2)(u_0, \ldots, u_{n-k})^* = 0$$

(29) $$-\tilde{\lambda}_0 u_{n+1-k+j} = 0 \qquad (j = 0, \ldots, 2k - n - 2)$$

(30) $$-\tilde{\lambda}_0 \left(u_{k+j} - \frac{\overline{\gamma}}{\tilde{\lambda}_0} u_j \right) = 0 \qquad (j = 0, \ldots, l - k - 1)$$

(31) $$-\tilde{\lambda}_0 \left(u_{l+j} - \frac{\overline{\gamma}}{\tilde{\lambda}_0} u_{l-k+j} - \frac{\overline{\delta}}{\tilde{\lambda}_0} u_j \right) = 0 \qquad (j = 0, \ldots, n - l)$$

which, by the above remarks,is equivalent to (25). We note that due to Steps 3, 4 we applied the transformation corresponding to (27) once to get (30), twice to get (31).

The solutions of (28) are given by (22) where now

$$\theta = \frac{\Pi}{2\left[\frac{n-k}{l-k}\right] + 3}$$

and $C_0, C_1, \ldots, C_{r'-1}$ are arbitrary constants, $C_{r'} = C_{r'+1} = \cdots = C_{k'-1} = 0$ if $r' > 0$ and $C_0, C_1, \ldots, C_{k'-1}$ are arbitrary constants if $r' = 0$.

From (29), (30), (31) by $\tilde{\lambda}_0 \neq 0$ it follows that

(32) $$u_{n+1-k+j} = 0 \qquad (j = 0, \ldots, 2k - n - 2)$$

(33) $$u_{k+j} = \frac{\overline{\gamma}}{\tilde{\lambda}_0} u_j \qquad (j = 0, \ldots, l - k - 1)$$

(34) $$u_{l+j} = \frac{\overline{\gamma}}{\tilde{\lambda}_0} u_{l-k+j} - \frac{\overline{\delta}}{\tilde{\lambda}_0} u_j \qquad (j = 0, \ldots, n - l).$$

Thus the solutions u_s $(s = 0, \dots, n)$ are given by (22) (with the above θ and constants) if $s = 0, \dots, n-k$, by (32) if $s = n-k+1, \dots, k-1$, by (33) if $s = k, \dots, l-1$ and by (34) if $s = l, \dots, n$.The extremal nonnegative trigonometric polynomial can then be calculated by (24).

The case of equality in the left hand side inequality of (15) can be settled similarly.

Acknowledgement. This research was supported by Hungarian National Science Foundation Operating Grant Number OTKA 1652.

References

1. R. Bellman, *Introduction to matrix analysis.* McGraw-Hill, New York, 1960.

2. E. Egerváry and O. Szász, *Einige Extremalprobleme im Bereiche der trigonometrischen Polynome.* Math. Z. **27** (1928), 641-652.

3. K. Fan, O. Taussky and J. Todd, *Discrete analogs of inequalities of Wirtinger.* Monatsh. Math. **59** (1955), 73-79.

4. L. Fejér, *Über trigonometrische Polynome.* J. Reine Angew. Math. **146** (1915), 53-82.

5. L. Losonczi, *On some discrete quadratic inequalities.* In: W. Walter (ed.) *General Inequalities 5* (pp. 73-85), Birkhäuser Verlag, Basel-Boston 1987.

6. L. Losonczi, *Eigenvalues and eigenvectors of some tridiagonal matrices.* to appear in Acta Math.Hung.

7. G.V. Milovanović and I.Ž. Milovanović, *On discrete inequalitites of Wirtinger's type.* J. Math. Anal. Appl. **88** (1982), 378-387.

8. G.V. Milovanović and I.Ž. Milovanović, *Some discrete inequalities of Opial's type.* Acta Sci. Math. (Szeged) **47** (1984), 413-417.

9. G. Szegő, *Koeffizientabschätzungen bei ebenen und räumlichen harmonischen Entwicklungen.* Math. Ann. **96** (1926-27), 601-632.

L. Losonczi, Institute of Mathematics, L. Kossuth University, H-4010 Debrecen, Pf. 12, Hungary

International Series of Numerical Mathematics, Vol. 103, © 1992 Birkhäuser Verlag Basel

On polynomial inequalities and extremal problems

Themistocles M. Rassias

Abstract

We shall give an account of some of the fundamental results concerning critical points of polynomials as well as some of the developments concerning extremal properties of polynomials with an analysis of some inequalities for polynomials and their derivatives as well as their zeros. We provide a counterexample to a theorem of M. Marden concerning the zeros of certain composite polynomials.

1. Zeros of polynomials

It is a main problem of algebra to find the solution of algebraic equations. The most important contributions of Carl Friedrich Gauß to the theory of algebraic equations are (cf. Van der Waerden [31, p. 89]):

I. The complete solution of the cyclotomic equation $x^n - 1 = 0$ by means of radicals.

II. The proof that every polynomial in one variable with real coefficients is a product of linear and quadratic factors. This theorem implies what we now call the *fundamental theorem of algebra*: Every polynomial $f(x)$ with complex coefficients is a product of linear factors.

S. Smale [27] has given a beautiful account of the old and new developments on the fundamental theorem of algebra as well as its relation to economics and numerical analysis. Smale [27] shows that a classical algorithm, Newton's method with a standard modification, is a tractable method for finding a zero of a complex polynomial. By "tractable" Smale means that the cost of finding a zero does not grow exponentially with the degree, in a certain statistical sense. Shub and Smale

---

This paper is in final form and no version of it will be submitted for publication elsewhere.

[26] proved that the number of iterations required for a certain fast algorithm to find a zero of a complex polynomial of degree d is linear in d, provided that an arbitrarily small set of problems is excluded. Smale [29] in his address to the International Congress of Mathematicians in Berkeley in 1986 gave an account of a global theory concerning the efficiency of algorithms for solving systems of equations. This approach introduces geometrical and topological ideas into numerical analysis (see also [28]).

2. Critical points of polynomials

The Gauß-Lucas theorem states that if S is the set of zeros of the polynomial p,

$$p(z) = \prod_{k=1}^{n}(z - z_k),$$

then every critical point (i.e. zero of the derivative) of p is contained in the smallest convex set that contains S. This is the well-known theorem which was implied in a note of Gauß (1836) and stated and proved explicitly by Lucas (1874) (cf. M. Marden [12, p. 22]). It has been investigated and sharpened in several ways; cf. M. Marden [12], A.W. Goodman, Q.I. Rahman and J.S. Ratti [7], G. Schmeisser [23, 24, 25], G.V. Milovanović, D.S. Mitrinović and Th.M. Rassias [18], Th.M. Rassias [21].

A conjecture of B. Sendov, better known as Ilieff's conjecture, states that if all zeros of a polynomial $p(z)$ lie in the unit disk $|z| \leq 1$ and if z_0 is any one such zero, then the disk $|z - z_0| \leq 1$ contains at least one zero of $p'(z)$. Proofs of *Sendov-Ilieff's conjecture* have been given for polynomials of degree at most five and in a number of other special cases for polynomials of any degree. For polynomials with three zeros the conjecture was proved by E.B. Saff and J.B. Twomey [22], D.A. Brannan [4], G. Schmeisser [25]. A nice elementary proof of the conjecture for all polynomials with at most four zeros was found by G.L. Cohen and G.H. Smith [5]. The conjecture was included in the short book of W.K. Hayman [8, Problem 4.5 on p. 25], which was published in 1967. M. Marden [13, p. 267] writes: "Since it had been brought to Hayman's attention by Professor Ilieff, it became known as *Ilieff's conjecture*. Actually, this conjecture was due to the Bulgarian mathematician B. Sendov who had acquainted me and probably others, with it in 1962 at the International Congress of Mathematicians held in Stockholm." M. Marden [13, p. 268] gave the following interpretation of this conjecture.

A. Physical interpretation. "...recall Gauß' Theorem that the critical points of a polynomial, which are not multiple zeros of the polynomial, are the equilibrium points in a certain force field. The field is due to particles placed at the zeros of the polynomial, the particles having masses equal to the multiplicity of the zeros and attracting with a force inversely proportional to the distance from the particle. These critical points therefore cannot be too close to any one zero since the force due to the particle at the zero would be relatively large. On the other hand, the Conjecture would imply that, if all the particles are of unit mass and situated on the disk $|z| \leq 1$, then at least one equilibrium point will lie within unit distance of each particle."

B. Geometric interpretation. "It is known [12, p. 9] that the critical points ζ_1 and ζ_2 ($\neq z_1, z_2, z_3$) of the polynomial

$$p(z) = (z - \zeta_1)^{m_1}(z - \zeta_2)^{m_2}(z - \zeta_3)^{m_3}$$

lie at the foci of the ellipse which touches the line segments (z_1, z_2), (z_2, z_3) and (z_3, z_1) in the points that divide these segments in the ratios m_1/m_2, m_2/m_3 and m_3/m_1, respectively. The Conjecture implies therefore that, if the vertices of this triangle z_1, z_2, z_3 all lie in the unit disk $|z| \leq 1$, each vertex is within unit distance from one of the foci of the inscribed ellipse. A similar interpretation can be given for the critical points of the polynomial

$$p(z) = (z - z_1)(z - z_2)\cdots(z - z_n)$$

as foci of a curve Γ of class n (see Marden [12, p. 11]), which is tangent to the sides of the polygon with vertices at the points z_j, $j = 1, 2, \ldots, n$. If true, the Conjecture would imply that within unit distance of each vertex lies at least one focus of Γ."

Let P be the set of all polynomials p,

$$p(z) = \prod_{\nu=1}^{n}(z - z_\nu) \text{ satisfying } |z_\nu| \leq 1 \text{ for } \nu = 1, 2, \ldots, n \text{ and } n \geq 2.$$

Denote by $H(p)$ the convex hull of the zeros of $p(z)$, that is, the smallest convex polygon containing all the zeros of $p(z)$. Sendov-Ilieff's conjecture suggests that close to every zero of $p(z)$ there should always lie a critical point. G. Schmeisser [25] asked whether Sendov-Ilieff's conjecture is valid for all polynomials in P having real coefficients and only real critical points. Also, if the conjecture holds true for all polynomials in P for which $H(p)$ is a quadrangular region. For the disk $\overline{D}(a; r) = \{z : |z - a| \leq r\}$, Schmeisser [24, problem 2] proposed to determine the smallest universal constant r^* such that for every polynomial in P, each of the disks $\overline{D}(z_\nu; r^*)$, $\nu = 1, 2, \ldots, n$, contains at least one of the zeros of $p'(z)$. From the Gauß-Lucas theorem it follows that $r^* \leq 2$. It is clear that $r^* \geq 1$ (just take $p(z) = z^n - 1$). The conjecture of Sendov-Ilieff can be stated as $r^* = 1$. Schmeisser [24, Satz 2] has obtained that $r^* < 1.568$. B.D. Bojanov, Q.I. Rahman and J. Szynal

[3] improved Schmeisser's bound so that $r^* \leq 1.08331641\ldots$ when $p(z)$ has all its zeros in $\overline{D}(0;1)$.

Remark. There is a problem which has occupied my interest during the last years.

Problem. Study the Gauß-Lucas theorem as well as the Sendov-Ilieff conjecture for polynomials in more than one variable.

3. Extremal problems for polynomials

The goal of this section is to explore the extremal problems of Markov's and Bernstein's type. It is interesting that the first result in this area was connected with some investigations of the well-known Russian chemist Mendeleev [17]. In mathematical terms, Mendeleev's problem was:

If $p(t)$ is an arbitrary quadratic polynomial defined on an interval $[a,b]$ with

$$\max_{t\in[a,b]} p(t) - \min_{t\in[a,b]} p(t) = L,$$

how large $p'(t)$ can be on $[a,b]$?

This problem can also be stated for polynomials of degree n in the following form:

If $p(t)$ is an arbitrary polynomial of degree n and $|p(t)| \leq 1$ on $[-1,1]$, how large can $|p'(t)|$ be on $[-1,1]$?

Such a problem was solved by A.A. Markov [14]. His brother V.A. Markov [15] investigated the upper bound of $|p^{(k)}(t)|$, where $k \leq n$. An analogue of Markov's theorem for the unit disk in the complex plane instead of the interval $[-1,1]$ was investigated by Bernstein [2]. Markov's and Bernstein's inequalities are fundamental for the proof of many inverse theorems in polynomial approximation theory (see Dzyadyk [6], Lorentz [10], Meinardus [16], Ivanov [9]). There are many results on Markov's and Bernstein's theorems and their generalizations in various norms and restricted classes of polynomials.

We begin this section by considering the following extremal problem:

Let P_n be the set of all algebraic polynomials p ($\not\equiv 0$) of degree at most n. For a given norm $||\cdot||$, determine the best constant A_n such that

$$||p'|| \leq A_n ||p|| \quad \text{if} \quad p \in P_n, \tag{3.1}$$

i.e.,

$$A_n = \sup_{p \in P_n} \frac{||p'||}{||p||}. \tag{3.2}$$

The first result in this area appeared in the year 1889. It was the well-known classical inequality of A.A. Markov [14]. Namely, for the maximum norm on $[-1,1]$, i.e.,

$$||p|| = ||p||_\infty = \max_{-1 \le t \le 1} |p(t)|, \tag{3.3}$$

Markov solved this extremal problem, giving a solution of Mendeleev's problem in a generalized form.

Theorem 3.1. *In the maximum norm of* (3.3), *we have*

$$||p'||_\infty \le n^2 ||p||_\infty \quad \text{if} \quad p \in P_n. \tag{3.4}$$

Equality holds only at ± 1 *and only when* $p(t) = cT_n(t)$, *where* T_n *is the Chebyshev polynomial of the first kind of degree* n *and* c *is an arbitrary constant.*

A natural question is how to get an upper bound for the k-th derivative of p? Iterating Markov's inequality (3.4) yields a crude result. The best possible inequality for the k-th derivative was found by V.A. Markov [15] in 1892. A German version of this remarkable paper was published in 1916.

Theorem 3.2. *For each* $k = 1, \ldots, n$, *the inequality*

$$||p^{(k)}||_\infty \le \frac{1}{(2k-1)!!} \prod_{i=0}^{k-1} (n^2 - i^2) ||p||_\infty \quad \text{if} \quad p \in P_n \tag{3.5}$$

holds. The extremal polynomial is T_n.

We note that the best constant in (3.5) is equal to

$$||T_n^{(k)}||_\infty = T_n^{(k)}(1).$$

Thus inequality (3.5) can be written in the form

$$||p^{(k)}||_\infty \le T_n^{(k)}(1) ||p||_\infty \quad \text{if} \quad p \in P_n.$$

Without loss of generality, we can suppose that

$$||p||_\infty = 1.$$

Markov's proof of this result is based on a complicated variational method. He investigated a more general problem:

If $\lambda_0, \lambda_1, \ldots, \lambda_n$ are given constants and $p(t) = \sum_{\nu=0}^{n} a_\nu t^\nu$ satisfies the condition $||p||_\infty = 1$, what is the precise bound for the linear form $\sum_{\nu=0}^{n} a_\nu \lambda_\nu$?

By suitably choosing the constants λ_ν the linear form can be made equal to any derivative of $p(t)$ at any preassigned point.

For a fixed k suppose that $t \mapsto \bar{p}(t)$ is an extremal polynomial, i.e., there exists $t^* \in [-1, 1]$ such that

$$|\bar{p}^{(k)}(t^*)| = \sup\{|p^{(k)}(t^*)| : p \in P_n,\ ||p||_\infty = 1\}.$$

It is easily shown that such a polynomial exists. Markov's variational approach was to show that $|\bar{p}^{(k)}(t^*)|$ must be equal to 1 at either n or $n+1$ different points

$$-1 < \tau_1 < \tau_2 < \cdots < \tau_m < 1 \quad \text{if } m = n \text{ or } m = n+1.$$

In the latter case $\pm\bar{p}(t)$ is the Chebyshev polynomial $T_n(t)$, whose derivatives satisfy (3.5). In the former case it can be proved that the extremal polynomial satisfies a differential equation of the form

$$1 - y^2 = \frac{(1-t^2)(t-b)(t-c)}{n^2(t-a)^2}(y')^2,$$

where a, b, c are real constants which depend upon one parameter. Markov thus proved that derivatives of this class of polynomials satisfy

$$|p^{(k)}(t)| \le \frac{n^2(n^2-1^2)(n^2-2^2)\cdots(n^2-(k-1)^2)}{1\cdot 3\cdot 5\cdots(2k-1)} \quad \text{if } -1 \le t \le 1.$$

Another type of these inequalities goes back to Bernstein [1] in 1912, who considered the following problem:

Let $p(z)$ be a polynomial of degree n and $|p(z)| \le 1$ in the unit disk $|z| \le 1$. Determine how large can $|p'(z)|$ be for $|z| \le 1$.

In other words, if we define $||f|| = \max_{|z|\le 1} |f(z)|$, this problem can be reduced to inequality (3.1). Thus, Bernstein's theorem can be stated in the following form:

Theorem 3.3. *Let $p \in P_n$, then*

$$||p'|| \leq n||p||.$$

Equality holds for $p(z) = cz^n$, $c = \text{const}$.

Bernstein's theorem (3.3) can be stated in several different forms. Since a polynomial $p(z)$ is an analytic function, it attains its maximum absolute value for $|z| \leq 1$ on the circumference $|z| = 1$. Therefore, it is enough to consider only the values of z such that $z = e^{i\theta}$, $-\pi < \theta \leq \pi$. Thus, we can define

$$||f|| = \max_{|z|=1} |f(z)| = \max_{-\pi<\theta\leq\pi} |f(e^{i\theta})|.$$

A polynomial $p(z)$, for $z = e^{i\theta}$, becomes a trigonometric polynomial $T(\theta)$ of degree n,

$$T(\theta) = p(e^{i\theta}) = \sum_{k=0}^{n}(a_k \cos k\theta + b_k \sin k\theta),$$

with complex coefficients in the general case.

An arbitrary trigonometric polynomial $T(\theta)$ of degree n, which is not identically zero, has the *Haar property*, i.e., it cannot have more than $2n$ distinct zeros in $(-\pi, \pi]$. Bernstein's theorem can also be stated in the following form for trigonometric polynomials.

Theorem 3.4. *Let $T(\theta)$ be a trigonometric polynomial of degree n and $|T(\theta)| \leq 1$, then*

$$|T'(\theta)| \leq n. \tag{3.6}$$

Equality holds for $T(\theta) = \gamma \sin n(\theta - \theta_0)$, where $|\gamma| = 1$.

Iterating Bernstein's inequality (3.6) we can obtain the corresponding inequality for the k-th derivative

$$|T^{(k)}(\theta)| \leq n^k.$$

This result can be stated in the following form:

Theorem 3.5. *Let $T(\theta)$ be a trigonometric polynomial of degree n. Then*

$$\max_{\theta} |T^{(k)}(\theta)| \leq n^k \max_{\theta} |T(\theta)| \quad \text{if } k = 1, 2, \ldots, \tag{3.7}$$

with equality only if $T(\theta) = a \cos \theta + b \sin \theta$, where a and b are constants.

Using the forward-difference operator Δ_h, defined by

$$\Delta_h f(\theta) = f(\theta + h) - f(\theta) \quad \text{if} \quad h > 0,$$

Stečkin [30] gave an interesting analogue of the Bernstein inequality (3.7).

Theorem 3.6. *Let* $0 < h < 2\pi/n$. *Then*

$$\max_\theta |T^{(k)}(\theta)| \leq \left(\frac{n}{2\sin\frac{nh}{2}}\right)^k \cdot \max_\theta |\Delta_h^k T(\theta)| \quad \textit{if } k = 1, 2, \ldots, \tag{3.8}$$

where

$$\Delta_h^k T(\theta) = \sum_{\nu=0}^{k} (-1)^{k-\nu} \binom{k}{\nu} T(\theta + \nu h).$$

Equality in (3.8) *is attained if and only if*

$$T(\theta) = a\cos\theta + b\sin\theta + c,$$

where a, b, c are constants.

This inequality as well as several similar inequalities with generalizations and applications will appear in [18].

Let $\mathbb{C}$ be the field of complex numbers and $f : \mathbb{C} \to \mathbb{C}$ be a complex-valued polynomial function of a complex variable z. Define

$$M_f = \max_{|z|=1} |f(z)|.$$

Th.M. Rassias [20] proved that the following inequality holds.

Theorem 3.7. *Let* $p_1, p_2, \ldots, p_n : \mathbb{C} \to \mathbb{C}$ *be polynomial functions of degrees* $d_1, d_2, \ldots, d_n$, *respectively, of a complex variable z. Then*

$$M_{p_1} \cdot M_{p_2} \cdots M_{p_n} \geq M_{p_1 \cdot p_2 \cdots p_n} \geq k \cdot M_{p_1} \cdot M_{p_2} \cdots M_{p_n},$$

where

$$k = \left(\sin\frac{2}{n}\frac{\pi}{8d_1}\right)^{d_1} \left(\sin\frac{2}{n}\frac{\pi}{8d_2}\right)^{d_2} \cdots \left(\sin\frac{2}{n}\frac{\pi}{8d_n}\right)^{d_n}. \tag{3.9}$$

Problem. Find examples which might show that the estimate on k is of the right order of magnitude. Maybe the worst case is when the p's form a complete set of factors of the Chebyshev polynomials, since those polynomials have the smallest sup norm.

4. A remark on a theorem of M. Marden

M. Marden [11] claimed to have proved the following theorem, which he also included in his book [12, Theorem (16.2)′, p. 70] in the following form:

If all the zeros of the polynomial f defined by

$$f(z) = \sum_{k=0}^{m} a_k z^k$$

lie in the ring R_0,

$$R_0: \quad 0 \le r_1 \le |z| \le r_2 \le \infty,$$

and if g is a polynomial of degree n,

$$g(z) = \sum_{k=0}^{n} b_k z^k,$$

then all the zeros of the polynomial h defined by

$$h(z) = \sum_{k=0}^{m} a_k g(k) z^k$$

lie in the ring

$$r_1 \min\left(1, \left|\frac{g(0)}{g(m)}\right|\right) \le |z| \le r_2 \max\left(1, \left|\frac{g(0)}{g(m)}\right|\right).$$

R. Peretz and Th.M. Rassias [19] provided the following counterexample to this theorem:

Counterexample. Let $f(z) = 1+2z+z^2$ and $g(z) = 1+2z-z^2$ be the polynomial of degree 2 that interpolates the coefficients of $f(z)$ at $z = 0, 1, 2$, i.e., $g(0) = 1$, $g(1) = 2$, $g(2) = 1$. Then $h(z) = 1 + 4z + z^2$. The polynomial $h(z)$ vanishes at $-2 \pm \sqrt{3}$ and $f(z)$ has a double root at -1. Thus we can take $r_2 = 1$. Then Marden's theorem implies

$$\left|-2 \pm \sqrt{3}\right| \le 1 \cdot \max\{1, 1\} = 1,$$

which is not correct.

Note. The theorem (16.2)′ is used for the proof of Theorem (35.1). Thus one has to correct all of Section 35 starting with Theorem (35.1) of the book.

In order to state a corrected version of Marden's theorem, R. Peretz and Th.M. Rassias [19] introduced the following

Definition. Let $g(z) = (\beta_1 - z)\cdots(\beta_n - z)$ and m a given positive integer. We define

$$g^+(z) = \prod_{\substack{1\le j\le n \\ \mathrm{Re}\,\{\beta_j\}\ge m/2}} (\beta_j - z)$$

and

$$g^-(z) = \prod_{\substack{1\le j\le n \\ \mathrm{Re}\,\{\beta_j\}< m/2}} (\beta_j - z)$$

with the understanding that g^+ or g^- takes the value 1 if one of the products is empty.

Theorem 4.1. *If all the zeros of the polynomial*

$$f(z) = \sum_{k=0}^{m} a_k z^k$$

lie in the ring R_0,

$$R_0: \quad 0 \le r_1 \le |z| \le r_2 \le \infty,$$

and if $g(z)$ *is a polynomial of degree* n, *then all zeros of*

$$h(z) = \sum_{k=0}^{m} a_k g(k) z^k$$

lie in the ring

$$r_1 \left| \frac{g^-(0)}{g^-(m)} \right| \le |z| \le r_2 \left| \frac{g^+(0)}{g^+(m)} \right| .$$

In order to state our second theorem we need

Definition. Let $g(z) = (\beta_1 - z)\cdots(\beta_n - z)$ and m a given positive integer. We define

$$g^{+\frac{1}{2}}(z) = \prod_{\substack{1\le j\le n \\ \mathrm{Re}\,\{\sqrt{\beta_j}\}\ge m(m-1)/2}} \left(\sqrt{\beta_j} - z\right)$$

and

$$g^{-\frac{1}{2}}(z) = \prod_{\substack{1\le j\le n \\ 0\le \mathrm{Re}\,\{\sqrt{\beta_j}\}<m(m-1)/2}} \left(\sqrt{\beta_j} - z\right)$$

such that if a product is empty, we take it to be equal to 1, and whenever we take the square roots $\sqrt{\beta_j}$ to be such that Re $\left\{\sqrt{\beta_j}\right\} \geq 0$ for $1 \leq j \leq n$.

Then we have

Theorem 4.2. *If all the zeros of the polynomial*

$$f(z) = \sum_{k=0}^{m} a_k z^k$$

lie in the ring R_0,

$$R_0: \quad 0 \leq r_1 \leq |z| \leq r_2 \leq \infty,$$

and if $g(z)$ *is a polynomial of degree* n, *then all zeros of*

$$h(z) = \sum_{k=0}^{m} a_k g(k(k-1)) z^k$$

lie in the ring

$$r_1 \left| \frac{g^{-\frac{1}{2}}(0)}{g^{-\frac{1}{2}}\left(\sqrt{m(m-1)}\right)} \right| \leq |z| \leq r_2 \left| \frac{g^{+\frac{1}{2}}(0)}{g^{+\frac{1}{2}}\left(\sqrt{m(m-1)}\right)} \right|.$$

References

1. S.N. Bernstein, *Sur l'ordre de la meilleure approximation des fonctions continues par des polynômes de degré donné.* Mémoires de l'Académie Royale de Belgique (2) **4** (1912), 1–103.

2. S.N. Bernstein, *Leçons sur les Propriétés Extrémales et la Meilleure Approximation des Fonctions Analytiques d'une Variable Réele.* Paris, 1926.

3. B.D. Bojanov, Q.I. Rahman and J. Szynal, *On a conjecture of Sendov about the critical points of a polynomial.* Math. Z. **190** (1985), 281–285.

4. D.A. Brannan, *On a conjecture of Ilieff.* Proc. Cambridge Philosophical Soc. **64** (1968), 83–85.

5. G.L. Cohen and G.H. Smith, *A proof of Ilieff's conjecture for polynomials with four zeros.* El. Math. **43** (1988), 18–21.

6. V.K. Dzyadyk, *Introduction to Uniform Approximation Theory by Polynomials* (Russian). Nauka, Moscow, 1977.

7. A.W. Goodman, Q.I. Rahman and J.S. Ratti, *On the zeros of a polynomial and its derivative.* Proc. Amer. Math. Soc. **21** (1969), 273–274.

8. W.K. Hayman, *Research Problems in Function Theory.* The Athlone Press, Univ. London, 1967.

9. V.I. Ivanov, *Some extremal properties of polynomials and inverse inequalities in approximation theory* (Russian). Trudy Mat. Inst. Steklov **145** (1979), 79–110.

10. G.G. Lorentz, *Approximation of functions.* Holt, Rinehart and Winston, New York, 1966.

11. M. Marden, *The zeros of certain composite polynomials.* Bull. Amer. Math. Soc. **49** (1943), 93–100.

12. M. Marden, *Geometry of Polynomials.* Math. Surveys, No. 3, Amer. Math. Soc., Providence, R.I., 1966.

13. M. Marden, *Conjectures on the critical points of a polynomial.* Amer. Math. Monthly **90** (1983), 267–276.

14. A.A. Markov, *On a problem of D.I. Mendeleev* (Russian). Zap. Imp. Akad. Nauk, St. Petersburg **62** (1889), 1–24.

15. V.A. Markov, *On functions deviating least from zero in a given interval* (Russian). Izdat. Imp. Akad. Nauk, St. Petersburg, 1892. [German Transl. Math. Ann. **77** (1916), 218–258.]

16. G. Meinardus, *Approximation of Functions: Theory and Numerical Methods*. Springer-Verlag, New York, 1967.

17. D. Mendeleev, *Investigation of aqueous solutions based on specific gravity* (Russian). St. Petersburg, 1887.

18. G.V. Milovanović, D.S. Mitrinović and Th.M. Rassias, *Topics in Polynomials: Extremal Problems, Inequalities, Zeros*. World Sci. Publ. Co., Singapore, London (to appear).

19. R. Peretz and Th.M. Rassias, *Some remarks on theorems of M. Marden concerning the zeros of certain composite polynomials*. Complex Variables (to appear).

20. Th.M. Rassias, *A new inequality for complex-valued polynomial functions*. Proc. Amer. Math. Soc. **97** (1986), 296–298.

21. Th.M. Rassias, *On certain properties of polynomials and their derivative*. In: Topics in Mathematical Analysis (ed. Th.M. Rassias), World Sci. Publ. Co., Singapore, London, 1989, pp. 758–802.

22. E.B. Saff and J.B. Twomey, *A note of the location of critical points of polynomials*. Proc. Amer. Math. Soc. **27** (1971), 303–308.

23. G. Schmeisser, *Bemerkungen zu einer Vermutung von Ilieff*. Math. Z. **111** (1969), 121–125.

24. G. Schmeisser, *Zur Lage der kritischen Punkte eines Polynoms*. Rend. Semin. Mat. Univ. Padova **46** (1971), 405–415.

25. G. Schmeisser, *On Ilieff's conjecture*. Math. Z. **156** (1977), 165–173.

26. M. Shub and S. Smale, *Computational complexity: on the geometry of polynomials and a theory of cost: Part I*. Ann. Sci. École Norm. Sup. (4) **18** (1985), 107–142.

27. S. Smale, *The fundamental theorem of algebra and complexity theory*. Bull. Amer. Math. Soc. (N.S.), **4** (1981), 1–36.

28. S. Smale, *On the efficiency of algorithms of analysis*. Bull. Amer. Math. Soc. (N.S.), **13** (1985), 87–121.

29. S. Smale, *Algorithms for solving equations.* Proc. ICM, Berkeley, CA, USA, 1986, Vol. I, pp. 172–195.

30. S.B. Stečkin, *Generalization of some inequalities of S.N. Bernstein.* Dokl. Acad. Nauk SSSR **60** (1948), 1511–1515.

31. B.L. Van der Waerden, *A History of Algebra, From al-Khwārizmī to Emmy Noether.* Springer-Verlag, Berlin, 1985.

Themistocles M. Rassias, Department of Mathematics, University of La Verne, P.O. Box 51105, Kifissia, Athens, Greece 14510

International Series of Numerical Mathematics, Vol. 103, © 1992 Birkhäuser Verlag Basel

An inequality between u^* and $|\text{grad } u|^*$

Giorgio Talenti

Abstract

We prove an inequality between the decreasing rearrangement of a function u and the decreasing rearrangement of the length of the gradient of u.

1. Introduction

Let u be a real-valued function defined in the Euclidean n-dimensional space $\mathbb{R}^n$. Suppose u is measurable and μ is the distribution function of u, i.e., for every nonnegative t

$$\{x \in \mathbb{R}^n : |u(x)| > t\}, \tag{1}$$

a typical level set of u, is measurable and $\mu(t)$ is the measure of such a level set. By definition, the *decreasing rearrangement* of u is the restriction to $[0, +\infty[$ of the distribution function of μ. Let decreasing rearrangements be denoted by superscript stars. Thus

$$\begin{aligned} u^*(s) &= \sup\{t : \mu(t) > s\} \\ &= \min\{t : \mu(t) \leq s\} \end{aligned} \tag{2}$$

for every nonnegative s; moreover, u^* is nonnegative, nonincreasing and right-continuous. Importantly,

$$\{s \geq 0 : u^*(s) > t\} = [0, \mu(t)[\tag{3}$$

for every t, so that the distribution function of u^* is μ. In other words, any real-valued measurable function and its decreasing rearrangement are equimeasurable or equidistributed. For more information on rearrangements see [12], [14], [19], [23].

This paper is in final form and no version of it will be submitted for publication elsewhere.

Rearrangements are closely related to the so-called embedding theorems, theorems which trace back to an important paper [22] by Sobolev and are subject of an extensive literature, see [1] and [17] for instance. Sharp forms of some embedding theorems heavily lie on rearrangement techniques, see [5], [15], [16], [18], [24]. In the case where the relevant norms are invariant under rearrangements, an embedding theorem à la Sobolev typically amounts to an inequality of the following form:

$$\frac{\text{a norm of } u^*}{\text{another norm of } |\text{grad } u|^*} \leq \text{Const. independent of } u, \tag{4}$$

where u is any member of a relevant function space. Here grad is a shorthand for gradient, thus $|\text{grad } u|^*$ is the decreasing rearrangement of $|\text{grad } u|$, the length of the gradient of u. Standard Sobolev embedding theorems involve customary Lebesgue norms, the simplest rearrangement invariant norms. Embedding theorems that involve rearrangement invariant norms other than Lebesgue norms appeared in a number of papers. See [2], [9], [21], where Lorentz spaces are considered. A non-conventional inequality of the form (4) appeared in [4], motivated by a problem of mechanics. Inequalities of the form (4) are also related to viscosity solutions of boundary value problems for first order partial differential equations, see [3], [10].

In this paper we offer an inequality, which may be a convenient tool in investigating inequalities of the form (4).

2. Main results

Theorem 1. *Assume u is real-valued and locally integrable, and the (distributional) gradient of u is integrable over $\mathbb{R}^n$. Assume the support of u has a finite measure, and let V be such a measure. Then*

$$u^*(s) \leq n^{-1}\kappa_n^{-1/n} \int_0^{V-s} |\text{grad } u|^*(t)(s+t)^{-1+1/n}\, dt \tag{5}$$

for every nonnegative s; here

$$\kappa_n = \frac{\pi^{n/2}}{\Gamma(n/2+1)},$$

the measure of the unit n-dimensional ball.

Equality holds in (5) if u is defined by

$$u(x) = 1 - |x| \quad \text{if } |x| < 1, \quad = 0 \quad \text{if } |x| \geq 1. \tag{6}$$

Remark. In the case where n, the dimension, is 1, Theorem 1 results in the following statement, which already appeared in [10]:

Let u be a real-valued function of one real variable. Assume u is absolutely continuous and the support of u has length V. Then

$$u^*(s) \leq \frac{1}{2} w(s),$$

where w is defined by

$$w(s) = 0 \quad \textit{if} \quad s \geq V,$$
$$-w'(V - s) = |u'|^*(s) \quad \textit{if} \quad 0 < s < V,$$

i.e.,

$$-w' = \textit{ the increasing rearrangement of } u'.$$

Theorem 1 has the following corollary.

Theorem 2. *Assume u is real-valued and locally integrable, and the (distributional) gradient of u is integrable over $\mathbb{R}^n$. Then*

$$u^*(s) \leq n^{-1} \kappa_n^{-1/n} \int_0^{+\infty} |\text{grad } u|^*(t)(s+t)^{-1+1/n}\, dt. \tag{7}$$

for every nonnegative s.

Proof of Theorem 1. By the Fleming-Rishel coarea formula [8],

$$\int_{\{x \in \mathbb{R}^n : a < |u(x)| \leq b\}} |\text{grad } u|\, dx = \int_a^b p\{x \in \mathbb{R}^n : |u(x)| > t\}\, dt, \tag{8}$$

whenever $0 \leq a < b$. By a form of the isoperimetric theorem [7],

$$p\{x \in \mathbb{R}^n : |u(x)| > t\} \geq n\kappa_n^{1/n} [m\{x \in \mathbb{R}^n : |u(x)| > t\}]^{1-1/n}. \tag{9}$$

Here m denotes the customary Lebesgue measure and p stands for perimeter à la De Giorgi; see [11], [25] for instance. Recall that μ, the distribution function of u, is given by

$$\mu(t) = m\{x \in \mathbb{R}^n : |u(t)| > t\}.$$

Thus we have

$$n\kappa_n^{1/n} \int_a^b [\mu(t)]^{1-1/n}\, dt \leq \int_{\{x \in \mathbb{R}^n : a < |u(x)| \leq b\}} |\text{grad } u|\, dx \tag{10}$$

whenever $0 \le a < b$.

A theorem by Hardy and Littlewood — see [12, Theorem 378] — says that

$$\int_E f(x)\,dx \le \int_0^{m(E)} f^*(s)\,ds,$$

provided f is nonnegative and measurable and E is a measurable subset of $\mathbb{R}^n$. Observe that

$$m\{x \in \mathbb{R}^n : a < |u(x)| \le b\} = \mu(a) - \mu(b).$$

Consequently, (10) implies

$$n\kappa_n^{1/n} \int_a^b [\mu(t)]^{1-1/n}\,dt \le \int_0^{\mu(a)-\mu(b)} |\text{grad } u|^*(s)\,ds \tag{11}$$

for any a and b such that $0 \le a < b$.

Now let us slightly abuse notations and replace a and b by $u^*(b)$ and $u^*(a)$, respectively. Equations (2) and (11) easily give

$$n\kappa_n^{1/n} \int_{u^*(b)}^{u^*(a)} [\mu(t)]^{1-1/n}\,dt \le \int_0^{b-a} |\text{grad } u|^*(s)\,ds \tag{12}$$

for any a and b such that $0 \le a < b$.

Equation (3) implies $s \ge \mu(u^*(s))$ for every nonnegative s. It also implies $\{s \ge 0 : u^*(s) \ge t\} = [0, \mu(t-)[$ for every t; here $\mu(t-)$ stands for the limit of μ at t from the left. Hence

$$\mu(u^*(s)) \le s \le \mu(u^*(s)-)$$

for every nonnegative s. We deduce

$$\int_a^b s^{1-1/n} \left[-\frac{du^*}{ds}(s)\right] ds = \int_{u^*(b)}^{u^*(a)} [\mu(t)]^{1-1/n}\,dt, \tag{13}$$

if a and b are such that $0 \le a < b$.

Inequality (12) and equation (13) give

$$n\kappa_n^{1/n} \int_a^b s^{1-1/n} \left[-\frac{du^*}{ds}(s)\right] ds \le \int_0^{b-a} |\text{grad } u|^*(s)\,ds \tag{14}$$

whenever $0 \le a < b$.

Clearly, $u^*(s) = 0$ if $s \ge V$. Hence

$$u^*(s) = \int_s^V \left[-\frac{du^*}{dt}(t)\right] dt$$

if $0 < s < V$. Partial integrations show

$$u^*(s) = V^{-1+1/n} \int_s^V t^{1-1/n} \left[-\frac{du^*}{dt}(t) \right] dt + \left(1 - \frac{1}{n}\right) \int_s^V dr\, r^{-2+1/n} \int_s^r t^{1-1/n} \left[-\frac{du^*}{dt}(t) \right] dt, \tag{15}$$

if $0 < s < V$.

Inequalities (14) and (15) yield

$$n\kappa_n^{1/n} u^*(s) \leq V^{-1+1/n} \int_0^{V-s} |\text{grad } u|^*(t)\, dt + \left(1 - \frac{1}{n}\right) \int_s^V dr\, r^{-2+1/n} \int_0^{r-s} |\text{grad } u|^*(t)\, dt \tag{16}$$

for $0 < s < V$. Partial integrations show that the right-hand side of (16) equals

$$\int_s^V dr\, r^{-1+1/n}\, |\text{grad } u|^*(r-s)\, dr.$$

In conclusion, we have the following inequality

$$n\kappa_n^{1/n} u^*(s) \leq \int_s^V dr\, r^{-1+1/n}\, |\text{grad } u|^*(r-s)\, dr \tag{17}$$

for $0 < s < V$, an equivalent form of (5).

Observe that the preceding arguments involve the absolute continuity of u^*; such an absolute continuity is a consequence of a well-known theorem on rearrangements, see [6] for instance.

The remaining part of Theorem 1 is easily checked by inspection.

The proof is complete. □

3. Applications

Theorem 3. *Assume u is real-valued and locally integrable in $\mathbb{R}^n$ and the (distributional) gradient of u belongs to the Lorentz space $L(p,q)$. Here $1 < p < n$ and $1 \leq q \leq +\infty$. Then u belongs to the Lorentz space $L\left(\frac{np}{n-p}, q\right)$.*

Recall that a real-valued measurable function f belongs to the Lorentz space $L(p,q)$ if and only if the decreasing rearrangement of f, f^*, satisfies

$$\left\{ \int_0^{+\infty} \left[s^{1/p} f^*(s) \right]^q \frac{ds}{s} \right\}^{1/q} < \infty;$$

the usual modification has to be made if $q = +\infty$. Lorentz spaces $L(p,q)$ have the following properties: $L(p,p) = L^p(\mathbb{R}^n)$, the customary Lebesgue space. If $q_1 < q_2$, $L(p,q_1)$ is a proper subset of $L(p,q_2)$; in particular, $L(p,q)$ is a proper subset of $L^p(\mathbb{R}^n)$ if $q < p$. For more information on Lorentz spaces see [13], [20], [23].

Theorem 3 is a significant improvement of a classical embedding theorem. Theorem 3 follows from [2], and special cases of it were remarked in [21] and [9]. Below we show that Theorem 3 can be deduced from Theorem 2.

A *Proof* of Theorem 3. If the following abbreviations

$$\phi(s) = n\kappa_n^{1/n} s^{1/p-1/n} u^*(s), \quad \psi(s) = s^{1/p} |\text{grad } u|^*(s) \tag{18}$$

are used, Theorem 2 gives

$$\phi(s) \le \int_0^{+\infty} K(s,t)\phi(t)\, dt, \tag{19}$$

where

$$K(s,t) = s^{1/p-1/n} t^{-1/p} (s+t)^{-1+1/n}.$$

Observe that

$$K(s,t) > 0,$$

$$K(r,rs) = \frac{1}{s} K(1,r),$$

$$\int_0^\infty K(1,r)\, dr = \frac{\Gamma(1-1/p)\,\Gamma(1/p-1/n)}{\Gamma(1-1/n)}.$$

A change of variables in the right-hand side of (19) gives

$$\phi(s) \le \int_0^{+\infty} K(1,r)\psi(rs)\, dr,$$

consequently the Minkowski inequality yields

$$\left\{ \int_0^{+\infty} \phi(s)^q \frac{ds}{s} \right\}^{1/q} \le \int_0^{+\infty} K(1,r) \left\{ \int_0^{+\infty} \psi(rs)^q \frac{ds}{s} \right\}^{1/q} dr.$$

The right-hand side of the last inequality equals

$$\int_0^{+\infty} K(1,r)\, dr \left\{ \int_0^{+\infty} \psi(s)^q \frac{ds}{s} \right\}^{1/q}.$$

In conclusion,

$$(20) \qquad \left\{ \int_0^{+\infty} \phi(s)^q \frac{ds}{s} \right\}^{1/q} \leq \frac{\Gamma(1-1/p)\,\Gamma(1/p-1/n)}{\Gamma(1-1/n)} \left\{ \int_0^{+\infty} \psi(s)^q \frac{ds}{s} \right\}^{1/q}.$$

Incidentally, the above argument is patterned on Theorem 319 of [12]. Theorem 3 follows. □

References

1. R.A. Adams, *Sobolev spaces.* Academic Press 1975.
2. A. Alvino, *Sulla disuguaglianza di Sobolev in spazi di Lorentz.* Boll. Unione Mat. It. (5) 14A (1977), 148–156.
3. A. Alvino, P.-L. Lions and G. Trombetti, *On optimization problems with prescribed rearrangements.* Nonlinear Analysis TMA **13** (1989), 185–220.
4. G. Aronsson and G. Talenti, *Estimating the integral of a function in terms of a distribution function of its gradient.* Boll. Unione Mat. It. (5) 18B (1981), 885–894.
5. T. Aubin, *Problèmes isopérimetriques et espaces de Sobolev.* J. Diff. Geometry **11** (1976), 573–598.
6. J. Brothers and W.P. Ziemer, *Minimal rearrangements of Sobolev functions.* J. Reine Angew. Math. **384** (1988), 153–179.
7. E. De Giorgi, *Sulla proprietà isoperimetrica dell'ipersfera.* Mem. Acc. Naz. Lincei (8) **5** (1958), 33–44.
8. W. Fleming and R. Rishel, *An integral formula for total gradient variation.* Arch. Math. **11** (1960), 218–222.
9. J.J.F. Fournier, *Mixed norms and rearrangements: Sobolev's inequality and Littlewood's inequality.* Ann. Mat. Pura Appl. (4) **148** (1987), 53–76.
10. E. Giarusso and D. Nunziante, *Symmetrization in a class of first-order Hamilton-Jacobi equations.* Nonlinear Analysis TMA **8** (1984), 289–299.
11. E. Giusti, *Minimal surfaces and functions of bounded variation.* Birkhäuser 1984.

12. G.H. Hardy, J.E. Littlewood and G. Pólya, *Inequalities.* Cambridge Univ. Press 1934.

13. A. Hunt, *On $L(p,q)$ spaces.* Enseignement math. **12** (1966), 249–276.

14. B. Kawohl, *Rearrangements and convexity of level sets in PDE.* Lecture Notes in Math., Vol. 1150, Springer-Verlag 1985.

15. H.A. Levine, *An estimate for the best constant in a Sobolev inequality involving three integral norms.* Ann. Mat. Pura Appl. **124** (1980), 181–197.

16. E.H. Lieb, *Sharp constants in the Hardy-Littlewood-Sobolev and related inequalities.* Ann. Math. **118** (1983), 349–374.

17. V.G. Maz'ja, *Sobolev spaces.* Springer-Verlag 1985.

18. J. Moser, *A sharp form of an inequality by N. Trudinger.* Indiana Math. J. **20** (1971), 1077–1092.

19. G.O. Okikiolu, *Aspects of the theory of bounded integral operators in L^p-spaces.* Academic Press 1971.

20. R. O'Neil, *Convolution operators and $L(p,q)$ spaces.* Duke Math. J. **30** (1963), 129–142.

21. S. Poornima, *An embedding theorem for the Sobolev space $W^{1,1}$.* Bull. Sci. Math. **107** (1983), 253–259.

22. S.L. Sobolev, *On a theorem of functional analysis.* Math. Sb. **4** (1938), 471–497.

23. E.M. Stein and G. Weiss, *Introduction to Fourier analysis on Euclidean spaces.* Princeton Univ. Press 1971.

24. G. Talenti, *Best constant in Sobolev inequality.* Ann. Mat. Pura Appl. **110** (1976), 353–372.

25. G. Talenti, *The standard isoperimetric theorem.* Handbook of convex geometry (P. Gruber and J.M. Wills editors), forthcoming.

Giorgio Talenti, Dipartimento di Matematica, viale Morgagni 67A, I-50134 Firenze, Italy

Inequalities of Functional Analysis

Valley and river Wolf

International Series of Numerical Mathematics, Vol. 103, © 1992 Birkhäuser Verlag Basel

Inequalities for summability factors

W. Beekmann and K. Zeller

Abstract

We consider two inequalities which are relevant for Cesàro summability factors. Combined use of these inequalities (in the realm of BK-spaces) leads to a rather short and lucid verification of Bosanquet's theorem.

1. Introduction

We consider factors which transform each C_α summable into a C_β summable series. Such factors have been investigated for a long time. At first one treated the cases where α and β are non-negative integers; partial summation leads to conditions involving differences of the factors. The extension to real non-negative parameters is rather delicate and has been completed by Bosanquet. But for the subcase $\alpha = \beta$ a functional analytic positivity argument yielded a rather short proof. This has been extended to the cases $\alpha \neq \beta$ where β is an integer. Here, besides the positivity inequality, we use an inequality connected with diaposivity (see below). In this way we get better insight and obtain a concise proof for the complete result (see Theorem 3b). In Section 9 we give more details and mention some other approaches.

2. Notation

In view of Bosanquet [3] we consider the following triangular (resp. diagonal) matrices:

$$S^\alpha \text{ with elements } S^\alpha_{nk} := \binom{n-k+\alpha-1}{n-k} \text{ for } 0 \leq k \leq n,$$

$$J_q := \operatorname{diag}\left(1/\binom{n+q}{n}\right),$$

$$C(\alpha, p) := J_{\alpha+p} S^{\alpha+1},$$

$$C_\alpha := C(\alpha, 0)$$

This paper is in final form and no version of it will be submitted for publication elsewhere.

and

$$A_\alpha := C(\alpha - 1, 1).$$

The latter two matrices are the series-to-sequence and sequence-to-sequence forms, respectively, of the Cesàro method C_α; we write $C_\alpha\text{-}\sum x_k := \lim C_\alpha x$, $C_\alpha\text{-}\lim w_k := \lim A_\alpha w$.

The null domain of $C(\alpha, p)$ — we denote it by $\mathcal{N}(\alpha, p)$ — is the set of elements $x = (x_k)$ for which $y := C(\alpha, p)x$ is a null sequence. It constitutes a BK-space with the norm

$$\|x\| := \sup |y_n|;$$

the continuous linear functionals are described by

$$L(x) = gy := \sum g_n y_n \qquad (\text{where } \sum |g_n| < \infty)$$

(each L has a g; each g determines an L). In the inner product gy we can consider g as a row vector and y as a column vector. Later (Theorem 3a) we use a representation

$$L(x) = C_\alpha\text{-}\sum f_k x_k.$$

For $p > 0$ the null domain contains the basic vectors $e^0 := (1, 0, 0 \ldots)$, $e^1 := (0, 1, 0, 0, \ldots)$, $e^2, \ldots$; for $p = 0$ it contains the vectors $e^k - e^{k+1}$. Thus

$$f_k = L(e^k) \text{ for } p > 0, \quad f_k - f_{k+1} = L(e^k - e^{k+1}) \text{ for } p = 0.$$

In the latter case the f_k are determined only up to a constant. (Recall that $C_\alpha\text{-}\sum x_k = 0$ in the null domain.) We take $f := gC(\alpha, p)$ in both cases.

3. Summability factors

The results described below (basic is the case $p = 0$) have a long history — see the references given in [8, **53** III]. In particular we mention Hardy, Bohr, Andersen, Lorentz–Zeller (Theorem 3a), Bosanquet, Jakimovski–Russell–Stieglitz (Theorem 3b), Beekmann–Zeller (general considerations).

Theorem 3a. *If f satisfies (for $\alpha, p \geq 0$ fixed)*

$$(L) \qquad f = gJ_{\alpha+p}S^{\alpha+1} \text{ with } \sum |g_n| < \infty,$$

(for a suitable g), then

$$(A) \qquad L(x) := C_\alpha\text{-}\sum f_k x_k \text{ exists for } x \in \mathcal{N}(\alpha, p)$$

(and hence describes a continuous linear functional L in $\mathcal{N}(\alpha, p)$.)
Conversely, if (A) is true for an f, $f_k = o(1)$ in the case $p = 0$, then (L) holds.

Thus we have a necessary and sufficient condition for C_α summability factors (modulo a constant in the case $p = 0$) and a description of the continuous linear functionals in $\mathcal{N}(\alpha, p)$.

Theorem 3b. *If f satisfies (L) (for $\alpha, p \geq 0$ fixed) and*

$$(M) \qquad f_k = \mathcal{O}(k^{\beta-\alpha-p})$$

(for a β with $0 \leq \beta \leq \alpha$), then

$$(B) \qquad L(x) := C_\beta\text{-}\sum f_k x_k \text{ exists for } x \in \mathcal{N}(\alpha, p)$$

(and hence describes a continuous linear functional L in $\mathcal{N}(\alpha, p)$.)
Conversely, if (B) is true for an f, $f_k = o(1)$ in the case $p = 0$, then (L) and (M) hold.

We remark that (L) implies $f_k = o(k^{-p})$, and hence (M) with $\beta - \alpha = 0$ is always satisfied in the presence of (L). The representation (L) can be inverted (with certain care) and leads to a difference condition for f (cf. Bosanquet [3], Lorentz–Zeller [7]). - The necessity part of both theorems can be verified by functional analytic considerations. Matrix summability factors in a BK–space determine a continuous linear functional; in view of the considerations in Section 2, this leads to (L). A matrix mapping of a BK–space into a BK–space is continuous; this yields (M).

4. Limitability factors

In the proof of Theorem 3b we are interested in factors achieving C-limitability. We state reasonable sufficient conditions (which could be refined to exact conditions). Here we put $w := Sx$ (partial sums of the series with terms x_k). Such a w is in the null domain of

$$C(\alpha - 1, p + 1) = J_{\alpha-1+p+1} S^\alpha = J_{\alpha+p} S^\alpha$$

if and only if x is in the null domain of $C(\alpha, p)$.

Proposition 4a. *If f satisfies (L) for some $\alpha \geq 0$ and some $p \geq 0$, then*

$$C_\alpha\text{-lim } f_n w_n = 0$$

for all w in the null domain of $C(\alpha - 1, p + 1)$.

This will follow from Theorem 3a (based on positivity) and the partial summation (P) described below (Section 7). In (P) the first term (series) will be $\boldsymbol{C}_\alpha$-summable, the third term (series) $\boldsymbol{C}_{\alpha-1}$-summable (resp. convergent if $\alpha - 1 < 0$). Thus the middle term (sequence) will be $\boldsymbol{C}_\alpha$-limitable (the change from f_n to f_{n+1} is irrelevant).

Proposition 4b. *If f satisfies (L) and (M) for some β and p with $0 \leq \beta \leq \alpha$ and $0 \leq p$, then*

$$\boldsymbol{C}_\beta\text{-}\lim f_n w_n = 0$$

for every w in the null domain of $C(\alpha - 1, p + 1)$.

The proof will be given in Sections 6 (for $0 \leq \beta \leq \alpha \leq 1, 0 \leq p$) and 8.

5. Positivity

The inequalities stated here and in the next section are basic for our considerations (in conjunction with summation by parts).

Lemma 5.1 *For $\alpha \geq 0$ and $m, n, r = 0, 1, \ldots,$*

$$\sum_{k=0}^{\infty} S_{mk}^{\alpha+1} S_{nk}^{\alpha+1} S_{kr}^{-\alpha-1} \geq 0$$

holds.

See [2], [5, Section 5] and the information given there.

We express Lemma 5.1 in terms of the matrices considered above:

$$J_{\alpha+p} S^{\alpha+1} \operatorname{diag}(C_{n0}^{\alpha}, \ldots, C_{nn}^{\alpha}, \ldots) S^{-\alpha-1} J_{\alpha+p}^{-1} \geq 0 \text{ (for } \alpha, p \geq 0 \text{ and } n = 0, 1, \ldots);$$

here we use the diagonal matrix formed by the elements of n-th row in C_α. Applying this positive matrix to $e := (1, 1, \ldots)$ we see that the row sums and hence the row norms are ≤ 1. This leads to the fact that the basic vectors $e^0, e^1, \ldots$ constitute a C_α-basis in the null domain resp. full domain of $C(\alpha, p)$ (for $p > 0$ resp. $p = 0$). Thus every continuous functional L in the corresponding BK-space can be written as a $\boldsymbol{C}$-sum, more precisely

$$L(x) = \mathbf{C}_\alpha\text{-}\sum f_k x_k \text{ with } f_k = L(e^k).$$

Comparison with the general form of L, which is

$$L(x) = g(C(\alpha, p)x) \quad \text{resp.} \quad L(x) = \mu\, \mathbf{C}_\alpha\text{-}\sum x_k + g(C(\alpha, 0)x)$$

(with a constant μ) in the cases $p > 0$ resp. $p = 0$, yields

$$f = gC(\alpha, p) \quad \text{resp.} \quad f = \mu e + gC(\alpha, p).$$

This shows the necessity of (L) in Theorem 3a (and similarly in Theorem 3b).

6. Diapositivity

The triangle $S^{-\alpha}$ (where $0 \le \alpha \le 1$) is diapositive (elements > 0 in the main diagonal, and ≤ 0 below it). A related property is present in S^β (where $0 \le \beta \le 1$): The elements S^β_{nk} in each row increase (in the wider sense) for $k = 0, \ldots, n$. This leads to the following inequality (cf. [1], [3, inequality (22)]):

Lemma 6.1 *For $0 \le \beta \le \alpha \le 1$, $0 \le r \le m \le n$ and suitable $K = K(\alpha, \beta)$,*

$$\left|\left(S^\beta I_m S^{-\alpha}\right)_{nr}\right| \le K \cdot S^\beta_{nr} \cdot S^{1-\alpha}_{mr}$$

holds (where I_m has entries 1 in $(0,0), \ldots, (m,m)$ and 0 elsewhere).

A lower bound is given by $S^{\beta-\alpha}_{nr}$ (consider $\left(S^\beta I_m S^{-\alpha}\right)_{nr} = \sum_{k=r}^m S^\beta_{nk} S^{-\alpha}_{kr}$ for the case $m = n$ and use diapositivity; a known asymptotic then leads to our estimate.) We obtain the upper bound (with $K = 1$) if we replace S^β_{nk} by S^β_{nr} (using monotonicity with respect to k).

Now we assume (L) and obtain the inequality (with $h := fS^{-1}$)

$$\left|hI_n S^{1-\alpha}\right| \le |h| S^{1-\alpha} \le |g| J_{\alpha+p} S^\alpha S^{1-\alpha} = |g| J_{\alpha+p} S,$$

$$\left|\left(\sum_{m=1}^{n} h_m S^\beta I_m S^{-\alpha}\right)\right| \le K\left(\sum |g_j|\right) S^\beta J_{\alpha+p}.$$

Partial summation yields

$$S^\beta \operatorname{diag}(f_j) S^{-\alpha} = \operatorname{diag}(f_{n+1}) S^{\beta-\alpha} + \sum_{m=1}^{n} h_m (S^\beta I_m S^{-\alpha}).$$

Here we treat the second term using (M) and the third term by the considerations above. We see that the row norms of $J_\beta S^\beta \operatorname{diag}(f_j) S^{-\alpha} J^{-1}_{\alpha+p}$ are bounded:

$$\mathcal{O}\left(n^{-\beta+(\beta-\alpha-p)+(\beta-\alpha+\alpha+p)}\right) + \mathcal{O}\left(n^{-\beta+\beta}\right) = \mathcal{O}(1);$$

here $J_{\alpha+\beta}^{-1}$ is just compensated in the second part, while the estimate $\mathcal{O}(n^{\alpha+p})$ in the first part is valid, since we have triangles; the "$\mathcal{O}$" indicates that the row norms of $S^{\beta-\alpha}$ are bounded. Moreover, our matrix has column limits zero. Thus it is null-preserving (it transforms every null sequence into a null sequence). Altogether we have verified Proposition 4b in the cases $0 \le \beta \le \alpha \le 1$ for all $p \ge 0$ (and Proposition 4a in the cases $0 \le \alpha \le 1$, $0 \le p$).

7. Partial summation

We use the standard formula

$$(P) \qquad \sum_{k=0}^{n} f_k x_k = f_{n+1} w_n + \sum_{k=0}^{n} h_k w_k \qquad (w := Sx;\ h := fS^{-1})$$

Suppose now that x is in the null domain of $C(\alpha, p)$ (where $\alpha, p \ge 0$) and that f satisfies (L) and (M). We'll show the desired $\boldsymbol{C}_\beta$–summability by induction ($\alpha = 1, 2, \ldots$). The last series in (P) is convergent (hence $\boldsymbol{C}_\beta$–summable) for $\alpha \le 1$, and we have already seen that Proposition 4b is true for $0 \le \alpha \le 1$ (and all $\beta, p \ge 0$ with $\beta \le \alpha$), i. e. the middle term in (P) has a $\boldsymbol{C}_\beta$–limit (equal to zero). Hence, by (P), Theorem 3b is true for these parameters. Next, we consider the cases $1 < \alpha \le 2$ (and so on). The last series causes no difficulties (it is covered by the cases in which Theorem 3b is already verified; cf. below).

The middle term is treated by the recursion described later (stepwise verification of Proposition 4b.)

For the recursion we again employ partial summation, but followed by taking differences:

$$f_n w_n = (f_{n+1} z_n - f_n z_{n-1}) + h_n z_n \qquad (z := Sw;\ h := fS^{-1};\ z_{-1} := 0).$$

Again the last term causes little difficulties, while more care is indicated for the middle term. We are going to explain the details.

8. Reduction

We have already proved Propositon 4b for $0 \le \alpha \le 1$ (and all $\beta, p \ge 0$ with $\beta \le \alpha$). Now we assume its truth for $0 \le \alpha \le r$ (where $r = 1, 2, \ldots$) and cover the cases $r < \alpha \le r+1$. Let f satisfy $(L), (M)$, let w be in the null domain of

$$C(\alpha - 1, p + 1) = J_{\alpha+p} S^{\alpha}.$$

Then z is in the null domain of

$$C(\alpha-2, p+2) = J_{\alpha+p} S^{\alpha-1}.$$

Moreover, h satisfies $(L), (M)$ with α, p replaced by $\alpha - 1, p+1$. Now we see

$$\boldsymbol{C}_{\alpha-1}\text{–lim } h_n z_n = 0, \quad \boldsymbol{C}_\beta\text{–lim } h_n z_n = 0$$

by Proposition 4a and (if $\beta < \alpha - 1$) Proposition 4b (induction).

The middle term will be transformed by the matrix

$$A_\beta = C(\beta - 1, 1) = J_\beta S^\beta = J_\beta S^{\beta-1} S.$$

The case $0 \le \beta \le 1$ is rather simple: $S^{\beta-1}$ has bounded row-norms, S cancels the differences, the transformed sequence is

$$\mathcal{O}(n^{-\beta}) \cdot \mathcal{O}(1) \cdot \mathcal{O}(n^{\beta-\alpha-p}) \cdot o(n^{\alpha+p}) = o(1).$$

This means that the $\boldsymbol{C}_\beta$–limit is zero.

Next, we consider the cases with $\beta \ge 1$. We find

$$[J_{(\alpha-1)+p} S^{\alpha-1} z]_n = o(n).$$

Our f satisfies (L), but then also (L) with α replaced by $\alpha - 1$. A similar remark applies to condition (M).

In the induction process we have already proved Proposition 4b in the case $\alpha - 1$; i. e. the matrix (triangle)

$$A_{\beta-1} \operatorname{diag}(f_k) S^{1-\alpha} J^{-1}_{(\alpha-1)+p}$$

transforms every null sequence into a null sequence. Now we see (using triangularity)

$$[J_{\beta-1} S^{\beta-1} \operatorname{diag}(f_k) z]_n = o(n).$$

The factor J_β (instead of $J_{\beta-1}$) changes this to $o(1)$. QED

9. Remarks

Bosanquet [3] uses classical methods in an expert way. Jakimovski, Russell and Stieglitz [6] apply functional analysis, but admit a restriction (β an integer). To a certain extent we have combined procedures from both papers. One could think of avoiding some of the partial summations and recursions used in the proof. Perhaps the following approaches and ideas are useful for this or another purpose.

Sectional summability allows comparison theorems based on factors (cf. [4]); these provide a bridge to summability factors. Tauberian theorems of the type $C_\alpha \to C_\beta$ with suitable conditions would be useful for the transition from Theorem 3a to 3b (and from Proposition 4a to 4b); we mention that Cesàro methods of different order are rather similar in positions away from the diagonal. Weak topologies based on continuous linear functionals satisfying (M) could give deeper insight into the structure of our problem. Special factors of the type $(n+1)^{-\delta}$ can be treated in advance and might be extracted (consider an algebra of factors).

References

1. A. F. Andersen, *Comparison theorems in the theory of Cesàro summability.* Proc. London Math. Soc. (2) **27** (1928), 39–71.

2. R. Askey, G. Gasper, M. E.-H. Ismail, *A positive sum from summability theory.* J. Approx. Th. **13** (1975), 413–420.

3. L. S. Bosanquet, *Note on convergence and summability factors (III).* Proc London Math. Soc. (2) **50** (1949), 482–496.

4. W. Beekmann, K. Zeller, *Positivity in summability.* ISNM **71** (1984), 111–117.

5. W. Beekmann, K. Zeller, *Positivity in absolute summability.* ISNM **80** (1987), 205–215.

6. A. Jakimovski, D. Russell, M. Stieglitz, *On multipliers in sequence spaces and a theorem of Bosanquet.* Analysis **9** (1989), 55–66.

7. G. G. Lorentz, K. Zeller, *Abschnittslimitierbarkeit und der Satz von Hardy-Bohr.* Archiv Math. **15** (1964), 208–213.

8. K. Zeller, W. Beekmann, *Theorie der Limitierungsverfahren.* Springer-Verlag, Berlin-Heidelberg-New York, 2nd edition 1970.

W. Beekmann, FB Mathematik und Informatik, FernUniversität – Gesamthochschule –, Postfach 940, D-5800 Hagen, Germany.

K. Zeller, Mathematische Fakultät, Universität Tübingen, Auf der Morgenstelle 10, D-7400 Tübingen, Germany.

International Series of Numerical Mathematics, Vol. 103, © 1992 Birkhäuser Verlag Basel

Seminorms and multiplicativity factors

Moshe Goldberg

Abstract

Let $\mathcal{A}$ be an algebra and let S be a seminorm on $\mathcal{A}$. In this note we discuss multiplicativity factors for S, i.e., constants $\mu > 0$ for which $S(xy) \leq \mu S(x)S(y)$ for all $x, y \in \mathcal{A}$. We begin by investigating these factors in terms of the kernel of S. We then specialize our study to function algebras and to seminorms generated by the sup norm. Finally, we study quasimonotonic seminorms on $\mathbb{C}^n$.

1. Introduction

Let $\mathcal{A}$ be an algebra over a field F where $F = \mathbb{R}$ or $F = \mathbb{C}$. Throughout this note we exclude the trivial case where all products in $\mathcal{A}$ are zero. As usual, a function $S : \mathcal{A} \to \mathbb{R}$ is called *seminorm* if for all $x, y \in \mathcal{A}$ and $\lambda \in F$:

$$S(x) \geq 0,$$

$$S(\lambda x) = |\lambda| \cdot S(x),$$

$$S(x + y) \leq S(x) + S(y).$$

If, in addition, S is positive definite, i.e.,

$$S(x) > 0 \quad \forall\, x \neq 0,$$

then S is a *norm.* We call a seminorm S *proper* if $S \neq 0$ and $S(x_0) = 0$ for some $x_0 \neq 0$. Finally, we say that S is (*sub-*) *multiplicative* if

$$S(xy) \leq S(x)S(y) \quad \forall\, x, y \in \mathcal{A}.$$

Of special interest are, of course, operator algebras. A well known example of a non-multiplicative norm on such an algebra is the numerical radius [10,8],

$$r(A) = \sup\,\{|(Ax, x)| : x \in H,\ (x, x) = 1\} \tag{1.1}$$

This paper is in final form and no version of it will be submitted for publication elsewhere.

defined on $\mathcal{B}(H)$, the algebra of bounded linear operators on a Hilbert space H over $\mathbb{C}$.

Another example of considerable interest is the l_p norm, $1 \leq p \leq \infty$, on the algebra $\mathbb{C}_{n \times n}$ of $n \times n$ complex matrices,

$$(1.2) \qquad ||A||_p = \left(\sum_{i,j=1}^{n} |a_{ij}|^p \right)^{1/p}, \quad A = (a_{ij}) \in \mathbb{C}_{n \times n},$$

which is multiplicative if and only if $1 \leq p \leq 2$, [13].

Given a seminorm S on an arbitrary algebra and a fixed constant $\mu > 0$, then obviously $S_\mu \equiv \mu S$ is a seminorm, too. Clearly, S_μ may or may not be multiplicative. If it is, we call μ a *multiplicativity factor* for S. That is, μ is a multiplicativity factor for S if and only if

$$(1.3) \qquad S(xy) \leq \mu S(x) S(y) \quad \forall\, x, y \in \mathcal{A}.$$

Evidently, if μ_0 is a multiplicativity factor for S, then so is any $\mu \geq \mu_0$. Thus, having a seminorm S, the question is whether S has multiplicativity factors; and if so, is there a best (least) one?

For example, it was shown in [11] and later in [5] that if H is a Hilbert space over $\mathbb{C}$ of dimension at least 2, and r is the numerical radius in (1.1), then the best multiplicativity factor for r is $\mu_{\min} = 4$, independently of dimension. Similarly, it was shown in [12] and later in [7] that the best multiplicativity factor for the l_p norm in (1.2), is

$$\mu_{\min} = \begin{cases} 1, & 1 \leq p \leq 2, \\ n^{1-2/p}, & 2 \leq p \leq \infty. \end{cases}$$

2. Multiplicativity factors and kernels

For arbitrary seminorms we have the following characterization of multiplicativity factors:

Theorem 2.1 [1]. *Let $S \neq 0$ be a seminorm on an algebra $\mathcal{A}$. Then:*

(a) *S has multiplicativity factors if and only if* Ker S *is an ideal in $\mathcal{A}$ and*

$$(2.1) \qquad \mu_{\inf} \equiv \sup\,\{S(x, y) : S(x) = S(y) = 1\} < \infty.$$

(b) *If S has multiplicativity factors and $\mu_{\inf} > 0$, then μ is a multiplicativity factor if and only if $\mu \geq \mu_{\inf}$.*

(c) *If S has multiplicativity factors and $\mu_{\inf} = 0$, then μ is a multiplicativity factor if and only if $\mu > 0$.*

Now, if S is a norm, then clearly $\mu_{\inf} > 0$, and Ker $S = \{0\}$ is an ideal in $\mathcal{A}$; hence in this case Theorem 2.1 says that μ is a multiplicativity factor for S if and only if $\mu_{\inf} < \infty$ and $\mu \geq \mu_{\inf}$. Further, a standard compactness argument implies that if $\mathcal{A}$ is finite dimensional and S is a norm, then $\mu_{\inf}$ in (2.1) is finite, so S has multiplicativity factors.

If $\mathcal{A}$ is a simple algebra, then by part (a) of Theorem 2.1, there are no multiplicative proper seminorms on $\mathcal{A}$. Since $F_{n\times n}$, the algebra of $n \times n$ matrices over F, is simple (e.g. [2]) we immediately obtain:

Theorem 2.2 [1]. *There are no multiplicative proper seminorms on $F_{n\times n}$.*

This result, proven directly by Goldberg and Straus in [5], shows that in the finite dimensional case, proper seminorms may fail to have multiplicativity factors. In fact, we can further prove:

Theorem 2.3 [1]. *Let $\mathcal{A}$ be a finite dimensional algebra, and let S be a seminorm on $\mathcal{A}$. Then S has multiplicativity factors if and only if* Ker S *is an ideal.*

In the infinite dimensional case, it was shown both in [6] and in [1] that norms as well as proper seminorms may or may not have multiplicativity factors. Such examples will be displayed at the end of Section 3.

3. A special class of seminorms

Let $\mathcal{A}$ be an algebra with a multiplicative seminorm S. Fix an element $c \in \mathcal{A}$ and define

$$S_c(x) = S(cx), \quad x \in \mathcal{A}. \tag{3.1}$$

Obviously, S_c is a seminorm on $\mathcal{A}$. In fact, it is quite evident that S_c is a norm on $\mathcal{A}$ if and only if S is a norm and c is not a zero divisor in $\mathcal{A}$. It is not hard to prove:

Theorem 3.1 [1]. *Let $\mathcal{A}$ be an algebra with a multiplicative norm S, and let S_c, $0 \neq c \in \mathcal{A}$ fixed, be the seminorm in (3.1). Then S_c has multiplicativity factors if either*

(a) *$\mathcal{A}$ has a unit and c is invertible*

or

(b) *c is in the center of $\mathcal{A}$ and $c = c^2d$ for some d in $\mathcal{A}$.*

Let us now specialize our discussion to certain function algebras where S in (3.1) is the sup norm. In this case we are able to give the following three characterizations of multiplicativity factors for S_c:

Theorem 3.2 [1]. *Let T be a set and let $\mathcal{A}$ be the algebra of bounded functions*

$$x : T \to F,$$

with the usual multiplication

$$xy(t) = x(t)y(t); \quad x, y \in \mathcal{A}; \quad t \in T.$$

For $0 \neq c \in \mathcal{A}$ define the seminorm

$$S_c(x) = \sup_{t \in T} |c(t)x(t)|. \tag{3.2}$$

Then:

(a) *S_c has multiplicativity factors if and only if*

$$\varepsilon \equiv \inf \{|c(t)| : t \in T,\ c(t) \neq 0\} > 0. \tag{3.3}$$

(b) *If S_c has multiplicativity factors, then μ is such a factor if and only if $\mu \geq \varepsilon^{-1}$.*

Theorem 3.3 [1]. *Let T be a topological space and let $\mathcal{A}$ be the algebra of bounded, continuous functions*

$$x : T \to F.$$

For $0 \neq c \in \mathcal{A}$ let S_c be the seminorm in (3.2). Then conclusions (a) and (b) of Theorem 3.2 hold.

Theorem 3.4 [1]. *Let T, $\mathcal{A}$, c and S_c be as in Theorem 3.3. Suppose T is connected. Then*

(a) *The following are equivalent:*

(i) S_c *has multiplicativity factors.*

(ii) $\delta \equiv \inf\{|c(t)| : t \in T\} > 0$.

(iii) c *is invertible.*

(b) *If* S_c *has multiplicativity factors then* S_c *is a norm in* $\mathcal{A}$, *and* μ *is a multiplicativity factor for* S_c *if and only if* $\mu \geq \delta^{-1}$.

For example, consider l^∞, the algebra of bounded sequences $x = \{\xi_j\}_{j=1}^{\infty}$ over F, with the usual Hadamard multiplication

$$xy = \{\xi_j \eta_j\}, \quad x = \{\xi_j\}, y = \{\eta_j\} \in l^\infty.$$

Fix an element $c = \{\gamma_j\} \in l^\infty$, $c \neq 0$, and define the seminorm

$$S_c(x) = \sup_j |\gamma_j \xi_j|, \quad x \in l^\infty.$$

Obviously, S_c is a norm on l^∞ if and only if

$$\gamma_j \neq 0, \quad j = 1, 2, 3, \ldots.$$

Otherwise S_c is a proper seminorm.

By Theorem 3.2 (here, $T = \mathbb{Z}^+ = \{1, 2, 3, \ldots\}$), S_c has multiplicativity factors if and only if

$$\varepsilon \equiv \inf_{\gamma_j \neq 0} |\gamma_j| > 0.$$

Thus, the four simple selections

$$\gamma_j = 1, \quad j = 1, 2, 3, \ldots,$$

$$\gamma_j = j^{-1}, \quad j = 1, 2, 3, \ldots,$$

$$\gamma_1 = 0, \quad \gamma_j = 1, \quad j = 2, 3, 4, \ldots,$$

$$\gamma_1 = 0, \quad \gamma_j = j^{-1}, \quad j = 2, 3, 4, \ldots$$

show, as indicated in the introduction, that in the infinite dimensional case, norms and proper seminorms may or may not have multiplicativity factors.

4. Quasimonotonic seminorms on $\mathbb{C}^n$

In the remainder of this note, let

$$S : \mathbb{C}^n \to \mathbb{R}, \quad n \geq 2,$$

be a seminorm on the space of complex n-tuples and let us study the function S^+ defined by

$$S^+(x) = S(x^+), \quad x \in \mathbb{C}^n,$$

where $x^+ = (|\xi_j|)$ is the vector obtained by taking the absolute values of the components of $x = (\xi_j)$.

Part of the motivation for studying S^+ lies in the fact that it is often considerably easier to compute the values of S^+ than those of S. In view of the results of [9], this is so, for example, in the case of the numerical radius which takes on $\mathbb{C}_{n\times n}$ the form

$$r(A) = \max\{|x^*Ax| : x \in \mathbb{C}^n,\ x^*x = 1\}. \tag{4.1}$$

Our purpose is to study the subadditivity and submultiplicativity properties of S^+ (when multiplication is defined in $\mathbb{C}^n$). This will be done under the assumption that S is *quasimonotonic*, i.e.,

$$x, y \in \mathbb{R}^n \text{ with } 0 \leq x \leq y \text{ implies } S(x) \leq S(y),$$

where the inequalities $0 \leq x \leq y$ are construed componentwise.

With this definition of quasimonotonicity, we can prove:

Theorem 4.1 [4]. *Let S be a seminorm (norm) on $\mathbb{C}^n$. Then S^+ is a seminorm (norm) if and only if S is quasimonotonic.*

For instance, it was shown in [3] and [4] that the numerical radius r in (4.1) is quasimonotonic on $\mathbb{C}_{n\times n}$. Since r is a norm, Theorem 4.1 implies that r^+ is a norm, too.

Now assume that $\mathbb{C}^n$ has been given a structure of an algebra over $\mathbb{C}$. For example, take $\mathbb{C}^n$ with the Hadamard multiplication

$$xy = (\xi_j \eta_j), \quad x = (\xi_j), y = (\eta_j) \in \mathbb{C}^n.$$

Another example is $\mathbb{C}_{n\times n}$ with the usual matrix multiplication. With this assumption we can state the following result that associates multiplicativity factors of S with those of S^+:

Theorem 4.2 [4]. *Let S be a quasimonotonic seminorm on $\mathbb{C}^n$ with a multiplicativity factor μ, and suppose that*

$$(xy)^+ \leq x^+y^+ \quad \forall\, x, y \in \mathbb{C}^n. \tag{4.2}$$

Then μ is a multiplicativity factor for S^+, and the least upper bounds for the multiplicativity factors of S and S^+ satisfy

$$\mu_{\inf}(S) \geq \mu_{\inf}(S^+).$$

We emphasize that (4.2) holds for all common multiplication rules on $\mathbb{C}^n$, including Hadamard's multiplication for which we have equality, and the usual matrix multiplication for which

$$(AB)^+ \leq A^+B^+ \quad \forall\, A, B \in \mathbb{C}_{n\times n}.$$

Since the numerical radius r in (4.1) is quasimonotonic on $\mathbb{C}_{n\times n}$ and its best multiplicativity factor is $\mu_{\min}(r) = 4$, Theorem 4.2, together with a simple calculation [4], imply that the best multiplicativity factor for r^+ is $\mu_{\min}(r^+) = 4$.

This example might lead us to conjecture that if S is a quasimonotonic norm on $\mathbb{C}^n$ then the best (least) multiplicativity factors for S and S^+ coincide. As shown in [4], this is quite false, and in fact the ratio

$$\mu_{\min}(S^+)/\mu_{\min}(S)$$

can be arbitrarily small.

References

1. R. Arens and M. Goldberg, *Multiplicativity factors for seminorms.* J. Math. Anal. Appl. **146** (1990), 169–181.

2. G. Birkhoff and S. MacLane, *A survey of Modern Algebra.* Macmillan, New York, 1977.

3. M. Goldberg, *A note on monotonic and semi-monotonic matrix functions.* Linear and Multilinear Algebra **24** (1989), 223–226.

4. M. Goldberg, *Quasimonotonic functions on $\mathbb{C}^n$ and the mapping $f \to f^+$.* Linear and Multilinear Algebra **27** (1990), 63–71.

5. M. Goldberg and E.G. Straus, *Norm properties of C-numerical radii.* Linear Algebra Appl. **24** (1979), 113–131.

6. M. Goldberg and E.G. Straus, *Operator norms, multiplicativity factors, and C-numerical radii.* Linear Algebra Appl. **43** (1982), 137–159.

7. M. Goldberg and E.G. Straus, *Multiplicativity factors of l_p norms for matrices.* Linear Algebra Appl. **52/53** (1983), 351–360.

8. M. Goldberg and E. Tadmor, *On the numerical radius and its applications.* Linear Algebra Appl. **42** (1982), 263–284.

9. M. Goldberg, E. Tadmor and G. Zwas, *Numerical radius of positive matrices.* Linear Algebra Appl. **12** (1975), 209–214.

10. P.R. Halmos, *A Hilbert Space Problem Book.* Van Nostrand, New York, 1967.

11. J.A.R. Holbrook, *Multiplicative properties of the numerical radius in operator theory.* J. Reine Angew. Math. **237** (1969), 166–174.

12. J.F. Maitre, *Norme composée et norme associée generalisée d'une matrice.* Numer. Math. **10** (1967), 132–141.

13. A.M. Ostrowski, *Über Normen von Matrizen.* Math. Z. **63** (1955), 2–18.

Moshe Goldberg, Department of Mathematics, Technion-Israel Institute of Technology, 32000 Haifa, Israel

International Series of Numerical Mathematics, Vol. 103, © 1992 Birkhäuser Verlag Basel

Some remarks on the Grothendieck inequality

Hermann König

Abstract

The exact value of the best constant in the Grothendieck inequality is still unknown. We sketch a method of proof yielding the best value known.

1. Introduction

The following inequality of Grothendieck [1] has important applications in the geometry of Banach spaces and, in particular, the theory of p-summing operators, cf. [5].

Theorem. *For any $n \in \mathbb{N}$ there is $c(n) > 1$ such that for any finite matrix $(a_{jk})_{j,k=1}^m$ with*

$$\left| \sum_{j,k=1}^m a_{jk} t_j s_k \right| \leq \sup_{1 \leq j,k \leq m} |t_j| \, |s_k|$$

and any sequence of vectors $(x_j)_{j=1}^m$, $(y_k)_{k=1}^m$ in an n-dimensional Hilbert space H_n one has

$$\left| \sum_{j,k=1}^m a_{jk} \langle x_j, y_k \rangle \right| \leq c(n) \sup_{1 \leq j,k \leq m} ||x_j|| \, ||y_k||. \tag{1}$$

Denoting the best constant $c(n)$ in the real or complex case by $K_G^{\mathbb{R}}(n)$ or $K_G^{\mathbb{C}}(n)$, respectively, one actually has $K_G^{\mathbb{K}} < K_G^{\mathbb{R}}(n)$. The best known values for the Grothendieck constants $K_G^{\mathbb{K}} := \sup_{n \in \mathbb{N}} K_G^{\mathbb{K}}(n)$, $\mathbb{K} \in \{\mathbb{R}, \mathbb{C}\}$, are due to Krivine and Haagerup

$$K_G^{\mathbb{R}} \leq \frac{\pi}{2 \ln (1 + \sqrt{2})} \approx 1.782 \qquad [4],$$

$$K_G^{\mathbb{C}} \leq \frac{8}{\pi (k_0 + 1)} \approx 1.405 \qquad [2],$$

where k_0 is the solution of a certain equation, see below.

This paper is in final form and no version of it will be submitted for publication elsewhere.

2. Sketch of the Proof

We now sketch a proof of the theorem yielding the best known values for $K_G(n)$ (except for $K_G^{\mathbb{R}}(2) = \sqrt{2}$ [4]). The resulting estimates for $K_G(n)$ will be stated at the end. We follow the method of [3].

Step 1. Let m denote the normalized Haar measure (Euclidean surface measure) on $S^{n-1}(\mathbb{K})$; in the complex case $S^{n-1}(\mathbb{C})$ is identified with $S^{2n-1}(\mathbb{R})$ for integration purposes. Averaging over the sphere shows for $x, y \in S^{n-1}(\mathbb{K})$

$$\int_{S^{n-1}(\mathbb{K})} \operatorname{sgn} \langle x, u\rangle \operatorname{sgn} \langle \bar{u}, y\rangle \, dm(u) = \Phi(\langle x, y\rangle), \tag{2}$$

where $\operatorname{sgn} z = z/|z|$ for $z \neq 0$ and $\operatorname{sgn} 0 = 0$, $\Phi(z) = \operatorname{sgn} z \cdot \varphi(|z|)$ with $\varphi : [0,1] \to \mathbb{R}$ given by

$$\varphi(s) = \left\{ \begin{array}{ll} \dfrac{2}{\pi} \sin^{-1} s, & \mathbb{K} = \mathbb{R} \\ s \displaystyle\int_0^{\pi/2} \frac{\cos^2 \theta}{\sqrt{1 - s^2 \sin^2 \theta \, d\theta}}, & \mathbb{K} = \mathbb{C} \end{array} \right\},$$

see [1], [2], [3]. Note that Φ does not depend on $n \in \mathbb{N}$; the averaging is essentially a two-dimensional problem. One easily checks that $\varphi : [0,1] \to [0,1]$ is bijective and increasing. Thus, letting $D = \{z \in \mathbb{K} | \ |z| \leq 1\}$, we have that $\Phi : D \to D$ is bijective and $\Phi^{-1}(z) = \operatorname{sgn} z\varphi^{-1}(|z|)$. To prove (1), it suffices to take $(x_j)_1^m, (y_k)_1^m \subset H_n$ with $||x_j|| = ||y_k|| = 1$, i.e., $x_j, x_k \in S^{n-1}(\mathbb{K})$. Take $t_j = \operatorname{sgn} \langle x_j, u\rangle$, $s_k = \operatorname{sgn} \langle \bar{u}, y_k\rangle$. By assumption on (a_{jk}), averaging using (2) yields

$$\left| \sum_{j,k=1}^m a_{jk} \Phi(\langle x_j, y_k\rangle) \right| \leq 1. \tag{3}$$

From (3), we will have to deduce that

$$\left| \sum_{j,k=1}^m a_{jk} \langle x_j, y_k\rangle \right| \leq c(n). \tag{4}$$

Step 2. For any $n \in \mathbb{N}$ we will find a Hilbert space H and $0 < c_n < 1$ and maps $f, g : S^{n-1} \to S(H) = \{x \in H | \ ||x|| = 1\}$ such that for all $x, y \in S^{n-1}$

$$\langle f(x), g(y)\rangle_H = \Phi^{-1}(c_n \langle x, y\rangle). \tag{5}$$

Having shown (5) and given $(x_j)_1^m, (y_k)_1^m \subseteq B(H_n) = S^{n-1}$, one applies (3) to the elements $f(x_j)$, $g(y_k)$ in the sphere of (a finite-dimensional subspace of) H to find

$$c_n \left| \sum_{j,k=1}^m a_{jk} \langle x_j, y_k\rangle \right| = \left| \sum_{j,k=1}^m a_{jk} \Phi(\langle f(x_j), g(y_k)\rangle) \right| \leq 1,$$

i.e., $K_G(n) \leq 1/c_n$. This will be the desired estimate. Thus it suffices to construct f and g satisfying (5).

Step 3. The addition formula of spherical harmonics states that for $n \in \mathbb{N}$ and all $k \in \mathbb{N}_0$ there are polynomials $p_k = p_k^{(n)}$ of degree k with $p_k(1) = 1$ and orthonormal (spherical) functions $(Y_{kl})_{l=1}^{N(n,k)} \leq L_2(S^{n-1}, m)$, $Y_{kl} : S^{n-1}(\mathbb{K}) \to \mathbb{K}$, such that for all $x, y \in S^{n-1}(\mathbb{K})$

$$N(n,k)p_k(\langle x,y\rangle) = \sum_{l=1}^{N(n,k)} Y_{kl}(x)\overline{Y_{kl}(y)}, \tag{6}$$

cf. [6]. Here $N(n,k)$ are certain positive integers. In fact, for odd orders, one has $p_{2k+1}(z) = z \cdot q_k(|z|)^2)$ where q_k is a polynomial of degree k, namely

$$q_k(t) := \left\{ \begin{array}{ll} P_k^{(\frac{n-3}{2},0)}(2t-1), & \mathbb{K} = \mathbb{R} \\ P_k^{(n-2,1)}(2t-1), & \mathbb{K} = \mathbb{C} \end{array} \right\}, \tag{7}$$

see [6], [3]. The $P_k^{\alpha,\beta}$ are the Jacobi polynomials, orthognal on $[-1,1]$ with respect to $(1-t)^\alpha(1+t)^\beta dt$, normalized by $P_k^{\alpha,\beta}(1) = 1$. $\varphi^{-1} : [0,1] \to [0,1]$ has a power series expansion $\varphi^{-1}(t) = \sum_{k=0}^{\infty} \beta_{2k+1}t^{2k+1}$. For $\mathbb{K} = \mathbb{R}$, $\varphi^{-1}(t) = \sin \frac{\pi}{2}t$, i.e., $\beta_{2k+1} = (-1)^k \frac{(\pi/2)^k}{2k+1)!}$ and for $\mathbb{K} = \mathbb{C}$, $\beta_1 = 4/\pi$, $\beta_3 = -8/\pi^3$, $\beta_5 = 0$ and all further $\beta_{2k+1} < 0$, see [2]. Take $0 < c < 1$ and expand $\varphi^{-1}(ct) = \sum_{k=0}^{\infty} \alpha_{2k+1}(c)p_{2k+1}(t)$, $0 \leq t \leq 1$. Then also $\Phi^{-1}(cz) = \sum_{k=0}^{\infty} \alpha_{2k+1}(c)p_{2k+1}(z)$, $|z| \leq 1$. We take H as the l_2-sum of the finite-dimensional $l_2^{N(n,2k+1)}$-spaces, $k \in \mathbb{N}_0$, fixing $n \in \mathbb{N}$.

Define $f, g : S(l_2^n) \to H$ by

$$f(x) := \left(\left(\operatorname{sgn} \alpha_{2k+1}(c) \sqrt{\frac{|\alpha_{2k+1}(c)|}{N(n,2k+1)}} Y_{2k+1,l}(x) \right)_{l=1}^{N(n,2k+1)} \right)_{k=0}^{\infty},$$

$$g(x) := \left(\left(\sqrt{\frac{|\alpha_{2k+1}(c)|}{N(n,2k+1)}} Y_{2k+1,l}(x) \right)_{l=1}^{N(n,2k+1)} \right)_{k=0}^{\infty}.$$

Then, using (6),

$$\langle f(x), g(y)\rangle_H = \sum_{k=0}^{\infty} \frac{\alpha_{2k+1}(c)}{N(n,2k+1)} \sum_{l=1}^{N(n,2k+1)} Y_{2k+1}(x)\overline{Y_{2k+1}(y)}$$

$$= \sum_{k=0}^{\infty} \alpha_{2k+1}(c) p_{2k+1}(\langle x, y\rangle) = \Phi^{-1}(c\langle x, y\rangle).$$

Similarly, with $p_{2k+1}(\langle x, x\rangle) = p_{2k+1}(1) = 1$,

$$\|f(x)\|^2 = \|g(x)\|^2 = \sum_{k=0}^{\infty} |\alpha_{2k+1}(c)|.$$

Thus f and g will map into the sphere of H, i.e., (5) is satisfied, provided $c = c_n > 0$ is determined (uniquely, in fact) by $\sum_{k=0}^{\infty} |\alpha_{2k+1}(c_n)| = 1$. Note that the p_k's and thus the α_{2k+1}'s actually depend on $n \in \mathbb{N}$. Then $K_G(n) \leq 1/c_n$.

Step 4. We evaluate the condition $\sum_{k=0}^{\infty} |\alpha_{2k+1}(c)| = 1$. The sign character of these coefficients is, in fact, different in the real and complex case. In the real case,

$$\varphi^{-1}(c_n t) = \sin c_n \frac{\pi}{2} t = \sum_{k=0}^{\infty} \alpha_{2k+1}(c_n) p_{2k+1}(t),$$

where $\operatorname{sgn} \alpha_{2k+1}(c_n) = (-1)^k$ and

$$(8_{\mathbb{R}}) \qquad 1 = \sum_{k=0}^{\infty} (-1)^k \alpha_{2k+1}(c) = \sum_{k=0}^{\infty} \frac{A_k^{(n)}}{(2k+1)!} \left(\frac{\pi}{2}\right)^{2k+1} c_n^{2k+1},$$

$$A_k^{(n)} := \frac{(n-4)(n-8)\cdots(n-4k)}{(n+2)(n+6)\cdots(n+4k-2)},$$

see [4]. For $n \to \infty$, $A_k^{(n)} \to 1$ ($k \in \mathbb{N}$) and $(8_{\mathbb{R}})$ implies $K_G^{\mathbb{R}} \leq 1/c$, $\sinh \pi/2c = 1$. In the complex case, expand $t^{2j+1} = \sum_{k=0}^{j} \gamma_{jk} p_{2k+1}(t)$. Calculation shows that all γ_{jk} are ≥ 0. We find

$$\varphi^{-1}(c_n t) = \sum_{j=0}^{\infty} \beta_{2j+1} c_n^{2j+1} t^{2j+1} = \sum_{k=0}^{\infty} \left(\sum_{j=k}^{\infty} \gamma_{jk} c_n^{2j+1} \beta_{2j+1} \right) p_{2k+1}(t),$$

i.e., $\alpha_{2k+1}(c_n) = \sum_{j=k}^{\infty} \gamma_{jk} c_n^{2j+1} \beta_{2j+1}$ with the coefficients β_{2j+1} of Step 3 (calculated in [2]). Since $\beta_1 = 4\pi$ and all other $\beta_{2k+1} \leq 0$ for $k \in \mathbb{N}$, we find the same for α_{2k+1}:

$$\alpha_1(c_n) > 0, \quad \alpha_{2k+1}(c_n) \leq 0, \quad k \in \mathbb{N}.$$

Hence

$$\begin{aligned} 1 &= \sum_{k=0}^{\infty} |\alpha_{2k+1}(c_n)| = 2\alpha_1(c_n) - \sum_{k=0}^{\infty} \alpha_{2k+1}(c_n) \\ &= 2\alpha_1(c_n) - \varphi^{-1}(c_n). \end{aligned}$$

With $\alpha_1(c_n) = \sum_{j=0}^{\infty} \gamma_{j0}\beta_{2j+1}c_n^{2j+1}$, $\gamma_{j0} = \dfrac{n!\,(j+1)!}{(n+1)!}$ and $A_j^{(n)} := 2\dfrac{n!\,(j+1)!}{(n+j)!} - 1$ this means

$$1 = \sum_{j=0}^{\infty} A_j^{(n)} \beta_{2j+1} c_n^{2j+1}. \tag{$8_{\mathbb{C}}$}$$

For $n \to \infty$, $A_j^{(n)} \to \left\{\begin{matrix} 1 & j = 0 \\ -1 & j > 0 \end{matrix}\right\}$ and $(8_{\mathbb{C}})$ means

$$1 = 2\beta_1 c - \varphi^{-1}(c).$$

Proposition [4], [2], [3]. *Determine c_n by the equations $(8_{\mathbb{R}})$ and $(8_{\mathbb{C}})$, respectively. Then*

$$K_G(n) \leq \frac{1}{c_n}.$$

For $n \to \infty$ we get

$$K_G \leq \begin{cases} \dfrac{\pi}{2\ln(1+\sqrt{2})}, & \mathbb{K} = \mathbb{R}, \\[2ex] \dfrac{8}{\pi(k_0+1)},\ \varphi(k_0) = \dfrac{\pi}{8}(1+k_0), & \mathbb{K} = \mathbb{C}. \end{cases}$$

There is exactly one such $c_n > 0$. Numerically, (8) can be solved to any degree of accuracy. One finds, e.g.,

n	$K_G^{\mathbb{R}}(n) \leq$	$K_G^{\mathbb{C}}(n) \leq$
2	$\sqrt{2}^*$	1.216
3	1.517	1.275
4	$\pi/2$	1.305
∞	1.782	1.405

* This particular estimate requires a separate proof, see [4].

References

1. A. Grothendieck, *Resumé de la théorie metrique des produits tensoriels topologiques.* Bol. Soc. Mat. Sao Paolo **8** (1956), 1–79.

2. U. Haagerup, *A new upper bound for the complex Grothendieck constant.* Israel J. Math. **60** (1987), 199–224.

3. H. König, *On the complex Grothendieck constant in the n-dimensional case.* Proc. Conf. Banach spaces (Strobl, 1989), Cambridge Univ. Press, 1991.

4. J.L. Krivine, *Constantes de Grothendieck et fonctions de type positif sur les sphères.* Adv. in Math. **31** (1979), 16–30.

5. J. Lindenstrauss and L. Tzafriri, *Classical Banach spaces II.* Springer, 1979.

6. C. Müller, *Spherical Harmonics.* Lect. Notes in Math. **17**, Springer, 1966.

Hermann König, Mathematisches Seminar, Universität Kiel, Olshausenstr. 40, D-2300 Kiel, Germany

Functional Equations and Inequalities

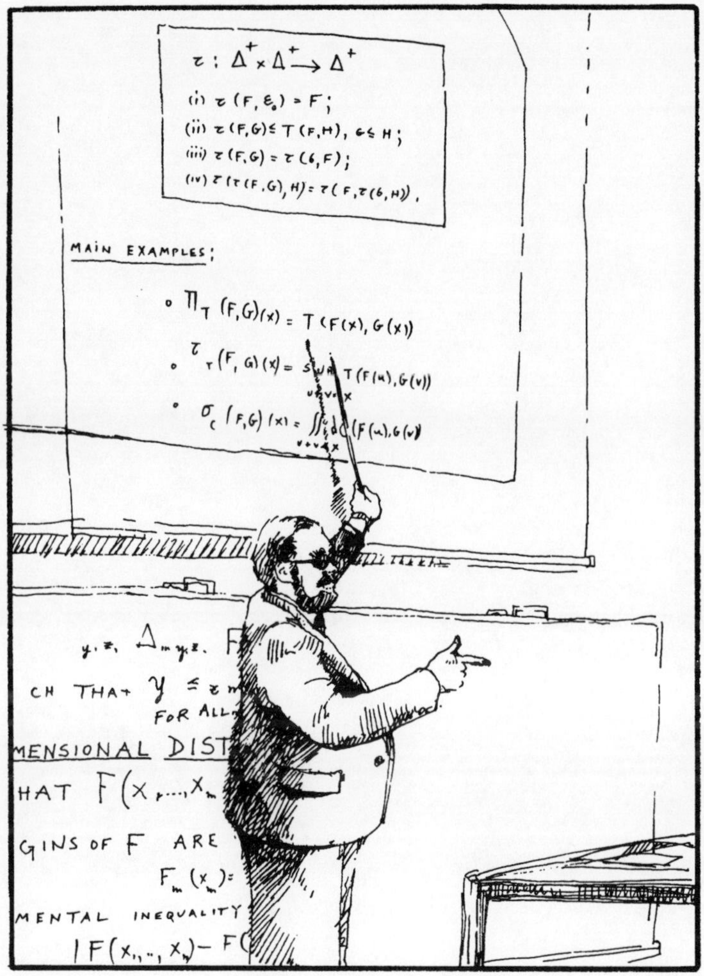

Functional inequalities

International Series of Numerical Mathematics, Vol. 103, © 1992 Birkhäuser Verlag Basel

Solutions of a functional equation, convex of higher order

János Aczél and Marek Kuczma

Abstract

We determine the general solutions of $f(kx) = k^\gamma f(x)$ ($k \in]0, \infty[\setminus\{1\}$, $\gamma \in \mathbb{R}$ are constants, $x \in]0, \infty[$ variable), in particular those which are convex of order γ.

It is known (mathematical folklore) that the restriction ($y = x$)

(1) $$f(2x) = 2f(x)$$

($x \in [0, \infty[$, say) of the Cauchy equation

$$f(x + y) = f(x) + f(y)$$

behaves very differently from the unrestricted Cauchy equation: Equation (1) has general, continuous, C^n, C^∞ or analytic solutions *depending on an arbitrary function* of the same regularity class. Continuity at 0 is almost automatic, but differentiability at 0 (from the right) or convexity guarantees the existence of a c such that $f(x) = cx$ [1].

P. Kahlig [2] found the somwhat more general equation

(2) $$f(kx) = k^\gamma f(x)$$

useful for applications to meteorology. Here γ is a real, $k \neq 1$ a positive constant. Without loss of generality *we can suppose* $0 < k < 1$. We may assume (2) to hold for all positive (or nonnegative) x but neighbourhoods of 0 (with 0 included or excluded) or of ∞ would do. Considering only negative x is equivalent to taking only positive ones and considering both positive and negative variables just splits the solution into two independent branches (with slight adjustment in case of differentiability at 0).

This paper is in final form and no version of it will be submitted for publication elsewhere.

Theorem 1. *The general solution* $f :]0,\infty[\to \mathbb{R}$ *of* (2) *on* $]0,\infty[$ *is given by*

$$f(x) = x^{\gamma} p(\log x/\log k) \quad (x \in]0,\infty[), \tag{3}$$

where $p : \mathbb{R} \to \mathbb{R}$ *is an arbitrary periodic function of period 1. The solution* f *is* C^n $(0 \le n \le \infty)$ *or analytic on* $]0,\infty[$ *if and only if* p *belongs to the same regularity class on* $\mathbb{R}$.

Proof. Define

$$p(t) = k^{-\gamma t} f(k^t).$$

Theorem 2. *Every solution of* (2) *on* $]0,\infty[$ *is of the form*

$$f(x) = cx^{\gamma} \quad (x \in]0,\infty[) \tag{4}$$

if and only if the following finite limit exists:

$$c := \lim_{x\to 0^+} (f(x)/x^{\gamma}). \tag{5}$$

Proof. "Only if" is obvious. For "if", define

$$F(x) = f(x)/x^{\gamma} \quad (x \in]0,\infty[). \tag{6}$$

By (2) and (5)

$$F(x) = F(kx) \tag{7}$$

and

$$c = \lim_{x\to 0^+} F(x),$$

respectively. Now, for an *arbitrary* $x \in]0,\infty[$,

$$F(x) = F(kx) = F(k^2 x) = \cdots = F(k^n x).$$

Since $0 < k < 1$, we get when $n \to \infty$

$$F(x) = c$$

which, with (6), proves Theorem 2.

Proposition 3. *If there exists a* $d > 0$ *such that* F, *defined by (6), is monotonic on* $]0,d[$ *(or on* $]d,\infty[$*) then there exists a* c *such that*

$$f(x) = cx^{\gamma} \quad (x \in]0,\infty[)$$

holds.

Proof. From (2) and (6) we again have (7). Take an arbitrary $x_0 \in]0, d[$ (or $\in]d, \infty[$) and define this time $c = F(x_0)$. Then

$$c = F(x_0) = F(kx_0) = F(k^2x_0) = \cdots \quad (x_0, kx_0, k^2x_0, \cdots \in]0, d[)$$

or

$$c = F(x_0) = F(x_0/k) = F(x_0/k^2) = \cdots \quad (x_0, x_0/k, x_0/k^2, \cdots \in]d, \infty[),$$

respectively. But F is *monotonic*, so it is constant on $]0, x_0[$ (or on $]x_0, \infty[$) and so, by (7) or its consequence $F(x) = F(x/k)$, everywhere. Thus again

$$F(x) = c, \quad f(x) = cx^\gamma \quad (x \in]0, \infty[,$$

as asserted.

Now, with γ *positive integer*, we come to the *solutions convex* (or *concave*) *of order* γ. We know (see e.g. [3]) that *functions convex (concave) of order* γ on an open interval *are* $C^{\gamma-1}$ *and* $f^{(j)}$ *is convex (concave) of order* $\gamma - j$ $(j = 1, 2, \ldots, \gamma - 1)$ there. In particular, $f^{(\gamma-1)}$ *is convex (concave).* We have

Lemma 4. *If* ϕ *is convex (concave) on an interval* $I \subseteq]0, \infty[$ *then* $x \mapsto \phi(x)/x$ *is either monotonic or unimodal (consisting of two monotonic parts) there.*

Proof. See the Note by J. Aczél and C.T. Ng in this volume.

And this leads to our final

Theorem 5. *Let* γ *be a positive integer,* $0 < k < 1$, *and* $f :]0, \infty[\to \mathbb{R}$ *a solution of*

(8) $$f(kx) = k^\gamma f(x) \quad (x \in]0, \infty[).$$

If there exists a D *such that* f *is convex (concave) of order* γ *on* $]0, D[$ *or on* $]D, \infty[$ *then there exists a* c *such that*

(4) $$f(x) = cx^\gamma \quad (x \in]0, \infty[).$$

Proof. By Lemma 4, there exists a d such that

(9) $$\phi(x) := f^{(\gamma-1)}(x)/x$$

is monotonic on $]0, d[$ (if f was convex of order γ on $]0, D[$, in this case $d \leq D$) or on $]d, \infty[$ (if f was convex of order γ on $]D, \infty[$, in that case $d \geq D$; the same holds for concavity). By differentiating (8) $(\gamma - 1)$ times we get, for $g = f^{(\gamma-1)}$,

$$g(kx) = kg(x)$$

which itself is an equation of the form (8) (with $\gamma = 1$). By (9), $\phi(x) = g(x)/x$, so ϕ corresponds for this equation to F in (6). But, as we just saw, ϕ is monotonic on $]0, d[$ or on $]d, \infty[$ so, by Proposition 3,

$$g(x) = f^{(\gamma-1)} = \tilde{c}x.$$

Therefore ,

$$f(x) = \frac{\tilde{c}}{\gamma!} x^\gamma + \sum_{j=0}^{\gamma-1} c_j x^j \tag{10}$$

and

$$\lim_{x \to 0+} f^{(j)}(x) = j!\, c_j \quad (j = 0, 1, \ldots, \gamma - 1). \tag{11}$$

If we differentiate (8), this time j times, we get

$$f^{(j)}(kx) = k^{\gamma-j} f^{(j)}(x), \quad (j = 0, 1, \ldots, \gamma - 1),$$

which is again of the form (8) with $(\gamma - j)$ in place of γ. By Theorem 1,

$$f^{(j)}(x) = x^{\gamma-j} p_j(\log x / \log k) \quad (j = 0, 1, \ldots, \gamma - 1), \tag{12}$$

where p_j is again periodic with period 1. Since $f^{(j)}$ is continuous (it is a polynomial by (10)), so is p_j, which therefore is bounded on every closed interval, in particular, say, on $[0, 1]$, and thus, by the periodicity of p_j, everywhere. So, by (12),

$$\lim_{x \to 0+} f^{(j)}(x) = 0$$

and, by (11),

$$c_j = 0 \quad (j = 0, 1, \ldots, \gamma - 1).$$

Now (10) with $c := \tilde{c}/\gamma!$ shows that indeed

$$f(x) = cx^\gamma.$$

Acknowledgement. The research of the first author was supported in part by the Natural Sciences and Engineering Research Council of Canada.

References

1. J. Aczél and J. Dhombres, *Functional equations in several variables*. Cambridge Univ. Press, Cambridge — New York, 1989, pp. 257–258.

2. P. Kahlig, *A general formula derived by dimensional analysis*. Ann. Geophys. **8** (1990), 167–170.

3. A.W. Roberts and D.E. Varberg, *Convex functions*. Academic Press, New York — London, 1973, pp. 4–5, 238.

János Aczél, Department of Pure Mathematics, University of Waterloo, Waterloo, Ontario, Canada N2L 3G1

Marek Kuczma, Institute of Mathematics, Silesian University, Bankowa 14, PL–40-007 Katowice, Poland

International Series of Numerical Mathematics, Vol. 103, © 1992 Birkhäuser Verlag Basel

On positive semidefinite strict t-norms

C. Alsina and M.S. Tomas

Abstract

We study and characterize positive semidefinite strict t-norms of orders 2 and 3.

1. Introduction

The theory of t-norms is a fundamental tool in probabilistic metric spaces (see [12]) as well as in the study of joint distribution functions with given marginals (see e.g. [13] and [5]). In this paper we will study first strict t-norms which are 2-positive semidefinite and their relations with copulas, sequences of exchangeable distribution functions and positive quadrant dependence. In fact positive semidefinite functions satisfy a less restrictive condition than that of total positivity, so our results may be placed into the framework of this theory initiated in [8] and [9] and developed extensively in the last decades in Matrices theory, Statistics, Probability and Majorization (see, e.g. [11], [2], [4] and [7]).

In the last part we will characterize positive semidefinite strict t-norms of order 3 and some interesting relations with bounded continuous semimetrics on real intervals will be obtained.

2. Preliminar notions

We begin with the following

Definition 2.1. A *strict t-norm* is a continuous two-place function T from $[0,1]\times[0,1]$ into $[0,1]$ which is associative, commutative, 1 is a unit and T is strictly increasing on $(0,1]\times(0,1]$.

This paper is in final form and no version of it will be submitted for publication elsewhere.

By the representation theorem for the solutions of the associativity equation (see [1]) we have that any strict t-norm can be represented in the form

$$T(x,y) = t^{-1}(t(x)+t(y)) \tag{1}$$

where t is a continuous strictly decreasing function from $[0,1]$ onto $[0,+\infty]$ such that $t(1)=0$.

The function t appearing in (1) is called the additive generator of T and it is unique up to a multiplicative constant. In the case that t is convex then the function T given by (1) is a copula, i.e., T satisfies the inequality

$$T(x_1,y_1)+T(x_2,y_2)-T(x_1,y_2)-T(x_2,y_1) \geq 0 \tag{2}$$

whenever $x_1 \leq x_2$ and $y_1 \leq y_2$.

Definition 2.2. If A is a subset of the real line $\mathbb{R}$, a function F from $A \times A$ into $\mathbb{R}$ is said to be *positive semidefinite of order* k if for all m, $1 \leq m \leq k$ and all $x_1 < \ldots < x_m$ in A we have

$$\begin{vmatrix} F(x_1,x_1) & \ldots & F(x_1,x_m) \\ \vdots & & \vdots \\ F(x_m,x_1) & \ldots & F(x_m,x_m) \end{vmatrix} \geq 0. \tag{3}$$

Note that (3) means that for all m, $1 \leq m \leq k$ and for all $x_1 < \ldots < x_m$ in A the corresponding symmetric matrices $(F(x_i,x_j))$ are positive semidefinite in the usual algebraic sense. In fact this notion is a particular case of the notion of that called *functions totally positive of order* k where for all m, $1 \leq m \leq k$ and for all $x_1 < \ldots < x_m$ and $y_1 < \ldots < y_m$ it is required that

$$\begin{vmatrix} F(x_1,y_1) & \ldots & F(x_1,y_m) \\ \vdots & & \vdots \\ F(x_m,y_1) & \ldots & F(x_m,y_m) \end{vmatrix} \geq 0.$$

Note that in this context a function $T(x,y)$ is a copula if and only if $\exp T(x,y)$ is totally positive of order 2.

Finally, we recall from [14] the following

Definition 2.3. A t-norm T_1 *weakly dominates* another t-norm T_2 if for all a,b,c in $[0,1]$ we have

$$T_1(a,T_2(b,c)) \geq T_2(T_1(a,b),c).$$

3. On strict t-norms positive semidefinite of order 2

For strict t-norms we have the following:

Theorem 3.1. *Let T be a strict t-norm. The following conditions are equivalent:*

(i) *T is totally positive of order 2;*

(ii) *T weakly dominates* Prod *(the t-norm $(x, y) \mapsto xy$);*

(iii) *T is positive semidefinite of order 2;*

(iv) *T is additively generated by $t(x) = f(-\ln x)$, where f is a continuous strictly increasing convex function from $\mathbb{R}^+$ onto $\mathbb{R}^+$ such that $f(0) = 0$.*

Proof. If T is totally positive of order 2, according to Definition 2.2 and the representation (1) we must have for $x_1 < x_2$ and $y_1 < y_2$

$$0 \leq \begin{vmatrix} T(x_1, y_1) & T(x_1, y_2) \\ T(x_2, y_1) & T(x_2, y_2) \end{vmatrix}, \tag{4}$$

so, in particular, in the case $x_2 = 1$, $z = \frac{y_1}{y_2}$, $y = y_2$, $x = x_1$ we obtain from (4) that T must satisfy the inequality

$$z \cdot T(x, y) \leq T(x, zy), \tag{5}$$

for all x, y, z in $[0, 1]$, i.e., Prod is weakly dominated by T. Now assume (ii). If we consider $u < v$, since $T(0, v) = 0 < u < v = T(1, v)$ and T is continuous there exists w in $[0, 1]$ such that $u = T(w, v)$.

If we apply (5) in the case $z = T(u, v)/T(v, v)$, $x = w$ and $y = T(v, v)$ we obtain

$$\frac{T(u, v)}{T(v, v)} T(w, T(v, v)) \leq T(w, T(u, v)), \tag{6}$$

and therefore by the associativity of T we have

$$\frac{T(v, v)}{T(u, v)} \geq \frac{T(w, T(v, v))}{T(w, T(u, v))} = \frac{T(T(w, v), v)}{T(u, T(w, v))} = \frac{T(u, v)}{T(u, u)},$$

or equivalently

$$0 \leq T(u, u) \cdot T(v, v) - T(u, v)^2 = \begin{vmatrix} T(u, u) & T(u, v) \\ T(u, v) & T(v, v) \end{vmatrix} \tag{7}$$

whence T is positive semidefinite of order 2. Now if (7) holds using (1) we obtain

$$t^{-1}(t(u)+t(v))^2 \leq t^{-1}(2t(u)) \cdot t^{-1}(2t(v)), \tag{8}$$

and introducing the variables $a = 2t(u)$, $b = 2t(v)$ and the function $f(x) = t(e^{-x})$ we immediately obtain from (8)

$$f^{-1}\left(\frac{a+b}{2}\right) \geq \frac{f^{-1}(a)+f^{-1}(b)}{2}, \tag{9}$$

i.e., f^{-1} is concave strictly increasing and continuous with $f^{-1}(0) = 0$ and therefore f is convex. Finally, if we have $t(x) = f(-\ln x)$ with f convex, and we consider $x_1 < x_2$ and $y_1 < y_2$ in $[0,1]$, then introducing $x = t(x_1)+t(y_1)$ and $y = t(x_2)+t(y_2)$ we have $y < t(x_1)+t(y_2) < x$ and there exists α in $[0,1]$ such that $t(x_1)+t(y_2) = \alpha x + (1-\alpha)y$ and $t(x_2)+t(y_1) = (1-\alpha)x + \alpha y$. Bearing in mind the concavity of $f^{-1}(x) = -\ln t^{-1}(x)$ we have

$$\begin{aligned} & f^{(-1)}(\alpha x + (1-\alpha)y) + f^{-1}((1-\alpha)x + \alpha y) \\ & \geq \alpha f^{-1}(x) + (1-\alpha)f^{-1}(y) + (1-\alpha)f^{-1}(x) + \alpha f^{-1}(y) \\ & = f^{-1}(x) + f^{-1}(y). \end{aligned}$$

or equivalently

$$\begin{aligned} 0 & \leq t^{-1}(x)t^{-1}(y) - t^{-1}(\alpha x + (1-\alpha)y)t^{-1}((1-\alpha)x + \alpha y) \\ & = \begin{vmatrix} t^{-1}(x) & t^{-1}(\alpha x + (1-\alpha)y) \\ t^{-1}((1-\alpha)x + \alpha y) & t^{-1}(y) \end{vmatrix} \\ & = \begin{vmatrix} T(x_1,y_1) & T(x_1,y_2) \\ T(x_2,y_1) & T(x_2,y_2) \end{vmatrix}, \end{aligned}$$

i.e., T is totally positive of order 2 and the proof of the theorem is complete.

From these results we can derive some interesting consequences.

Corollary 3.1. *Any strict t-norm T which is positive semidefinite of order 2 is a copula and $T \geq$* Prod.

Of course not all copulas are positive semidefinite of order 2. This corollary yields that if two random variables x and y have an associative copula C_{xy} which is strictly increasing and positive semidefinite of order 2 then x and y will be *positively quadrant dependent*, i.e., $H_{xy}(u,v) = C_{xy}(F_x(u), F_y(v)) \geq F_x(u) \cdot F_y(v)$ (see, e.g. [3], [6] and [10]).

Corollary 3.2. *If a strict t-norm T can be represented in the form (1) with additive generator t two-times differentiable then T is definite positive of order 2 if and only if $t'(x)x$ is increasing. In this case all generators t can be represented in the form $t(x) = \int_1^x g(t)/t\, dt$ with g an increasing function.*

Corollary 3.3. *If a strict t-norm T is positive semidefinite of order 2 and $T =$ Prod on $[0,1] \times [0,\epsilon]$ for some $\epsilon > 0$, then $T =$ Prod on $[0,1] \times [0,1]$.*

Proof. By Corollary 3.1 we known that $T(x,y) \geq x \cdot y$. On the other hand by (ii) of Theorem 3.1 we have that T weakly dominates Prod, so $T(x,y) \leq T(x,zy)/z$ and since $T =$ Prod on $[0,1] \times [0,\epsilon]$ we have $xy = \lim_{z \to 0} \frac{T(x,zy)}{z} \geq T(x,y)$, whence $T =$ Prod on $[0,1]^2$.

Now we will prove the following

Theorem 3.2. *If T is a strict t-norm representable in the form (1) and t^{-1} is completely monotone on $[0,\infty)$ then T is positive semidefinite of order 2.*

Proof. If t^{-1} is completely monotone, i.e., t^{-1} is continuous and has derivatives of all orders on $(0,\infty)$ which alternate successively in sign: $(-1)^n (t^{-1})^{(n)} \geq 0$, $n = 0,1,2,\ldots$ then by using a special form of Hölder's inequality it was proved in [5] that for all $x, y \geq 0$ and positive real m and n satisfying $\frac{1}{m} + \frac{1}{n} = 1$ we must have

$$t^{-1}(x+y) \leq [t^{-1}(x+my)]^{1/m} [t^{-1}(x)]^{1/n} \tag{10}$$

or, equivalently,

$$-\ln t^{-1}(x+y) \geq \frac{1}{m}[-\ln t^{-1}(x+my)] + \frac{1}{n}[-\ln t^{-1}(x)]. \tag{11}$$

Given any $b \geq a \geq 0$ and α in $[0,1]$ we have by (11) with $x = a$, $m = \frac{1}{\alpha}$ and $y = \alpha(b-a)$ that

$$-\ln t^{-1}\left((1-\alpha)a + \alpha b\right) \geq \alpha \left[-\ln t^{-1}(b)\right] + (1-\alpha)\left[-\ln t^{-1}(a)\right],$$

i.e., the function $f^{-1}(x) = -\ln t^{-1}(x)$ must be concave, i.e., $t = f(-\ln x)$ with f convex and by Theorem 3.1 necessarily T is positive semidefinite of order 2.

Corollary 3.4. *Any strict t-norm with additive generator t such that $t^{-1}(x)$ can be represented in the form*

$$t^{-1}(x) = \int_0^\infty e^{-xt}\, d\alpha(t),$$

where $\alpha(t)$ is bounded and nondecreasing (and the integral converges for $0 \leq x < \infty$) will be positive semidefinite of order 2.

Corollary 3.5. *A strict t-norm is positive semidefinite of order 2 if and only if for all a, b in $[0,1]$ we have*

$$T(a,b)^2 \leq T(a, T(b,b))a. \tag{12}$$

Proof. By Theorem 3.1, if T is positive semidefinite of order 2 then $f(x) = t^{-1}(e^{-x})$ is convex so (10) holds. In particular for $m = n = 2$, $t^{-1}(x) = a$ and $t^{-1}(y) = b$ we deduce (12). Conversely, if (12) holds then we obtain that (10) is valid for $x, y \geq 0$ and $m = n = 2$, i.e., $-\ln t^{-1}(x)$ is mid-point concave but since it is continuous then it is full concave and therefore T is positive semidefinite of order 2.

Corollary 3.6. *If a sequence of random variables is T-admissible respect to a strict t-norm T then T is semidefinite positive or order 2.*

Proof. Following [5] we have that a sequence of random variables $\{X_n\}$, with corresponding continuous distribution functions F_n, is *admissible* or *exchangeable under* T if the joint distribution function $F_{i_1,\ldots,i_m}$ of any m subset $\{X_{i_1}, \ldots, X_{i_m}\}$ is given by $F_{i_1,\ldots,i_m}(x_1, \ldots, x_m) = t^{-1}\left(\sum_{j=1}^{n} t(F_{ij}(x_j))\right)$ and this situation implies the complete monotonicity of t^{-1} on $[0, \infty)$. Thus in view of Corollary 3.2, T must be positive semidefinite of order 2.

4. On strict t-norms positive semidefinite of order 3

Let T be a strict t-norm representable in the form (1). If T is positive semidefinite of order 3 by Definition 2.2 it is also positive semidefinite of order 2 and consequently all results presented in Theorem 3.1 can be taken into account. Now we begin our study with the following numerical inequality:

Lemma 4.1. *For a, b, c in $[0,1]$ the inequality*

$$a^2 + b^2 + c^2 \leq 1 + 2abc \tag{13}$$

holds, if and only if,

$$\operatorname{Max}\left(ab - \sqrt{(1-a^2)(1-b^2)}, 0\right) \leq c \leq ab + \sqrt{(1-a^2)(1-b^2)}. \tag{14}$$

Proof. If (14) holds then $|c-ab| \leq [(1-a^2)(1-b^2)]^{1/2}$ and squaring the inequality we have (13). Conversely, if we assume (13) by considering the quadratic function $P(t) = t^2 - 2abt + a^2 + b^2 - 1$ it is immediate to show that $\{t \in [0,1] |\ P(t) \leq 0\} = \left[\text{Max}(ab - \sqrt{(1-a^2)(1-b^2)}, 0), ab + \sqrt{(1-a^2)(1-b^2)}\right]$, and (14) follows.

Theorem 4.1. *A strict t-norm T, with additive generator t, is positive semidefinite of order 3 if and only if there exists a continuous semimetric $d : (0,1]^2 \to \left[0, \frac{\pi}{2}\right)$ such that*

$$t^{-1}\left(\frac{t(x)+t(y)}{2}\right) = \sqrt{xy}\cos d(x,y).$$

Proof. If T is positive semidefinite of order 3 and representable in the form (1) then we have for all x,y,z in $\mathbb{R}^+$:

$$\begin{vmatrix} t^{-1}(x) & t^{-1}\left(\frac{x+y}{2}\right) & t^{-1}\left(\frac{x+z}{2}\right) \\ t^{-1}\left(\frac{x+y}{2}\right) & t^{-1}(y) & t^{-1}\left(\frac{y+z}{2}\right) \\ t^{-1}\left(\frac{x+z}{2}\right) & t^{-1}\left(\frac{y+z}{2}\right) & t^{-1}(z) \end{vmatrix} \geq 0 \tag{15}$$

If we substitute in (15) $t^{-1}(x) = u$, $t^{-1}(y) = v$, $t^{-1}(z) = w$ and we define M from $(0,1] \times (0,1]$ into $\mathbb{R}$ by means of:

$$M(a,b) = \frac{t^{-1}\left(\frac{t(a)+t(b)}{2}\right)}{\sqrt{ab}}, \tag{16}$$

then by Theorem 3.1 we have that M takes values in $(0,1]$ and (15) becomes:

$$1 + 2M(u,v)M(u,w)M(v,w) \geq M(u,v)^2 + M(u,w)^2 + M(v,w)^2, \tag{17}$$

for all u, v, w in $(0,1]$. In view of Lemma 4.1 the inequality (17) is equivalent to

$$\begin{aligned} &\text{Max}(M(u,v)M(u,w) - (1-M(u,v)^2)^{1/2}(1-M(u,w)^2)^{1/2}, 0) \\ &\leq M(v,w) \\ &\leq M(u,v)M(u,w) + (1-M(u,v)^2)^{1/2}(1-M(u,w)^2)^{1/2}. \end{aligned} \tag{18}$$

Let us introduce now the continuous function $d : (0,1]^2 \to \left[0, \frac{\pi}{2}\right)$ defined by

$$d(x,y) = \arccos M(x,y). \tag{19}$$

Since $M(x,y) = \cos d(x,y)$ we can rewritte (18) in the form

$$\begin{aligned}\text{Max}\,(\cos d(u,v)\cos d(u,w) - \sin d(u,v)\sin d(u,w), 0)\\ \le \cos d(v,w)\\ \le \cos d(u,v)\cos d(u,w) + \sin d(u,v)\sin d(u,w),\end{aligned}$$

and these inequalities yield

$$\text{Max}\,(\cos(d(u,v)+d(u,w)), 0) \le \cos d(v,w) \le \cos(d(u,w)-d(u,v)) \tag{20}$$

Since $-\frac{\pi}{2} \le -d(u,v) \le d(u,w) - d(u,v) \le d(u,w) \le \frac{\pi}{2}$, from the second inequality in (20) we obtain

$$-d(v,w) \le d(u,w) - d(u,v) \le d(v,w)$$

and therefore

$$|d(u,w) - d(u,v)| \le d(v,w) \tag{21}$$

and condition (21) is equivalent to the fact that d is a semimetric. We note that the first inequality in (20) is also equivalent to (21). The theorem is proved.

Corollary 4.1. *A strict t-norm T is positive semidefinite of order 3 if and only if there exists a continuous semimetric $d : (0,1]^2 \to [0, \frac{\pi}{2})$ such that*

$$T(x,y) = T\left(\sqrt{xy}\cos d(x,y), \sqrt{xy}\cos d(x,y)\right), \tag{22}$$

or equivalently

$$T\left(\sqrt{x}\cos d(x,1), \sqrt{y}\cos d(y,1)\right) = \sqrt{xy}\cos d(x,y).$$

In view of (22) we have a geometrical interpretation of the property of positive semidefinition of order 3: the surface $z = T(x,y)$ of the strict t-norm takes the same value on a point (x,y) than in the point of the diagonal $(\sqrt{xy}\cos d(x,y), \sqrt{xy}\cos d(x,y))$. In particular for $y = 1$ we get from (22) that

$$x = T(\sqrt{x}\cos d(x,1), \sqrt{x}\cos d(x,1)), \tag{23}$$

which in turn implies that the diagonal of T is completely determined because the function $\sqrt{x}\cos d(x,1)$ maps $[0,1]$ onto $(0,1]$.

Since by theorem 4.1 we have translated the condition of positive semidefinite of order 3 to the relation

$$(24) \qquad t^{-1}\left(\frac{t(x)+t(y)}{2}\right) = \sqrt{xy}\cos d(x,y),$$

we need to study the existence and uniqueness of solutions of (24) when d is a given continuous semimetric, $0 \le d(x,y) \le \arccos[\operatorname{Min}(x,y)/\operatorname{Max}(x,y)]^{1/2}$, and t is to be found.

Theorem 4.2. *Given a continuous semimetric* $d : (0,1]^2 \to [0, \frac{\pi}{2})$ *such that* $0 \le d(x,y) \le \arccos(\operatorname{Min}(x,y)/\operatorname{Max}(x,y))^{1/2}$ *if there exists a solution t of (24) then it is unique up to a multiplicative positive constant.*

Proof. If for the same d there would exist two functions t_1 and t_2 satisfying (24) then $(t_1 \circ t_2^{-1})(x)$ would satisfy the Jensen equation and therefore $t_1(x) = at_2(x)+b$. Since $t_1(1) = t_2(1) = 0$, $b = 0$ and $t_1 = at_2$, with $a > 0$.

We observe now the following examples:

Example 4.1. If $d \equiv 0$ is the zero semimetric then by (24) we obtain $t(x) = -\ln x$, i.e., $T = \text{Prod}$.

Example 4.2. If $d(x,y) = \frac{\pi}{2}|x-y|$ then we will show that there exists no function t generating a t-norm and satisfying (24). If such t would exist then since the quasiarithmetic mean $m(x,y) = t^{-1}\left(\frac{t(x)+t(y)}{2}\right)$ would satisfy the bisymmetry equation, in particular:

$$m(m(x,y), m(x,z)) = m(x, m(y,z)),$$

and taking $x = 1/2$, $y = 1$ we obtain a contradiction.

Example 4.3. If $d(x,y) = \arccos[\operatorname{Min}(x,y)/\operatorname{Max}(x,y)]^{1/2}$ we have a continuous semimetric on $(0,1]^2$ and (24) yields that necessarily $t^{-1}\left(\frac{t(x)+t(y)}{2}\right) = \operatorname{Min}(x,y)$, whence there is no function t solution of (24).

Example 4.4. Let us consider the mapping

$$(25) \qquad d(x,y) = \arccos\frac{\sqrt{xy}}{\frac{x+y}{2}}.$$

It is easy to show that d is a semimetric on $(0,1]^2$. Moreover any solution t of (24) has to be of the form $t(x) = k\left(\frac{1}{x} - 1\right)$ with $k > 0$. In this case we obtain that the strict t-norm $T(x,y) = \frac{xy}{x+y-xy}$ is positive semidefinite of order 3.

In relation with this Example 4.4 we can give the following characterization

Theorem 4.3. *Let $T(x,y) = t^{-1}(t(x)+t(y))$ be a strict t-norm positive semidefinite of order 3. Then the semimetric*

$$d(x,y) = \arccos\left(\frac{t^{-1}\left(\frac{t(x)+t(y)}{2}\right)}{\sqrt{xy}}\right) \tag{26}$$

satisfies the conditions $d(\lambda u, \lambda v) = d(u,v)$ and $d(1,1/2) = \frac{2\sqrt{2}}{3}$ if and only if $T(x,y) = \frac{xy}{x+y-xy}$.

Proof. If d as given by (26) is a semimetric satisfying $d(\lambda u, \lambda v) = d(u,v)$, i.e., d is homogeneous of degree 0, then we define $D(x,y) = -\ln\cos d(e^{-x}, e^{-y})$, i.e.

$$D(x,y) = f^{-1}\left(\frac{f(a)+f(b)}{2}\right) - \frac{a+b}{2}, \tag{27}$$

where $f(x) = t(e^{-x})$ is convex and $f(0) = 0$ (because of the positive semidefinition of order 2). D will be invariant under translations: $D(x+z, y+z) = D(x,y)$ and by (27) we will obtain immediately:

$$f^{-1}\left(\frac{f(x+z)+f(y+z)}{2}\right) = f^{-1}\left(\frac{f(x)+f(y)}{2}\right) + z. \tag{28}$$

It is well-known (see [1]) that the only strictly increasing convex solution f of (28) satisfying $f(0) = 0$ are $f(x) = e^{kx} - 1$, with $k > 0$, i.e., $t(x) = \frac{1}{x^k} - 1$ and $T(x,y) = \frac{xy}{(x^k+y^k-x^ky^k)^{1/k}}$. Since in addition we have fixed a particular value $d(1,1/2) = \frac{2\sqrt{2}}{3}$ then necessarily $k = 1$ and we have obtained a characterization of the functions appearing in Example 4.4.

All previous examples make clear that given the semimetric d it is not true, in general, that we can find t. It seems an interesting open question to find, whenever possible, explicit representations for t involving d.

References

1. J. Aczél, *Lectures on Functional Equations and Their Applications*. Academic Press, New York, 1966.

2. F.R. Gantmacher, *Theory of Matrices*, Vol. I, II. Chelsea, Bronx, New York, 1959.

3. W. Hoefding, *Maßstabinvariante Korrelationstheorie*. Schriften Math. Inst. Univ. Berlin. **5** (1940), 181–233.

4. S. Karlin, *Total Positivity*, vol. I. Stanford Univ. Press, Stanford, California, 1968.

5. C.H. Kimberling, *Exchangeable events and completely monotonic sequences*. Rocky Mountain J. Math. **3** (1973), 565–574.

6. E.L. Lehmann, *Some concepts of dependence*. Ann. Math. Statist. **37** (1966), 1137–1153.

7. A.W. Marshall and I. Olkin, *Inequalities: Theory of Majorization and Its Applications*. Academic Press, New York, 1979.

8. G. Pólya, *Algebraische Unterschungen über ganze Funktionen vom Geschlechte Null und Eins*. J. für Mathematik **145** (1915), 224–249.

9. G. Pólya, *Über Annäherung durch Polynome mit lauter reellen Wurzeln*. Rend. Circ. Mat. Palermo **36** (1913), 1–17.

10. J.J. Quesada, *On independence in some families of multivariate distributions*. Stochastica X, **3** (1986), 213–222.

11. I.J. Schoenberg, *On Pólya frequency functions, I. The totally positive functions and their Laplace transforms*. J. Analyse Mat. **1** (1951), 331–374.

12. B. Schweizer and A. Sklar, *Probabilistic Metric Spaces*. North Holland Elsevier, New York, 1983.

13. A. Sklar, *Fonctions de répartition à n dimensions et leur marges*. Publ. Inst. Statist. Univ. Paris. **8** (1959), 229–231.

14. R.M. Tardiff, *On a functional inequality arising in the construction of the product of several metric spaces*. Aequat. Math. **20** (1980), 51–58.

C. Alsina and M.S. Tomas, Sec. Matemàtiques i Informàtica (ETSAB), Universitat Politècnica de Catalunya, Diagonal 649, E-08028 Barcelona, Spain

International Series of Numerical Mathematics, Vol. 103, © 1992 Birkhäuser Verlag Basel

On functional inequalities stemming from stability questions

Roman Ger

Abstract

A subadditive (resp. subquadratic) functional f is called a control function for a vector-valued mapping F on a semigroup, group or Banach space into a normed linear space whenever

$$||F(x+y) - F(x) - F(y)|| \leq f(x) + f(y) - f(x+y)$$

(resp.

$$||F(x+y)+F(x-y)-2F(x)-2F(y)|| \leq 2f(x)+2f(y)-f(x+y)-f(x-y)).$$

We look for conditions in order for a map F controlled by f to be asymptotically close to an additive mapping A (resp. quadratic mapping Q) in the sense that

$$||F(x) - A(x)|| \leq c \cdot f(x) \qquad (\text{resp. } ||F(x) - Q(x)|| \leq c \cdot f(x)),$$

where c is a nonnegative constant.

1. Introduction

Themistocles M. Rassias has proved in [9] that, given two Banach spaces $(X, ||\cdot||)$, $(Y, ||\cdot||)$, an $\varepsilon > 0$, a number $p \in [0, 1)$ and a mapping $F : X \to Y$ such that

$$(1) \qquad ||F(x+y) - F(x) - F(y)|| \leq \varepsilon(||x||^p + ||y||^p)$$

for all $x, y \in X$, there exists a unique additive mapping $A : X \to Y$ such that

$$(2) \qquad ||F(x) - A(x)|| \leq \varepsilon_p ||x||^p$$

for all $x \in X$, where $\varepsilon_p := 2\varepsilon(2 - 2^p)^{-1}$.

This paper is in final form and no version of it will be submitted for publication elsewhere.

For $p = 0$ this result reduces to the classical D.H. Hyers theorem [6] concerning the stability of the linear functional equation.

Z. Gajda [2] has recently shown that Rassias' result carries over to all exponents $p \in \mathbb{R} \setminus \{1\}$ (with ε_p replaced by $-\varepsilon_p$ in (2) provided that $p > 1$). The value $p = 1$ turns out to be exceptional: there exists a function $F : \mathbb{R} \to \mathbb{R}$ such that $|F(x+y) - F(x) - F(y)| \le |x| + |y|$, for all $x, y \in \mathbb{R}$, but for any additive mapping $A : \mathbb{R} \to \mathbb{R}$ the function $\mathbb{R} \setminus \{0\} \ni x \mapsto \frac{1}{|x|} \cdot |F(x) - A(x)|$ remains unbounded (see also Gajda [2] and a recent paper of Th.M. Rassias and P. Šemrl [10] where more explicit counterexamples are given).

On the other hand, D. Yost repeatedly asked (cf. [13] and [14]) whether a homogeneous mapping F from a Banach space into another one and such that

$$||F(x+y) - F(x) - F(y)|| \le ||x|| + ||y|| - ||x+y|| \tag{3}$$

has to be linear (the latter problem has also an equivalent reformulation in terms of some geometric properties of the complex plane). Since this problem still remains open, we might ask the following "intermediate" question: Does inequality (3) of Yost imply that F is asymptotically additive in the sense that the function

$$x \mapsto \frac{1}{||x||} \cdot ||F(x) - A(x)|| \tag{4}$$

is bounded for some additive mapping A? Bearing Gajda's counterexample in mind we realize that inequality (obtained by setting $p = \varepsilon = 1$ in (1))

$$||F(x+y) - F(x) - F(y)|| \le ||x|| + ||y|| \tag{5}$$

is too "weak" for F to enforce its asymptotic additivity. Since (3) is definitely stronger than (5) one might expect to get the desired assertion (i.e. the boundedness of the map (4)) in this case. In what follows we shall try to answer this question. Inspired by the dissertation of L. Veselý and L. Zajiček [12], where functional inequalities of the type

$$\left|\left| F\left(\frac{x+y}{2}\right) - \frac{F(x)+F(y)}{2} \right|\right| \le \frac{f(x)+f(y)}{2} - f\left(\frac{x+y}{2}\right)$$

are considered with a given "control function" f, we shall finally deal with inequality

$$||F(x+y) - F(x) - F(y)|| \le f(x) + f(y) - f(x+y) \tag{3f}$$

with a subadditive control function f. Obviously, (3f) reduces to (3) provided that $f = ||\cdot||$. A "quadratic version" of inequality (3f) with a subquadratic control function f will also be considered.

2. Scalar equivalents of inequality (3f)

The following theorem admits arbitrary semigroups as possible domains of the functions considered. We shall keep the additive notation for the underlying binary law although the corresponding semigroup is not necessarily commutative.

Theorem 1. *Let $(S,+)$ be a semigroup and let $(Y, \|\cdot\|)$ be a real normed linear space. Let further $f : S \to \mathbb{R}$ be a subadditive functional. Then, for every mapping $F : S \to Y$, the following statements are equivalent:*

(i) *Inequality* (3f) *holds true for all $x, y \in S$;*

(ii) *For every bounded linear functional y^* from the conjugate space Y^* the functional $y^* \circ F + \|y^*\| \cdot f$ is subadditive;*

(iii) *For every $y^* \in Y^*$ the functional $y^* \circ F - \|y^*\| \cdot f$ is superadditive;*

(iv) *For every $y^* \in Y^*$ such that $\|y^*\| = 1$ the functional $y^* \circ F - f$ is superadditive;*

(v) *For every $y^* \in Y^*$ such that $\|y^*\| = 1$ the functional $y^* \circ F + f$ is subadditive.*

Proof. (i) implies (ii). Let $y^* \in Y^*$. Then, for every $x, y \in S$,

$$\begin{aligned} |y^*(F(x+y) - F(x) - F(y))| &\leq \|y^*\| \, \|F(x+y) - F(x) - F(y)\| \\ &\leq \|y^*\|(f(x) + f(y) - f(x+y)), \end{aligned}$$

whence

$$y^*(F(x+y)) + \|y^*\| \cdot f(x+y) \leq y^*(F(x)) + \|y^*\| \cdot f(x) + y^*(F(y)) + \|y^*\| \cdot f(y),$$

i.e., $y^* \circ F + \|y^*\| \cdot f$ is subadditive.

(ii) implies (iii). Fix a $y^* \in Y^*$. Then $-y^* \in Y^*$ whence, by (ii), the functional $-y^* \circ F + \|y^*\| \cdot f$ is subadditive. Therefore, the functional $y^* \circ F - \|y^*\| \cdot f$ is superadditive.

(iii) implies (iv). Trivial.

(iv) implies (v). Fix a $y^* \in Y^*$ with $\|y^*\| = 1$. Then $-y^* \in Y^*$ and $\|-y^*\| = 1$ whence, by (iv), $-y^* \circ F - f$ is superadditive. Thus, $y^* \circ F + f$ is subadditive.

(v) implies (i). Let $y^* \in Y^*$ be such that $\|y^*\| = 1$. Then, for any $x, y \in S$, one has

$$y^*(F(x+y)) + f(x+y) \leq y^*(F(x)) + f(x) + y^*(F(y)) + f(y),$$

whence

$$y^*(F(x+y)-F(x)-F(y)) \leq f(x)+f(y)-f(x+y).$$

Replacing y^* by $-y^*$ one gets similarly that

$$-y^*(F(x+y)-F(x)-F(y)) \leq f(x)+f(y)-f(x+y).$$

for all $x, y \in S$. Consequently, the estimation

$$|y^*(F(x+y)-F(x)-F(y))| \leq f(x)+f(y)-f(x+y), \quad x, y \in S,$$

holds true for any functional y^* from the unit sphere in Y^*. Thus, for all $x, y \in S$,

$$\begin{aligned} ||F(x+y)-F(x)-F(y)|| &= \sup_{||y^*||=1} |y^*(F(x+y)-F(x)-F(y))| \\ &\leq f(x)+f(y)-f(x+y), \end{aligned}$$

i.e., inequality (i) holds true.

3. Some asymptotic additivity results

For the case where the target space is finite-dimensional we have the following result.

Theorem 2. *Let $(S,+)$ be an Abelian semigroup and let $(Y, ||\cdot||)$ be an n-dimensional real normed linear space. Let further $f: S \to \mathbb{R}$ be a nonnegative subadditive functional on S and let $F: S \to Y$ be a mapping such that inequality* (3f) *holds true for all $x, y \in S$. Then there exists an additive mapping $A: S \to Y$ such that*

$$(6) \qquad ||F(x)-A(x)|| \leq n \cdot f(x)$$

for all $x \in S$. The mapping A is unique provided that

$$(7) \qquad \liminf_{k\to\infty} \frac{f(kx)}{k} = 0 \quad \textit{for all } x \in S.$$

Proof. Let $\{e_1, \ldots, e_n\}$ be a basis of Y such that $||e_i|| = 1$ for all $i \in \{1, \ldots, n\}$. Let further $y_i^*: Y \to \mathbb{R}$ be a projection onto the i-th axis, i.e., $y_i^*(y_1 e_1 + \cdots + y_n e_n) := y_i$ for $(y_1, \ldots, y_n) \in \mathbb{R}^n$, $i \in \{1, \ldots, n\}$. Clearly, $y_i^* \in Y^*$ and $||y_i^*|| = 1$ for all $i \in \{1, \ldots, n\}$. Put $p_i := y_i^* \circ F + f$ and $q_i^* := y_i^* \circ F - f$, $i = 1, \ldots, n$. In view of the nonnegativity of f we have $q_i \leq p_i$ and, on account of Theorem 1 (assertions (iv) and (v)), q_i is superadditive and p_i is subadditive, $i = 1, \ldots, n$. Consequently, in view of P. Kranz's theorem from [7], there exist additive functionals $a_i : S \to \mathbb{R}$ such that

$$q_i \le a_i \le p_i \quad \text{for all } i \in \{1, \dots, n\}.$$

Hence, by means of the definitions of p_i and q_i, we obtain

$$(8) \qquad |y_i^*(F(x)) - a_i(x)| \le f(x), \quad x \in S,$$

valid for all $i \in \{1, \dots, n\}$.

Define a mapping $A : S \to Y$ by the formula:

$$A(x) := a_1(x)e_1 + \cdots + a_n(x)e_n, \quad x \in S.$$

Plainly, A yields a homomorphism of the semigroup $(S, +)$ into the additive group $(Y, +)$ of the linear space $(Y, ||\cdot||)$. On the other hand,

$$\begin{aligned} ||F(x) - A(x)|| &= \left|\left| \sum_{i=1}^{n} (y_i^*(F(x)) - a_i(x))e_i \right|\right| \\ &\le \sum_{i=1}^{n} |y_i^*(F(x)) - a_i(x)| \, ||e_i|| \le n \cdot f(x) \end{aligned}$$

for all $x \in S$, in view of (8) and the fact that the basis $\{e_1, \dots, e_n\}$ was taken from the unit sphere in $(Y, ||\cdot||)$.

To prove uniqueness (under the assumption that (7) holds true) suppose that $B : S \to Y$ is another additive mapping fulfilling the inequality

$$||F(x) - B(x)|| \le n \cdot f(x), \quad x \in S.$$

This jointly with (6) implies that

$$||A(x) - B(x)|| \le 2n \cdot f(x), \quad x \in S,$$

whence, for any positive integer k, we get

$$k||A(x) - B(x)|| = ||A(kx) - B(kx)|| \le 2n \cdot f(kx), \quad x \in S,$$

which, by means of (7), gives $A(x) = B(x)$, $x \in S$, and finishes the proof.

Remark 1. If one drops the additional assumption (7) then the uniqueness part of the assertion fails to hold. In fact, take $Y = S = \mathbb{R}$ and $F(x) := \frac{1}{2}|x|$, $x \in \mathbb{R}$. Then, for all $x, y \in \mathbb{R}$, one has

$$|F(x+y) - F(x) - F(y)| = \frac{1}{2}(|x| + |y| - |x+y|) \le |x| + |y| - |x+y|,$$

and the additive mappings $A, B : \mathbb{R} \to \mathbb{R}$ given by the formulae: $A(x) = 0$, $B(x) = \frac{1}{2}x$, $x \in \mathbb{R}$, satisfy the inequalities $\left|\frac{1}{2}|x| - A(x)\right| \le |x|$, $\left|\frac{1}{2}|x| - \frac{1}{2}x\right| \le |x|$ for all $x \in \mathbb{R}$.

The function $f(x) = \sqrt{x}$, $x \in [0, \infty)$, yields an example of a nonnegative mapping satisfying condition (5).

Remark 2. To get the uniqueness of an additive mapping A fulfilling (6) it suffices to assume (7) not necessarily for all $x \in S$ but merely for $x \in E \subset S$, where E is any set generating the semigroup S.

Theorem 3. *Let $(X, ||\cdot||)$ be a real Banach space and let $(Y, ||\cdot||)$ be a real normed linear space. Suppose that $f : X \to \mathbb{R}$ is a nonnegative subadditive functional on X and $F : X \to Y$ is a mapping such that inequality* (3f) *holds true for all $x, y \in S$. If f and F have a common continuity point (or, more generally, if the function $X \ni x \mapsto ||F(x)|| + f(x)$ is bounded above on a second category Baire set), then there exists a nonnegative constant c such that*

$$||F(x)|| \le c \cdot ||x|| + f(x), \quad x \in X. \tag{9}$$

Proof. Fix arbitrarily $y^* \in Y^*$ such that $||y^*|| = 1$. Then, by virtue of Theorem 1 (assertions (iv) and (vi)) and the nonnegativity of f, we infer that a superadditive functional $y^* \circ F - f$ is majorized by subadditive one: $y^* \circ F + f$. Therefore (cf. a "sandwich theorem" of P. Kranz [7]) there exists an additive functional $a_{y^*} : X \to \mathbb{R}$ such that

$$y^* \circ F - f \le a_{y^*} \le y^* \circ F + f,$$

whence

$$|y^*(F(x)) - a_{y^*}(x)| \le f(x) \quad \text{for all } x \in X. \tag{10}$$

Thus

$$|a_{y^*}(x)| \le |y^*(F(x))| + f(x) \le ||y^*||\, ||F(x)|| + f(x) = ||F(x)|| + f(x) \tag{11}$$

for all $x \in X$, and in view of the hypothesis assumed, there exists a second category Baire set $E \subset X$ and a constant $c(y^*) > 0$ such that $|a_{y^*}(x)| \le c(y^*)$ for all $x \in E$. Plainly, $|a_{y^*}(x)| \le 2c(y^*)$ for $x \in E + E$, and since int $(E + E) \ne \emptyset$ (see, e.g., M. Kuczma [8]) the functional a_{y^*} is continuous. Therefore, $a_{y^*} \in X^*$ for all

$y^* \in Y^*$ with $||y^*|| = 1$, and in view of (11) with the aid of the Banach-Steinhaus uniform boundedness principle we obtain the existence of a $c \geq 0$ such that

$$||a_{y^*}|| \leq c \quad \text{for all } y^* \in Y^*,\ ||y^*|| = 1.$$

Consequently, for all $y^* \in Y^*$ such that $||y^*|| = 1$, relation (10) implies that

$$|y^*(F(x))| \leq |a_{y^*}(x)| + f(x) \leq ||a_{y^*}||\,||x|| + f(x) \leq c \cdot ||x|| + f(x), \quad x \in X.$$

Now,

$$||F(x)|| = \sup_{||y^*||=1} |y^*(F(x))|| \leq c \cdot ||x|| + f(x) \quad \text{for all } x \in X,$$

which completes the proof.

Corollary 1. *Let $(X, ||\cdot||)$ be a real Banach space and let $(Y, ||\cdot||)$ be a real normed linear space. Assume that $F : X \to Y$ satisfies inequality (3) and has the property that a function $X \ni x \mapsto ||F(x)||$ is bounded above on a second category Baire set. Then F is continuous and there exists an $\alpha > 0$ such that*

$$||F(x)|| \leq \alpha ||x||, \quad x \in X, \tag{12}$$

i.e. F is asymptotically additive with the approximating additive mapping vanishing identically.

Proof. Take $f := ||\cdot||$ and put $\alpha := c + 1$, where c is a constant occurring in the assertion of Theorem 3. Then estimation (9) becomes (12). Relation (12) implies the continuity of F at zero. Fix arbitrarily an $x_0 \in X$. Then, for any $h \in X$, one gets

$$\begin{aligned} ||F(x_0 + h) - F(x_0)|| &\leq ||F(x_0 + h) - F(x_0) - F(h)|| + ||F(h)|| \\ &\leq ||x_0|| + ||h|| - ||x_0 + h|| + ||F(h)||, \end{aligned}$$

because of (3); obviously, the right hand side of the latter inequality tends to zero as h approaches zero. So, F is continuous at x_0, which was to be proved.

4. The asymptotic behaviour of quadratic mappings

It is well known (Jordan-von Neumann theorem) that the norm $||\cdot||$ in a linear space comes from an inner product if and only if the functional $f := ||\cdot||^2$ satisfies the functional equation

$$f(x + y) + f(x - y) = 2f(x) + 2f(y) \tag{13}$$

(of quadratic functionals) which not merely for this reason belongs to one of the best known and important functional equations (cf., e.g., the book of J. Aczél and J. Dhombres [1]). In what follows we are going to show that all the results of the previous sections have their counterparts for quadratic mappings.

Let $(G,+)$ be a group (not necessarily commutative). A functional $f : G \to \mathbb{R}$ is said to be subquadratic (resp. superquadratic) whenever relation (13) holds with equality sign replaced by $\le$ (resp. $\ge$) for all $x, y \in G$. The following analogue of Theorem 1 holds true.

Theorem 4. *Let $(G,+)$ be a group and let $(Y, \|\cdot\|)$ be a real normed linear space. Let further $f : G \to \mathbb{R}$ be a subquadratic functional. Then, for every mapping $F : G \to Y$, the following statements are equivalent:*

(i)$_2$ $\|F(x+y)+F(x-y)-2F(x)-2F(y)\| \le 2f(x)+2f(y)-f(x+y)-f(x-y)$ *for all $x, y \in G$;*

(ii)$_2$ *For every $y^* \in Y^*$ the functional $y^* \circ F + \|y^*\| \cdot f$ is subquadratic;*

(iii)$_2$ *For every $y^* \in Y^*$ the functional $y^* \circ F - \|y^*\| \cdot f$ is superquadratic;*

(iv)$_2$ *For every $y^* \in Y^*$ such that $\|y^*\| = 1$ the functional $y^* \circ F - f$ is superquadratic;*

(v)$_2$ *For every $y^* \in Y^*$ such that $\|y^*\| = 1$ the functional $y^* \circ F + f$ is subquadratic.*

We omit the proof since it is almost literally the same proof as that of Theorem 1.

Theorem 5. *Let $(G,+)$ be an Abelian group and let $(Y, \|\cdot\|)$ be an n-dimensional real normed linear space. Let further $f : G \to \mathbb{R}$ be a nonnegative subquadratic functional on G and let $F : G \to Y$ be a mapping such that inequality (i)$_2$ holds true. Then there exists a solution $Q : G \to Y$ of equation (13) such that*

$$\|F(x) - Q(x)\| \le n \cdot f(x) \tag{14}$$

for all $x \in G$. The mapping Q is unique provided that

$$\liminf_{k\to\infty} \frac{f(kx)}{k^2} = 0 \quad \textit{for all } x \in G. \tag{15}$$

Proof. We shall use the notation applied in the proof of Theorem 2. On account of Theorem 4 (assertions $(\mathrm{iv})_2$ and $(\mathrm{v})_2$) the functionals q_i and p_i are superquadratic and subquadratic, respectively, for $i \in \{i, \dots, n\}$. In view of the nonnegativity of f we have also $q_i \le p_i$ for all $i \in \{1, \dots, n\}$. Consequently, in view of W. Smajdor's theorem from [11] (the 2-divisibility assumption on the group $(G, +)$ occurring in [11] turns out to be inessential), there exist quadratic functionals $Q_i : G \to \mathbb{R}$ such that

$$q_i \le Q_i \le p_i \quad \text{for } i \in \{1, \dots, n\}.$$

Hence, by means of the definitions of p_i and q_i, we obtain the inequalities

$$|y_i^*(F(x)) - Q_i(x)| \le f(x), \quad x \in G, \tag{16}$$

valid for all $i \in \{1, \dots, n\}$.

Define a mapping $Q : G \to Y$ by the formula:

$$Q(x) := Q_1(x)e_1 + \dots + Q_n(x)e_n, \quad x \in G.$$

Plainly, Q yields a solution to (13). On the other hand,

$$\begin{aligned} ||F(x) - Q(x)|| &= \left\| \sum_{i=1}^{n} (y_i^*(F(x)) - Q_i(x))e_i \right\| \\ &\le \sum_{i=1}^{n} |y_i^*(F(x)) - Q_i| \cdot ||e_i|| \\ &\le n \cdot f(x), \end{aligned}$$

for all $x \in G$, in view of (16).

The proof of uniqueness (under the assumption that (15) holds true) is analogous to the corresponding part of the proof of Theorem 2. We only need to apply the property: $Q(kx) = k^2 Q(x)$, $x \in G$, $k \in \mathbb{N}$, valid for any quadratic mapping Q. This completes the proof.

Remark 3. The uniqueness fails to hold whenever assumption (15) is removed. In fact, take any function $F : \mathbb{R} \to \mathbb{R}$ such that $|F(x)| \le \frac{1}{3}$ for $x \in \mathbb{R}$ and put $f(x) := x^2 + 1$, $x \in \mathbb{R}$. Then, for all $x, y \in \mathbb{R}$,

$$2f(x) + 2f(y) - f(x+y) - f(x-y) = 2; \tag{17}$$

in particular, f is subquadratic and nonnegative. We have also

$$|F(x+y) + F(x-y) - 2F(x) - 2F(y)| \le 6\,\frac{1}{3} = 2, \quad x, y \in \mathbb{R},$$

whence, in view of (17), inequality $(\mathrm{i})_2$ holds true with $||\cdot||$ being the absolute value norm on $\mathbb{R}$. Nevertheless, there exist two different quadratic functionals, namely $Q_1 := 0$ and $Q_2(x) := x^2$, $x \in \mathbb{R}$, such that $|F(x) - Q_i(x)| \le f(x)$, $x \in \mathbb{R}$, $i \in \{1, 2\}$.

Remark 4. However, in many instances, to guarantee the uniqueness of a quadratic mapping satisfying (14), assumption (15) may considerably be weakened. For example, in the case where the domain $(G, +)$ is a divisible Abelian group (i.e., it forms a vector space over the field of rationals) it suffices to assume (15) not necessarily for all $x \in G$ but merely for $x \in \frac{1}{2}(H + H)$, where H stands for any Hamel basis over the $\mathbb{Q}$-vector space $(G, +)$ (see [3, Theorem 1]).

Remark 5. Nonnegative subquadratic functionals occur naturally in a paper of Hudzik [5] where self-mappings φ of $[0, \infty)$ are considered such that $\varphi(0) = 0$ and $\varphi \circ \sqrt{\ }$ is convex. All such φ's are necessarily subquadratic and that is what the author of [5] actually needs. One can also easily check that the even extension of any nondecreasing concave and vanishing at zero self-mapping of $[0, \infty)$ yields a nonnegative subquadratic function on $\mathbb{R}$.

To prove an analogue of Theorem 3 we need first to prove a "bilinear version" of the Banach-Steinhaus uniform boundedness principle.

Proposition 1. *Let $(E, ||\cdot||)$ be a Banach space and let $(F, ||\cdot||)$ be a normed space. Let further $\{\Phi_\alpha : \alpha \in T\}$ be a nonempty family of bilinear symmetric and continuous operators on $E \times E$ into F. If, for every $(x, y) \in E \times E$, the set $\{\Phi_\alpha(x, y) : \alpha \in T\}$ is bounded in F, then*

$$\sup_{\alpha \in T} ||\Phi_\alpha|| < \infty.$$

Proof. Fix arbitrarily a $y \in E$ and put $\Phi_{\alpha,y}(x) := \Phi_\alpha(y, x)$, $x \in X$, $\alpha \in T$. The family $\{\Phi_{\alpha,y} : \alpha \in T\}$ of bounded linear operators on E into F satisfies the assumptions of the classical Banach-Steinhaus uniform boundedness principle. Therefore, there exists a positive constant $m(y)$ such that $||\Phi_{\alpha,y}|| \le m(y)$. Consequently, for all pairs $(x, y) \in E \times E$, one has that

$$||\Phi_{\alpha,y}(x, y)|| = ||\Phi_{\alpha,y}(x)|| \le ||\Phi_{\alpha,y}|| \, ||x|| \le m(y) \cdot ||x||,$$

which, owing to the symmetry of Φ_α, implies that

$$||\Phi_\alpha(y, x)|| \le m(x) \cdot ||y||, \quad x, y \in E. \tag{18}$$

Consider the family $\mathcal{F} := \{\Phi_{\alpha,y} : \alpha \in T, \ ||y|| \leq 1\}$. In view of (18), one has $||\Phi_{\alpha,y}(x)|| \leq m(x)$, $x \in E$, for all $\Phi_{\alpha,y} \in \mathcal{F}$. Applying the Banach-Steinhaus principle once more we infer the existence of a positive constant M such that $||\Phi_{\alpha,y}|| \leq M$ for all operators $\Phi_{\alpha,y} \in \mathcal{F}$. Finally, for every $\alpha \in T$, one has

$$\begin{aligned} ||\Phi_\alpha|| &= \sup_{||y||\leq 1} \left(\sup_{||x||\leq 1} ||\Phi_\alpha(y,x)|| \right) \\ &= \sup_{||y||\leq 1} \left(\sup_{||x||\leq 1} ||\Phi_\alpha(x)|| \right) \\ &\leq \sup_{||y||\leq 1} ||\Phi_{\alpha,y}|| \leq M, \end{aligned}$$

which was to be proved.

Remark 6. Obviously, one might consider n-linear symmetric continuous forms on E^n into F as well, getting a similar uniform boundedness principle. The result is perhaps known but not having any reference we have presented the detailed proof.

Theorem 6. *Let $(X, ||\cdot||)$ be a real Banach space and let $(Y, ||\cdot||)$ be a real normed linear space. Suppose that $f : X \to \mathbb{R}$ is a nonnegative subquadratic functional on X and $F : X \to Y$ is a mapping such that inequality* $(\mathrm{i})_2$ *holds true for all $x, y \in X$. If f and F have a common continuity point, then there exists a nonnegative constant c such that*

$$||F(x)|| \leq c \cdot ||x||^2 + f(x), \quad x \in X. \tag{19}$$

Proof. For any $y^* \in Y^*$ such that $||y^*|| = 1$ Theorem 3 (assertions $(\mathrm{iv})_2$ and $(\mathrm{v})_2$), nonnegativity of f and W. Smajdor's "sandwich theorem" from [11] imply that there exists a quadratic functional $Q_{y^*} : X \to \mathbb{R}$ such that

$$y^* \circ F - f \leq Q_{y^*} \leq y^* \circ F + f,$$

whence

$$|y^*(F(x)) - Q_{y^*}(x)| \leq f(x) \quad \text{for all } x \in X. \tag{20}$$

In particular,

$$|Q_{y^*}(x)| \leq |y^*(F(x))| + f(x) \leq ||F(x)|| + f(x), \quad x \in X, \tag{21}$$

and, in view of the hypothesis assumed, Q_{y^*} is bounded on a nonempty open subset of X. Therefore (see, e.g., Hille-Phillips [4, Theorem 26.24]), Q_{y^*} is continuous and there exists a continuous bilinear symmetric functional $\Phi_{y^*} : X \times X \to \mathbb{R}$ such that

$$(22) \qquad Q_{y^*}(x) = \Phi_{y^*}(x, x), \quad x \in X,$$

([4, Theorem 26.2.6]). Since

$$\Phi_{y^*}(x, y) = \frac{1}{4}\left(Q_{y^*}(x + y) - Q_{y^*}(x - y)\right), \quad (x, y) \in X \times X$$

(cf., e.g., Aczél-Dhombres [1, p. 166]), relation (21) implies that

$$|\Phi_{y^*}(x, y)| \leq \frac{1}{4}(||F(x + y)|| + f(x + y) + ||F(x - y)|| + f(x - y)), \quad (x, y) \in X \times X,$$

for all $y^* \in Y^*$, $||y^*|| = 1$. Hence, the family $\{\Phi_{y^*} : ||y^*|| = 1\}$ satisfies all the assumptions of Proposition 1, whence

$$c := \sup\,\{||\Phi_{y^*}|| : ||y^*|| = 1\} < \infty.$$

Relation (22) gives now:

$$|Q_{y^*}(x)| = |\Phi_{y^*}(x, y)| \leq ||\Phi_{y^*}||\,||x||^2 \leq c \cdot ||x||^2, \quad x \in X,$$

for every functional $y^* \in Y^*$ such that $||y^*|| = 1$. Hence, by (20), for all $x \in X$ on obtains

$$\begin{aligned} ||F(x)|| &= \sup_{||y^*||=1} |y^*(F(x))| \\ &\leq \sup_{||y^*||=1} (|Q_{y^*}(x)| + f(x)) \leq c \cdot ||x||^2 + f(x), \end{aligned}$$

which completes the proof.

Corollary 2. *Let $(X, ||\cdot||)$ be a real Banach space and let $(Y, ||\cdot||)$ be a real normed linear space. Assume that $F : X \to Y$ satisfies the inequality*

$$||F(x + y) + F(x - y) - 2F(x) - 2F(y)|| \leq 2||x|| + 2||y|| - ||x + y|| - ||x - y||$$

for all $x, y \in X$, and has the property that a function $X \ni x \mapsto ||F(x)||$ has a point of continuity. Then there exists a constant $c \geq 0$ such that

$$||F(x)|| \leq c \cdot ||x||^2 + ||x||, \quad x \in X.$$

Proof. The functional $f := ||\cdot||$ is nonnegative and subquadratic.

Unfortunately, there exists no satisfactory analogue of Corollary 1. Indeed, one might ask whether the inequality

$$(23)\quad ||F(x+y)+F(x-y)-2F(x)-2F(y)|| \leq 2||x||^2+2||y||^2-||x+y||^2-||x-y||^2,$$

$x, y \in X$, (with, say, F continuous at a point) implies the existence of an $\alpha > 1$ such that

$$(24)\qquad ||F(x)|| \leq \alpha||x||^2, \quad x \in X.$$

However, such (at first glance more natural) question becomes trivial, in a way. Indeed, first of all, inequality (23) implies that the functional $||\cdot||^2$ has to be subquadratic. On the other hand, we have the following

Proposition 2. *Let $(X, ||\cdot||)$ be a normed space. If the functional $||\cdot||^2$ is subquadratic (superquadratic), then it is quadratic (and hence the norm comes from inner product).*

Proof. Assume that, for all $x, y \in X$, one has

$$(25)\qquad ||x+y||^2 + ||x-y||^2 \leq 2||x||^2 + 2||y||^2,$$

and fix arbitrarily $u, v \in X$. Put $x := \dfrac{u+v}{2}$ and $y := \dfrac{u-v}{2}$ in (25). Then

$$||u||^2 + ||v||^2 \leq 2\left|\left|\frac{u+v}{2}\right|\right|^2 + 2\left|\left|\frac{u-v}{2}\right|\right|^2 = \frac{||u+v||^2 + ||u-v||^2}{2} \leq ||u||^2 + ||v||^2,$$

which proves that (25) holds with equality sign. In the case where $||\cdot||^2$ is superquadratic the proof is literally the same.

Consequently, (23) forces F to be quadratic. Continuity at one point implies that there exists a symmetric continuous bilinear form $\Phi : X \times X \to Y$ such that $F(x) = \Phi(x, x)$, $x \in X$, whence (24) is satisfied with $\alpha := ||\Phi||$.

References

1. J. Aczél and J. Dhombres, *Functional equations in several variables.* Cambridge Univ. Press, Cambridge – New York – Sydney, 1989.

2. Z. Gajda, *On stability of additive mappings.* Internat. J. Math. Sci. (to appear).

3. R. Ger, *Some remarks on quadratic functionals.* Glas. Mat. Ser. III **23** (43) (1988), 315–330.

4. E. Hille and R.S. Phillips, *Functional analysis and semi-groups.* Amer. Math. Soc., Providence, R.I., 1957.

5. H. Hudzik, *Lower and upper estimations of the modulus of convexity in some Orlicz spaces.* Arch. Math. (to appear).

6. D.H. Hyers, *On the stability of the linear functional equation.* Proc. Nat. Acad. Sci. U.S.A. **27** (1941), 222–224.

7. P. Kranz, *Additive functionals on abelian semigroups.* Comment. Math. Prace Mat. **16** (1972), 239–246.

8. M. Kuczma, *An introduction to the theory of functional equations and inequalities.* Polish Scientific Publishers and Uniwersytet Śląski, Warszawa – Kraków – Katowice, 1985.

9. Th.M. Rassias, *On the stability of the linear mapping in Banach spaces.* Proc. Amer. Math. Soc. **72** (1978), 297–300.

10. Th.M. Rassias and P. Šemrl, *On the behaviour of mappings which do not satisfy Hyers-Ulam stability.* Proc. Amer. Math. Soc. (to appear).

11. W. Smajdor, *Subadditive and subquadratic set-valued functions.* Prace Nauk. Uniw. Śląsk. 889, Katowice, 1987.

12. L. Veselý and L. Zajiček, *Delta-convex mappings between Banach spaces and applications.* Dissertationes Math. (Rozprawy Mat.) 285, Polish Scientific Publishers, Warszawa, 1989.

13. D. Yost, Talk at the 17-th Winter School on Abstract Analysis (Section of Analysis). January, 1989, Šrni, Czechoslovakia.

14. D. Yost, Talk at the 18-th Winter School on Abstract Analysis (Section of Analysis). January, 1990, Šrni, Czechoslovakia.

Roman Ger, Institute of Mathematics, Silesian University, Bankowa 14, 40-007 Katowice, Poland

Inequalities for Differential Operators

Lunch time

International Series of Numerical Mathematics, Vol. 103, © 1992 Birkhäuser Verlag Basel

Interpolation inequalities and nonoscillatory differential equations

Mark Ashbaugh, Richard Brown, and Don Hinton

Abstract

Weighted interpolation inequalities of the form

$$\left(\int_I |x^\beta u^{(j)}|^p\,dx\right)^{1/p} \le K\left(\int_I |x^\gamma u|^q\,dx\right)^{(1-\lambda)/q}\left(\int_I |x^\alpha u^{(n)}|^r\,dx\right)^{\lambda/r}$$

are applied to the quadratic form associated with a symmetric differential operator to obtain nonoscillation criteria. These criteria are in the form of L^s integral conditions on the coefficients. Additionally some remarks are made on connections between these inequalities and conditions for disconjugacy as well as the problem of counting the number of negative eigenvalues of a self-adjoint operator arising from mathematical physics.

1. Introduction

Let p_i, $i = 0, \ldots, n-1$, be real, locally Lebesgue integrable functions defined on an interval I where I is one of $(0,\infty)$, $[b,\infty)$, or $(0,b]$ with $b > 0$; additionally let $p_n(x) = x^\Delta$ where Δ is a real number. Quasi-derivatives for a function y are defined by

$$y^{[i]} = y^{(i)} \quad (i = 0, \ldots, n-1), \qquad y^{[n]} = p_n y^{(n)},$$
$$y^{[n+i]} = -(y^{[n+i-1]})' + p_{n-i}y^{(n-i)} \quad (i = 1, \ldots, n-1),$$

and the symmetric differential operator L is defined by

$$L(y) = -(y^{[2n-1]})' + p_0 y. \tag{1.1}$$

In case the coefficients are sufficiently smooth, L has the familiar form

$$L(y) = \sum_{i=0}^{n}(-1)^i(p_i y^{(i)})^{(i)}.$$

This paper is in final form and no version of it will be submitted for publication elsewhere.

This paper has two major purposes: One is to establish new and quite concrete nonoscillation criteria for L which take the form of requiring the finiteness of weighted integrals of the sth power of the coefficients, where $1 \le s \le \infty$. The case $s = \infty$ is to be interpreted as a sup norm, i.e., $\left(\int_I |p(x)|^s \, dx\right)^{1/s}$ for $s = \infty$ is $\sup\{|p(x)| : x \in I\}$. This is done in Theorems 3.2–3.4 of Section 3. Another is to call attention by our methods to the usefulness in spectral theory of a new class of weighted interpolation inequalities derived in [3] and which are presented here in Section 2. Towards this end the final section of the paper consists of some general remarks designed to illustrate further connections between inequalities and the spectral analysis of L. We focus on two areas of current interest: (i) the discovery of conditions guaranteeing the disconjugacy of L (a concept related to but not the same as nonoscillation); (ii) the estimation of upper bounds on the number of negative eigenvalues in certain second order examples of L arising in mathematical physics from the Schrödinger operator in three dimensions with spherically symmetric potential.

For the special cases $s = 1, \infty$ (the latter however yielding just a pointwise bound) many integral criteria for nonoscillation are known; see for example [10] and [16]. For intermediate values of s the literature is sparse. For operators defined on compact intervals, Fink [6, 7], Melentsova [14] ($1 < s < \infty$), and Melentsova and Mil'shtein [13] ($s = \infty$) have given conditions for disconjugacy in terms of the length of the interval and the L^s norms of the coefficients. In the singular case known results seem to be limited to second order operators. For example Glasser, Martin, Grosse, and Thirring [8] have studied the operator

$$L(y) = -y'' + [l(l+1)x^{-2} + q(x)]y, \quad 0 < x < \infty, \quad l \ge 0. \tag{1.2}$$

They gave upper bounds on the number of negative eigenvalues of a self-adjoint realization of L in terms of a weighted L^s integral of q; on the same problem see also Simon [18] and Glasser, Grosse, and Martin [9]. We will discuss some connections between our approach and this work in Section 4.

The following notation and conventions are used in the paper: given an interval $J := (c, d)$, where $0 \le c < d \le \infty$, $AC^n(J)$ denotes the space of real functions u defined on J such that $u, \ldots, u^{(n-1)}$ are absolutely continuous on compact subintervals in the interior of J; we further define

$$AC^n_{LR}(J) = \left\{u \in AC^n(J) : \lim_{x \to c+} u^{(i)}(x) = 0 = \lim_{x \to d-} u^{(i)}(x),\ i = 0, \ldots, n-1\right\}.$$

Finally $L^p(J)$ is the real Banach space of (equivalence class of) real functions f on J satisfying $\int_J |f|^p \, dx < \infty$, and for a real function p, $p^+(x) = \max\{0, p(x)\}$ and $p^- = p(x) - p^+(x)$.

2. A weighted interpolation inequality

In [3] we proved an interpolation inequality of the form

$$(2.1)\qquad \left(\int_I |x^\beta u^{(j)}|^p\,dx\right)^{1/p} \le K\left(\int_I |x^\gamma u|^q\,dx\right)^{(1-\lambda)/q}\left(\int_I |x^\alpha u^{(n)}|^r\,dx\right)^{\lambda/r},$$

$u \in AC^n_{LR}(I)$, with $0 \le j \le n-1$, and where the constant K may depend on I, α, β, γ, p, q, r, n and j, but does not depend on u. This result, which is needed for Section 3, is as follows.

Theorem 2.1. *Suppose $0 \le j \le n-1$, and $1 \le r, p, q \le \infty$. Let S_n denote the positive integers not exceeding n, let α, β, γ be real numbers, and let $\mathcal{D}^n_{LR}(I)$ be the subspace of $AC^n_{LR}(I)$ for which the two integrals on the right-hand side of (2.1) are finite. Further assume $p \ge p_0$ where p_0 satisfies*

$$\frac{n}{p_0} = \frac{n-j}{q} + \frac{j}{r},$$

and $\lambda_0 \le \lambda \le 1$ where

$$\lambda_0 := \frac{j + q^{-1} - p^{-1}}{n - r^{-1} + q^{-1}} \qquad (\lambda_0 = 0 \text{ if } n = r = 1 \text{ and } q = \infty).$$

Then inequality (2.1) holds for all $u \in \mathcal{D}^n_{LR}(I)$ with finite constant K in the following cases.

(i) $I = (0,\infty)$, $\quad \alpha + r^{-1} \notin S_n$, $\quad p \ge r$,
$\beta + p^{-1} = \lambda(\alpha + r^{-1} - n + j) + (1-\lambda)(\gamma + q^{-1} + j)$.

(ii) $I = [b,\infty)$ *and either*

(a) $\alpha + r^{-1} \notin S_n$, $p \ge r$, $\beta + p^{-1} \le \lambda(\alpha + r^{-1} - n + j) + (1-\lambda)(\gamma + q^{-1} + j)$

or

(b) $\{\alpha + r^{-1} \in S_n$ *or* $p < r\}$, $\beta + p^{-1} < \lambda(\alpha + r^{-1} - n + j) + (1-\lambda)(\gamma + q^{-1} + j)$.

(iii) $I = (0,b]$ *and either*

(a) $\alpha + r^{-1} \notin S_n$, $p \ge r$, $\beta + p^{-1} \ge \lambda(\alpha + r^{-1} - n + j) + (1-\lambda)(\gamma + q^{-1} + j)$

or

(b) $\{\alpha + r^{-1} \in S_n$ *or* $p < r\}$, $\beta + p^{-1} > \lambda(\alpha + r^{-1} - n + j) + (1-\lambda)(\gamma + q^{-1} + j)$.

Remark 2.1. In the unweighted case ($\beta = \gamma = \alpha = 0$) (2.1) is called a Landau inequality. Only in this case and then only for a few combinations of the parameters has the best constant K been determined. For a survey of the best constant problem see [12].

Remark 2.2. If $\lambda = 1$ and $j = 0$, then (2.1) becomes the Hardy-type inequality

$$\left(\int_I |x^\beta u|^p \, dx\right)^{1/p} \le K \left(\int_I |x^\alpha u^{(n)}|^r \, dx\right)^{1/r}. \tag{2.2}$$

The proof of Theorem 2.1 for $\lambda = 1$ basically depends on iterating (2.2) for $n = 1$ which in turn is a special case of Opic and Kufner [11, Theorem 8.2]. We also note that this special case does not require that $p \ge p_0$, and that the conditions given in [11] are necessary and sufficient.

3. Nonoscillation criteria

The operator L in (1.1) is said to be *oscillatory* on $[c, d]$ provided there is a nontrivial solution y of $L(y) = 0$ and numbers f and g, $c \le f < g \le d$, such that

$$y(f) = \cdots = y^{(n-1)}(f) = 0 = y(g) = \cdots = y^{(n-1)}(g).$$

The operator L is also called oscillatory at ∞ (respectively, 0) provided for each $M > 0$ ($\delta > 0$) there is an interval $[c, d] \subset (M, \infty)$ ($[c, d] \subset (0, \delta)$) such that L is oscillatory on $[c, d]$.

Associated with L is a quadratic function Q where

$$Q(y) := \sum_{i=0}^{n} \int_I p_i(x)|y^{(i)}(x)|^2 \, dx$$

for appropriate y in $AC^n(I)$. Let

$$\mathcal{A}_n(c, d) := \left\{u \in AC^n_{LR}(c, d) : u^{(n)} \in L^2(c, d)\right\}.$$

The following classical result (see e.g. [10, 17]) relates the oscillatory properties of L to an inequality condition on Q.

Theorem 3.1. *The operator L is oscillatory on $[c, d] \subset I$ if and only if there is a nontrivial $y \in \mathcal{A}_n(c, d)$ such that $Q(\tilde{y}) \le 0$ where $\tilde{y}(x) = y(x)$ on $[c, d]$ and 0 otherwise.*

We can now state and prove our first application of Theorem 2.1.

Theorem 3.2. *Suppose $I = [b, \infty)$ (respectively, $I = (0, b]$) and $(\Delta+1)/2 \notin \{1, \ldots, n\}$. Then L is nonoscillatory at $\infty(0)$ if there exists $s_0, \ldots, s_{n-1}$ such that for each i, $1 \le s_i \le \infty$, and with $\phi_i := \Delta - 2(n-i) + s_i^{-1}$, we have*

$$\int_I |x^{-\phi_i} p_i^-(x)|^{s_i}\, dx < \infty \quad \textit{if } s_i < \infty \tag{3.1}$$

and

$$x^{-\phi_i} p_i^-(x) \to 0 \quad \textit{as} \quad x \to \infty(0) \quad \textit{if } s_i = \infty.$$

Proof. We consider only the case $I = [b, \infty)$; the proof for $(0, b]$ is similar. In Theorem 2.1 (ii) take $r = 2$, $\alpha = \Delta/2$, $p = 2/(1 - s_i^{-1})$, $\lambda = 1$, $j = 0$, and $\beta = \alpha + \frac{1}{2} - n + i - p^{-1} = \phi_i/2$. Then the conditions (ii) (a) hold so there is a least number K_i such that

$$\left(\int_I |x^\beta u|^p\, dx\right)^{1/p} \le K_i \left(\int_I |x^\alpha u^{(n-i)}|^2\, dx\right)^{1/2}. \tag{3.2}$$

for all $u \in AC_{LR}^{n-i}(I)$ for which the right side integral of (3.2) is finite. By (3.1) there is a number $a \ge b$ so that

$$\sum_{i=0}^{n-1} K_i^2 \left(\int_a^\infty |x^{-\phi_i} p_i^-|^{s_i}\, dx\right)^{1/s_i} < 1. \tag{3.3}$$

Suppose now $[c, d] \subset [a, \infty)$ and y is a nontrivial member of $\mathcal{A}_n(c, d)$. Then (recall $\tilde{y}(x) = y(x)$ on $[c, d]$ and 0 otherwise)

$$Q(\tilde{y}) = \sum_{i=0}^{n} \int_c^d p_i(x)|y^{(i)}|^2\, dx \ge \int_c^d x^\Delta |y^{(n)}|^2\, dx - \sum_{i=0}^{n-1} \int_c^d |p_i^-|\, |y^{(i)}|^2\, dx. \tag{3.4}$$

Hölder's inequality followed by (3.2) with $u = y^{(i)}$ next yields that

$$\begin{aligned} &\int_c^d |p_i^-|\, |y^{(i)}|^2\, dx \\ &\quad \le \left(\int_c^d |x^{-\phi_i} p_i^-|^{s_i}\, dx\right)^{1/s_i} \left(\int_c^d |x^{\phi_i/2} y^{(i)}|^{2/(1-s_i^{-1})}\, dx\right)^{1-s_i^{-1}} \\ &\quad \le K_i^2 \left(\int_a^\infty |x^{-\phi_i} p_i^-|^{s_i}\, dx\right)^{1/s_i} \left(\int_c^d x^\Delta |y^{(n)}|^2\, dx\right). \end{aligned} \tag{3.5}$$

Finally, substitution of (3.5) into (3.4) and using (3.3) shows that $Q(\tilde{y}) > 0$; application of Theorem 3.1 completes the proof.

Remark 3.1. For the operator $T(y) := (x^\Delta y')' + q(x)y$ on $[b,\infty)$, Theorem 3.2 implies that T is nonoscillatory at ∞ if either $\int_b^\infty x^{(2-\Delta)s-1}[q^+(x)]^s\,dx < \infty$ for some s, $1 \le s < \infty$, or $x^{2-\Delta}q^+(x) \to 0$ as $x \to \infty$. Comparison of T with the Euler operator $(x^\Delta y')' + cx^{\Delta-2}y$, which is oscillatory for $c > (\Delta-1)^2/4$, shows that the exponent on x, namely $(2-\Delta)s-1$ for $1 \le s < \infty$ and $2-\Delta$ for $s=\infty$, is optimal and cannot be decreased.

In the proof of Theorem 3.2 we have used only the simple special case of Theorem 2.1 when $\lambda = 1$ and consequently (2.1) collapses into the Hardy-type inequality (2.2). In this approach the negative parts of the quadratic function Q can be compared only to the positive leading term. The full generality of Theorem 2.1 however allows comparison of p_i^- with positive terms p_j^+, p_k^+ with $j > i \ge k$. In the last two theorems of the section we illustrate how this type of result can be applied to fourth-order and second-order operators. The proofs are similar to that of Theorem 3.2 but require the introduction of the true interpolation inequality (2.1). In Theorem 3.3 we compare the middle term of a fourth-order operator to the end terms while in Theorem 3.4 we make a decomposition of the coefficient of y for a second-order operator.

Let Δ, μ, A be real numbers with $A > 0$, and let $q(x)$ be a real locally Lebesgue integrable function on I. Set

$$L_4(y) = [(x^\Delta y'')' + q(x)y']' + Ax^\mu y. \tag{3.6}$$

Theorem 3.3. *Suppose $I = [b,\infty)$ (respectively, $I = (0,b]$) and $\Delta \notin \{1,3\}$. Then L_4 is nonoscillatory at $\infty(0)$ if there exists s, $1 \le s \le \infty$, such that with $\phi := (\Delta/2)(1+1/2s) + (\mu/2)(1-1/2s)$,*

$$\int_I |x^{-\phi}q^+(x)|^s\,dx < \infty \quad \textit{if } s < \infty \tag{3.7}$$

and

$$x^{-\phi}q^+(x) \to 0 \quad \textit{as } x \to \infty(0) \quad \textit{if } s = \infty.$$

Proof. As in Theorem 3.2 only the case $I = [b,\infty)$ is considered. In Theorem 2.1 (ii) take $r = q = 2$, $n = 2$, $j = 1$, $p = 2/(1-s^{-1})$, $\alpha = \Delta/2$, $\beta = \phi/2$, $\gamma = \mu/2$ and $\lambda = \lambda_0 = (1+1/2s)/2$. Then conditions (ii) (a) hold so that there is a least number K such that

$$\left(\int_I |x^\beta u'|^p\,dx\right)^{1/p} \le K\left(\int_I x^\mu |u|^2\,dx\right)^{(1-\lambda)/2}\left(\int_I x^\Delta |u''|^2\,dx\right)^{\lambda/2}, \quad u \in \mathcal{D}^2_{LR}(I).$$

By (3.7), there is a number a so that

$$K^2 \left(\int_a^\infty |x^{-\phi} q^+(x)|^s \, dx \right)^{1/s} < B := \frac{A^{1-\lambda}}{\lambda^\lambda (1-\lambda)^{1-\lambda}}. \tag{3.8}$$

Thus if $[c,d] \subset [a,\infty)$ and y is a nontrivial member of $\mathcal{A}_2(c,d)$, we may proceed as in the proof of Theorem 3.2 showing that

$$\begin{aligned} \int_c^d q^+(x)\,|y'|^2 \, dx \; &\le \; \left(\int_c^d |x^{-\phi} q^+(x)|^s \, dx \right)^{1/s} \left(\int_c^d |x^{\phi/2} y'|^p \, dx \right)^{2/p} \\ &< \; B \left(\int_c^d x^\mu |y|^2 \, dx \right)^{1-\lambda} \left(\int_c^d x^\Delta |y''|^2 \, dx \right)^{\lambda}. \end{aligned}$$

Hence

$$\begin{aligned} Q(\tilde{y}) \; &= \; \int_c^d [x^\Delta |y''|^2 - q(x)|y'|^2 + A x^\mu |y|^2] \, dx \\ &> \; \int_c^d x^\Delta |y''|^2 \, dx - B \left(\int_c^d x^\mu |y|^2 \, dx \right)^{1-\lambda} \left(\int_c^d x^\Delta |y''|^2 \, dx \right)^{\lambda} \\ &\quad + A \int_c^d x^\mu |y|^2 \, dx. \end{aligned} \tag{3.9}$$

It is known (cf. [4] or [15, p. 30]) that the inequality

$$X^Q - BXY + AY^R \ge 0 \tag{3.10}$$

holds if $X \ge 0$, $Y \ge 0$, $1 < Q$, $R < \infty$, $Q^{-1} + R^{-1} = 1$, $B > 0$, $A > 0$, and $B \le A^{1/R} Q^{1/Q} R^{1/R}$. The right hand side of (3.9) is the same as (3.10) with $X = \left(\int_c^d x^\Delta |y''|^2 \, dx \right)^\lambda$, $Q = 1/\lambda$, and $Y = \left(\int_c^d x^\mu |y|^2 \, dx \right)^{1-\lambda}$. The definition of B in (3.8) implies then $Q(\tilde{y}) > 0$, and the proof is complete.

In both Theorems 3.2 and 3.3 the proofs show that L is actually nonoscillatory on I, i.e., no nontrivial solution y of $Ly = 0$ has a pair of n-fold zeros on I, if the weighted L^s norms of the coefficients are sufficiently small. For example if (3.8) holds for some s, $1 \le s \le \infty$, and with a replaced by b, then L_4 is nonoscillatory on $[b,\infty)$. Our last theorem is stated in this fashion.

Let Δ, μ, A and q be as in (3.6) and set

$$L_2(y) := -(x^\Delta y')' + [Ax^\mu + q(x)]y.$$

The three cases $I = (0,b], [b,\infty), (0,\infty)$ are all allowed. Note that L_2 nonoscillatory on I means no nontrivial solution y of $L_2 y = 0$ has two zeros on I.

Theorem 3.4. *Suppose $\Delta \neq 1$, $\phi := \mu + (\Delta - \mu)/2s$, $\lambda := 1/2s$, and $p := 2/(1-s^{-1})$. Then L_2 is nonoscillatory on I if for some s, $1 \leq s < \infty$,*

$$K^2 \left(\int_I |x^{-\phi} q^-(x)|^s \, dx \right)^{1/s} < B := \frac{A^{1-\lambda}}{\lambda^\lambda (1-\lambda)^{1-\lambda}}, \tag{3.11}$$

where K is the best constant of the inequality

$$\left(\int_I |x^{\phi/2} u|^p \, dx \right)^{1/p} \leq K \left(\int_I x^\mu |u|^2 \, dx \right)^{(1-\lambda)/2} \left(\int_I x^\Delta |u'|^2 \, dx \right)^{\lambda/2}, \tag{3.12}$$

$u \in \mathcal{D}^1_{LR}(I)$.

Proof. First note that Theorem 2.1 implies the existence of K with $q = r = 2$, $\gamma = \mu/2$, $\alpha = \Delta/2$ and $\beta = \phi/2$. If y is a nontrivial member of $\mathcal{A}_1(c,d)$, then

$$Q(\tilde{y}) = \int_c^d [x^\Delta |y'|^2 + (Ax^\mu + q(x))|y|^2] \, dx.$$

As in the proof of Theorem 3.3, we find that

$$\begin{aligned} \int_c^d q(x)|y'|^2 \, dx \ &\geq \ -\left(\int_c^d |x^{-\phi} q^-(x)|^s \, dx \right)^{1/s} \left(\int_c^d |x^{\phi/2} y|^p \, dx \right)^{2/p} \\ &> \ -B \left(\int_c^d x^\mu |y|^2 \, dx \right)^{1-\lambda} \left(\int_c^d x^\Delta |y'|^2 \, dx \right)^{\lambda}. \end{aligned}$$

Using this lower bound in $Q(\tilde{y})$ leads to $Q(\tilde{y}) > 0$ as in Theorem 3.3; hence the proof is complete.

Remark 3.2. As noted above in the introduction many examples of criteria similar to Theorems 3.3 and 3.4 are known in the two extreme cases $s = 1, \infty$. The novelty of our approach is that all values $1 \leq s \leq \infty$ are allowed as well as comparison of one negative term in the function Q with one or more positive terms. Furthermore, different values of s for each coefficient can be used.

Note also that Theorem 3.4 yields a stronger result for L_2 than Theorem 3.2 when the exponent on x in (3.11), $-\phi = -\mu - (\Delta - \mu)/2s$, is smaller than the exponent of Theorem 3.2, namely $(2 - \Delta) - 1/s$. This is equivalent to $\mu > \Delta - 2$.

4. Related results

In the previous sections of the paper we have seen that inequalities like (2.1) or (2.2) are a powerful tool in proving facts about nonoscillation properties of operators. We have omitted discussions of other questions like discreteness of the spectrum, boundedness of the operators, or the derivation of limit-point or limit circle criteria. In all these topics Hardy or interpolation type inequalities have also proven useful. The literature on such applications is voluminous; see, e.g., the books [4], [10], [16], or their references.

In this section we informally sketch how some other inequalities related to (2.1) or (2.2) have been applied recently to two areas of spectral theory.

Disconjugacy. Recall that the classical operator L defined by

$$L(y) := y^{(n)} + p_1(t)y^{(n-1)} + \cdots + p_n(t)y \tag{4.1}$$

is said to be disconjugate on the compact interval $J := [c, d]$ if no nontrivial solution of $L(y) = 0$ has more than $n - 1$ zeros on $[c, d]$ counting multiplicity; hence this concept is the same as "nonoscillation" only for second order operators. An important tool in the investigation of the disconjugacy of L are inequalities of the form

$$\left(\int_J |u|^p\,dx\right)^{1/p} \le C L^{m+1/p-1/r} \left(\int_J |u^{(m)}|^r\,dx\right)^{1/r}, \quad L := d - c, \tag{4.2}$$

defined on the class of functions u in $AC^m(J)$ such that $u^{(m)} \in L^r(J)$ and u has at least $i \ge m$ zeros on J counting multiplicity. Although (4.2) seems a different inequality than those presented here, for some cases it is actually equivalent to a Hardy-type inequality like (2.2) on $I = [1, \infty)$: in the $m = 1$, $i = 2$ case this can be seen by normalizing first to the interval $[0, 1]$ followed by the change of variables $x = 1/t$. Applications of (4.2) to disconjugacy may be found in Fink [6, 7], Melentsova [14], Melentsova and Mil'shtein [13], and Willett [20]. ([13, [14] apply methods of optimal control to determine the length of the maximal interval of disconjugacy.) For work on (4.2) alone see Brink [2], Fink [6, 7], and Talenti [19]. The best constant C in (4.2) is unknown in almost all cases (see however [2, 5, 6] for general bounds), the case $m = 1$ being an exception. Then the inequality

$$\left(\int_J |u|^p\,dx\right)^{1/p} \le L^{1+1/p-1/r} \frac{p(1 + r'/p)^{1/r}\,\Gamma(1/p + 1/r')}{2(1 + p/r')^{1/p}\,\Gamma(1/p)\,\Gamma(1/r')} \left(\int_J |u'|^r\,dx\right)^{1/r} \tag{4.3}$$

holds (cf. [1], [14], or [19]); here u is a real absolutely continuous function on J such that $u(c) = u(d) = 0$, $1 \le p < \infty$, $1 < r < \infty$, $r' = r/(r-1)$, and Γ is the gamma function. Inequality (4.3) in turn leads directly to a nonoscillation criterion for $L(y) = y'' + q(x)y$. This is (cf. [14]): L is nonoscillatory on J if for some s, $1 < s < \infty$,

$$\left(\int_c^d q^+(x)^s\,dx\right) < \frac{k}{L^{2s-1}}, \quad L := d - c, \tag{4.4}$$

where

$$k = k_0 := \frac{(s-1)^{s+1}}{(2s-1)s^s}\left[\frac{\Gamma((s-1)/2s)\,\Gamma(1/2)}{\Gamma((2s-1)/2s)}\right]^{2s}.$$

Moreover the criterion fails if k in (4.4) is greater than k_0.

Another application of (4.4) is to the calculation of a lower bound of the first conjugate point $\nu(c)$ for $y'' + q(x)y = 0$ by computing the value of d which makes (4.4) an equality. To illustrate this consider $y'' + xy = 0$ with $c = 0$. The conjugate point $\nu(0)$ is 2.666. A few lower bounds d for this example corresponding to choices of s implied by (4.4) are:

s	d
1	2
2	2.396
∞	2.145

Taking another example, the equation $y'' + a(\sin x)y = 0$ is nonoscillatory on $[0,\pi]$ if and only if $a < a_0$, where the critical exponent $a_0 \approx 1.1724$. The lower bound $\tilde{a}$ for a_0 implied by (4.4) is

s	$\tilde{a}$
1	.6366
2	1.1375
3	1.1719
4	1.1681
∞	1

Bounds on the number of negative eigenvalues. The operator given by (1.2) arises from the separation of variables of the Schrödinger operator in three dimensions with a spherically symmetric potential. The real functions $q(x)$ is locally L^1 on $(0,\infty)$ and its behavior at $x = 0$ is such to allow the existence of solutions y_0, y_1 of $L(y) = 0$ with the asymptotic behavior $y_0(x) \sim x^{l+1}$ and $y_1(x) \sim x^{-l}$ as $x \to 0$. Furthermore, if $q(x)$ is sufficiently well behaved at ∞, e.g., $q(x) = O(x^2)$, then L is of limit-point type at ∞. Under these circumstances it is possible to associate a unique self-adjoint operator A_l with L (for $l = 0$, the boundary condition $y(0) = 0$ is imposed), and a problem of interest is to estimate the number of negative eigenvalues of A_l. Such estimates frequently use the counting principle: If y_0 is the solution of $L(y) = 0$ with $y_0(x) \sim x^{l+1}$ as $x \to 0$, then the number of eigenvalues of A_l in $(-\infty, 0)$ is equal to the number of zeros of y_0 in $(0, \infty)$; see [18] for a discussion of this theorem.

If $n_l(q)$ denotes the number of negative eigenvalues of A_l, then it has been proved by Glasser, Martin, Grosse, and Thirring [8] that for all s, $1 \le s < \infty$,

$$n_l(q) \le \frac{(s-1)^{s-1}\,\Gamma(2s)}{(2l+1)^2 s^s\,\Gamma(s)^2} \int_0^\infty x^{2s-1}|q(x)|^s\,dx. \tag{4.5}$$

To see how (4.5) is connected to the inequalities of Section 2, we have, following [8], that if c, d are consecutive zeros of y_0, then $(1/s + 1/s' = 1)$

$$\begin{aligned} &\int_c^d [|y_0'|^2 + x^{-2}l(l+1)|y_0|^2]\,dx \\ &\quad = \int_c^d -q(x)|y_0|^2\,dx \\ &\quad \le \left(\int_c^d x^{2s-1}|q|^s\,dx\right)^{1/s} \left(\int_c^d x^{-s'-1}|y_0|^{2s'}\,dx\right)^{1/s'} \\ &\quad \le k_1^{-1}\left(\int_c^d x^{2s-1}|q|^s\,dx\right)^{1/s} \int_c^d [|y_0'| + x^{-2}l(l+1)|y_0|^2]\,dx, \end{aligned} \tag{4.6}$$

where k_1 is the infimum of the function

$$J_1(y) = \int_0^\infty [|y'|^2 + x^{-2}l(l+1)|y|^2]\,dx \left(\int_0^\infty x^{-s'-1}|y|^{2s'}\,dx\right)^{-1/s'} \tag{4.7}$$

over all absolutely continuous real y with $y', x^{-1}y \in L^2(0,\infty)$. From (4.6) it follows that $1 \le k_1^{-s}\int_c^d x^{2s-1}|q|^s\,dx$ and summing over the consecutive pairs of zeros of y_0 we find by the counting principle that $n_l(q) \le k_1^{-s}\int_0^\infty x^{2s-1}|q|^s\,dx$. The determination of k_1^{-1} is accomplished in [8] by a variational argument. If we make the transformation $y(x) = x^{l+1}u(x)$, then we obtain after some calculation that

$$J_1(y) = J_2(u) = \int_0^\infty x^{2l+2}|u'|^2\,dx \left(\int_0^\infty x^{s'(2l+1)-1}|u|^{2s'}\,dx\right)^{-1/s'}$$

Thus the minimization of J_1 is equivalent to that of J_2. Note the minimization of J_2 is equivalent to a Hardy-like inequality similar to (2.2). Further the domain of J_2 is all absolutely continuous real u such that $x^l u, x^{l+1}u' \in L^2(0,\infty)$. These two conditions are equivalent to $x^{l+1}u' \in L^2(0,\infty)$, $x^{2l+1}u^2 \to 0$ as $x \to \infty$, which in turn are equivalent to $x^{l+1}u' \in L^2(0,\infty)$, $u \to 0$ as $x \to \infty$. The latter equivalence follows from

$$u(x)^2 = \left(\int_x^\infty u'\right)^2 \leq \left(\int_x^\infty \tau^{-2l-2}\, d\tau\right)\left(\int_x^\infty \tau^{2l+2}|u'|^2\, d\tau\right).$$

Such a u is said to belong to $AC^1_{\mathbb{R}}(0,\infty)$. That the infimum of $J_2(u)$ over $AC^1_{\mathbb{R}}(0,\infty)$ exists and is positive follows from the examples of Section 6.7, Chapter 1 of [11].

References

1. M.S. Ashbaugh and E. Harrell, II, *Maximal and minimal eigenvalues and their associated nonlinear equations.* J. Math. Phys. **28** (1987), 1770–1786.

2. J. Brink, *Inequalities involving* $||f||_p$ *and* $||f^{(n)}||_q$ *for* f *with* n *zeros.* Pacific J. Math. **42** (1972), 289–311.

3. R.C. Brown and D.B. Hinton, *Interpolation inequalities with power weights for functions in* $\mathbb{R}$. J. Math. Anal. Appl., to appear.

4. D.E. Edmunds and W.D. Evans, *Spectral theory and differential operators.* Oxford Univ. Press, Oxford, 1987.

5. W.N. Everitt, *On the spectrum of a second-order linear differential equation with a p-integrable coefficient.* Applicable Analysis **2** (1972), 143–160.

6. A.M. Fink, *Conjugate inequalities for functions and their derivatives.* SIAM J. Math. Anal. **5** (1974), 399–411.

7. A.M. Fink, *Differential inequalities and disconjugacy.* J. Math. Anal. Appl. **49** (1975), 758–772.

8. V. Glasser, A. Martin, H. Grosse, and W. Thirring, *A family of optimal conditions for the absence of bound states in a potential.* In: Studies in Mathematical Physics (E.H. Lieb, B. Simon, A.S. Wightman, eds.), Princeton Univ. Press, 1976, pp. 169–194.

9. V. Glasser, H. Grosse, and A. Martin, *Bounds on the number of eigenvalues of the Schrödinger operator.* Comm. Math. Phys. **59** (1978), 197–212.

10. I.M. Glazman, *Direct methods of qualitative spectral analysis of singular differential operators.* Israel Program for Scientific Translation, Jerusalem, 1965.

11. A. Kufner and B. Opic, *Hardy-type inequalities.* Pitman Research Notes in Mathematics, Longman, London, 1990.

12. M.K. Kwong and A. Zettl, *Norm inequalities for derivatives.* In: Lecture Notes in Mathematics, Vol. 846, Springer, Berlin, 1981, pp. 229–243.

13. Y.A. Melentsova and G.N. Mil'shtein, *Optimal estimation of the nonoscillation interval for linear differential equations with bounded coefficients.* Diff. Eqs. **17** (1981), 1368–1379.

14. Y.A. Melentsova, *A best-possible estimate of the nonoscillation interval for a linear differential equation with coefficients bounded in L_r.* Diff. Eqs. **13** (1977), 1236–1243.

15. D.S. Mitrinović, *Analytic Inequalities.* Springer-Verlag, Berlin, 1970.

16. E. Müller-Pfeiffer, *Spectral theory of ordinary differential operators.* Ellis Horwood, Chichester, 1981.

17. W.T. Reid, *Riccati matrix differential equations and nonoscillation criteria for associated systems.* Pacific J. Math. **13** (1963), 665–685.

18. B. Simon, *On the number of bounded state of two-body Schrödinger operators.* In: Studies in Mathematical Physics (E.H. Lieb, B. Simon, A.S. Wightman, eds.), Princeton Univ. Press, 1976, pp. 305–326.

19. G. Talenti, *Best constant in Sobolev inequality.* Annali di Mat. Pura ed Appl. **110** (1976), 353–372.

20. D. Willett, *Generalized De La Vallée Poussin disconjugacy tests for linear differential equations.* Canad. Math. Bull. **14** (1971), 419–428.

Mark Ashbaugh, Mathematics Department, University of Missouri, Columbia, MO 65211, USA

Richard Brown, Mathematics Department, University of Alabama, Tuscaloosa, AL 35487, USA

Don Hinton, Mathematics Department, University of Tennessee, Knoxville, TN 37996, USA

International Series of Numerical Mathematics, Vol. 103, © 1992 Birkhäuser Verlag Basel

Optimal bounds for the spectrum of a one-dimensional Schrödinger operator

C. Bennewitz and E.J.M. Veling

Abstract

The lowest point σ_0 of the spectrum of the eigen-value problem (0.1) is considered for the case of a p-integrable potential q. Optimal upper and lower bounds for σ_0 in terms of the L^p-norm of q are deduced.

0. Introduction

Consider the simple eigen-value problem

$$\begin{cases} -u'' + qu = \lambda u & \text{in } [0, \infty) \\ u(0)\cos\beta + u'(0)\sin\beta = 0 \end{cases} \tag{0.1}$$

where β is given, $0 < \beta \leq \pi$, and q is real-valued and locally integrable. With appropriate conditions on q this gives rise to a selfadjoint operator in $L^2(0, \infty)$. For example, it is well known that this is the case if the negative part of q is in $L^p(0, \infty)$ for some p, $1 \leq p \leq \infty$.

The question we set out to answer is the following: Given the L^p-norm $\|q\|_p$ of q, what is the possible range of σ_0, the lowest point of the spectrum? The answer to the question is contained in Theorem 1.1. Note that if $q \in L^p(0, \infty)$, $1 \leq p < \infty$, then the continuous spectrum is $[0, \infty)$ so $\sigma_0 \leq 0$, and if < 0, then σ_0 is an eigenvalue.

Similar questions have been treated repeatedly in the literature. The case when the interval is bounded was considered, e.g., by Essén [3] and by Talenti [6]. Egnell [2] considered the case of an elliptic operator on a bounded domain in $\mathbb{R}^n$. Everitt [4] gave a lower bound for the spectrum in the cases $\beta = \pi/2$ and $\beta = \pi$; this was improved to an optimal bound by Veling [8]. It should be remarked that our technique can be used to deal with far more general cases; higher order equations,

---

This paper is in final form and no version of it will be submitted for publication elsewhere.

elliptic partial differential equations, etc. However, in such cases the bounds will be far less explicit since they are expressed in terms of solutions of non-linear differential equations. In the present case these solutions can be found explicitly.

Acknowledgement. The results of this paper were obtained in early summer 1984 while both authors were visiting the University of Birmingham, England, under the auspices of the Science and Engineering Research Council. Various circumstances have delayed the publication, but although late we wish to express our appreciation to our host, professor W. N. Everitt, for providing us with this opportunity, and to SERC for generous support.

1. Results

In the sequel we will always assume that the potential q is such that (0.1) generates a selfadjoint operator in $L^2(0,\infty)$ (the 'limit-point' case), and that the lowest point σ_0 of the spectrum is given by

$$\sigma_0 = \sigma_0(\beta, q) = \inf\{\int_0^\infty (|u'|^2 + q|u|^2) - \cot\beta|u(0)|^2\} , \tag{1.1}$$

infimum being taken over those u with $\int_0^\infty |u|^2 = 1$ which are in $H^1(0,\infty)$. As usual $H^1(0,\infty)$ denotes the set of locally absolutely continuous functions which together with their derivatives are in $L^2(0,\infty)$. Similarly $H_0^1(0,\infty)$ is the subset of $H^1(0,\infty)$ of functions vanishing at 0. In the case $\beta = \pi$ one has to drop the $\cot\beta$-term and instead consider $u \in H_0^1(0,\infty)$. It is well known that (1.1) holds, e.g., if the potential q has a negative part $q_- = \max(0,-q)$ which is in $L^p(0,\infty)$ for some p, $1 \le p \le \infty$. We will only discuss the case $1 \le p < \infty$ since the optimal potentials in the case $p = \infty$ are obviously constant. Our bounds for σ_0 are given in Theorem 1.1. In order to state the theorem we need to introduce some notation. For $1 \le p < \infty$ let

$$a_0 = a_0(p,\beta) = \begin{cases} p(p-1)^{\frac{1-p}{p}}(2p-1)^{-\frac{1}{p}}(\cot\beta)^{\frac{2p-1}{p}}, & 0 < \beta \le \pi/2 \\ -p(p-1)^{\frac{1-p}{p}}B(p,\tfrac{1}{2})^{\frac{1}{p}}(-\cot\beta)^{\frac{2p-1}{p}}, & \pi/2 < \beta < \pi \\ -\infty & \beta = \pi \end{cases}$$

where $B(\cdot,\cdot)$ is the classical beta-function. For $p = 1$ this is interpreted as $a_0 = \cot\beta$ for $0 < \beta \le \pi/2$ and $a_0 = 2\cot\beta$ for $\pi/2 < \beta < \pi$. Also put

$$\lambda(p,\beta,a) = \begin{cases} -(-a)^{\frac{2p}{2p-1}}\left(\frac{p^p}{(p-1)^{p-1}}\int_C^\infty \cosh^{-2p} t\,dt\right)^{\frac{2}{1-2p}} & \text{for } a < 0 \\ -(\cot\beta\tanh C)^2 & \text{for } a > 0\,. \end{cases} \tag{1.2}$$

Here we have set $C = C(p,\beta,a) = -\infty$ if $a_0(p,\beta) \le a < 0$; for other $a < 0$, $C(p,\beta,a)$ is the unique solution of

$$(-a)^{\frac{p}{1-2p}} \cot\beta = \left(\frac{p^p}{(p-1)^{p-1}} \int_C^\infty \cosh^{-2p} t\,dt\right)^{\frac{1}{1-2p}} \tanh C$$

(for $p = 1$ we interprete $(p-1)^{p-1}$ as 1). For $0 < a \le a_0(p,\beta)$, $C(p,\beta,a)$ is the unique (positive) solution of

$$a^{\frac{p}{1-2p}} \cot\beta = \left(\frac{p^p}{(p-1)^{p-1}} \int_C^\infty \sinh^{-2p} t\,dt\right)^{\frac{1}{1-2p}} / \tanh C\ .$$

For other $a > 0$ we set $C(p,\beta,a) = 0$. Note that for $p = 1$ we have $\lambda(1,\beta,a) = 0$ for $a \ge \max(0,\cot\beta)$ and $\lambda(1,\beta,a) = -a^2/4$ if $2\cot\beta < a < 0$. Otherwise $\lambda(1,\beta,a) = -(\cot\beta - a)^2$. Finally, we set

$$A(p,\beta,a) = \frac{\sqrt{-\lambda(p,\beta,a)}}{p-1}\ .$$

We use the notation $q_+ = \max(q,0)$ and $q_- = \max(-q,0)$. The $L^p(0,\infty)$-norm, $1 \le p \le \infty$, is denoted by $\|\cdot\|_p$. Our main result is the following.

Theorem 1.1. *The following estimates for the lowest point σ_0 of the spectrum for the problem* (0.1) *are optimal:*

$$\lambda(p,\beta,-\|q_-\|_p) \le \sigma_0(\beta,q) \le \lambda(p,\beta,\|q_+\|_p)\ . \tag{1.3}$$

(1) *If $p > 1$ and $\|q_-\|_p > -a_0(p,\beta)$, then the lower bound is realized if and only if*

$$q(x) = -p(p-1)A^2\cosh^{-2}(Ax + C)$$

where $A = A(p,\beta,-\|q_-\|_p)$ and $C = C(p,\beta,-\|q_-\|_p)$. The corresponding eigenfunctions are all multiples of $\cosh^{1-p}(Ax + C)$.

(2) *In all other cases the lower bound in (1.3) is still optimal in the sense that there are potentials q with $\|q_-\|_p > 0$ given which give $\sigma_0(\beta,q)$ arbitrarily close to $\lambda(p,\beta,-\|q_-\|_p)$, but there are no potentials for which they are equal.*

(3) *Similarly, if $p > 1$ and $\|q_+\|_p < a_0(p,\beta)$ the upper bound is realized if and only if*

$$q(x) = p(p-1)A^2\sinh^{-2}(Ax + C)$$

where $A = A(p,\beta,\|q_+\|_p)$ and $C = C(p,\beta,\|q_+\|_p)$ (> 0). The corresponding eigenfunctions are all multiples of $\sinh^{1-p}(Ax + C)$.

(4) *If* $\|q_+\|_p = a_0(p,\beta)$ *and* $p > \frac{3}{2}$, *the upper bound is* 0. *This is realized with* 0 *as an eigenvalue if and only if* $q(x) = p(p-1)(x + \frac{p-1}{\cot\beta})^{-2}$. *The corresponding eigenfunctions are all multiples of* $(x + \frac{p-1}{\cot\beta})^{1-p}$.

(5) *In the remaining cases the upper bound is still optimal in the following sense. If* $\lambda(p,\beta,\|q_+\|_p) = 0$, *then any positive potential of the required norm gives* $\sigma_0(\beta,q) = 0$, *but there is no choice of potential for which this is an eigenvalue. If* $p = 1$ *and* $\|q_+\|_p < a_0(\beta,q)$, *then there are potentials* q *which give* $\sigma_0(\beta,q)$ *arbitrarily close to* $\lambda(p,\beta,\|q_+\|_p)$, *but there are no potentials for which they are equal.*

The *proof* of Theorem 1.1 is based on the following inequality.

Theorem 1.2. *If* $r = \frac{2p}{p-1}$ $(= \infty$ *for* $p = 1)$, $0 < \beta \le \pi$ *and* $a \in \mathbb{R}$, *then* $\lambda(p,\beta,a)$, *as defined in* (1.2), *is the best (largest) value of* λ *for which the inequality*

$$\lambda\|u\|_2^2 \le \|u'\|_2^2 - \cot\beta|u(0)|^2 + a\|u\|_r^2 \tag{1.4}$$

is valid for all $u \in H^1(0,\infty)$. *In case* $\beta = \pi$ *the* $\cot\beta$*-term should be dropped and the inequality is valid for* $u \in H_0^1(0,\infty)$. *Setting* $A = A(p,\beta,a)$ *and* $C = C(p,\beta,a)$ *the cases of equality are as follows:*

(1) *If* $0 < a < a_0(p,\beta)$, *then if* $p > 1$ *there is equality for all multiples of* $\sinh^{1-p}(Ax + C)$, *and if* $p = 1$ *for all multiples of* $\exp(-x(\cot\beta - a))$.

(2) *If* $a = a_0(p,\beta) > 0$ *and* $p > \frac{3}{2}$, *then there is equality for all multiples of* $(x + \frac{p-1}{\cot\beta})^{1-p}$.

(3) *If* $a < \min(a_0(p,\beta), 0)$, *then if* $p > 1$ *there is equality for all multiples of* $\cosh^{1-p}(Ax + C)$, *and if* $p = 1$ *for all multiples of* $\exp(-x(\cot\beta - a))$.

(4) *If* $p = 1$, $a = a_0(1,\beta) = 2\cot\beta < 0$, *and* B *arbitrary, then there is equality for all multiples of* $\exp(|x - B|\cot\beta)$.

(5) *There are no other cases of equality.*

If (1.4) has an extremal and $p > 1$, then it must satisfy the Euler equation

$$\begin{cases} -u'' + a\|u\|_r^{2-r}u^{r-1} = \lambda u \\ u(0)\cos\beta + u'(0)\sin\beta = 0\,. \end{cases} \tag{1.5}$$

To see this, note that (some non-zero multiple of) u is assumed to minimize $\|u'\|_2^2 - \cot\beta|u(0)|^2 + a\|u\|_r^2$ under the constraint $\|u\|_2^2 = 1$. That the Lagrange multiplier then equals the optimal constant λ in (1.4) follows easily upon multiplying (1.5) by u and integrating. Note that since $u \in H^1(0,\infty)$ and by (1.5) u, u' and u'' are all in $L^2(0,\infty)$ so that $u'u \to 0$ at ∞. Hence $-\int_0^\infty u''u = \int_0^\infty |u'|^2$. We have here assumed $u \geq 0$, as we may, since replacing u by $|u|$ in (1.4) leaves the left hand side unchanged but decreases the right hand side strictly if the argument of u changes between zeros of u. Using this and Theorem 1.2 it is easy to prove Theorem 1.1.

Proof of Theorem 1.1. Using Hölder's inequality and then (1.4) with $a = -\|q_-\|_p$ we obtain ($r = \frac{2p}{p-1}$ as before)

$$\begin{aligned}\|u'\|_2^2 - \cot\beta|u(0)|^2 + \int_0^\infty q|u|^2 &\geq \|u'\|_2^2 - \cot\beta|u(0)|^2 - \int_0^\infty q_-|u|^2 \\ &\geq \|u'\|_2^2 - \cot\beta|u(0)|^2 - \|q_-\|_p\|u\|_r^2 \\ &\geq \lambda(p,\beta,-\|q_-\|_p)\|u\|_2^2\end{aligned}$$

which is the desired lower bound. There is equality in the first inequality if $q_+ \equiv 0$. In the last inequality we may choose a normalised u to get either equality, if (1.4) admits equality, or else equality with an arbitrarily small relative error. If $p > 1$ we may finally choose q_- as the appropriate positive multiple of $|u|^{\frac{2}{p-1}}$ to give equality in the second inequality. If $p = 1$ we may similarly choose q_- to have its support in an arbitrarily small neighbourhood of a point where $|u|$ assumes its maximum. This will give equality with arbitrarily small error in the second inequality. It is clear, however, in the case $p = 1$, that even if (1.4) admits equality we can not choose q_- to give equality in the second inequality since an extremal of (1.4) only assumes its maximum in a single point. This completes the proof for the lower bound.

Note that if we, in the case $p = 1$, allow q more generally to be a measure with given total variation we can find an optimal potential in all cases where (1.4) admits equality. The optimal potential will then be an appropriate multiple of the δ-function supported at the point where the extremal of (1.4) achieves its maximum.

Turning now to the upper bound we similarly have

$$\begin{aligned}\|u'\|_2^2 - \cot\beta|u(0)|^2 + \int_0^\infty q|u|^2 &\leq \|u'\|_2^2 - \cot\beta|u(0)|^2 + \int_0^\infty q_+|u|^2 \\ &\leq \|u'\|_2^2 - \cot\beta|u(0)|^2 + \|q_+\|_p\|u\|_r^2\end{aligned}$$

which shows that $\sigma_0(\beta,q) \leq \lambda(p,\beta,\|q_+\|_p)$ for any q. If $p > 1$ it is clear from (1.5) that $\lambda(p,\beta,a)$ is an eigenvalue of (0.1) precisely if there is an extremal u for (1.4) and we choose $q = a\|u\|_r^{2-r}|u|^{r-2}$. The corresponding eigenfunction is u, and, since u has no zeros, $\lambda(p,\beta,a)$ is the *smallest* eigenvalue. In the case $p = 1$ it is clear that $\lambda(1,\beta,a)$ can not be an eigenvalue even if there are extremals for (1.4) since this would require the support of q to be contained in the set of points where the extremal assumes its maximum absolute value; but this is a single point according to Theorem 1.2.

It remains to show the optimality of the estimate in the cases when $p = 1$ or there are no extremals for (1.4). If $p = 1$ but $\lambda(1, \beta, \|q_+\|_1) < 0$ this follows by letting q_p with $\|q_p\|_p = \|q_+\|_1$ be the optimal potential for $p > 1$, but p sufficiently close to 1. The smallest eigenvalue is then $\lambda(p, \beta, \|q_+\|_1) \to \lambda(1, \beta, \|q_+\|_1)$ as $p \to 1$. It is easily seen that $\|q_p\|_1 \to \|q_+\|_1$ as $p \to 1$. The optimality of the estimate follows, since it is obvious that the best constant in (1.4) depends continuously on a.

Finally, if there are no extremals for (1.4), so that $\|q_+\|_p \geq a_0(p, \beta)$, the estimate is still optimal since it is obvious that $\sup_{\|q_+\|_p = a} \sigma_0$ increases with a and $\lambda(p, \beta, a) \uparrow 0$ as $a \uparrow a_0(p, \beta)$. The proof is complete.

The rest of the paper is devoted to proving Theorem 1.2.

2. The Euler equations

In the sections that follow, as before, the numbers p, $1 \leq p < \infty$, and r, $2 < r \leq \infty$, are always assumed to be connected by $r = \frac{2p}{p-1}$, $p = \frac{r}{r-2}$. We first give a result to restrict the possible range of λ in (1.4).

Lemma 2.1. *The best constant λ of (1.4) is ≤ 0 for $0 < \beta \leq \pi$, $1 \leq p < \infty$ and all $a \in \mathbb{R}$.*

Proof. Let $v \in C_0^\infty(0, \infty)$ have $L^2(0, \infty)$-norm 1 and put $v_\varepsilon(x) = \sqrt{\varepsilon}\, v(\varepsilon x)$. Then $\|v_\varepsilon\|_2 = 1$ and

$$\|v_\varepsilon'\|_2^2 - \cot \beta |v_\varepsilon(0)|^2 + a\|v_\varepsilon\|_r^2 = \varepsilon^2 \|v'\|_2^2 + \varepsilon^{1/p} \|v\|_r^2 \to 0 \text{ as } \varepsilon \to 0 .$$

The lemma follows.

In the rest of this section we will only consider the case $2 < r < \infty$, *i.e.*, $1 < p < \infty$. By Lemma 2.1 we have $\lambda \leq 0$ and we will first consider the case $\lambda < 0$. If we multiply the equation (1.5) by $2u'$ and integrate the result is

$$(u')^2 = \frac{2a}{r} \|u\|_r^{2-r} u^r - \lambda u^2 .$$

The constant of integration is 0 since all terms above are integrable (by the usual Soboleff estimates $L^r(0, \infty) \subset H^1(0, \infty)$). This equation is separable and its solutions are the multiples of $\sinh^{1-p}(Ax + C)$ if $a > 0$ and of $\cosh^{1-p}(Ax + C)$ if $a < 0$.

From the differential equation we have $\lambda = -(p-1)^2 A^2$ and

$$A = \left(\tfrac{a}{p(p-1)}\right)^{\frac{p}{2p-1}} \left(\int_C^\infty \sinh^{-2p} t\, dt\right)^{\frac{1}{1-2p}} \qquad \text{if } a > 0\,,$$
$$A = \left(\tfrac{-a}{p(p-1)}\right)^{\frac{p}{2p-1}} \left(\int_C^\infty \cosh^{-2p} t\, dt\right)^{\frac{1}{1-2p}} \qquad \text{if } a < 0\,.$$

Inserting this in the boundary condition it follows that

$$a^{\frac{p}{1-2p}} \cot\beta = \left(\frac{p^p}{(p-1)^{p-1}} \int_C^\infty \sinh^{-2p} t\, dt\right)^{\frac{1}{1-2p}} / \tanh C \quad \text{if } a > 0\,,$$
$$(-a)^{\frac{p}{1-2p}} \cot\beta = \left(\frac{p^p}{(p-1)^{p-1}} \int_C^\infty \cosh^{-2p} t\, dt\right)^{\frac{1}{1-2p}} \tanh C \quad \text{if } a < 0\,.$$

As functions of C the right hand sides of these equations are strictly increasing, defined for $C \geq 0$ if $a > 0$ and for all $C \in \mathbb{R}$ if $a < 0$. Their ranges are $[((p-1)^{p-1}p^{-p}(2p-1))^{\frac{1}{2p-1}}, +\infty)$ and $(-(p^p(p-1)^{1-p}B(p,\frac{1}{2}))^{\frac{1}{1-2p}}, +\infty)$ respectively. It follows that there can be no extremals for (1.4) except as stated in Theorem 1.2. It still remains to show that the solutions of (1.5) actually are extremals, which we will do in the next section. It also remains to discuss the case $p = 1$, which will be done in the last section.

3. Proof of optimality

We must now show that the values of λ calculated in Section 2 are in fact optimal. First note that by standard Soboleff estimates the expression

$$\|u'\|_2^2 - \cot\beta |u(0)|^2 + a\|u\|_r^2 + K\|u\|_2^2\,, \tag{3.1}$$

K large, is a norm-square equivalent to the standard H^1-norm. It follows that $\lambda > -\infty$ in all cases. We first consider the simpler case of $a > 0$.

Lemma 3.1. *For* $1 < p < \infty$ *and* $a > 0$ (1.4) *has an extremal as soon as* $\lambda(p, \beta, a) < 0$.

Proof. Let u_j, $j = 1, 2\ldots$, be a normalised ($\|u_j\|_2 = 1$) minimizing sequence for (1.4). By (3.1) it follows that u_j, $j = 1, 2, \ldots$, is a bounded sequence in $H^1(0, \infty)$. We may thus assume it to be weakly convergent in H^1 and so pointwise convergent. If the limit is u we have, by Fatou's lemma, that $\|u\|_r^2 \leq \|u_j\|_r^2 + o(1)$ and $\|u\|_2^2 \leq$

$\|u_j\|_2^2 + o(1)$. Also $u_j(0) \to u(0)$ and, by weak convergence in H^1, $\|u'\|_2^2 = \|u_j'\|_2^2 - \|u' - u_j'\|_2^2 + o(1) \le \|u_j'\|_2^2 + o(1)$. Since $\lambda < 0$ we therefore get

$$\begin{aligned}\|u'\|_2^2 - \cot\beta|u(0)|^2 + a\|u\|_r^2 &\le \|u_j'\|_2^2 - \cot\beta|u_j(0)|^2 + a\|u_j\|_r^2 + o(1)\\ &= \lambda\|u_j\|_2^2 + o(1) \le \lambda\|u\|_2^2 + o(1)\end{aligned}$$

so that u is optimal unless identically 0. On the other hand, from $\|u_j\|_2 = 1$ follows $\cot\beta|u_j(0)|^2 \ge -\lambda + o(1)$ so $\cot\beta|u(0)|^2 \ge -\lambda > 0$. This excludes $u \equiv 0$. Note that the proof also shows that $u_j \to u$ strongly in $H^1(0,\infty)$.

By the results of Section 2 it follows that for $a > 0$, $\lambda(p,\beta,a)$ as given in (1.2) is the best constant in (1.4), and that the cases of equality are as stated. This is clear if $\lambda(p,\beta,a) < 0$, but by Lemma 2.1, and since the best constant clearly increases with a, the only possible value for the remaining cases is 0.

We leave the case $p = 1$ to the last section and turn now to the case $a < 0$. The following lemma by Brezis and Lieb [1] is crucial, and will serve a similar purpose in the present case as did Fatou's lemma in the proof of Lemma 3.1.

Lemma 3.2 (Brezis and Lieb). *Suppose u_j, $j = 1, 2, \ldots$ is a bounded sequence in $L^r(0,\infty)$, $0 < r < \infty$, such that $u_j \to u$ a.e. Then*

$$\|u_j\|_r^r - \|u - u_j\|_r^r \to \|u\|_r^r \quad \text{as } j \to \infty\ .$$

The *proof*, as given in [1] of a slightly more general statement, is an ingenious application of dominated convergence. Using Lemma 3.2 we can now prove the following result.

Lemma 3.3. *If $a < 0$, $1 < p < \infty$ and (1.4) has no extremal, then every minimizing sequence, normalised with respect to $\|\cdot\|_2$, tends to 0 weakly in $H^1(0,\infty)$.*

Proof. Let u_j, $j = 1, 2, \ldots$, be a minimizing sequence for (1.4) normalized by $\|u_j\|_2 = 1$. Like in Lemma 3.1 it is clear that this sequence is bounded in $H^1(0,\infty)$ so we may as well assume it to converge weakly; $u_j \rightharpoonup u$, say. As in the proof of Lemma 3.1 we have $\|u'\|_2^2 = \|u_j'\|_2^2 - \|u' - u_j'\|_2^2 + o(1)$ and $\|u\|_2^2 = \|u_j\|_2^2 - \|u - u_j\|_2^2 + o(1)$. By Lemma 3.2 we have $\|u - u_j\|_r^r + \|u\|_r^r = \|u_j\|_r^r + o(1)$ so, since $r > 2$, $\|u\|_r^2 \ge \|u_j\|_r^2 - \|u - u_j\|_r^2 + o(1)$. If we denote the right hand side of (1.4) by $F(u)$ we therefore obtain:

$$F(u) \le F(u_j) - F(u - u_j) + o(1) \le \lambda\|u_j\|_2^2 - \lambda\|u - u_j\|_2^2 + o(1) = \lambda\|u\|_2^2 + o(1)\ .$$

Hence, either $u \equiv 0$ or u is an extremal of (1.4).

From Lemma 3.3 follows easily the following central result.

Lemma 3.4. *Under the assumptions of Lemma 3.3 there is a minimizing sequence for* (1.4) *in* $H_0^1(0,\infty)$.

Proof. Let u_j, $j = 1, 2, \ldots$, be a normalized minimizing sequence. According to Lemma 3.3, $u_j \rightharpoonup 0$ weakly in $H^1(0,\infty)$. In particular $u_j(0) \to 0$. Now define

$$v_j(x) = \begin{cases} xu_j(0) & \text{for } 0 \le x \le 1 \\ u_j(x-1) & \text{for } x > 1 \end{cases}.$$

Then it is clear that $v_j \in H_0^1(0,\infty)$ and that v_j, $j = 1, 2, \ldots$ is minimizing.

The final lemma we need to complete the proof of Theorem 1.2 for $p > 1$ is the result of Veling [8], which in turn is based on an inequality of Nagy [7]. For the sake of completeness we give a proof, which is essentially that of Nagy.

Lemma 3.5. *For* $p > 1$, $a < 0$ *and* $\beta = \pi$ *the best constant of* (1.4) *is* $\lambda(p, \pi, a)$.

Proof. Every element $u \in H_0^1(0,\infty)$ tends to 0 at ∞ so $|u|$ assumes its maximum at some strictly positive point m. Since (1.4) is homogeneous we need only consider the case $|u(m)| = 1$. If as before $r = \frac{2p}{p-1}$, it is easily seen that $4\int_0^1 \sqrt{t^2 - t^r}\,dt = B(p, \frac{1}{2})$, so that we obtain

$$\begin{aligned} B(p, \tfrac{1}{2}) = 4\int_0^1 \sqrt{t^2 - t^r}\,dt &= 2\int_0^m |u|'\sqrt{|u|^2 - |u|^r} - 2\int_m^\infty |u|'\sqrt{|u|^2 - |u|^r} \\ &\le 2\int_0^\infty |u'|\sqrt{|u|^2 - |u|^r} \le \|u'\|_2^2 + \|u\|_2^2 - \|u\|_r^r \end{aligned}$$

by the geometric-arithmetic inequality. We may rewrite this inequality as

$$-\|u\|_2^2 \le \|u'\|_2^2 - (\|u\|_r^{r-2} + B(p, \tfrac{1}{2})\|u\|_r^{-2})\|u\|_r^2 \le \|u'\|_2^2 + a_1\|u\|_r^2$$

where $a_1 = \max_{t>0}(-(t^{r-2} + B(p, \frac{1}{2})t^{-2})) = -p(p-1)^{\frac{p-1}{p}}(B(p, \frac{1}{2}))^{\frac{1}{p}}$. Applying this inequality to $v(x) = u(\varepsilon x)$ and choosing $\varepsilon = (\frac{a_1}{a})^{\frac{p}{2p-1}}$ gives the inequality (1.4) for $\beta = \pi$ with $\lambda = \lambda(p, \pi, a)$. This estimate is optimal. To see this, put $u_j(x) = \cosh^{1-p}(\frac{1}{p-1}(\frac{a}{a_1})^{\frac{p}{2p-1}}x + j)$ for $x \ge 1$ and $u_j(x) = xu_j(1)$ for $0 \le x < 1$. This is easily seen to be a minimizing sequence.

Lemma 3.4 means that if (1.4) has no extremal for certain $p > 1$, β and $a < 0$, then the best constant $\lambda = \lambda(p, \pi, a)$. On the other hand, if there is no extremal, then $\lambda < \lambda(p,\beta,a)$ for any p, β and a for which the Euler equation (1.5) has solutions. Since obviously $\lambda(p,\beta,a) \le \lambda(p,\pi,a)$ by (1.2), it is clear that all the solutions calculated in Section 2 actually are extremals. This completes the proof of Theorem 1.2 for $p > 1$.

4. The case $p = 1$

For a given $u \in H^1(0,\infty)$ it is clear that $\|u\|_r \to \|u\|_\infty$ as $r \to \infty$. Thus letting $p \downarrow 1$ in (1.4) shows that the best constant λ for $p = 1$, β and a given, satisfies $\lambda \ge \lim_{p\downarrow 1} \lambda(p,\beta,a) = -(\cot\beta - a)^2$ if $a < a_0(1,\beta)$. Similarly $\lambda \ge -a^2/4$ if $a_0(1,\beta) \le a < 0$. On the other hand, $u(x) = \exp(-x(\cot\beta - a))$ gives the opposite inequality for $a < a_0(1,\beta)$. Similarly, if $a_0(1,\beta) \le a < 0$ we may set

$$u_j(x) = \begin{cases} x\exp(a(j-1)/2) & \text{for } 0 \le x \le 1 \\ \exp(a|x-j|/2) & \text{for } x \ge 1 \end{cases} .$$

Clearly $u_j \in H_0^1(0,\infty)$ and a simple calculation shows that $\lambda(1,\beta,a)$ is indeed the best constant in (1.4) if $a_0(1,\beta) \le a < 0$. Finally, since the best constant increases with a and $\lambda(1,\beta,a) \to 0$ as $a \to \max(a_0(1,\beta),0)$ we have verified that (1.2) indeed gives the optimal constants for (1.4) also if $p = 1$. To determine the cases of equality requires a more detailed analysis.

Assume that $p = 1$ and u is an extremal for (1.4). Let $\phi \in C_0^\infty(\mathbb{R})$ have support which does not intersect the (closed) set where $|u|$ assumes its maximum. It is then clear that $\|u+\delta\phi\|_\infty$ is independent of δ for $|\delta|$ sufficiently small. It therefore follows that u satisfies the Euler equation

$$\begin{cases} -u'' = \lambda u \\ u(0)\cos\beta + u'(0)\sin\beta = 0 \end{cases} \tag{4.1}$$

except in points where it assumes its maximum. We may as before assume that $u \ge 0$. This immediately excludes the possibility of extremals for $\lambda = 0$.

If $2\cot\beta < a < 0$ we have $\lambda(1,\beta,a) = \lambda(1,\pi,a)$. Any extremal u then satisfies $u(0) = 0$, since otherwise (1.4) would be violated for any β' such that $\cot\beta < \cot\beta' < a$, keeping $p = 1$ and the same a. Since u therefore does not assume its maximum at 0, it also satisfies the boundary condition; but this implies $u \equiv 0$ by (4.1). There are therefore no extremals for $2\cot\beta < a < 0$.

If $a \le 2\cot\beta < 0$ or $0 \le a < \cot\beta$, then any extremal u must, after the last time it assumes its maximum, be a multiple of $\exp(-\sqrt{-\lambda}\,x)$ (note that $u(x) \to 0$ as $x \to \infty$ since $u \in H^1(0,\infty)$). If u assumes its maximum in more than one point, let b be the first and c the last such point. Now replace u in (1.4) by v where $v(x) = u(x)$ for $0 \le x \le b$ and $v(x) = u(x-b+c)$ for $x \ge b$. In effect, we have 'removed' the part of u which is between b and c. This will clearly strictly increase the left hand side and decrease the right hand side of (1.4), so that an extremal must assume its maximum in a single point. This shows that we have found all extremals in the case $a = 2\cot\beta < 0$, $p = 1$.

In all other cases we claim that an extremal must assume its maximum at 0 and nowhere else. To see this, we now remove the part of v before b, *i.e.*, we replace v by $v(x-b)$. In the case $0 \le a < \cot\beta$ we will again strictly increase the left hand side and decrease the right hand side of (1.4) unless $b = 0$. In the case $a < 2\cot\beta < 0$, we will have removed $\lambda \int_0^b |u|^2$ from the left hand side and

$$\int_0^b |u'|^2 + \cot\beta(|u(b)|^2 - |u(0)|^2)$$

from the right hand side. Integrating by parts and using (4.1) this equals $\lambda \int_0^b |u|^2 + u(b)(u(b)\cot\beta + u'(b))$. We claim that $u(b)\cot\beta + u'(b) > 0$ unless $b = 0$, which will show that an extremal can only assume its maximum at 0, and therefore verifies that we have no other cases of equality than those stated in Theorem 1.2.

To show that $u(b)\cot\beta + u'(b) > 0$, set $w(x) = u(x)\cot\beta + u'(x)$. It is clear that w satisfies the equation $-w'' = \lambda w$. We also have $w(0) = 0$ since u satisfies $u(0)\cos\beta + u'(0)\sin\beta = 0$. Finally, $w'(0) = -(\cot^2\beta + \lambda)u(0) > 0$ since $\cot^2\beta + \lambda = a(2\cot\beta - a) < 0$. Therefore w is a positive multiple of $\sinh(\sqrt{-\lambda}\,x)$ and thus $w(b) > 0$. We have now verified all claims of Theorem 1.2.

References

1. H. Brezis and E. Lieb, *A relation between pointwise convergence of functions and convergence of functionals.* Proc. Amer. Math. Soc. **88** (1983), 486–490.

2. H. Egnell, *Extremal properties of the first eigenvalue of a class of elliptic eigenvalue problems.* Ann. della Scuola Norm. Sup. di Pisa **XIV** (1987), 1–48.

3. M. Essén, *On estimating eigenvalues of a second order linear differential operator.* General Inequalities 5 (ed. by W. Walter), 347–366, Birkhäuser 1987.

4. W. N. Everitt, *On the spectrum of a second order linear differential equation with a p-integrable coefficient.* Applicable Anal. **2** (1972), 143–160.

5.a E. M. Harrell II, *Hamiltonian operators with maximal eigenvalues*. J. of Math. Phys. **25** (1984), 48–51.

5.b E. M. Harrell II, *Erratum to* [5a]. J. of Math. Phys. **27** (1986), 419.

6. G. Talenti, *Estimates for eigenvalues of Sturm-Liouville problems*. General Inequalities 4 (ed. by W. Walter), 341–350, Birkhäuser 1984.

7. B. Szőkefalvi-Nagy, *Ueber Integralungleichungen zwischen einer Funktion und ihrer Ableitung*. Acta Sci. Math. (Szeged) **10** (1941), 64–74.

8. E.J.M. Veling, *Optimal lower bounds for the spectrum of a second order linear differential equation with a p-integrable coefficient*. Proc. Roy. Soc. Edinburgh **92A** (1982), 95–101.

C. Bennewitz, Deptartment of Mathematics, University of Alabama at Birmingham, Birmingham, AL 35 294, U.S.A.

E.J.M. Veling, Nat. Inst. of Public Health & Environmental Protection, P. O. Box 1, 3720 BA Bilthoven, The Netherlands

International Series of Numerical Mathematics, Vol. 103, © 1992 Birkhäuser Verlag Basel

HELP integral and series inequalities

B.M. Brown, W.D. Evans and W.N. Everitt

Abstract

A survey of recent work on the Hardy Everitt Littlewood and Pólya (HELP) inequality and analogous series inequalities investigated by Brown and Evans is presented. Included is a description of the numerical techniques devised by Brown, Kirby and Pryce to determine the best constants in the inequalities.

1. Introduction

The original study of the Hardy Everitt Littlewood and Pólya (hence HELP) integral inequalities began in 1932 with the paper [20] by Hardy and Littlewood. In this work the Hardy-Littlewood inequality was first stated and proved; see [20, Theorem 6] which gives:

If y and y'' are L^2 in $(0,\infty)$ then

$$\left(\int_0^\infty y'^2\right)^2 < 4\int_0^\infty y^2 dx \int_0^\infty y''^2 dx \tag{1.1}$$

unless $y(x) = AY(Bx)$ where $B > 0$ and

$$Y(x) = e^{-\frac{x}{2}} \sin(x \sin\gamma - \gamma) \quad \left(\gamma = \frac{\pi}{3}\right) \tag{1.2}$$

when there is equality.

We note that the conditions on y force y' to be in $L^2(0,\infty)$; also that the number 4 in (1.1) is best possible. The form of the equalising function (1.2) is due to the connection between the inequality (1.1) and a problem in the calculus of variations for which the function Y in (1.2) is the extremal solution of the corresponding Euler-Lagrange differential equation.

This paper is in final form and no version of it will be submitted for publication elsewhere.

A detailed discussion of the Hardy-Littlewood inequality (1.1) which includes three different proofs of this result, is given in Hardy, Littlewood and Pólya [21, Section 7.8].

Since this paper is also concerned with inequalities for infinite series we give here the series analogue of (1.1) which was first considered by Copson [12] in 1979. Let $\mathbb{N}_0 = \{0, 1, 2, ...\}$ represent the non-negative integers and $\mathbb{N} = \mathbb{N}_0 \setminus \{0\}$. For any sequence of real numbers $\{a_n : n \in \mathbb{N}_0\}$ let the (first) difference sequence $\{\Delta a_n : n \in \mathbb{N}_0\}$ be determined by

$$\Delta a_n = a_{n+1} - a_n.$$

The second difference sequence $\{\Delta^2 a_n : n \in \mathbb{N}_0\}$ is then given by

$$\Delta^2 a_n \equiv \Delta(\Delta a_n) = a_{n+2} - 2a_{n+1} + a_n.$$

The Copson series inequality may now be stated in the form:

If $\{a_n : n \in \mathbb{N}_0\}$ is a sequence of real numbers such that both the infinite series

$$\sum_{n=0}^{\infty} a_n^2 \quad \textit{and} \quad \sum_{n=0}^{\infty} (\Delta^2 a_n)^2$$

are convergent, then

$$\left(\sum_{n=0}^{\infty} (\Delta a_n)^2 \right)^2 < 4 \sum_{n=0}^{\infty} a_n^2 \sum_{n=0}^{\infty} (\Delta^2 a_n)^2, \tag{1.3}$$

with equality only when $a_n = 0$ $(n = 0, 1, 2, ..)$.

As before we note that the conditions on the sequence $\{a_n\}$ force the infinite series $\sum_{n=0}^{\infty} (\Delta a_n)^2$ to be convergent. It is shown in [12] that the number 4 in (1.3) is best possible although in this case there are no equalising sequences which are equivalent to the extremal functions (1.2) for the integral inequality. Nevertheless the proof of the series inequality (1.3) is adapted from the results in [21, Section 7.8] for the integral inequality.

The HELP extension of the Hardy-Littlewood inequality was first considered by Everitt [19] in 1971, and then in a generalised form by Evans and Everitt [14] in 1982. These ideas were revised and extended by Bennewitz [6] in 1984.

We give here the general form of the HELP integral inequality as considered in [14]:

(i) Let [a,b) be an interval of the real line $\mathbb{R}$ with $-\infty < a < b \leq \infty$,

(ii) let the coefficients $p, q, w : [a, b) \to \mathbb{R}$ and satisfy the local Lebesgue integrability conditions

$$p^{-1}\left(\equiv \frac{1}{p}\right), q, w \in L^1_{loc}[a, b),$$

(iii) let the weight coefficient w satisfy

$$w(x) > 0 \quad \text{(almost all (Lebesgue) } x \in [a, b)),$$

(iv) let $L^2(a, b; w)$ denote the Hilbert function space of equivalence classes of complex-valued functions on [a,b) which satisfy

$$\int_a^b w(x)|f(x)|^2 dx < \infty,$$

and for which the inner-product is determined by

$$(f, g)_w = \int_a^b w(x) f(x) \bar{g}(x) dx,$$

(v) let the linear manifold Δ of $L^2(a, b; w)$ be determined by

$$\begin{aligned} \Delta \;=\; & \{f : [a, b) \to \mathbb{R} : \\ & f, pf' \in AC_{loc}[a, b); \quad f, w^{-1}(-(pf')' + qf) \in L^2(a, b; w)\}. \end{aligned}$$

Then the general HELP inequality takes the form

$$\begin{aligned} (1.4) \qquad & \left(\int_a^{\to b} (pf'^2 + qf^2) dx\right)^2 \\ & \leq K \int_a^b w f^2 dx \int_a^b w(w^{-1}(-(pf')' + qf))^2 dx \quad (f \in \Delta) \end{aligned}$$

where the notation indicates that the integral on the left hand side may be only conditionally convergent.

Unlike the original Hardy-Littlewood inequality (1.1) taking $f \in \Delta$ does not, in general, imply that the integral on the left of (1.4) is convergent or even conditionally convergent. To ensure that all terms in (1.4) are finite it is sufficient:

(1) to take $f \in \Delta$ when both terms on the right-hand side are finite,

(2) to make the technical assumption that the quasi-differential expression, $-(pf')' + qf$ with $f \in \Delta \subset L^2(a, b; w)$, is in the so-called strong limit-point condition at the end-point b, i.e.

$$(1.5) \qquad \lim_{x \to b} (pf')(x) g(x) = 0 \quad (f, g \in \Delta).$$

For a discussion of the significance and consequences of (1.5) see Evans and Everitt [14, Section 3]. Here it is sufficient to note that :

(a) the condition (1.5) gives conditional convergence of the integral on the left of (1.4), for each $f \in \Delta$,

(b) the condition (1.5) forces the end-point b to be a singular point of the Sturm-Liouville differential equation

$$-(py')' + qy = \lambda wy \text{ on } [a, b) \tag{1.6}$$

where $\lambda \in \mathbb{C}$, the complex number field, is the spectral parameter,

(c) there is an extensive literature giving explicit conditions on the coefficients p, q and w which ensure that the strong limit-point condition (1.5) is satisfied.

These definitions and conditions provide the framework in which the general HELP integral inequality (1.4) may be considered. The problem now is to decide if (1.4) is a valid inequality, i.e. to decide if there is a positive number K, which must be independent of $f \in \Delta$, for which the the inequality holds for all $f \in \Delta$. If such a number can be found then we denote the best possible number (that is the smallest such number) for which (1.4) holds, by the symbol K.

If, in this sense, the inequality is not valid then we write $K = \infty$ for (1.4), although this is of course a symbolic notation for the invalidity of the inequality.

In the case of a valid inequality with K so determined, or possibly characterised in some form, there is then the problem of describing all cases of equality in the resulting inequality.

All this may now be seen in the light of the original Hardy-Littlewood example given by (1.1). In the general case (1.4) take

$$a = 0, \quad b = \infty; \quad p(x) = w(x) = 1, \quad q(x) = 0 \ \ (x \in [0, \infty)).$$

It may be shown that the strong limit-point condition is satisfied, i.e. if f, g, f'', g'' are all in $L^2(0, \infty)$ then it follows that $\lim_{x \to \infty} f'(x)g(x) = 0$. In this case the analysis in [20] and [21], alternatively see [14, Section 9, Example 1], shows that $K = 4$ and that all cases of equality are given by (1.2).

It may be asked why the general inequality (1.4) should be studied as an extension of the original Hardy-Littlewood inequality (1.11). The main reason is that the connection with the Sturm - Liouville differential equation (1.6) allows for a general analysis of the inequality (1.4) firstly in terms of the classical theory of the differential equation, and secondly in terms of the spectral theory of the differential operators associated with (1.6) and defined in the Hilbert function space

$L^2(a,b;w)$. It proves to be possible to obtain certain general results about the HELP inequality (1.4), and in certain explicit examples to use a combination of spectral theory and properties of special functions to characterise the best possible number K, and then to identify all cases of equality. All this is best seen in terms of the analysis and examples in Evans and Everitt [14], and in the subsequent account given by Evans and Everitt [16].

In Section 2 we report on a recent paper by Evans and Everitt [17] on the extension of the HELP integral inequality to the case when the end-point b in (1.4) is not in the limit-point classification, i.e. to the cases when this end-point is either regular, or in the limit-circle classification in $L^2(a,b;w)$.

There is a corresponding extension of the Copson series inequality (1.3) to the general second-order Sturm-Liouville difference equation considered on the set $\mathbb{N}_0$ of non-negative integers. Details of this extension are given in Section 3 below, which draws results from the recent paper by Brown and Evans [8].

The analytic difficulties in analysing any particular example of the HELP integral inequality may be extreme; the seemingly simple examples do sometimes yield to the techniques of showing that the inequality is valid, but the subsequent search for further information about the best possible number K may be beyond the reach of presently known methods. For this reason recent efforts have been made to apply numerical analytic techniques to the determination of accurate numerical values for the best possible number K. Success in this direction has been reported on in the recent Ph.D. thesis of Kirby [24] and in the recent publications of Brown, Kirby and Pryce [10], and [11]. In Section 4 we report on these results and the remarkable implications for the future study of the integral and series inequalities.

Finally in this introductory section we outline the analytic techniques for the analysis of the HELP integral inequality (1.4) to pave the way for the results reported on in Sections 2, 3 and 4 below. Full details of the proof of the results given here, together with a wide range of examples, are to be found in Evans and Everitt [16]. Earlier results, but with some additional conditions on the coefficients p, q and w in (1.4) are given in Everitt [19], and Evans and Zettl [18].

We start with the pair of solutions, in the now standard notation, $\theta(x,\lambda)$ and $\phi(x,\lambda)$ ($x \in [a,b)$, $\lambda \in \mathbb{C}$, of the differential equation (1.6) determined by the following initial conditions at the regular end point a, and this for all $\lambda \in \mathbb{C}$;

$$\theta(a,\lambda) = 0, \quad (p\theta')(a,\lambda) = 1,$$

$$\phi(a,\lambda) = -1, \quad (p\phi')(a,\lambda) = 0. \tag{1.7}$$

The standard existence theorem, with the above given conditions on the coefficients p, q and w, imply the unique existence of θ and ϕ on $[a,b) \times \mathbb{C}$; also that $\theta(x,\cdot)$ and $\phi(x,\cdot)$ are holomorphic functions on the whole plane $\mathbb{C}$.

The Titchmarsh-Weyl theory for the differential equation (1.6), see [14, Section 4], implies the existence of a unique pair of Cauchy analytic functions $\{m_+, m_-\}$ with the following properties, here we use the notation $\mathbb{C}_+ = \{\lambda \in \mathbb{C} : \operatorname{im}(\lambda) > 0\}$, $\mathbb{C}_- = \{\lambda \in \mathbb{C} : \operatorname{im}(\lambda) < 0\}$,

(i) $m_\pm : \mathbb{C}_\pm \to \mathbb{C}_\pm$,

(ii) $m_\pm$ is holomorphic on $\mathbb{C}_\pm$,

(iii) $\bar{m}_\pm(\lambda) = m_\mp(\bar{\lambda})$ $(\lambda \in \mathbb{C}_\pm)$,

(iv) if $\psi_\pm(x, \lambda)$ is the solution of (1.6) defined by

$$\psi_\pm(x,\lambda) = \theta(x,\lambda) + m_\pm(\lambda)\phi(x,\lambda) \quad (x \in [a,b),\ \lambda \in \mathbb{C}_\pm),$$

then

(i) $\psi_\pm(\cdot, \lambda) \in L^2(a, b; w)$ $(\lambda \in \mathbb{C}_\pm)$,
(ii) $\int_a^b w(x)|\psi_\pm(x,\lambda)|^2 dx = \operatorname{im} m_\pm(\lambda))/\operatorname{im}(\lambda)$ $(\lambda \in \mathbb{C}_\pm)$.

The Titchmarsh-Weyl m-coefficient is then defined by

$$m(\lambda) = m_\pm(\lambda) \quad (\lambda \in \mathbb{C}_\pm)$$

and $m(\cdot)$ is holomorphic on $\mathbb{C} \setminus \mathbb{R} = \mathbb{C}_+ \cup \mathbb{C}_-$. The behaviour of the m-coefficient near the real axis of $\mathbb{C}$ is linked with the spectral properties of the self-adjoint operator generated by the differential equation (1.6) in the Hilbert function space $L^2(a, b; w)$.

The uniqueness of the m-coefficient for (1.6) as given here, follows from the technical assumptions (1.5) that the coefficients p, q and w have been chosen to give the strong limit-point condition at the singular end-point b.

The connection between the m-coefficient for the differential equation (1.6) and the HELP integral inequality may be stated as follows:

(i) let $\lambda = r\exp(i\theta)$ with $r \in (0, \infty)$ and $\theta \in [0, 2\pi)$,

(ii) for $\theta \in [0, \frac{1}{2}\pi]$ let the line segments $L_\pm(\theta)$ be defined by $L_+(\theta) := \{r\exp(i\theta) : r \in (0,\infty)\}$, $L_-(\theta) := \{r\ exp\,(i(\theta + \pi)) : r \in (0, \infty)\}$,

(iii) let $\theta_\pm \in [0, \frac{1}{2}\pi]$ be defined by $\theta_\pm := \inf\{\theta \in (0, \frac{\pi}{2}]$: for all $\phi \in [\theta, \frac{\pi}{2}]$, $\mp\operatorname{im}[\lambda^2 m_\pm(\lambda)] \geq 0$ $(\lambda \in L_\pm(\phi))\}$,

(iv) define $\theta_0 \in [0, \frac{1}{2}\pi]$ by $\theta_0 := \max\{\theta_+, \theta_-\}$,

(v) define the sets $E_\pm$ and the functions $Y_\pm$ by $E_\pm := \{r \in (0,\infty) : \lambda \in L_\pm(\theta_0)$ and $\mathrm{im}\,(\lambda^2 m_\pm(\lambda)) = 0\}$ and $Y_\pm(x;r) := \mathrm{im}\,(\lambda\psi_\pm(x,\lambda))$ $(x \in [0,\infty);\ \lambda \in L_\pm(\theta_0))$.

Then we have the following results:

(1) in all cases $\theta_0 \neq 0$, i.e. $\theta_0 \in (0, \frac{1}{2}\pi]$,

(2) the HELP integral inequality (1.4) is valid if and only if $\theta_0 \in (0, \frac{1}{2}\pi)$, i.e.

$$\theta_0 \neq \frac{1}{2}\pi, \tag{1.8}$$

(3) when (1.8) is satisfied the best possible number K in the inequality (1.4) is given by

$$K = \sec^2(\theta_0), \tag{1.9}$$

(4) when (1.8) is satisfied and K is given by (1.9) then all $f \in \Delta$ which give equality in (1.4) are determined by the three following mutually exclusive conditions:

(i) $f = 0$, the zero function on $[a, b)$,

(ii) there exists $f \in \Delta$ with $f \neq 0$, $-(pf')' + qf = 0$, either $f(a) = 0$ or $(pf')(a) = 0$ (but not both) in which case both sides of (1.4) are equal to zero,

(iii) $E_+ \cup E_- \neq \emptyset$ and then, for $A \in \mathbb{C}$ with $A \neq 0$,

$$f(x) = AY_\pm(x;r) \quad (x \in [a,b),\ \ r \in E_\pm)$$

in which case the two sides of (1.4) are equal but not to zero.

The effectiveness of this connection between the HELP integral inequality and the Titchmarsh-Weyl m-coefficient may be seen in the application to the sets of examples considered in Evans and Everitt [14, Section 9] and [16, Section 4]. In all these examples information on the best possible number K and on the cases of equality has been obtained by analytic methods. However many of the remaining examples of interest for qualitative and general properties of the HELP inequality are so intractable analytically that numerical techniques have been introduced in the search for additional information. In general numerical methods cannot replace analytical techniques but in the case of the HELP inequality they have provided results of extraordinary value.

To illustrate the interplay between analytical and numerical results we give a general case of the HELP integral inequality which commands special attention from both view points.

In addition to the general conditions on coefficiants p, q and w given above we now add two special conditions:

(i) $p(x) \geq 0 \quad (x \in [a, b))$,

(ii)

$$\lim_{x \to b} q(x) = \infty, \tag{1.10}$$

Consider then the Dirichlet (D) and Neumann (N) boundary value problems for the differential equation

$$-(py')' + qy = \lambda wy \quad \text{on} \quad [a, b)$$

with one or other of the two boundary conditions at the regular end-point a

$$\mathrm{D} : y(a) = 0, \quad \mathrm{N} : (py')(a) = 0.$$

For both of these problems the spectrum is discrete, simple and bounded below; here the additional conditions (1.10) play a significant role. Let the spectra of the two problems be denoted by

$$\mathrm{D} : \{\mu_n : n \in \mathbb{N}_0\}, \qquad \mathrm{N} : \{\lambda_n : n :\in \mathbb{N}_0\}.$$

Then general results show that these two sets of real numbers satisfy the properties:

(i) $\lambda_n < \mu_n < \lambda_{n+1} < \mu_{n+1} \quad (n \in \mathbb{N}_0)$,

(ii) $\lim_{n\to\infty} \mu_n = \lim_{n\to\infty} \lambda_n = \infty$.

Consider now the HELP inequality for the coefficients $p, q - \tau w, w$ on $[a, b)$, where $\tau \in \mathbb{R}$ is a real parameter, i.e.

$$\begin{aligned} &\left[\int_a^b (pf'^2 + (q - \tau w)f^2)dx\right]^2 \\ &\leq K(\tau) \int_a^b wf^2 dx \int_a^b w(w^{-1}(-(pf')' + (q - \tau w)f))^2 dx \quad (f \in \Delta(\tau)) \end{aligned} \tag{1.11}$$

where the dependence of K and Δ on the parameter τ is shown. Then from results in [9, Section 17] it may be shown that the following properties hold

(i)

$$\Delta(\tau) = \Delta(0) = \Delta \quad (\tau \in \mathbb{R}),$$

(ii) the inequality (1.11) is valid if and only if

$$\text{for some } n \in \mathbb{N}_0 \quad \tau = \mu_n \quad \text{or} \quad \tau = \lambda_n \tag{1.12}$$

provided that additionally the coefficients p and w satisfy technical conditions to give an asymptotic form to the m-coefficient of the kind given in [14, Section 10, (10.4)].

This last result (1.12) is typical of the type of qualitative result which can be proved using the connection between the HELP inequality and the m-coefficient.

In respect of the technical conditions to be satisfied by p and w in order that the result (1.12) follows, it should be noted that these conditions are always satisfied when $p(x) \geq 0$ $(x \in [a,b))$ and $w(x) = 1$ $(x \in [a, a+\delta])$ for some $\delta > 0$.

To illustrate these results, in particular (1.12), we consider the case, for some $\alpha \in (0, \infty)$.

$$a = 0, \quad b = \infty; \quad p(x) = w(x) = 1, \quad q(x) = x^\alpha \quad (x \in [0, \infty)) \tag{1.13}$$

for which we designate the best possible constants K by

$$K_\alpha(\lambda_n) \quad (n \in \mathbb{N}_0), \qquad K_\alpha(\mu_n) \quad (n \in \mathbb{N}_0)$$

to indicate their dependence on the particular α. (Note that the eigenvalues $\{\lambda_n\}$ and $\{\mu_n\}$ depend also on $\alpha \in (0, \infty)$.)

The only case of (1.13) for which the HELP inequality has been completely analysed analytically is when $\alpha = 2$. Full details are given in [16, Section 4, Example 7] and [15]; we have

$$\lambda_n = 4n + 1 \quad (n \in \mathbb{N}_0), \qquad \mu_n = 4n + 3 \quad (n \in \mathbb{N}_0),$$

$$K_2(\lambda_n) = 4 \quad (n \in \mathbb{N}_0),$$

$$K_2(\mu_n) > 4 \quad (n \in \mathbb{N}_0),$$

$$K_2(\mu_n) = 4 + \frac{3}{\pi^2 n^2} + O(\frac{1}{n^3}) \quad (n \to \infty).$$

All other information concerning this example when $q(x) = x^\alpha$ $(x \in [0, \infty))$ is available only as a result of the numerical analysis of the problem as considered in [10], [11], [24]. However the now established authority of these numerical results gives compelling reasons to believe that the following properties of these examples gives a correct analytic viewpoint:

(i) for all $\alpha > 0$ $K_\alpha(\lambda_n) = 4$ $(n \in \mathbb{N}_0)$,

(ii) for $\alpha = 1$ $K_1(\mu_n) = 4$ $(n \in \mathbb{N}_0)$,

(iii) for all $\alpha > 0$ but $\alpha \neq 1$ $K_\alpha(\mu_n) > 4$ $(n \in \mathbb{N}_0)$,

(iv) $\lim_{\alpha\to\infty} K_\alpha(\mu_n)$ exists and is finite, for all $n \in \mathbb{N}_0$, and

$$\lim_{\alpha\to\infty} K_\alpha(\mu_n) > 4 \quad (n \in \mathbb{N}_0).$$

Additional information about these numerical results is given in Section 4 below.

2. Limit-circle and regular HELP integral inequalities

In Section 1 above we outlined the connection between the HELP integral inequality and the Titchmarsh-Weyl m-coefficient, presupposing that the associated differential equation is regular at one end-point and singular but strong limit-point at the other end-point, of the interval of definition. In this section we report on the cases when the differential equation remains regular at the one end-point, but is either non-oscillatory limit-circle or regular at the other end-point.

The significant references are Bennewitz [6], which considers an alternative approach to the analysis of the HELP integral inequality, Bennewitz [7], Evans and Everitt [16, Section 5], and the recent paper Evans and Everitt [17]. Our account here is based on the results in [17] but the earlier work of Bennewitz is particularly significant and inflenced the results presented in both the papers [16] an [17].

A general account of the HELP integral inequality for various classifications of the two end-points of the interval (a, b) of the associated Sturm-Liouville differential equation, is given in [16, Section 5].

For the case considered in this Section, one regular end-point and either regular or limit-circle at the other end-point, one significant remark should be made at this stage. In the circumstances any self-adjoint differential operator, defined from the associated Sturm-Liouville differential expression in the weighted integrable-square space $L^2(a, b; w)$, always has a discrete spectrum. This spectral result implies that the associated Titchmarsh-Weyl m-coefficient is a meromorphic function on the complex plane $\mathbb{C}$ with a countable infinity of simple poles on the real axies $\mathbb{R}$ of $\mathbb{C}$, but holomorphic on the open upper and lower half planes $\mathbb{C}_+$ and $\mathbb{C}_-$ of $\mathbb{C}$. This leads then as in Section 1 (see (1.10) onwards) to consideration of the translated

HELP integral inequality

$$(2.1)\qquad \left(\int_a^{\to b}(pf'^2+(q-\tau w)f^2)dx\right)^2 \le K(\tau)\int_a^b wf^2dx\int_a^b w(w^{-1}(-(pf')'+(q-\tau w)f))^2dx \quad (f\in\Delta(\tau))$$

where $\tau \in \mathbb{R}$ is a real parameter. In the regular/non-oscillatory limit-circle case under consideration in this section we find also that the inequality (2.1) is valid on a well determined, countable set of values of the parameter τ.

Proceeding now with the case in hand the definition and form of the HELP inequality in this section is taken from the previous definitions given in Section 1. Again we consider the inequality (1.4) over the interval $[a, b)$ of the real line $\mathbb{R}$ with the end-point a given as regular. We now impose the additional conditions:

(1) $p(x) \geq 0$ (almost all $x \in [a, b)$),

(2) the end point b is *either* regular, i.e. $-\infty < a < b < \infty$ and $p^{-1}, q, w \in L^1(a, b)$ *or* b is non-oscillatory limit-circle in which case the following conditions hold:

(i) $-\infty < a < b \leq \infty$ and $p^{-1}, q, w \in L^1_{loc}(a, b)$,

(ii) if $b < \infty$, $\int_a^b\{p^{-1} + |q| + w\} = \infty$,

(iii) all non-null solutions of the Sturm-Liouville differential equation

$$(2.2)\qquad -(py')' + qy = \mu wy \text{ on } [a, b),$$

for some (and then for all) real number μ lie in the space $L^2(a, b; w)$ (the limit-circle requirement) and have only a finite number of zeros in the interval $[a, b)$ (the non oscillatory requirement). (Note that these properties of the solution given in (iii) for the non-oscillatory limit-circle case are automatically satisfied in the regular case).

There are known conditions on the coefficients p, q and w which ensure that the non-oscillatory limit-circle property holds, for example see [17, Sections 3 and 5].

Under the above given classification for the end points a and b it is known that for any $\mu \in \mathbb{R}$ the differential equation (2.2) has a pair of real-valued linearly independent solutions u, v with the properties, see [17, Section 2]

(1) $p(x)\{u(x)v'(x) - u'(x)v(x)\} = 1 \quad (x \in [a, b))$,

(2) for some point $c \in (a,b)$

$$u(x), v(x) \neq 0 \quad (x \in [c,b)), \tag{2.3}$$

(3) $\lim_{x\to b} u(x)/v(x) = 0$,

(4) $\int_c^b (1/pu^2) = \infty, \quad \int_c^b (1/pv^2) < \infty$,

(5) $\int_c^b wu^2 / \int_c^b wv^2 = O(u(x)^2/v(x)^2) = o(1) \quad (x \to b)$.

The solution u, with the above properties, is called the *principal* solution of the differential equation (2.3) at the end-point b. The solution v is called a *non-principal* solution; given v, clearly $v + \alpha u$ $(\alpha \in \mathbb{R})$ is also a non principal solution at the end-point b. However up to linear dependence the principal solution u is unique.

In the case when the equation (2.2) is regular at b, the principal solution u is determined on $[a,b]$ from the initial conditions

$$u(b) = 0, \qquad (pu')(b) = 1. \tag{2.4}$$

In this case, for technical convenience, we determine the non-principal solution v from the initial conditions

$$v(b) = -1, \qquad (pv')(b) = 0. \tag{2.5}$$

We now consider the extension of the HELP integral inequality (1.4) to these regular and non-oscillatory limit-circle cases. This requires one additional technical condition which is essential to ensure that the integral on the left-hand side of (1.4) is at least conditionaly convergent: given all the above notations definitions and conditions let

$$\lim_{x\to b} p(x)v'(x)v(x) = 0 \tag{2.6}$$

which, if failing, let

$$\lim_{x\to b} p(x)u'(x)u(x) = 0. \tag{2.7}$$

We have presented the condition in this form since it may be shown that (2.6) implies (2.7), see [17, Section 3, Remark 3.13], but that in general in the limit-circle case (2.7) may hold whilst (2.6) is not satisfied, see [17, Section 3, Example 3.6].

Note that both (2.6) and (2.7) always hold in the case when the end-point b is regular, provided that the non-principal solution v is determined by the initial conditions (2.5).

As in Section 1 let Δ denote the maximal linear manifold in $L^2(a,b;w)$ associated with the differential equation (2.3), i.e.

$$\Delta = \{f : [a,b) \to \mathbb{R} : \quad f, pf' \in AC_{loc}[a,b); \quad f, w^{-1}(-(pf')' + qf) \in L^2(a,b;w)\}. \tag{2.8}$$

Unlike the limit-point case in Section 1 it is rare for the HELP inequality to be valid on Δ in the regular/limit-circle case. The underlying reason is that in the latter case all the solutions of the differential equation

$$-(py')' + qy = 0 \text{ on } [a,b)$$

belong to the domain Δ; such solutions may invalidate the inequality by rendering the right-hand side of (1.4) to be zero, and yet give a non-zero left-hand-side. Indeed, in general, it proves to be necessary to reduce the domain Δ to a linear sub-manifold in order to eliminate such elements of Δ to obtain a valid inequality of the form (1.4).

It can happen that the inequality (1.4) is valid on the maximal domain Δ in the regular/limit-circle case but such circumstances are exceptional; for a discussion of such possibilities see Bennewitz [6] and [7], and [17, Section 1].

For the regular and non-oscillatory limit-circle cases as considered above we define two sub-domains of Δ, first choosing, for some real μ, u and v as principal and non-principal solutions of the differential equation (2.3) so that

(i) Property 1 of (2.3) is satisfied

(ii)

(2.9) in the limit-circle case at b (2.6) holds or, if failing, (2.7) holds.

Now define (the general theory shows that the limits given exist and are finite)

$$\Delta_u = \{f \in \Delta : \lim_{x\to b} p(x)(f(x)u'(x) - f'(x)u(x)) = 0\}, \tag{2.10}$$

$$\Delta_v = \{f \in \Delta : \lim_{x\to b} p(x)(f(x)v'(x) - f'(x)v(x)) = 0\}. \tag{2.11}$$

It can be shown, see [17, Section 3], that with (2.7) and (2.10) holding for the principal solution u (respectively (2.6) and (2.11) holding for the non-principal solution v) the left-hand side of inequality (1.4) is always finite for all $f \in \Delta_u$ ($f \in \Delta_v$).

At this stage one point should be noted. In the limit-point case discussed in Section 1 the HELP inequality is considered on the maximal domain Δ; in particular there is no boundary condition at either the regular end-point a or the singular end-point b. In this regular or limit-circle case at b it turns out to be essential to place a boundary condition at the singular end-point and this is precisely the effect of the conditions (2.10) or (2.11). However as with the limit-point case in Section 1 no boundary condition is placed at the regular end-point a when the inequality domains Δ_u and Δ_v are determined.

As discussed in [17, Sections 3 and 4] the above conditions also give the existence of a uniquely determined Titchmarsh-Weyl m-coefficient here denoted by m_u (respectivly m_v). The definition of m_u is given in [17, Section 3, (3.17)], and there is an equivalent definition for m_v. In terms of the special solutions θ, ϕ determined by the initial conditions (1.7) we now define the solutions

$$\psi_u(x,\lambda) = \theta(x,\lambda) + m_u\phi(x,\lambda) \tag{2.12}$$

$$\psi_v(x,\lambda) = \theta(x,\lambda) + m_v\phi(x,\lambda) \tag{2.13}$$

for all $x \in [a,b]$ and for all $\lambda \in \mathbb{C} \setminus \mathbb{R}$.

All the properties of the m-coefficient and the solutions ψ given in Section 1 now apply to m_u and ψ_u, m_v and ψ_v. Additionally in this case, since as mentioned above the spectra of the associated differential operators are discrete, the m-coefficients m_u and m_v are meromorphic on $\mathbb{C}$.

With all this to hand the connection between the general HELP inequality (1.4) and the m-coefficient and ψ-solutions follows entirely the same results as given in Section 1. The critical angle θ_0 and the extremal functions $Y_\pm$ are determined in the same way; the proof of these results follows the argument given in [14, Sections 7 and 9] and can be found in [17, Sections 3 and 4]; see in particular [17, Theorems 3.8, 3.14 and 4.1].

We are now in a position to look at the translated HELP inequality (2.1). As in Section 1 there is a requirement for an additional technical condition on the coefficients p and w to give an asymptotic form to the m-coefficients m_u and m_v of the kind given in [14, Section 10, (10.4)]. We suppose in giving the results below, that such a condition is placed on the coefficients p and w.

Consider now the HELP inequality (2.1) on the domain Δ_u. With all the information to hand as given above it is known that m_u has an infinity of poles (all simple) on the real axis and that these are bounded below; if these poles are at the points $\{\lambda_n : n \in \mathbb{N}_0\}$ then we have

$$\lambda_n < \lambda_{n+1} \quad (n \in \mathbb{N}_0) \text{ and } \lim_{x\to\infty} \lambda_n = \infty.$$

It may be then shown that m_u has an infinity of simple zeros $\{\mu_n : n \in \mathbb{N}_0\}$ on the real axis, and that these zeros relate to the poles of m_u by

$$\lambda_n < \mu_n < \lambda_{n+1} < \mu_{n+1} \quad (n \in \mathbb{N}_0).$$

Following the results given in Section 1 it may be shown that the following properties hold:

(i) $\Delta_u(\tau) = \Delta_u(0) = \Delta_u \ (\tau \in \mathbb{R})$,

(ii) the inequality (2.1) is valid *on the domain* Δ_u if and only if

$$\text{for some } n \in \mathbb{N}_0, \quad \tau = \mu_n \text{ or } \lambda_n.$$

Entirely similar results hold for the HELP inequality (2.1) on the domain Δ_v. Note that the position of the simple poles and zeros $\{\lambda_n : n \in \mathbb{N}_0\}$ and $\{\mu_n : n \in \mathbb{N}_0\}$ depend on u and v but the same notation suffices and the form of the results is the same.

Examples of the results presented in this Section are given in [17 , Section 5]. In particular the numerical methods reported on in that Section are given in [11, Section 5]. In particular the numerical methods reported on in Section 4 below have been applied to the examples and a clear view is obtained of the magnitude of the best possible constant K, and the cases of equality both for regular and non-oscillatory limit-circle end-points.

3. Extensions of Copson's inequality

3.1 Introduction

The series analogue of the HELP inequality (1.4) considered by Brown and Evans in [8] is of the form

$$\left\{\sum_{n=1}^{\infty}(p_n|\Delta x_n|^2 + q_n|x_n|^2) + p_0|\Delta x_0|^2\right\}^2 \le K\sum_{n=0}^{\infty} w_n|x_n|^2 \sum_{n=1}^{\infty} w_n\left|\frac{1}{w_n}Mx_n\right|^2 \tag{3.1}$$

where M is the difference expression

$$Mx_n := -\Delta(p_{n-1}\Delta x_{n-1}) + q_n x_n \quad (n \in \mathbb{N}) \tag{3.2}$$

and $(p_n)_0^\infty$, $(q_n)_1^\infty$, and $(w_n)_0^\infty$ are real sequences which satisfy the conditions

$$p_n \neq 0, \quad w_n > 0 \quad (n \in \mathbb{N}_0). \tag{3.3}$$

The inequality (3.1) is required to hold for all sequences $(x_n)_0^\infty$ in the linear manifold

$$(3.4)\qquad \mathcal{D} := \left\{ (x_n)_0^\infty : \sum_{n=0}^\infty w_n|x_n|^2 < \infty, \quad \sum_{n=1}^\infty w_n|\frac{1}{w_n}Mx_n|^2 < \infty \right\}.$$

The Copson inequality (1.3) is the special case $p_n = w_n = 1$, $q_n = 0$, when $Mx_n = -\Delta^2 x_{n-1}$.

An important feature of Everitt's result in [19] is the intimate relation between the integral inequality (1.4) and the Titchmarsh-Weyl m-function for the differential equation

$$-(py')' + qy = \lambda wy \quad (\lambda \in \mathbb{C}).$$

As was described in Section 1, the validity of the inequality (1.4), the value of the best constant K and the equalising functions (if they exist) are determined by properties of m. A similar connection was established in [8] between the series inequality (3.1) and the analogue of the m-function for the recurrence relation

$$(3.5)\qquad Mx_n = \lambda w_n x_n \quad (n \in \mathbb{N},\ \lambda \in \mathbb{C})$$

introduced simultaneously by Hellinger and Nevanlinna in [22], [26]. This function is defined as follows. Let $\phi(\lambda) = (\phi_n(\lambda))_0^\infty$, $\theta(\lambda) = (\theta_n(\lambda))_0^\infty$ be the solutions of (3.5) which satisfy the initial conditions

$$(3.6)\qquad \phi_0 = 0, \quad p_0\phi_1 = 1, \quad \theta_0 = 1, \quad p_0\theta_1 = 0,$$

and set

$$(3.7)\qquad f_k(\lambda, z) := -\frac{p_{k-1}\theta_k(\lambda) + z\theta_{k-1}(\lambda)}{p_{k-1}\phi_k(\lambda) + z\phi_{k-1}(\lambda)} \quad (k \in \mathbb{N},\ z \in \mathbb{C},\ \lambda \in \mathbb{C}_+).$$

The function $z \mapsto f_k(\lambda, z)$ maps $\mathbb{R}$ onto a circle $C(k,\lambda)$ in $\mathbb{C}_+$ and maps $\mathbb{C}_-$ onto the disc $D(k,\lambda)$ interior to $C(k,\lambda)$. As $k \to \infty$ the circles $C(k,\lambda)$ are nested and converge either to a point $m(\lambda)$ or a circle $C(\infty,\lambda)$ in $\mathbb{C}_+$: these alternatives are known as the limit-point and limit-circle cases respectively. There are similar alternatives when $\lambda \in \mathbb{C}_-$ in (2.7) the disks $D(k,\lambda)$ now being the images of $\mathbb{C}_+$ under $z \mapsto f_k(\lambda, z)$. Morover the limit-point, limit-circle classification of M is independent of $\lambda \in \mathbb{C} \setminus \mathbb{R}$.

It is proved by Hinton and Lewis in [23] that M is in the limit-point case if and only if

$$(3.8)\qquad \lim_{k\to\infty} \{p_k(\overline{x}_{k+1}\Delta y_k - y_{k+1}\Delta\overline{x}_k)\} = 0$$

for all $(x_k)_0^\infty, (y_k)_0^\infty \in \mathcal{D}$. To consider (3.1), Brown and Evans in [8] imposed the condition

$$\lim_{k\to\infty} p_k \bar{x}_{k+1} \Delta y_k = 0 \quad ((x_k)_0^\infty \in \mathcal{D}); \tag{3.9}$$

when this is satisfied M is said to be in the strong limit-point case. This ensures that the series on the left-hand side of (3.1) is convergent (possible conditionally) for all $(x_k)_0^\infty \in \mathcal{D}$. Furthermore (3.9) obviously implies (3.8) and hence the strong limit-point condition implies the limit-point condition. The converse need not be true, as is the case for the corresponding notions in Section 1 when M is the differential expression given by $My = -(py')' + qy$.

If M is in the limit-point case, it follows from the Hellinger-Nevanlinna theory that there exists a pair of functions $\{m_+, m_-\}$ with the following properties:

(i) $m_\pm : \mathbb{C}_\pm \to \mathbb{C}_\pm$,

(ii) $m_\pm$ is holomorphic in $\mathbb{C}_\pm$,

(iii) $\bar{m}_\pm(\lambda) = m_\mp(\bar{\lambda})$ $(\lambda \in \mathbb{C}_\pm)$.

The m function is defined by

$$m(\lambda) = m_\pm(\lambda) \quad (\lambda \in \mathbb{C}_\pm).$$

It is a Nevanlinna function, being holomorphic in $\mathbb{C}_+ \cup \mathbb{C}_-$ and mapping $\mathbb{C}_+$ and $\mathbb{C}_-$ into themselves. The sequence $\psi(\lambda) = (\psi_n(\lambda))_0^\infty$ defined by

$$\psi_n(\lambda) = \theta_n(\lambda) + m(\lambda)\phi_n(\lambda) \quad (\lambda \in \mathbb{C} \setminus \mathbb{R}) \tag{3.10}$$

satisfies

$$\sum_{n=1}^{\infty} w_n |\psi_n(\lambda)|^2 = \frac{\text{im}\,[m(\lambda)]}{\text{im}\,[\lambda]} \quad (\lambda \in \mathbb{C} \setminus \mathbb{R}) \tag{3.11}$$

and, indeed, is the unique solution of (3.5) for $(\lambda \in \mathbb{C} \setminus \mathbb{R})$ which satisfies $\sum_{n=1}^\infty w_n |x_n|^\infty < \infty$.

The result proved in [8] is as follows. Let

$$\theta_\pm := \inf\{\theta \in (0, \frac{\pi}{2}] : \text{for all } \phi \in [\theta, \frac{\pi}{2}],\ \mp\text{im}[\bar{\lambda}^2\{m(\lambda) - p_0 + \lambda w_0\}] \geq 0 \quad (\lambda \in L_\pm(\phi))\},$$

$$\theta_0 := \max(\theta_+, \theta_-),$$

$$E_\pm := \{r \in (0,\infty) : \lambda \in L_\pm(\theta_0) \text{ and } im\,[\bar{\lambda}^2\{m(\lambda) - p_0 + \lambda w_0\}] = 0\},$$

$$Y_n(r) := \text{im}[\bar{\lambda}\psi_n(\lambda)]\ (\lambda \in L_\pm(\theta_0), n \in \mathbb{N}_0).$$

Then

(1) $\theta_0 \in (0, \frac{\pi}{2}]$,

(2) the inequality (3.1) is valid if and only if $\theta_0 \in (0, \frac{\pi}{2})$,

(3) if $\theta_0 \in (0, \frac{\pi}{2})$ then the best possible constant K in (3.1) is $K = \sec^2 \theta_0$,

(4) if $\theta_0 \in (0, \frac{\pi}{2})$ and $K = \sec^2 \theta_0$ then the non-trivial sequences $(x_n)_0^\infty \in \mathcal{D}$ which give equality in (3.1) are determined by the following distinct conditions:

 (i) there exist $(x_n)_0^\infty \not\equiv 0$ such that $\sum_{n=1}^\infty w_n |x_n|^2 < \infty$, $Mx_n = 0$ $(n \in \mathbb{N}_0)$ and either $x_0 = 0$ or $\Delta x_0 = 0$ in which case both sides of (3.1) are equal to zero,

 (ii) $E_+ \cup E_- \neq \emptyset$ and $x_n = AY_n(r)$ $(r \in E_\pm, n \in \mathbb{N}_0)$ with $A \in \mathbb{C} \setminus \{0\}$, in which case both sides of (3.1) are equal but not zero.

The proof of this result in [8] is based on the following observations and is motivated by the operator theoretic proof given by Evans and Zettl in [18] of Everitt's result for the inequality (1.4) in [19]. Let

$$\mathcal{D}_0' := \{\mathbf{x} = (x_n)_0^\infty : x_0 = 0 \text{ and } x_n = 0 \\ \text{for all but a finite number of values of } n\ \}$$

and denote by l_w^2 the space of sequences $(x_n)_0^\infty$ such that $\sum_{n=0}^\infty w_n |x_n|^2$ is convergent; l_w^2 is a Hilbert space with inner product

$$(\mathbf{x}, \mathbf{y}) = \sum_{n=0}^\infty w_n x_n \bar{y}_n, \quad (\mathbf{x} = (x_n)_0^\infty, \mathbf{y} = (y_n)_0^\infty).$$

Define the linear operator $T_0' : \mathcal{D}_0' \mapsto l_w^2$ by $T_0' : \mathbf{x} \mapsto \mathbf{y}$, where

$$w_n y_n = \begin{cases} Mx_n, & n \in \mathbb{N}, \\ -p_0 x_1, & n = 0, \end{cases} \tag{3.12}$$

and let G_0' denote its graph in $l_w^2 \times l_w^2$, namely,

$$G_0' := \{< \mathbf{x}, T_0' \mathbf{x} >: \mathbf{x} \in \mathcal{D}_0'\}.$$

Then, the closure G_0 of G_0' is a closed symmetric relation in l_w^2 and it transpires that its adjoint $G_0^* = (G_0')^*$ is given by

$$G_0^* = G := \{< \mathbf{x}, \mathbf{x}^* >: w_n x_n^* = Mx_n \text{ for } n \in \mathbb{N}, \quad \mathbf{x}, \mathbf{x}^* \in l_w^2\};$$

see [5] for background information on linear relations.

Bennewitz in [5, Theorem 1.4] gives an analogue for symmeteric relations of the von Neumann formula for symmetric operators and their adjoints. This is the direct sum decomposition

$$G = G_0 \oplus N_\lambda \oplus N_{\bar{\lambda}} \ (\text{ im } \lambda \neq 0), \tag{3.13}$$

where

$$\begin{aligned} N_\lambda &:= \{< \mathbf{x}, \lambda\mathbf{x} > \in G\} \\ &= \{< \mathbf{x}, \lambda\mathbf{x} >: \mathbf{x} \in l_w^2 \text{ and } Mx_n = \lambda w_n x_n, n \in \mathbb{N}\}. \end{aligned}$$

In fact the sum is an orthogonal sum with respect to the inner product

$$(< \mathbf{x}, \mathbf{x}^* >, < \mathbf{y}, \mathbf{y}^* >)_\lambda := (\mathbf{x}^* - \mu\mathbf{x}, \mathbf{y}^* - \mu\mathbf{y}) + \nu^2(\mathbf{x}, \mathbf{y})$$

on G, where $\lambda = \mu + i\nu$. Since we are assumuing that M satisfies the limit-point condition, N_λ and $N_{\bar{\lambda}}$ are one-dimensional and are the linear spans of $< \psi(\lambda), \lambda\psi(\lambda) >$ and $< \bar{\psi}(\lambda), \bar{\lambda}\bar{\psi}(\lambda) >$ respectively.

It is proved that any $< \mathbf{y}, \mathbf{y}^* > \in G_0$ satisfies $y_0 = 0, w_0 y_0^* = -p_0 y_1$ and on using this fact in (3.13) it follows that for any real $< \mathbf{x}, \mathbf{x}^* > \in G$, there exist $< \mathbf{y}, \mathbf{y}^* > \in G_0$ and $A_\lambda \in \mathbb{C}$ such that

$$< \mathbf{x}, \mathbf{x}^* > = < \mathbf{y}, \mathbf{y}^* > + 2 \text{ re } [A_\lambda < \psi(\lambda), \lambda\psi(\lambda) >];$$

note that it is sufficient to prove (3.1) for real sequences only.

If $D[\mathbf{x}]$ now denotes the sum on the left-hand side of (3.1), namely

$$D[\mathbf{x}] := \sum_{n=1}^{\infty} \{p_n |\Delta x_n|^2 + q_n |x_n|^2\} + p_0 |\Delta x_0|^2,$$

we obtain

$$\begin{aligned} (\mathbf{x}, \mathbf{x}^*) &= \sum_{n=1}^{\infty} x_n M x_n + w_0 x_0 x_0^* \\ &= D[\mathbf{x}] + x_0(w_0 x_0^* + p_0 x_1 - p_0 x_0) \\ &= D[\mathbf{x}] + 4 \text{ re } [A_\lambda]\{\text{re } [A_\lambda(m(\lambda) + \lambda w_0 - p_0)]\}. \end{aligned}$$

On substituting this in the identity

$$\begin{aligned} || < \mathbf{x}, \mathbf{x}^* >_\lambda ||^2 &= ||\mathbf{x}^* - \mu\mathbf{x}||^2 + \nu^2 ||\mathbf{x}||^2 \\ &= ||\mathbf{x}^*|| - 2\mu(\mathbf{x}, \mathbf{x}^*) + |\lambda|^2 ||\mathbf{x}||^2 \end{aligned}$$

and noting that

$$\begin{aligned} || < \mathbf{x}, \mathbf{x}^* >_\lambda ||^2 &= || < \mathbf{y}, \mathbf{y}^* >_\lambda ||^2 + 4|A_\lambda|^2 || < \psi(\lambda), \lambda\psi(\lambda) >_\lambda || \\ &= || < \mathbf{y}, \mathbf{y}^* >_\lambda ||^2 + 4\nu^2 |A_\lambda|^2 ||\psi(\lambda)||^2 \\ &= || < \mathbf{y}, \mathbf{y}^* >_\lambda ||^2 + 4\nu^2 |A_\lambda|^2 \{ \frac{\mathrm{im}\,[\mathrm{m}(\lambda)]}{\nu} + w_0 \} \end{aligned}$$

it follows that on setting $A_\lambda = \alpha_\lambda + i\beta_\lambda$ and $n(\lambda) := m(\lambda) + \lambda w_0 - p_0 = n_1(\lambda) + i\, n_2(\lambda)$,

(3.14)

$$\begin{aligned} ||\mathbf{x}^*||^2 &+ |\lambda|^2 ||\mathbf{x}||^2 - 2\mu D[\mathbf{x}] \\ &= || < \mathbf{y}, \mathbf{y}^* > ||_\lambda^2 + 4\nu(\alpha_\lambda^2 + \beta_\lambda^2) n_2(\lambda) + 8\mu\alpha_\lambda \operatorname{Re}[A_\lambda n(\lambda)] \\ &= || < \mathbf{y}, \mathbf{y}^* >_\lambda ||^2 + 4\{(\nu\beta_\lambda^2 - \mu\alpha_\lambda)^2 \frac{n_2(\lambda)}{\nu} + (-\frac{\mathrm{im}\,[\bar{\lambda}^2 n(\lambda)]}{\nu})\alpha_\lambda^2\}. \end{aligned}$$

The validity of (3.1) turns out to be equivalent to the non-negativness of the quadratic form (3.14) and this fact yields the result. See [8] for further details.

3.2 Connection with orthogonal polynomials

In the examples associated with the classical orthogonal polynomials discussed in Section 3.3 below the polynomials $\Phi_n(\lambda)$ ($n \in \mathbb{N}_0$) say, satisfy a three term recurrence relation of the form (3.5) and initial conditions

$$\Phi_0 = 1, \quad p_0\Phi_1 = (b_0 - \lambda w_0), \tag{3.15}$$

where $b_0 = p_0 + q_0$. Let $\Theta_n(\lambda)$ ($n \in \mathbb{N}_0$) satisfy (3.5) and

$$\Theta_0 = 0, \quad p_0\Theta_1 = -1.$$

Associated with the pair of sequences $(\Phi_n(\lambda))_0^\infty$, $(\Theta_n(\lambda))_0^\infty$ is another $m(\lambda)$ function, $\mathcal{M}(\lambda)$ say, which is obtained from consideration of the mapping

$$F_k(\lambda, z) := -\frac{p_{k-1}\Theta_k(\lambda) + z\Theta_{k-1}(\lambda)}{p_{k-1}\Phi_k(\lambda) + z\Phi_{k-1}(\lambda)} \quad (k \in \mathbb{N},\ z \in \mathbb{C},\ \lambda \in \mathbb{C}_+).$$

As before with the map $z \mapsto f_k(\lambda, z)$, $F_k(\lambda, z)$ defines circles in $\mathbb{C}_\pm$ and these converge to $\mathcal{M}(\lambda)$ when M is in the limit-point case. The function $\mathcal{M}(\cdot)$ is a Nevanlinna function and the sequence $(\Psi_n(\lambda))_0^\infty$, where $\Psi_n(\lambda) = \Theta_n(\lambda) + \mathcal{M}(\lambda)\Phi_n(\lambda)$, lies in l_w^2 and hence $\Psi_n(\lambda) = A\psi_n(\lambda)$ ($n \in \mathbb{N}_0$). We infer from (3.6) and (3.15) that

$$m(\lambda) = -\frac{1}{\mathcal{M}(\lambda)} + b_0 - \lambda w_0. \tag{3.16}$$

It is proved by Brown, Evans and Littlejohn in [9] that

$$\limsup_{\nu\to\infty} |\nu \mathcal{M}(i\nu)| < \infty$$

(a property shared by $m(\lambda)$) and this is well-known to ensure that for the Nevanlinna function $\mathcal{M}$, there exists a non-decreasing function ρ of bounded variation on $\mathbb{R}$ such that

$$\mathcal{M}(\lambda) = \int_{-\infty}^{\infty} \frac{d\rho(t)}{t-\lambda} \quad (\lambda \in \mathbb{C} \setminus \mathbb{R}). \tag{3.17}$$

The function ρ is unique in the sense that any other function can only differ from ρ by a constant at points of continuity. We also have the orthogonality relation

$$\int_{-\infty}^{\infty} \Phi_n(t)\Phi_m(t)d\rho(t) = w_n^{-1}\delta_{m,n} \tag{3.18}$$

where $\delta_{m,n}$ is the Kronecker delta. A full treatment of these and other fascinating facts about the $\mathcal{M}(\cdot)$ function and its role in the solution of the Hamburger moment problem may be found in Akhiezer [1] and Atkinson [2]; for our purpose in this article, (3.17) and (3.18) provide the means of determining the $\mathcal{M}(\cdot)$ function as we shall show in Section 3.3 below.

3.3 Examples

3.2.1 Copson's inequality. This is the case $p_n = w_n = 1$ and $q_0 = 0$, so that $Mx_n = -\Delta^2 x_{n-1}$. Note that in Copson's inequality it is enough to assume that $\sum_{n=0}^{\infty} |x_n|^2 < \infty$, for this implies that $\sum_{n=0}^{\infty} |\Delta^2 x_n|^2 < \infty$.

The recurrence relation (3.5) in this case, that is

$$-\Delta^2 x_{n-1} \equiv -x_{n+1} - x_{n-1} + 2x_n = \lambda x_n \quad (n \in \mathbb{N}),$$

has the two linearly independent solutions $(\alpha_\pm^n(\lambda))_0^\infty$, where

$$\alpha_\pm(\lambda) := 1 - \frac{\lambda}{2} \pm \frac{1}{2}\sqrt{\lambda(\lambda-4)}; \tag{3.19}$$

the square root of a complex number $z = r\exp(i\theta), 0 \le \theta \le 2\pi$ will be the branch defined by $z^{\frac{1}{2}} = r^{\frac{1}{2}}\exp(i\frac{\theta}{2})$ and the square root in (3.19) is taken to be the single-valued function defined in $\mathbb{C} \setminus [0,4]$. It is proved in [8] that for all $\lambda \in \mathbb{C}_+$, $|\alpha_+(\lambda)| < 1$ and $|\alpha_-(\lambda)| > 1$ and from this it follows that

$$m_\pm(\lambda) = \alpha_\pm(\lambda) \quad (\lambda \in \mathbb{C}_\pm). \tag{3.20}$$

Also M is in the strong limit-point case. Furthermore, it is shown in [8] that $\theta_+ = \frac{\pi}{3}$, $\theta_- = 0$ and hence $K = \sec^2 \frac{\pi}{3} = 4$. There are no non-trivial cases of equality and Copson's result is confirmed.

3.2.2. The Legendre problem. The Legendre polynomials $P_n(\lambda)$ $(n \in \mathbb{N}_0)$ satisfy the recurrence relation

$$MP_n := (n+1)P_{n+1} + nP_{n-1} = \lambda(2n+1)P_n \quad (n \in \mathbb{N}),$$

and

$$P_0 = 1, P_1 = \lambda.$$

Hence, this is the case $p_n = -(n+1)$, $w_n = q_n = 2n+1$ of (3.5) and (3.15) is satisfied. The orthogonality relation (3.17) is now

$$\frac{1}{2}\int_{-1}^{1} P_n(t)P_m(t)dt = (2n+1)^{-1}\delta_{n,m}$$

and so from (3.17) we obtain

$$\mathcal{M}(\lambda) = \frac{1}{2}\int_{-1}^{1} \frac{dt}{t-\lambda} = \frac{1}{2}\left\{\log\left(\frac{1-\lambda}{1+\lambda}\right) - \pi i\right\} \quad (\lambda \in \mathbb{C}_\pm). \tag{3.21}$$

It is an easy matter to prove that M is in the strong limit-point case.

The following results are proved in [9]:

$$\theta_+ = \frac{1}{2}\tan^{-1}\left(\frac{2}{\pi}\right), \quad \theta_0 = \theta_- = \frac{\pi}{2} - \frac{1}{2}\tan^{-1}\left(\frac{2}{\pi}\right),$$

$$K = 2\frac{\sqrt{\pi^2+4}}{\sqrt{\pi^2+4}-\pi},$$

$$E_+ = E_- = \emptyset.$$

Also there are no trivial cases of equality. In this case for real sequences the inequality takes the form

$$\left\{2\left[\sum_{n=0}^{\infty}(n+1)x_n x_{n+1} - x_0^2\right]\right\}^2$$
$$\leq \left(\frac{2\sqrt{\pi^2+4}}{\sqrt{\pi^2+4}-\pi}\right)\sum_{n=0}^{\infty}(2n+1)x_n^2 \sum_{n=1}^{\infty}(2n+1)^{-1}[(n+1)x_{n+1} + nx_n]^2.$$

3.3.3 The Hermite problem. The Hermite polynomials $H_n(\lambda)$ $(n \in \mathbb{N}_0)$ satisfy

$$H_{n+1} + 2nH_{n-1} = 2\lambda H_n \quad (n \in \mathbb{N})$$

and this can be written in the symmetric form of (3.5) as

$$MH_n := -\Delta\left(\frac{-1}{2^n(n-1)!}\Delta H_{n-1}\right) + \frac{2n+1}{2^{n+1}n!}H_n = \frac{\lambda}{2^n n!}H_n.$$

Also

$$H_0 = 1, H_1 = 2\lambda.$$

It follows again that M is in the strong limit-point case.

The orthogonality relation (3.18) is

$$\int_{-\infty}^{\infty} H_n(t)H_m(t)e^{-t^2}\,dt = \sqrt{\pi}2^n n!\delta_{n,m}$$

and by (3.17) this yields

$$\mathcal{M}(\lambda) = \frac{1}{\sqrt{\pi}}\int_{-\infty}^{\infty}\frac{e^{-t^2}dt}{t-\lambda} \quad (\lambda \in \mathbb{C}_\pm). \tag{3.22}$$

This problem is considered in [9] but an analytic proof is not found. However strong numerical evidence is presented for the following conjectures:

$$\theta_+ = \frac{1}{2}\tan^{-1}\left(\frac{2}{\sqrt{\pi}}\right), \quad \theta_0 = \theta_- = \frac{\pi}{2} - \frac{1}{2}\tan^{-1}\left(\frac{2}{\sqrt{\pi}}\right),$$

$$K = 2\frac{\sqrt{\pi+4}}{\sqrt{\pi+4}-\pi},$$

$$E_+ = E_- = \emptyset,$$

and there are no non-trivial cases of equality.

3.3.4 The Laguerre problem. The symmeteric form (3.5) of the recurrence relation satisfied by the Laguerre polynomials $L_n^\alpha(\lambda)$ ($n \in \mathbb{N}_0$, $\alpha > -1$) is

$$ML_n^\alpha := -\Delta\left(\frac{n!}{\Gamma(\alpha+n)}\Delta L_{n-1}^\alpha\right) = \frac{n!}{\Gamma(\alpha+n+1)}\lambda L_n^\alpha \quad (\alpha > -1,\ n \in \mathbb{N})$$

and

$$L_0^\alpha = 1, \quad \frac{1}{\Gamma(\alpha+1)}L_1^\alpha = b_0^\alpha - \frac{\lambda}{\Gamma(\alpha+1)}$$

where $b_0^\alpha = \frac{1}{\Gamma(\alpha+1)} + \frac{1}{\Gamma(\alpha)}$ if $\alpha \neq 0$ and $b_0^0 = 1$. The proof that M is in the strong limit-point case is no longer easy as in the preceding examples; see [9, Lemma 6.1].

The orthogonality relation is

$$\int_0^\infty L_n^\alpha(t)L_m^\alpha(t)t^\alpha e^{-t}dt = \frac{\Gamma(\alpha+n+1)}{n!}\delta_{n,m}$$

and so (3.17) yields

$$\mathcal{M}(\lambda) = \int_0^\infty \frac{t^\alpha e^{-t}dt}{t-\lambda} \quad (\lambda \in \mathbb{C}_\pm). \tag{3.23}$$

It is proved in [9] that there is no valid inequality for $\alpha \in (-1, 0]$; for $\alpha > 0$ no proof is given but strong numerical evidence is presented to indicate that the inequality is valid. In all cases $\theta_- = 0$ and hence $\theta_0 = \theta_+$.

Figure 1 illustrates the predicted dependence of K on α.

Figure 1. Graph for best constant K as a function of α

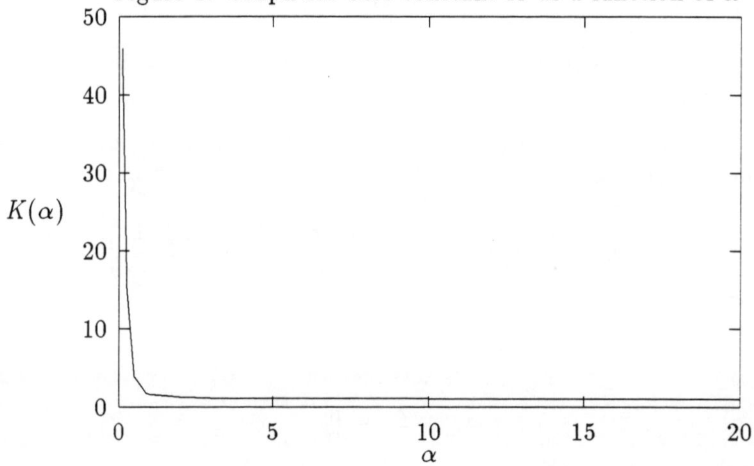

4. Numerical Estimates for the best constant in the integral HELP inequality

4.1 Introduction

In view of the difficulties in the analysis of examples of the inequality (1.4) discussed in Section 1, in this section we describe a numerical method for providing convincing estimates of the best constant K. In order to illustrate the approach of Brown, Kirby and Pryce in [10], we confine ourselves to the case, discussed at the end of Section 1 above, namely

$$x \in [0,\infty), \quad p(x) = 1, \quad w(x) = 1, \quad q(x) = x^\alpha - \tau, \quad (\alpha \geq 0, \ \tau \text{ is a parameter}).$$

However the methods discussed below, are capable of being used in a more general context; indeed they have been used to estimate the best constant K when the coefficient $p(x)$ is allowed to change sign. Full details may be found in [10], [11], [24]. We continue to write $\lambda = r\exp(i\theta)$ where $0 \leq \theta < 2\pi$.

In order to discribe the method of estimating K we remind the reader of the criterion described in Section 1, which determines the existence of, and establishes the value of the best constant. The criterion is given in terms of the behaviour of the Titchmarsh-Weyl $m(\lambda)$ coefficient in the first and third quadrants of the complex plane. The criterion leads to the determination of two numbers θ_+, θ_- and to the determination of the number $\theta_0 = \max(\theta_+, \theta_-)$. The number θ_0 satisfies $0 < \theta_0 \leq \pi/2$. In order that there shall be a finite number K, it is required that $\theta_0 \neq \pi/2$ and in this case the value of the best possible K is $\sec^2\theta_0$. Full details of the criterion are to be found in [14, Section 6].

It is shown in [10] that the following is an equivalent definition of θ_0. Let $L_\pm$ be defined as in Section 1, i.e.

$$L_+ = \{re^{i\theta} : r \in (0,\infty)\}, \quad L_- = \{re^{i\theta+\pi} : r \in (0,\infty)\}.$$

Then, it follows from [14, Corollary 7.1] that

$$\theta_\pm = \sup\{\theta \in (0,\pi/2] : \text{ im }[\lambda^2 m(\lambda)] = 0 \text{ for some } \lambda \in L_\pm(\theta)\}.$$

Thus θ_0 may be be found, at least in principal, by solving im $[\lambda^2 m(\lambda)] = 0$ for all $|\lambda| > 0$ and for $\theta \in [0,\pi/2] \cup [\pi, 3\pi/2]$ and then taking the supremum of all such θ's.

This result forms the basis for a numerical algorithm to estimate θ_0. In order to develop such an algorithm we first need an algorithm to estimate $m(\lambda)$ at points away from the real line; in the following sub-section we show how such an algorithm may be developed.

4.2 The $m(\lambda)$ algorithm

In the classical limit-point, limit-circle theory of Weyl [27] it is shown that the mapping $z \mapsto F_X(\lambda, z)$ defined by

$$F_X(\lambda, z) = -\frac{\theta(X,\lambda)z - \theta'(X,\lambda)}{\phi(X,\lambda)z - \phi'(X,\lambda)} \quad (\lambda \in \mathbb{C}\pm),$$

maps the real line onto the boundary of a circle $C_X \equiv C_X(\lambda)$. In the above the functions $\theta(x,\lambda)$ and $\phi(x,\lambda)$ are solutions of

$$-y'' + qy = \lambda y, \quad x \in [0,x], \ \lambda \in \mathbb{C} \setminus \mathbb{R}$$

which satisfy

$$\theta(0) = 0, \quad \theta'(0) = 1, \quad \phi(0) = -1, \quad \phi'(0) = 0.$$

The diameter d_X of the circle C_X is given by

$$d_X = |\text{im}\,\lambda \int_0^X |\phi|^2 dx|^{-1}, \quad (\text{im}\,\lambda \neq 0). \tag{4.1}$$

Weyl showed also that in the limit-point case, when $Y > X$ then $C_X \subset C_Y$ and as $X \to \infty$ C_X tends to a point in the plane which serves to define the $m(\lambda)$ function. These results imply that if η is any point interior to C_X then $|m(\lambda) - \eta| < d_X$. The final ingredient for our algorithm is provided by a result of Atkinson [3] which relates points on the Weyl circles C_X to solutions of an initial value problem. Atkinson's result is the following:

Let y be a solution of $-y'' + qy = \lambda y$, $x \in [0, X]$, $\lambda \in \mathbb{C} \setminus \mathbb{R}$, with $-\dfrac{y(X)}{y'(X)} \in \mathbb{R}$. *Then $-y(0)/y'(0)$ lies on the bounday of C_X.*

Thus our algorithm for estimating $m(\lambda)$ consists of using (4.1) to find an X which makes the diameter of the Weyl circle smaller than the required numerical accuracy, then using the above result of Atkinson and a differential equation solver to find an approximation to $m(\lambda)$.

This suggests a computational algorithm on the following general lines. Integrate the differential equation $-y'' + qy = \lambda y$ $x \in [0, \infty)$ with initial values $y(0) = -1$, $y'(0) = 0$ accumulating $\int_0^X |y|^2 dx$ as X increases, until the resulting estimates of d_X drops below a given tolerance ϵ. Then take X as being sufficiently close to infinity, and integrate backwards from X using the initial values $y(X) = 1$, $y'(X) = 0$. If both integrations could be done exactly we would end up with values $y(0)$, $y'(0)$ such that $\hat{m}(\lambda) = -y(0)/y'(0)$ lies on a Weyl circle of diameter less then ϵ and which contains $m(\lambda)$ within it. Thus $\hat{m}(\lambda)$ approximates $m(\lambda)$ with absolute error less than ϵ.

The method of solving the differential equation which was used in [10] is the fourth order implicate Runge-Kutta method of England [13]. In practice errors are committed in the integration whose effect must be estimated and controlled. Error estimation is performed by the one step two half steps method with Richardsons extrapolation which yields a sixth order accurate result.

Numerical tests have shown that the resulting Fortran Code is capable of delivering 10-15 places of accuracy at modest cost in certain regions of the plane and for functions $q(x) = x^\alpha - \tau$; such coefficients are, in a well-determined sense, not too difficult to handle. The solution y becomes progressively easier to compute as α increases, and the regions of the complex plane on which it is efficient to compute

$\hat{m}(\lambda)$ are those in which $|\text{im}\,\lambda|$ is large. Other cases give rise to solutions which are highly oscillatory and therefore expensive to compute.

We illustrate these features of the performance of the algorithm with results in Tables 1 and 2 below. In all the illustrations that follow we give, for differing values of λ, the computed value of $\hat{m}(\lambda)$ as returned by our algorithm and the 'true' value of $m(\lambda)$ as known from analytic results and evaluated by an independent algorithm. The calculations are accurate to within a tolerance of 10^{-5} (in view of the errors from the integration process this means that the diameter of the Weyl circle is set, at most, to 10^{-8}). We give also the number of integration steps needed to compute $\hat{m}(\lambda)$, this is a measure of the computational complexity of the algorithm. In Table 1 we give results for the case when $q = 0$ ($x \in [0, \infty)$). Here it can be seen that compared to the 'true value' of $m(\lambda) = i/\sqrt{\lambda}$ [14, Section 9, Example 2] and computed by the Fortran double complex 'zsqrt' function, the algorithm is able to return results of high accuracy. It can also be seen from the table, that the complexity of the computation, as measured by the number of steps needed to compute the solution, can be great for those regions of the plane in which λ is near the real axis.

Table 1 $\quad p = w = 1,\ q = 0$

λ	computed value of $m(.)$	true value of $m(.)$	steps	X
$10^{-6} + i10^{-6}$	$321.7971264507 + i776.8869870208$	$321.7971264527 + i776.8869870108$	91	28931
$10^{-3} + i10^{-3}$	$10.1761186469 + i24.5673236304$	$10.1761186408 + i24.5673236351$	78	788
$1 + i$	$0.3217971176 + i0.7768869863$	$0.3217971264 + i0.7768869870$	65	21
$10 + i10$	$0.1017611955 + i0.245673288$	$0.1017611864 + i0.245673263$	61	7
$100 + i100$	$0.0321797035 + i0.0776886941$	$0.0321797126 + i0.0776886987$	57	2
$1 + i10^{-1}$	$0.0496899459 + 0.9962771250$	$0.0496899401 + 0.9962771201$	859	195
$1 + i10^{-2}$	$0.049996803 + i0.9999624996$	$0.049996875 + i0.9999625027$	17726	1937

A further indication of the numerical accuracy of the code can be found by considering the example $q(x) = x^2$. In this case it is known that $m(\lambda)$ is given by

$$m(\lambda) = \frac{\Gamma(\frac{1}{4} - \frac{1}{4}\lambda)}{2\Gamma(\frac{3}{4} - \frac{1}{4}\lambda)}$$

where Γ is the classical gamma function, see [17, Section 4, Example 7]. In this example, with error bound less than 5×10^{-5}, the results returned by the code are compared to the 'true value' of $m(\lambda)$ as calculated from the algorithm of Kuki [25]. As can be seen from the tables, and is extensivly discussed in [11] and [24] the estimate $\hat{m}(\lambda)$ gets harder to compute for values of λ with small imaginary part.

Table 2 $\quad p = w = 1,\ q = x^2$

λ	computed value of $m(.)$	true value of $m(.)$	steps	X
$10^{-6} + i10^{-6}$	$1.4703387213 + i0.000001618$	$1.4703387214 + i0.000001618$	71	6
$10^{-3} + i10^{-3}$	$1.48.4994260 + i0.001164138$	$1.48.4994260 + i0.001164138$	51	6
$1 + i$	$0.3815889884 + i1.1744183613$	$0.3815889841 + i1.1744183522$	35	5
$10 + i10$	$0.104525163 + i0.2457963859$	$0.104525158 + i0.2457963850$	61	7

However with $q(x) = x^\alpha$ the number of steps needed to compute $\hat{m}(\lambda)$ (for a fixed λ), decreases as the value of α increases, and is less than when q is a constant.

4.3 An algorithm to find K

After describing the algorithm for estimating $m(\lambda)$ we now return to the problem of estimating θ_0 and hence the best constant K. The methods used in Brown, Kirby and Pryce [10] to solve $\mathrm{im}\,(\lambda^2 m(\lambda)) = 0$ are Newton's method and the method of bisection. The equation is solved for a selection of values of $|\lambda|$ (> 0), less than some empiric value $|\lambda_{max}|$, say, with $m(\lambda)$ estimate by the above algorithm.

Solving $\mathrm{im}\,(\lambda^2 m(\lambda)) = 0$ produces a set of ordered pairs $(|\lambda|, \theta)$ which are examined to find the number θ_+, obtained by choosing λ in the first quadrant of the complex plane, and then taking θ_+ as the greatest of the values θ so obtained. The number θ_- is found in a similar manner, but this time λ is chosen to lie in the third quadrant of the plane.

In the examples that follow we display the null sets of $\mathrm{im}\,(\lambda^2 m(\lambda))$ as Cartesian graphs, with θ values along the vertical axis and $|\lambda|$ values along the horizontal axis. The number θ_+ (respectively θ_-) can then be read off as the maximum value of θ attained on the graph. Thus θ_0 and hence the best constant $K = \sec^2 \theta_0$ may be estimated.

4.4 Examples

We now turn from general considerations of the method to some specific examples. As stated in Section 1 we confine ourselves to the cases $q(x) = x^\alpha - \tau$, $\alpha \in [0, 8]$, $\tau \in \mathbb{R}$. We are grateful to Dr. V.G. Kirby for providing the numerical data contained in this section.

4.4.1 $q(x) = -\tau$

The case $\tau = 0$ is of course the Hardy Littlewood example.

In order to illustrate this example we consider $\tau = 0, 1, -1$. The results for θ_+ and θ_- are displayed in Figures 2 and 3 below.

From these figures it can be seen that when $\tau \geq 0$ $\theta_+ = \pi/3$, the graph for θ_- is zero for all $|\lambda|$ and is not shown, while if $\tau < 0$, $\theta_+ = \pi/2$, $\theta_- = \pi/4$. Thus when $\tau \geq 0$ $\theta_0 = \pi/3$ which gives a best constant $K = 4$. However, when $\tau < 0$ $\theta_0 = \pi/2$ and there is no valid inequality. This numerical result confirms the analysis: see [16, Examples 1, 2].

Figure 2. Graph for θ_+; λ in first quadrant of $\mathbb{C}$; $q(x) = -\tau$

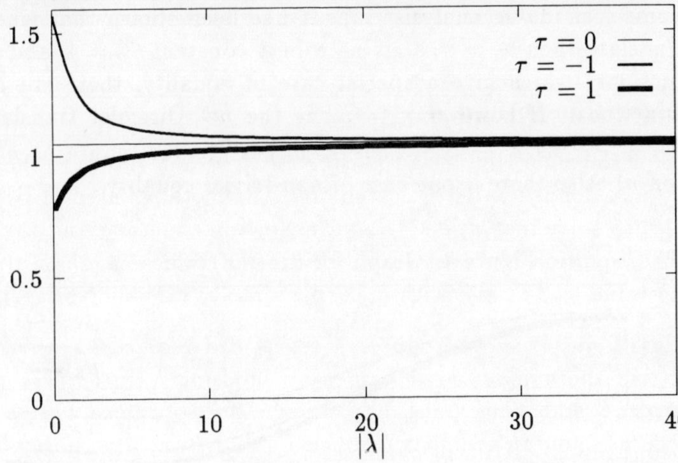

Figure 3. Graph for θ_-; λ in third quadrant of $\mathbb{C}$; $q(x) = -\tau$

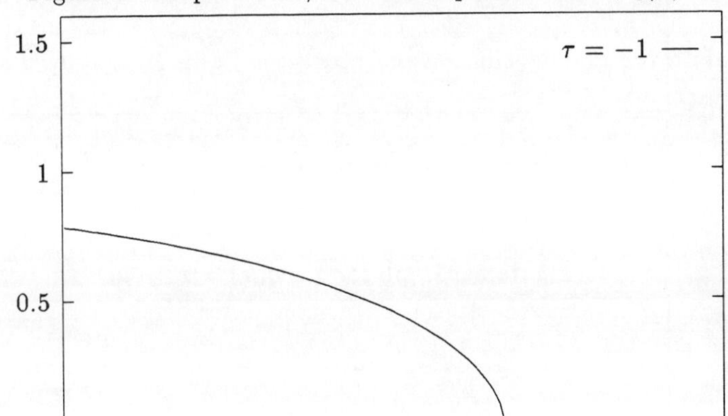

4.4.2 $q(x) = x^\alpha - \tau \qquad (\alpha > 0)$

This example is the one that was discussed in outline in Section 1. We remind the reader that it is shown in [19] that in order to obtain a valid inequality when $q(x) = x^\alpha - \tau$ it is both necessary and sufficient that τ be an eigenvalue of $-y'' + qy = \lambda y$ with either Neumann or Dirichlet boundary values at $x = 0$. (In terms of the notation of Section 1 τ is λ_n or μ_n).

As stated in Section 1 in the special case $\alpha = 2$ the example has been the subject of some formidable analysis [15]. It has been shown that when τ is any Neumann translate then $\theta_0 = \pi/3$ giving a best constant $K = 4$, and apart from the eigenfunctions, which give a special case of equality, there are no cases of equalising functions. If however $\tau = \mu_n$ is the nth Dirichlet translate $(\theta_0)_n = \frac{\pi}{3} - \frac{\sqrt{3}}{8\pi^2(2n+1)^2} + \frac{O(1)}{n^3}$ as $n \to \infty$. Thus $K(\mu_n)$ is greater than 4 as $n \to \infty$ and depends upon n. Also there is one case of non-trivial equality.

Figure 4. Graph for θ_+; $q(x) = x^2 - \mu_n$

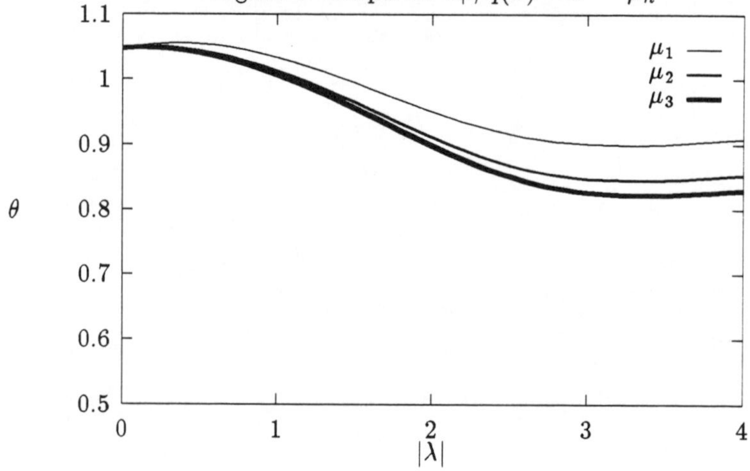

Figure 5. Graph for θ_-; $q(x) = x^2 - \mu_n$

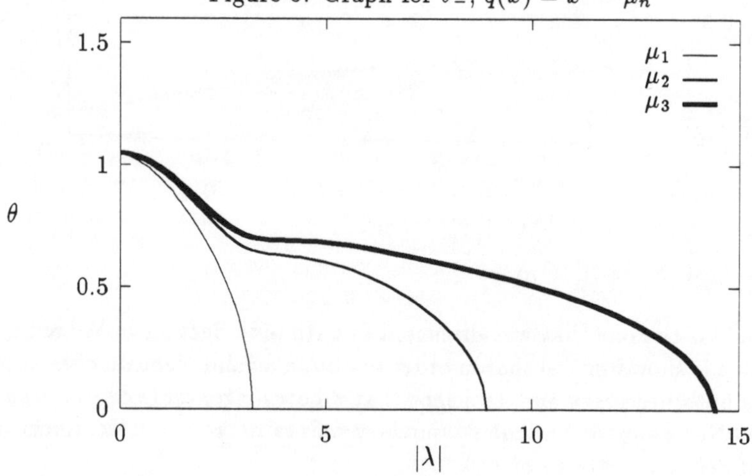

Estimates have been computed for θ_+, θ_- for this case ($\alpha = 2$). Our estimates, displayed in Figures 4 and 5 above, confirm all the features of the above analysis.

We now discuss results for other examples of $\alpha \in (0, 8]$. The reader will recall that in order to obtain a valid inequality the translates must be known. In the examples that follow the values τ of the translates are not explicitly known. In [10] they are estimated using the computer program SLEIGN [4] to 10 decimal places of accuracy.

In order to discuss the results in a structured manner, it will be convenient to discuss the Dirichlet and Neumann translate cases separatly. We commence by taking τ to be a Dirichlet translate.

We first discuss the case $\alpha > 1$. In this example the first Dirichlet translate is used and the graphs of θ_+, θ_- are shown in Figures 6 and 7. The graphs of the null set of $\operatorname{im}(\lambda^2 m(\lambda))$ for λ in the first quadrant and $1 < \alpha \leq 8$ all display the same general shape. They tend to $\pi/3$ both as $|\lambda| \to 0$ and a as $|\lambda| \to \infty$. The curve increases to a maximum as $|\lambda|$ moves away from 0 then decreases to a minimum before tending to $\pi/3$ at ∞. Thus θ_+ is given by the value of θ attained at the maximum which we note is at some value of $|\lambda| \neq 0$.

Figure 6. Graph for θ_+; $q(x) = x^\alpha - \mu_1$ $(\alpha > 1)$

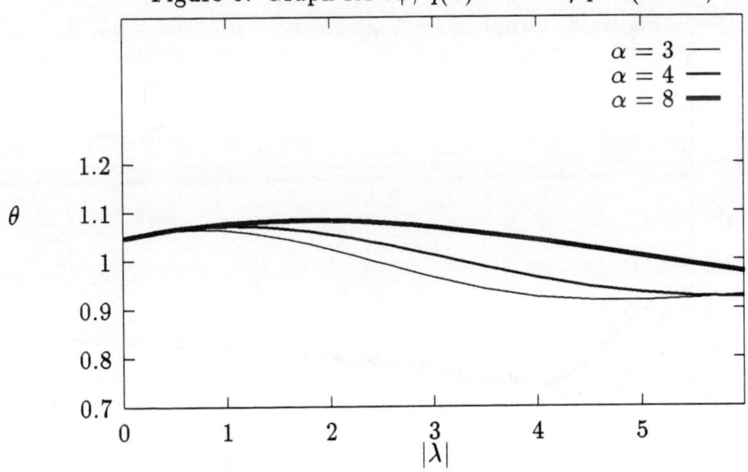

In order to estimate θ_-, λ is chosen to lie in the third quadrant of the complex plane; the null set of $\operatorname{im}(\lambda^2 m(\lambda))$ is shown in Figure 7. It can be seen in this case that $\theta_- = \pi/3$. This means that $\theta_0 = \theta_+ > \pi/3$ and since θ_+ is attained at one value of $|\lambda| \neq 0$, E_+ contains a single point.

Figure 7. Graph for θ_-; $q(x) = x^\alpha - \mu_1 \quad (\alpha > 1)$

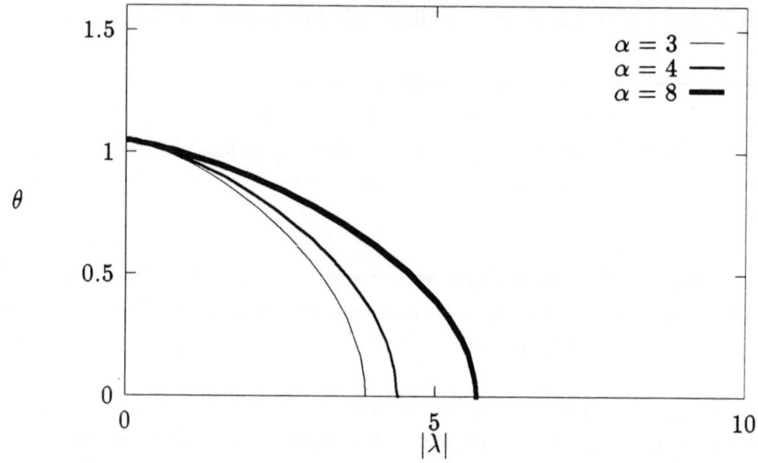

We next consider the case $0 < \alpha < 1$. The null sets of im $(\lambda^2 m(\lambda))$ are shown in Figures 8 and 9. In the case that λ is chosen from the first quadrant, the graph in Figure 8 indicates that the value of θ_+ is attained at $|\lambda| = 0$ and is $\pi/3$.

Figure 8. Graph for θ_+; $q(x) = x^\alpha - \mu_1 \quad (\alpha < 1)$

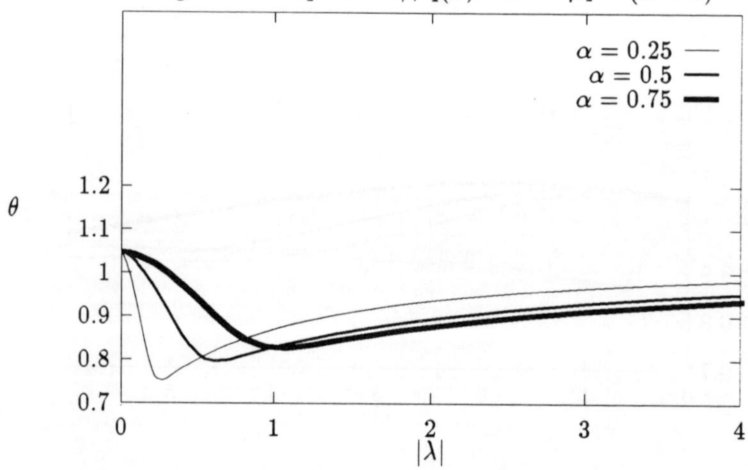

However an examination of Figure 9 indicates that the value of θ_- is attained when $|\lambda| \neq 0$ and is greater than $\pi/3$.

Figure 9. Graph for θ_-; $q(x) = x^\alpha - \mu_1 \quad (\alpha < 1)$

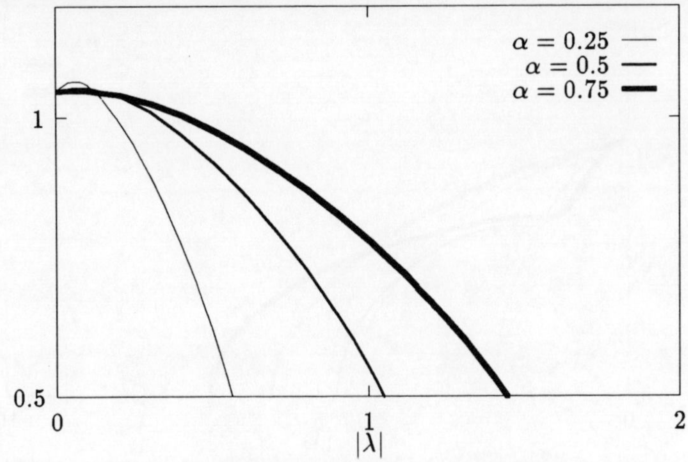

In this case $\theta_0 = \theta_- > \pi/3$ and E_- contains a single point. This, we believe, is the first example known in which $\theta_- > \theta_+$.

When $\alpha = 1$ the null sets of im $(\lambda^2 m(\lambda))$ are shown in Figures 10 and 11. It can be seen that in this case both θ_+ and θ_- take the values $\pi/3$, this value being attained at $|\lambda| = 0$. Thus $\theta_0 = \frac{\pi}{3}$, $K = 4$ and there are no equalising functions apart from the eigenfunctions, i.e. $E_+ = E_- = \emptyset$.

Figure 10. Graph for θ_+; $q(x) = x - \tau$

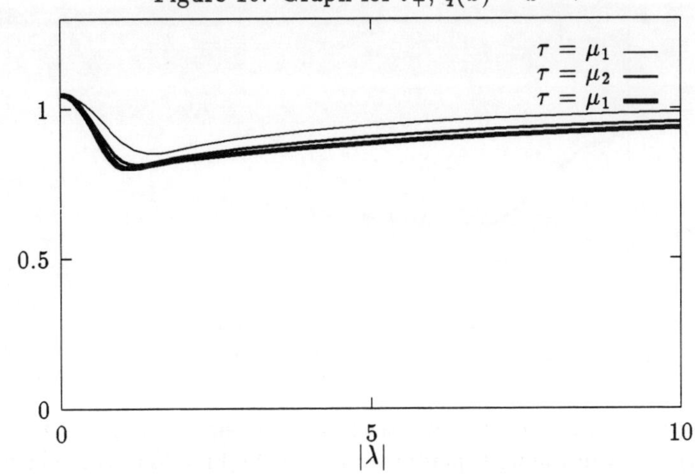

Figure 11. Graph for θ_-; $q(x) = x - \tau$

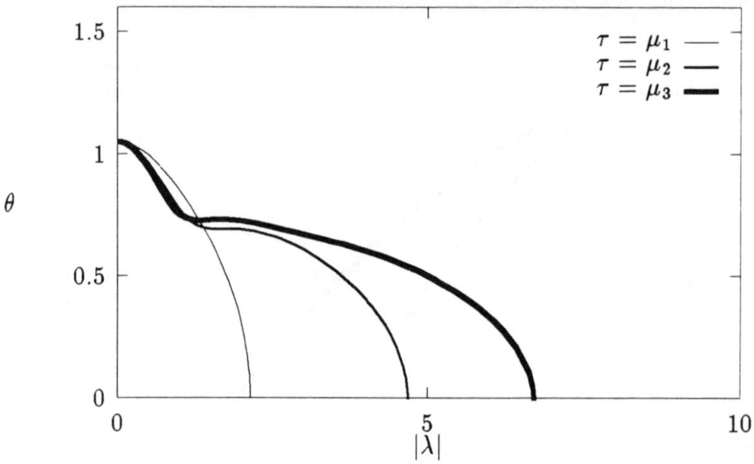

The remarkable behaviour of this example is, we predict, due to the concavity of the function $q(x)$; for $0 < \alpha < 1$ the function is concave up, while for $1 < \alpha < 8$ the function is concave down, with $q(x)$ linear when $\alpha = 1$.

Finally for this example we display, in Figure 12 the graph of the estimates of the best constant $K(\alpha)$ plotted against values of α. The evidence displayed on this graph enables us to make the conjecture at the end of Section 1 above.

Figure 12. Graph for best constant K as a function of α

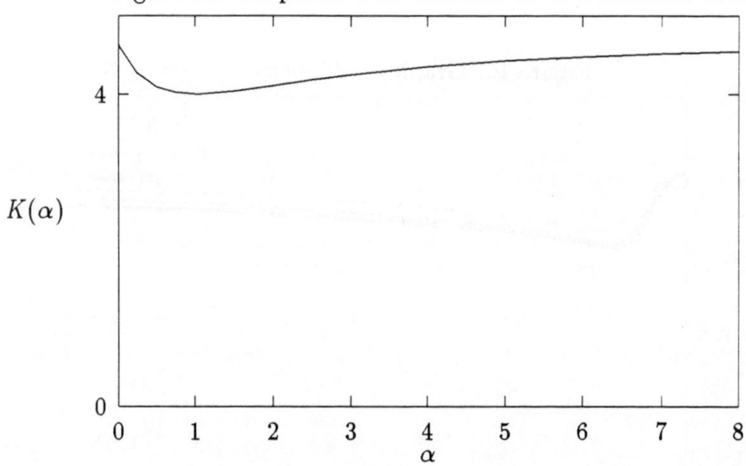

Turning to the case of Neumann translates we show in Figure 13 the null set of $\mathrm{im}\,(\lambda^2 m(\lambda))$ for λ in the first quadrant and for the first Neumann eigenvalue. The appropriate graph, when θ lies in the third quadrant, is zero, and is therefore not shown. If τ is not the first eigenvalue then $\theta_- \neq 0$, however $\theta_- < \theta_+$. In all cases $\theta_+ = \pi/3$ (with the value attained as $|\lambda| \to \infty$) and $\theta_- < \frac{\pi}{3}$. Thus $\theta_0 = \pi/3$ and $K = 4$ for all values on n and $E_+ = \emptyset$: hence there are no cases of equality apart from the eigenfunctions.

Figure 13. Graph for θ_+; $q(x) = x^\alpha - \lambda_1$

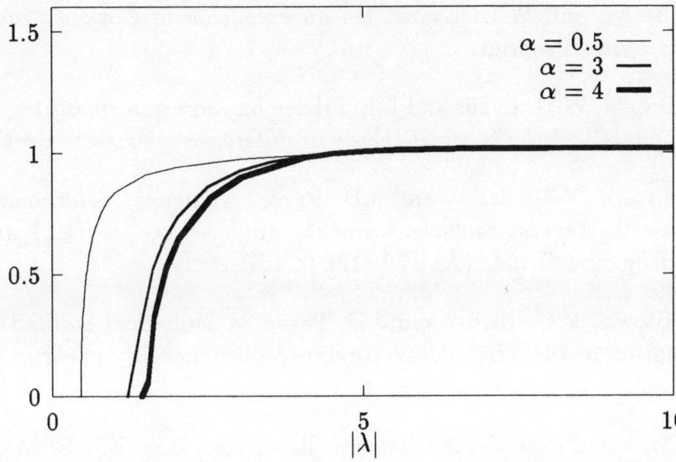

References

1. N.I. Akhiezer, *The Classical Moment Problem.* Oliver and Boyd, Edinburgh and London, 1965. (Translated from the Russian edition of 1960.)

2. F.V. Atkinson, *Discrete and Continuous Boundary Problems.* Academic Press, New York, 1964.

3. F.V. Atkinson, *On the location of the Weyl circles.* Proc. Roy. Soc. Edinb. A **88** (1981), 345–356.

4. P.B. Baily, M.K. Gordon and L.F. Shampine, *Automatic solution of the Sturm-Liouville problem.* ACM Trans Math. Software **4** (1978), 193–207.

5. C. Bennewitz, *Spectral theory for pairs of differential operators.* Arkiv. för mat **15** (1977), 33–61.

6. C. Bennewitz, *A general version of the Hardy-Littewood -Pólya-Everitt (HELP) inequality.* Proc. Roy. Soc. Edinb (A) **97** (1984), 9–20.

7. C. Bennewitz, *The HELP inequality in the regular case.* General Inequalities 5 (Proc. Oberwolfach, 1986; edited by W. Walter; Birkhäuser-Verlag, Basel, 1987).

8. B.M. Brown and W.D. Evans, *On an extension of Copson's inequality for infinite series.* Preprint.

9. B.M. Brown, W.D. Evans and L.L. Littlejohn, *Series inequalitites, orthogonal polynomials and the spectral theory of difference operators.* Preprint.

10. B.M. Brown, V.G. Kirby, and J.D. Pryce, *Numerical determination of the Titchmarsh-Weyl m-coefficient and its applciations to HELP inequalities.* Proc. Roy. Soc. Lond. (A) **426** (1989), 167–188.

11. B.M. Brown, V.G. Kirby, and J.D. Pryce, *A numerical method for the determination of the Titchmarsh-Weyl m-coefficient.* To appear in Proc. Roy. Soc. Lond.

12. E.T. Copson, *Two series inequalities.* Proc. Roy. Soc. Edinb. (A) **83** (1979), 109–114.

13. R. England, *Error estimates for Runga-Kutta-type solutions to systems of ordinary differential equations.* Compt. J. **12** (1969), 166–170.

14. W.D. Evans and W.N. Everitt, *A return to the Hardy-Littlewood inequality.* Proc. Roy. Soc. Lond. (A) **380** (1982), 447–486.

15. W.D. Evans, W.N. Everitt, W.K. Hayman and S. Ruscheweyh, *On a class of integral inequalities of Hardy-Littlewood type.* J. d'Anal. Math. **46** (1986), 118–147.

16. W.D. Evans and W.N. Everitt, *Hardy-Littlewood integral inequalities.* Lecture Notes in Pure and Applied Mathematics **129** (1991), 29–51 (Marcel Dekker, Inc., New York, 1991; edited by W.N. Everitt).

17. W.D. Evans and W.N. Everitt, *HELP inequalities for limit-circle and regular problems.* Proc. Roy. Soc. Lond. A **432** (1991), 367–390.

18. W.D. Evans and A. Zettl, *Norm inequalities involving derivatives.* Proc. Roy. Soc. Edinb. (A) **82** (1978), 51–70.

19. W.N. Everitt, *On an extension to an integro-differential inequality of Hardy, Littlewood and Pólya.* Proc. Roy. Soc. Edinb. A **69** (1971/72), 295–333.

20. G.H. Hardy and J.E. Littlewood, *Some inequalities connected with the calculus of variations.* Quart. J. Math. (Oxford) (2) **3** (1932), 241–252.

21. G.H. Hardy, J.E. Littlewood and G. Pólya, *Inequalitites.* Cambridge Univ Press, 1934.

22. E. Hellinger, *Zur Stieltjesschen Kettenbruchtheorie.* Math. Ann. **86** (1922), 18–29.

23. D.B.Hinton and R.T.Lewis, *Spectral analysis of second-order difference equations.* J. Math. Anal. and Appl. **63** (1978), 421–438.

24. V.G. Kirby, *A numerical method for determining the Titchmarsh-Weyl m-coefficient, and its applications to certain integro-differential inequalities.* Ph.D. Thesis, University of Wales College of Cardiff, 1990.

25. H. Kuki, *Complex gamma functions with error control.* Comm. ACM **15** (1972), 262–267.

26. R. Nevanlinna, *Asymptotische Entwickelungen beschränkter Funktionen und das Stieltjessche Momentenproblem.* Ann. Acad. Sci. Fenn. A **18** (5) (1922), 1–52.

27. H. Weyl, *Über gewöhnliche Differentialgleichungen mit Singularitäten und die zugehörigen Entwicklungen willkürlicher Funktionen.* Math. Ann. **68** (1910), 220–269.

B.M. Brown, Department of Computing Mathematics, University of Wales College of Cardiff, Senghennydd Road, Cardiff, CF2 4YN U.K.

W.D. Evans, School of Mathematics, University of Wales College of Cardiff, Senghennydd Road, Cardiff, CF2 4AG, U.K.

W.N. Everitt, School of Mathematics and Statistics, The University of Birmingham, B15 2TT, U.K.

International Series of Numerical Mathematics, Vol. 103, © 1992 Birkhäuser Verlag Basel

Gewöhnliche Differentialgleichungen mit quasimonoton wachsenden rechten Seiten in geordneten Banachräumen

Alice Chaljub-Simon, Roland Lemmert,
Sabina Schmidt und Peter Volkmann

Herrn Ray Redheffer, einem Liebhaber der Quasimonotonie,
zum 70. Geburtstage gewidmet

Abstract

In *General Inequalities 5*, Ray Redheffer and two authors of the present paper have given an existence theorem for ordinary differential equations, where the right hand side is assumed to be monotone increasing. The main objective here is to prove the same result with monotonicity replaced by quasimonotonicity. Moreover, we consider variable order cones.

1. Einleitung

Vorliegende Note ist eine Erweiterung von [6], und wie [6] ist sie ein Beitrag zur Existenztheorie für gewöhnliche Differentialgleichungen in Banachräumen. Einige grundlegende Tatsachen dieser Theorie, die hier nicht wiederholt werden, wurden 1987 in [6] erwähnt. Seitdem hat es weitere Resultate gegeben, z.B. [2], [5], [7], [11], [12]; vgl. auch den Übersichtsartikel [16]. Ausgehend von [5] spielen dabei verstärkt Quasimonotonie-Bedingungen für die rechte Seite der Differentialgleichung eine Rolle. In der Vergangenheit war Quasimonotonie in erster Linie ein wichtiges Hilfsmittel in der Theorie der Differentialungleichungen; vgl. das Buch von W. Walter [17] für den endlich-dimensionalen Fall und die Arbeiten von R.M. Redheffer [9] sowie R.M. Redheffer und W. Walter [10] (und die in diesen Artikeln zitierte Literatur) für den Fall allgemeinerer Räume. Natürlich hat man mit Hilfe von Quasi-

This paper is in final form and no version of it will be submitted for publication elsewhere.

monotonie schon früher die Existenz von Maximal- und Minimalintegralen für Anfangswertprobleme nachgewiesen. Bei einem unendlichen System von Differentialgleichungen haben das z.B. W. Mlak und C. Olech [8] getan, aber die Existenz einer Lösung als solche ist (auch ohne Quasimonotonie) bei diesem System bereits nach A. Tychonoff [13] gesichert. Im folgenden wird die Existenz von Lösungen überhaupt erst durch die Quasimonotonie garantiert.

2. Grundlegendes über Quasimonotonie

Es sei E ein reeller Banachraum, und es sei $K \subseteq E$ ein Kegel, d.h. K ist abgeschlossen, konvex, $\lambda K \subseteq K$ $(\lambda \geq 0)$ und $K \cap (-K) = \{0\}$. Dann wird in E durch

$$x \leq y \Longleftrightarrow y - x \in K \quad (x, y \in E) \tag{1}$$

eine (reflexive, antisymmetrische, transitive) Ordnung erzeugt. Ferner setzen wir

$$x \ll y \Longleftrightarrow y - x \in \operatorname{Int} K \quad (x, y \in E) \tag{2}$$

und

$$K^* = \{\varphi |\, \varphi \in E^*,\ \varphi(x) \geq 0 \quad (x \in K)\}$$

(dabei bezeichnen Int K und E^* das Innere von K bzw. den topologischen Dual von E; alle bisherigen Begriffsbildungen findet man bei M.G. Kreĭn und M.A. Rutman [4]).

Eine Funktion $f : D \to E$ mit $D \subseteq E$ nennen wir *monoton wachsend*, wenn sie der folgenden Bedingung genügt:

$$x, y \in D,\ x \leq y \Longrightarrow f(x) \leq f(y).$$

Wir nennen f *quasimonoton wachsend* (vgl. [14]), wenn gilt:

$$x, y \in D,\ x \leq y,\ \varphi \in K^*,\ \varphi(x) = \varphi(y) \Longrightarrow \varphi(f(x)) \leq \varphi(f(y)).$$

Offenbar ist jede monoton wachsende Funktion f auch quasimonoton wachsend.

Im folgenden wollen wir im Banachraume E Ordnungen $\leq$, $\ll$ betrachten, welche von einem Parameter t abhängen. Dazu sei $[0, T]$ (mit $T > 0$) ein Segment der reellen Geraden, und für jedes $t \in [0, T]$ sei K_t ein Kegel in E. Die entsprechend (1), (2) erzeugten Relationen werden mit $\leq_t$ bzw. $\ll_t$ bezeichnet, und wir setzen noch voraus:

$$0 \leq s \leq t \leq T \Longrightarrow K_s \subseteq K_t. \tag{3}$$

Lemma 1. *Es gelte* (3), *es sei* $\Delta \subseteq [0,T] \times E$, *und es habe* $f : \Delta \to E$ *die Eigenschaft, daß* $f(t,\cdot)$ *für jedes* $t \in (0,T]$ *bezüglich des Kegels* K_t *quasimonoton wächst. Ferner seien* $v, w : [0,T] \to E$ *stetige Funktionen, deren linksseitige Ableitungen* $v'_-(t)$, $w'_-(t)$ *für* $0 < t \leq T$ *existieren. Dann folgt aus*

$$(4) \qquad v'_-(t) - f(t, v(t)) \ll_t w'_-(t) - f(t, w(t)) \quad (0 < t \leq T),$$

$$(5) \qquad v(0) \ll_0 w(0)$$

die Ungleichung

$$(6) \qquad v(t) \ll_t w(t) \quad (0 \leq t \leq T).$$

Beweis. Wir nehmen an, (6) sei falsch. Dann ergibt sich aus (3), (5) leicht die Existenz eines $\tau \in (0,T]$, so daß

$$(7) \qquad v(t) \ll_t w(t) \quad (0 \leq t < \tau)$$

und

$$(8) \qquad w(\tau) - v(\tau) \in \partial K_\tau$$

ausfällt. Wegen (8) existiert ein $\varphi \in K^*_\tau$, $\varphi \neq 0$ mit

$$(9) \qquad \varphi(w(\tau) - v(\tau)) = 0$$

(Satz von Hahn/Banach). Die Quasimonotonie der Funktion $f(\tau,\cdot)$ liefert

$$(10) \qquad \varphi(f(\tau, v(\tau))) \leq \varphi(f(\tau, w(\tau))).$$

Aus (3), (7) folgt $v(t) \ll_t w(t)$ $(0 \leq t < \tau)$, also gilt

$$\varphi(w(t) - v(t)) > 0 \quad (0 \leq t < \tau).$$

Beachtet man (9), so hat demnach die reellwertige Funktion $\varphi(w(t) - v(t))$ an der Stelle $t = \tau$ eine linksseitige Ableitung ≤ 0, d.h.

$$(11) \qquad \varphi(w'_-(\tau) - v'_-(\tau)) \leq 0.$$

Andererseits ist nach (4) (für $t = \tau$) und (10) $\varphi(v'_-(\tau)) < \varphi(w'_-(\tau))$; das ist ein Widerspruch zu (11). Damit ist das Lemma bewiesen.

Im Spezialfalle eines von t unabhängigen Kegels $K_t = K$ $(0 \le t \le T)$ wurde Lemma 1 bereits in [14] bewiesen, und später [7] erhielt man damit einen Existenzsatz für gewöhnliche Differentialgleichungen in Banachräumen E, welche durch reguläre Kegel geordnet sind. Es sei daran erinnert, daß ein Kegel $K \subseteq E$ regulär heißt, wenn aus

$$x_1 \le x_2 \le x_3 \le \cdots \le y \quad \text{in } E$$

die Existenz eines $x \in E$ mit $\lim_{n\to\infty} ||x - x_n|| = 0$ folgt. In Verallgemeinerung des eben erwähnten Existenzsatzes gilt nun:

Satz 1. *Es sei E ein reeller Banachraum, und für $0 \le t \le T$ sei K_t ein regulärer Kegel in E mit* Int $K_t \ne \emptyset$*; außerdem werde* (3) *vorausgesetzt. Ferner sei $a \in E$, und es sei*

$$f : [0,T] \times E \to E$$

eine stetige, beschränkte Funktion, so daß für $0 < t \le T$

$$f(t,\cdot) : E \to E$$

bezüglich K_t quasimonoton wächst. Dann besitzt das Anfangswertproblem

$$u(0) = a, \quad u'(t) = f(t,u(t)) \quad (0 \le t \le T) \tag{12}$$

ein Maximalintegral $\bar{u} : [0,T] \to E$ und ein Minimalintegral $\underline{u} : [0,T] \to E$, d.h. $\bar{u}$, $\underline{u}$ lösen (12), *und für jede Lösung $u : [0,T] \to E$ gilt*

$$\underline{u}(t) \le_t u(t) \le_t \bar{u}(t) \quad (0 \le t \le T).$$

Ist noch $v : [0,T] \to E$ differenzierbar mit

$$v(0) \le_0 a, \quad v'(t) \le_t f(t,v(t)) \quad (0 < t \le T), \tag{13}$$

so folgt

$$v(t) \le_t \bar{u}(t) \quad (0 \le t \le T). \tag{14}$$

Zum *Beweise* dieses Satzes werde zunächst $p \in \text{Int } K_0$ gewählt. Dann kann in E eine äquivalente Norm $|| \cdot ||$ derart eingeführt werden, daß

$$\{x |\ x \in E,\ ||x|| \le 1\} = \{x |\ x \in E,\ -p \le_0 x \le_0 p\}$$

gilt, und anschließend ergibt sich die Existenz von $\underline{u}$, $\bar{u}$ (unter Verwendung von Lemma 1) ähnlich wie in [7]. Insbesondere erhält man

$$\bar{u}(t) = \lim_{n\to\infty} u_n(t) \quad (0 \le t \le T)$$

mit einer Folge stetig differenzierbarer Funktionen $u_n : [0,T] \to E$, für welche

$$a \ll_0 u_n(0), \quad f(t, u_n(t)) \ll_t u_n'(t) \quad (0 < t \leq T)$$

gilt. Aus diesen Ungleichungen folgt mit (13)

$$v(0) \ll_0 u_n(0), \quad v'(t) - f(t, v(t)) \ll_t u_n'(t) - f(t, u_n(t)) \quad (0 < t \leq T),$$

und Lemma 1 liefert $v(t) \ll_t u_n(t)$ $(0 \leq t \leq T)$. Durch den Grenzübergang $n \to \infty$ ergibt sich dann (14).

Wir betrachten nun Banachräume $E = l_\infty(E_\alpha; \alpha \in A)$, die wie folgt erklärt sind: A ist eine beliebige Menge, und für jedes $\alpha \in A$ ist E_α ein reeller Banachraum. Wir setzen

$$\begin{aligned} &l_\infty(E_\alpha; \alpha \in A) = \\ &\{x \mid x = (x_\alpha)_{\alpha \in A},\ x_\alpha \in E_\alpha\ (\alpha \in A),\ \|x\| = \sup_{\alpha \in A} \|x_\alpha\| < \infty\}. \end{aligned} \tag{15}$$

Sind die E_α durch Kegel K_α geordnet, so wird der Kegel $K = l_\infty(K_\alpha; \alpha \in A)$ in E definiert durch

$$l_\infty(K_\alpha; \alpha \in A) = \{x \mid x = (x_\alpha)_{\alpha \in A} \in E,\ x_\alpha \in K_\alpha\ (\alpha \in A)\}. \tag{16}$$

Für eine Funktion $f : E \to E$ bezeichne dann $f_\alpha : E \to E_\alpha$ die α-te Koordinate, d.h. es ist $f(x) = (f_\alpha(x))_{\alpha \in A}$. Jede der Funktionen f_α (und auch f selbst) kann als Funktion der Variablen $x_\beta \in E_\beta$ (mit $\beta \in A$) aufgefaßt werden.

Lemma 2. *Der Banachraum $E = l_\infty(E_\alpha; \alpha \in A)$ sei geordnet durch den Kegel $K = l_\infty(K_\alpha; \alpha \in A)$, und es sei $f : E \to E$ quasimonoton wachsend. Dann gilt (für $\alpha \in A$):*

I) *$f_\alpha : E \to E_\alpha$ ist quasimonoton wachsend bezüglich $x_\alpha \in E_\alpha$.*

II) *$f_\alpha : E \to E_\alpha$ ist monoton wachsend bezüglich der Variablen $x_\beta \in E_\beta$ mit $\beta \neq \alpha$ $(\beta \in A)$.*

Beweis. Zu I): Zu zeigen ist, daß aus

$$x, y \in E,\ x_\beta = y_\beta\ (\beta \in A,\ \beta \neq \alpha),\ x_\alpha \leq y_\alpha,\ \varphi_\alpha \in K_\alpha^*,\ \varphi_\alpha(x_\alpha) = \varphi_\alpha(y_\alpha) \tag{17}$$

folgt:

$$\varphi_\alpha(f_\alpha(x)) \leq \varphi_\alpha(f_\alpha(y)). \tag{18}$$

Es sei also (17) erfüllt. Dann werde $\varphi : E \to \mathbb{R}$ definiert durch

$$(19) \qquad \varphi(z) = \varphi_\alpha(z_\alpha) \quad (z \in E).$$

Mit (17) folgt dann

$$(20) \qquad x, y \in E,\ x \le y,\ \varphi \in K^*,\ \varphi(x) = \varphi(y),$$

und die Quasimonotonie von $f : E \to E$ liefert $\varphi(f(x)) \le \varphi(f(y))$, d.h. es gilt (18).

Zu II): Es sei $\beta \in A$, $\beta \neq \alpha$ fixiert; dann ist zu zeigen, daß aus

$$(21) \qquad x, y \in E,\ x_\gamma = y_\gamma\ (\gamma \in A,\ \gamma \neq \beta),\ x_\beta \le y_\beta$$

folgt:

$$(22) \qquad f_\alpha(x) \le f_\alpha(y).$$

Wir setzen nun (21) voraus und nehmen an, (22) sei falsch. Dann existiert nach dem Satz von Hahn-Banach ein $\varphi_\alpha \in K_\alpha^*$ mit

$$(23) \qquad \varphi_\alpha(f_\alpha(x)) > \varphi_\alpha(f_\alpha(y)).$$

Wieder wird $\varphi : E \to \mathbb{R}$ durch (19) erklärt; dann ergibt sich (20) aus (21), und die Quasimonotonie von $f : E \to E$ liefert erneut (18). Damit erhält man einen Widerspruch zu (23).

3. Ein Existenzsatz in $l_\infty(E_\alpha; \alpha \in A)$

Wie bereits in [6] verwenden wir einen Fixpunktsatz von N. Bourbaki [1] zum Beweise eines Existenzsatzes für gewöhnliche Differentialgleichungen. Für unsere Zwecke am besten geeignet ist die nachstehende Folgerung dieses Fixpunktsatzes, die man z.B. bei E. Zeidler ([18], Satz 11.D auf S. 506 und Korollar 11.23 auf S. 507) findet:

Fixpunktsatz. *Es sei $\le$ eine (reflexive, antisymmetrische, transitive) Ordnung in einer Menge Ω, und es sei $\Phi : \Omega \to \Omega$ monoton wachsend, d.h.*

$$\omega^1, \omega^2 \in \Omega,\ \omega^1 \le \omega^2 \Longrightarrow \Phi\omega^1 \le \Phi\omega^2.$$

I) Ω *habe folgende Eigenschaft:*

(P) *Ist C eine Kette in Ω (d.h. $C \subseteq \Omega$ und für $\omega^1, \omega^2 \in C$ ist $\omega^1 \leq \omega^2$ oder $\omega^2 \leq \omega^1$), $C \neq \emptyset$, so existiert* $\sup C \in \Omega$.

Gilt dann $v \in \Omega$, $v \leq \Phi v$, so besitzt Φ in der Menge

$$\{\omega |\, \omega \in \Omega,\ v \leq \omega\}$$

einen kleinsten Fixpunkt.

II) *Dual zu* (P) *habe Ω folgende Eigenschaft:*

(P′) *Ist C eine Kette in Ω, $C \neq \emptyset$, so existiert* $\inf C \in \Omega$.

Gilt dann $w \in \Omega$, $\Phi w \leq w$, so besitzt Φ in

$$\{\omega |\, \omega \in \Omega,\ \omega \leq w\}$$

einen größten Fixpunkt.

Satz 2. *Im Banachraume $E = l_\infty(E_\alpha; \alpha \in A)$ (vgl.* (15)*) sei das Anfangswertproblem*

$$u(0) = a, \quad u'(t) = f(t, u(t)) \quad (0 \leq t \leq T) \tag{12}$$

vorgelegt, wobei $a \in E$ und $f : [0,T] \times E \to E$ stetig und beschränkt ist. Für jedes $t \in [0,T]$ sei $K_t = l_\infty(K_{t,\alpha}; \alpha \in A)$ (analog zu (16)*) ein Kegel in E mit* $\operatorname{Int} K_t \neq \emptyset$, *und es gelte*

$$0 \leq s \leq t \leq T \Longrightarrow K_s \subseteq K_t. \tag{3}$$

Für $\alpha \in A$ seien die Kegel $K_{t,\alpha}$ regulär im Banachraume E_α, und schließlich sei $f(t,\cdot) : E \to E$ für jedes $t \in (0,T]$ quasimonoton wachsend bezüglich des Kegels K_t. Dann besitzt (12) *ein Minimalintegral $\underline{u} : [0,T] \to E$ und ein Maximalintegral $\bar{u} : [0,T] \to E$.*

Beweis. 1. Nach Voraussetzung gibt es ein $\mu \geq 0$, so daß

$$||f(t,x)|| \leq \mu \quad (0 \leq t \leq T,\ x \in E) \tag{24}$$

ausfällt. Dann kann $p \in \operatorname{Int} K_0$ so gewählt werden, daß gilt:

$$y \in E,\ ||y|| \leq \mu \Longrightarrow y \leq_0 p.$$

Mit (24) folgt

$$f(t,x) \leq_0 p \quad (0 \leq t \leq T,\ x \in E),$$

und somit wird für jede Lösung $u : [0,T] \to E$ von (12)

$$u(t) \leq_0 a + pt \quad (0 \leq t \leq T).$$

Setzt man

(25) $$w(t) = a + pt \quad (0 \leq t \leq T)$$

und beachtet man $K_0 \subseteq K_t$ für $0 \leq t \leq T$ (vgl. (3)), so erhält man also

(26) $$u \text{ löst } (12) \Longrightarrow u(t) \leq_t w(t) \quad (0 \leq t \leq T).$$

Mit

$$M = \max\{\mu, \|p\|\}$$

bilde man die Funktionenmenge

$$\Omega = \{\omega |\ \omega : [0,T] \to E,\ \omega(0) = a,\ \|\omega(s) - \omega(t)\| \leq M|s-t|\ (s,t \in [0,T])\}.$$

Dann gilt

$$w \in \Omega$$

und wegen (24)

(27) $$u \text{ löst } (12) \Longrightarrow u \in \Omega.$$

Für $\omega^1, \omega^2 \in \Omega$ werde nun gesetzt

(28) $$\omega^1 \leq \omega^2 \Longleftrightarrow \omega^1(t) \leq_t \omega^2(t) \quad (0 \leq t \leq T).$$

Dann ist $\leq$ eine (reflexive, antisymmetrische, transitive) Ordnung in Ω, und (26), (27) besagen

(29) $$u \text{ löst } (12) \Longrightarrow u \in \{\omega |\ \omega \in \Omega,\ \omega \leq w\}.$$

Unter Heranziehung einer geeigneten Funktion $\Phi : \Omega \to \Omega$ werden wir nun zeigen, daß es in

(30) $$\{\omega |\ \omega \in \Omega,\ \omega \leq w\}$$

Lösungen u von (12) und unter diesen ein größtes Element $\bar{u}$ gibt. Wegen (29) ist dann $\bar{u}$ Maximalintegral von (12); die Existenz eines Minimalintegrals kann analog bewiesen werden.

Es sei schon hier bemerkt, daß Ω bezüglich der durch (28) gegebenen Ordnung die im Fixpunktsatze genannten Eigenschaften (P), (P′) besitzt. Der Beweis zu (P) kann durch geringfügige Änderungen des Beweises des Lemmas 2 in [6] gegeben werden; (P′) gilt analog.

2. Wir definieren nun den Operator $\Phi : \Omega \to \Omega$. Dazu sei $\omega \in \Omega$; für $0 \le t \le T$ ist dann $\omega(t) = (\omega_\alpha(t))_{\alpha \in A}$ mit $\omega_\alpha(t) \in E_\alpha$ $(\alpha \in A)$, und entsprechend ist $a = (a_\alpha)_{\alpha \in A}$ mit $a_\alpha \in E_\alpha$ $(\alpha \in A)$. Für $\alpha \in A$ werde

$$r^\alpha : E \times E_\alpha \to E$$

erklärt durch

$$r^\alpha(x, \xi_\alpha) = (r^\alpha_\beta(x, \xi_\alpha))_{\beta \in A} \quad (x \in E,\ \xi_\alpha \in E_\alpha)$$

mit

$$r^\alpha_\beta(x, \xi_\alpha) = \begin{cases} \xi_\alpha & (\beta = \alpha) \\ x_\beta & (\beta \neq \alpha). \end{cases}$$

Dann wird durch

$$(t, \xi_\alpha) \to f_\alpha(t, r^\alpha(\omega(t), \xi_\alpha))$$

eine stetige, beschränkte Funktion

$$[0, T] \times E_\alpha \to E_\alpha$$

gegeben; diese ist nach Teil I) von Lemma 2 bei fixiertem $t \in (0, T]$ in der Variablen ξ_α bezüglich des Kegels $K_{t,\alpha}$ quasimonoton wachsend, also besitzt nach Satz 1 das in E_α vorgelegte Anfangswertproblem

$$\text{(31)} \qquad \xi_\alpha(0) = a_\alpha, \quad \xi'_\alpha(t) = f_\alpha(t, r^\alpha(\omega(t), \xi_\alpha(t))) \quad (0 \le t \le T)$$

ein Maximalintegral $\bar{\xi}_\alpha : [0, T] \to E_\alpha$. Mit diesen $\bar{\xi}_\alpha$ $(\alpha \in A)$ kann $\bar{\xi} : [0, T] \to E$ erklärt werden durch

$$\bar{\xi}(t) = (\bar{\xi}_\alpha(t))_{\alpha \in A} \quad (0 \le t \le T).$$

Man sieht leicht: $\bar{\xi} \in \Omega$. Setzt man also für $\omega \in \Omega$

$$\Phi\omega = \bar{\xi},$$

so gilt $\Phi : \Omega \to \Omega$.

3. Analog zu (26) erhält man für alle Lösungen von (31)

$$\xi_\alpha(t) \le_t w_\alpha(t) \quad (0 \le t \le T)$$

(hier bedeutet $\leq_t$ die durch $K_{t,\alpha}$ in E_α erzeugte Ordnung, und es ist $w_\alpha(t)$ die α-te Koordinate der Funktion (25)). Insbesondere wird für jedes $\omega \in \Omega$

$$\bar{\xi}_\alpha(t) \leq_t w_\alpha(t) \quad (0 \leq t \leq T,\ \alpha \in A),$$

d.h.

$$\bar{\xi}(t) \leq_t w(t) \quad (0 \leq t \leq T),$$

$$\bar{\xi} \leq w,$$

also gilt

$$\Phi\omega \leq w \quad (\omega \in \Omega) \tag{32}$$

und insbesondere

$$\Phi w \leq w. \tag{33}$$

4. Jetzt zeigen wir, daß $\Phi : \Omega \to \Omega$ monoton wächst. Dazu seien $\omega^1, \omega^2 \in \Omega$, $\omega^1 \leq \omega^2$. Unser Ziel ist

$$\Phi\omega^1 \leq \Phi\omega^2. \tag{34}$$

Aus $\omega^1 \leq \omega^2$ folgt zunächst mit Teil II) von Lemma 2 die Ungleichung

$$f_\alpha(t, r^\alpha(\omega^1(t), \xi_\alpha)) \leq_t f_\alpha(t, r^\alpha(\omega^2(t), \xi_\alpha)) \quad (0 < t \leq T,\ \alpha \in A,\ \xi_\alpha \in E_\alpha). \tag{35}$$

Setzen wir $\bar{\xi}^1 = \Phi\omega^1$, $\bar{\xi}^2 = \Phi\omega^2$, $\bar{\xi}^1(t) = (\bar{\xi}^1_\alpha(t))_{\alpha\in A}$, $\bar{\xi}^2(t) = (\bar{\xi}^2_\alpha(t))_{\alpha\in A}$, so gilt für $\alpha \in A$

$$\bar{\xi}^2_\alpha(0) = a_\alpha, \quad (\bar{\xi}^2_\alpha)'(t) = f_\alpha(t, r^\alpha(\omega^2(t), \bar{\xi}^2_\alpha(t))) \quad (0 \leq t \leq T), \tag{36}$$

und es gilt entsprechendes für $\bar{\xi}^1_\alpha$, also infolge von (35)

$$\bar{\xi}^1_\alpha(0) = a_\alpha, \quad (\bar{\xi}^1_\alpha)'(t) \leq_t f_\alpha(t, r^\alpha(\omega^2(t), \bar{\xi}^1_\alpha(t))) \quad (0 < t \leq T).$$

Da $\bar{\xi}^2_\alpha$ das Maximalintegral von (36) ist, folgt mit Satz 1

$$\bar{\xi}^1_\alpha(t) \leq_t \bar{\xi}^2_\alpha(t) \quad (0 \leq t \leq T,\ \alpha \in A),$$

und das beweist (34).

5. Für $\Phi : \Omega \to \Omega$ sind nun alle Voraussetzungen von Teil II) des Fixpunktsatzes erfüllt; insbesondere gilt (33). Damit besitzt Φ in der Menge (30) einen größten Fixpunkt $\bar{u}$, und aus $\Phi\bar{u} = \bar{u}$ folgt, daß $\bar{u}$ (12) löst. Ist u irgendeine Lösung von (12), so sieht man leicht: $u \leq \Phi u$. Nach Teil I) des Fixpunktsatzes besitzt Φ einen

Fixpunkt $\tilde{u}$ mit $u \leq \tilde{u}$. Wegen (32) ist $\tilde{u} = \Phi\tilde{u} \leq w$, also liegt $\tilde{u}$ in der Menge (30), und es folgt $\tilde{u} \leq \bar{u}$. Insgesamt hat man

$$u \leq \tilde{u} \leq \bar{u},$$

und daher ist $\bar{u}$ Maximalintegral des Anfangswertproblems (12).

Bemerkungen. 1. In folgendem Spezialfalle wurde Satz 2 in [2] bewiesen: $E_\alpha = \mathbb{R}$, $K_\alpha = [0, \infty)$ für alle $\alpha \in A$, also $E = l_\infty(A)$, geordnet durch seinen natürlichen positiven Kegel $l_\infty^+(A)$. Auch in [2] wurde der hier verwendete Operator Φ benutzt; es sei aber betont, daß dieser Operator bereits auf W. Mlak und C. Olech [8] zurückgeht. Da die Verhältnisse in $l_\infty(A)$ etwas einfacher als in den hier betrachteten Banachräumen $l_\infty(E_\alpha; \alpha \in A)$ sind, brauchte man in [2] nicht den Fixpunktsatz von Bourbaki, sondern man kam mit dem Fixpunktsatz von Tarski aus.

2. In folgendem Spezialfalle wurde Satz 2 in [6] bewiesen: K_t unabhängig von t, also $K_t = K$ $(0 \leq t \leq T)$, und $f(t, \cdot) : E \to E$ für jedes $t \in [0, T]$ monoton wachsend bezüglich K. In Wirklichkeit wurde in [6] nur $f : [0, T] \times U \to E$ mit einer Umgebung U von a vorausgesetzt und dann die Existenz einer lokalen Lösung von (12) gezeigt. Entsprechend kann auch der hier gegebene Satz 2 verfeinert werden.

4. Nichtexistenz in Räumen stetiger Funktionen

Ist M ein metrischer Raum, so bezeichnet $C(M)$ den Banachraum der stetigen, beschränkten Funktionen $x : M \to \mathbb{R}$, also

$$C(M) = \{x \mid x : M \to \mathbb{R} \text{ stetig}, \ ||x|| = \sup_{\alpha \in M} |x_\alpha| < \infty\}$$

(für $x : M \to \mathbb{R}$ bedeutet x_α den Wert von x an der Stelle $\alpha \in M$). Ferner sei

$$C^+(M) = \{x \mid x \in C(M), \ x = (x_\alpha)_{\alpha \in M} \text{ mit } x_\alpha \geq 0 \ (\alpha \in M)\}$$

der natürliche Ordnungskegel in C(

Satz 3. *Ist M ein metrischer Ra[...] ssen Metrik nicht diskret ist (d.h. in M gibt es mindestens einen Häufung[...]) und wird $E = C(M)$ geordnet durch $K = C^+(M)$, so gibt es eine stet[...] hränkte, monoton wachsende Funktion $f : E \to E$ und ein $a \in E$ derart, [...] Anfangswertproblem*

(37) $$u(0) = a, \quad u' \ [...] (t)) \quad (0 \leq t \leq T)$$

für jedes $T > 0$ unlösbar ist.

Beweis. $\varphi : \mathbb{R} \to \mathbb{R}$ sei definiert durch

$$\varphi(\xi) = \begin{cases} 2 & (\xi \geq 4) \\ \sqrt{\xi} & (0 \leq \xi \leq 4) \\ 0 & (\xi \leq 0). \end{cases}$$

Dann kann $f : E \to E$ erklärt werden durch $f(x) = (f_\alpha(x))_{\alpha \in M}$ mit $f_\alpha(x) = \varphi(x_\alpha)$ ($x \in E$, $\alpha \in M$). Man sieht leicht, daß $f : E \to E$ stetig, beschränkt und monoton wachsend ist. Es sei α_0 ein Häufungspunkt von M. Dann existiert in M eine Folge

$$\alpha_1, \alpha_2, \alpha_3, \ldots \to \alpha_0$$

mit $\alpha_m \neq \alpha_n$ ($m \neq n$). Die Menge $\{\alpha_n |\ n = 0, 1, 2, \ldots\}$ ist abgeschlossen in M, und auf ihr wird durch

$$a(\alpha_0) = 0, \quad a(\alpha_n) = (-1)^n/n \quad (n = 1, 2, 3, \ldots) \tag{38}$$

eine stetige, reellwertige Funktion definiert. Diese läßt sich nach Tietze/Urysohn zu einer stetigen, beschränkten Funktion $a : M \to \mathbb{R}$ fortsetzen. Wir nehmen nun an, daß (37) für ein $T \in (0, 2]$ eine Lösung $u : [0, T] \to E$ besitze. Schreiben wir $u(t) = (u_\alpha(t))_{\alpha \in M}$ mit $u_\alpha : [0, T] \to \mathbb{R}$, so folgt für $\alpha \in M$

$$u_\alpha(0) = a(\alpha), \quad u'_\alpha(t) = \varphi(u_\alpha(t)) \quad (0 \leq t \leq T).$$

Damit wird nach (38) für $\alpha = \alpha_n$, n ungerade (also $a(\alpha_n) < 0$)

$$u_{\alpha_n}(t) = -1/n \quad (0 \leq t \leq T), \tag{39}$$

und für $\alpha = \alpha_n$, $n \geq 2$ gerade (also $a(\alpha_n) > 0$) wird

$$u_{\alpha_n}(t) = \left(\frac{t}{2} + \frac{1}{\sqrt{n}}\right)^2 \quad (0 \leq t \leq T). \tag{40}$$

Grenzübergang $n \to \infty$ in (39) bzw. (40) liefert

$$u_{\alpha_0}(t) = 0, \quad u_{\alpha_0}(t) = t^2/4.$$

Das ist für $t > 0$ unmöglich.

Bemerkungen. 1. Satz 3 wurde im Falle $M = [-1, 1]$ (also $E = C[-1, 1]$) in [15] bewiesen. Der hier gegebene Beweis ist ähnlich, und er beruht auf einer Modifikation eines berühmten Beispiels von J. Dieudonné [3].

2. Satz 3 ist nicht richtig, wenn der metrische Raum M als diskret vorausgesetzt wird. Dann nämlich gilt $C(M) = l_\infty(M)$, und die Lösbarkeit von (37) folgt z.B. aus Satz 2 (aber auch aus [15], Satz 1).

Literatur

1. N. Bourbaki, *Éléments de mathématique I: Théorie des ensembles, Fascicule de résultats*. Hermann & Cie., Paris, 1939.

2. A. Chaljub-Simon und P. Volkmann, *Un théorème d'existence et de comparaison pour des équations différentielles dans les espaces de fonctions bornées*. C.R. Acad. Sci., Paris, Sér. I **311** (1990), 515–517.

3. J. Dieudonné, *Deux exemples singuliers d'équations différentielles*. Acta Sci. Math. **12 B** (1950), 38–40.

4. M.G. Kreĭn und M.A. Rutman, *Lineĭnye operatory, ostavljajuščie invariantnym konus v prostranstve Banaha*. Uspehi Mat. Nauk **3** (1948), Nr. 1(23), 3–95.

5. R. Lemmert, *Existenzsätze für gewöhnliche Differentialgleichungen in geordneten Banachräumen*. Funkcial. Ekvac., Ser. Internac. **32** (1989), 243–249.

6. R. Lemmert, R.M. Redheffer und P. Volkmann, *Ein Existenzsatz für gewöhnliche Differentialgleichungen in geordneten Banachräumen*. General Inequalities 5 (Oberwolfach 1986), 381–390, International Series of Numerical Mathematics 80, Birkhäuser, Basel, 1987.

7. R. Lemmert, S. Schmidt und P. Volkmann, *Ein Existenzsatz für gewöhnliche Differentialgleichungen mit quasimonoton wachsender rechter Seite*. Erscheint in Math. Nachr.

8. W. Mlak und C. Olech, *Integration of infinite systems of differential inequalities*. Ann. Polon. Math. **13** (1963), 105–112.

9. R.M. Redheffer, *Gewöhnliche Differentialungleichungen mit quasimonotonen Funktionen in normierten linearen Räumen*. Arch. Rational Mech. Analysis **52** (1973), 121–133.

10. R.M. Redheffer und W. Walter, *Remarks on ordinary differential equations in ordered Banach spaces*. Monatshefte Math. **102** (1986), 237–249.

11. S. Schmidt, *Existenzsätze für gewöhnliche Differentialgleichungen in Banachräumen*. Diss. Univ. Karlsruhe 1989; erscheint in Funkcial. Ekvac., Ser. Internac.

12. S. Schmidt, *Properties of integral operators corresponding to ordinary differential equations in Banach spaces.* Erscheint in Ann. Univ. Mariae Curie-Skłodowska, Sect. A.

13. A. Tychonoff, *Ein Fixpunktsatz.* Math. Ann. **111** (1935), 767–776.

14. P. Volkmann, *Gewöhnliche Differentialungleichungen mit quasimonoton wachsenden Funktionen in topologischen Vektorräumen.* Math. Z. **127** (1972), 157–164.

15. P. Volkmann, *Équations différentielles ordinaires dans les espaces des fonctions bornées.* Czechoslovak Math. J. **35 (110)** (1985), 201–211.

16. P. Volkmann, *Quelques résultats récents sur les équations différentielles dans les espaces de Banach.* Erscheint in Actes Colloque Internat. Théorie Point Fixe Appl. Marseille-Luminy 1989.

17. W. Walter, *Differential and integral inequalities.* Springer, Berlin, 1970. [Übersetzung aus dem Deutschen.]

18. E. Zeidler, *Nonlinear functional analysis and its applications I: Fixed-point theorems.* Springer, New York, 1986. [Übersetzung aus dem Deutschen.]

Alice Chaljub-Simon, Département de Mathématiques, Université d'Orléans, BP 6759, F-45067 Orléans Cedex 2, Frankreich.

Roland Lemmert, Sabina Schmidt und Peter Volkmann, Mathematisches Institut I, Universität Karlsruhe, Postfach 6980, W-7500 Karlsruhe 1, Deutschland.

International Series of Numerical Mathematics, Vol. 103, © 1992 Birkhäuser Verlag Basel

Comparison theorem and its applications

G.S. Ladde, M. Sambandham and S. Sathananthan

Abstract

By developing comparison theorems for stochastic integro-differential equations, stability and error estimate problems are investigated. The obtained results provide the information about random environmental perturbations. In particular, the presented work provides a partial solution to the problem of "Stochastic versus Deterministic" in the modelling of dynamic systems in biological, engineering, and physical sciences. Examples are given to show the scope of the results.

1. Introduction

In modelling dynamical systems, one of the fundamental problems is the study of the effects of randomness (versus nonrandomness) in the system. In recent years problems of this nature has been intensively studied [5, 6, 7]. This was motivated by the wide applications in stochastic optimal control, life sciences, queuing theory, various branches of engineering and physics [1, 2, 3, 4].

Although stochastic differential systems have been the subject of several investigations, there is only very little work in stochastic integro-differential systems. Quite recently, some attempts have been made to study the existence and uniqueness of solutions [1], the generalized variation of constant formulas [6], and various stability results [6, 7] for stochastic integro-differential equations of Itô-type.

It is well known that Lyapunov's second method has played a very significant role in the qualitative and quantitative analysis of systems of stochastic integro-differential equations. In this paper we have attempted to develop comparison results by employing the concept of random differential and integro-differential inequalities. This important technique permits us to reduce the study of stochastic integro-differential equations to stochastic differential equations in the context of the minimal sets of functions [8]. These comparison theorems have many applica-

---

This paper is in final form and no version of it will be submitted for publication elsewhere.

tions in the theory of error estimates and stability analysis of stochastic integro-differential systems. Stability analysis and error estimates are developed using these comparison theorems. Examples are worked out to demonstrate the fruitfulness of the results.

2. Preliminaries

Let us consider the stochastic integro-differential equation (SIDE)

$$(2.1)\qquad y' = f(t, y, Ty(\omega), \omega), \qquad y(t_0, \omega) = y_0(\omega)$$

and the corresponding deterministic initial value problem (DIVP)

$$(2.2)\qquad m' = \hat{f}(t, m, \hat{T}m), \qquad m(t_0) = m_0 = E[y_0(\omega)],$$

which is obtained by ignoring the random disturbance in the system described by (2.1). In our subsequent analysis we will utilize the following random initial value problem (RIVP)

$$(2.3)\qquad x' = \hat{f}(t, x, \hat{T}x), \qquad x(t_0, \omega) = x_0(\omega).$$

In (2.1), (2.2) and (2.3) $y, m, x \in \mathbb{R}^n$, $x_0, y_0 \in R[\Omega, \mathbb{R}^n]$; E stands for the expected value of a random variable; $\Omega \equiv (\Omega, F, P)$ is a complete probability space, $f \in M[\mathbb{R}^+ \times S_\rho \times \mathbb{R}^n, R[\Omega, \mathbb{R}^n]]$, $S_\rho = \{x \in \mathbb{R}^n : |x| < \rho\}$, $Tx(\omega) = \int_{t_0}^{t} K(t, s, x(s, \omega), \omega)\, ds$ and $K \in C[\mathbb{R}^+ \times \mathbb{R}^+ \times S_\rho, R[\Omega, \mathbb{R}^n]]$. Further $f(t, 0, 0) \equiv 0$, $K(t, s, 0) \equiv 0$ and f satisfies desired regularity conditions that (2.1) possesses the trivial sample solution processes. Our notations are those of [4].

We need the following results in our subsequent study. These results are direct extension of the results in [12, 8].

3. Comparison results

Lemma 3.1. *Assume that*

(H_0) $V \in C[\mathbb{R}^+ \times S_\rho, R[\Omega, \mathbb{R}^n]]$, *$V(t, x, \omega)$ is locally Lipschitzian in x with probability 1, and*

$$b(||x||) \leq \sum_{i=1}^{n} V_i(t, x, \omega) \leq a(t, ||x||, \omega),$$

$(t, x) \in \mathbb{R}^+ \times S_\rho$, *where $b \in K$, $a(t, \cdot, \omega) \in K$ and $a \in C[\mathbb{R}^+ \times [0, \rho]$, $R[\Omega, \mathbb{R}^+]]$.*

(H_1) *$g \in C[\mathbb{R}^+ \times \mathbb{R}^n, R[\Omega, \mathbb{R}^n]]$ is quasi-monotone non-decreasing in u for each t and $r(t, t_0, u_0(\omega))$ is the right maximal solution process of*

$$u' = g(t, u, \omega), \qquad u(t_0, \omega) = u_0(\omega) \geq 0$$

existing on $[t_0, \infty)$.

(H_2) $D_-V(t, x, Tx(\omega), \omega) \equiv \liminf_{h \to 0^-} \frac{1}{h}[V(t+h, x+hf(t, x, Tx(\omega), \omega), \omega) - V(t, x, \omega)] \leq g(t, V(t, x, \omega))$ *on Ω, where $\Omega = \{x \in C[\mathbb{R}^+, \mathbb{R}^n] : V(s, x(s), \omega) \leq V(t, x(t), \omega),\ t_0 \leq s \leq t\}$.*

Let $x(t, t_0, x_0(\omega))$ be any solution of (2.1) such that $V(t_0, x_0(\omega), \omega) \leq u_0(\omega)$. Then

$$V(t, x(t, t_0, x_0(\omega)), \omega) \leq r(t, t_0, u_0(\omega), \omega), \qquad t \geq t_0.$$

One can formulate the *proof* by following the proofs of results in [4, 8].

Theorem 3.1. *Assume that the hypotheses H_0 and H_1 of Lemma 3.1 hold. Further the hypothesis H_2 of Lemma 3.1 is replaced by*

(H_2)

$$\begin{aligned} &D^+V(s, x(t, s, y), Tx(t, s, y), \omega) \\ &\equiv \limsup_{h \to 0^+} \frac{1}{h}[V(s+h, x(t, s+h, y+hf(s, y, Ty, \omega)), \omega) \\ &\quad - V(t, x(t, s, y), \omega)] \\ &\leq g(s, V(t, x(t, s, y, \omega), \omega)) \end{aligned} \tag{3.1}$$

on Ω, where $\Omega = \{x \in C[\mathbb{R}^+, \mathbb{R}^n] : V(r, x(t, r, y), \omega) \leq V(s, x(t, s, y), \omega),\ t_0 \leq r \leq s \leq t\}$; $x(t, s, y)$ is a solution process of (2.3) through (s, y);

(H_3) *$f \in M[\mathbb{R}^+ \times S_\rho \times \mathbb{R}^n, R[\Omega, \mathbb{R}^n]]$ and f is smooth enough to guarantee the existence and uniqueness of a sample process $y(t, \omega) = y(t, t_0, y_0(\omega), \omega)$ of (2.1) for $t \geq t_0$;*

(H_4) *$\hat{f} \in C[\mathbb{R}^+ \times S_\rho \times \mathbb{R}^n, \mathbb{R}^n]$, the solution process $x(t, \omega) = x(t, t_0, x_0(\omega))$ of (2.3) exists for all $t \geq t_0$, it is unique and sample continuous with respect to initial data, and is locally Lipschitzian in x_0 with probability 1.*

Let $x(t,t_0,x_0(\omega))$ with $x_0(\omega) = y_0(\omega)$ with $||x|| < \rho$, and $y(t,\omega) = y(t,t_0,y_0(\omega),\omega)$ are solution processes of (2.3) and (2.1), respectively. Then, we have

$$V(t,y(t,\omega),\omega) \le r(t,t_0,u_0(\omega),\omega), \quad t \ge t_0, \tag{3.2}$$

provided that $V(t_0,x(t,\omega),\omega) \le u_0(\omega)$.

Proof. Set $t_0 \le s \le t$, $W(s,\omega) = V(s,x(t,s,y(s,\omega)),\omega)$,

$$W(t_0,\omega) = V(t_0,x(t,\omega),\omega). \tag{3.3}$$

Since $x(t,s,y(s,\omega))$ and $y(s,\omega)$ are sample solution processes of (2.3) and (2.1) through $(s,y(s,\omega))$ and $(t_0,y_0(\omega))$, respectively, and $V \in C[\mathbb{R}^+ \times \mathbb{R}^n, R[\Omega,\mathbb{R}^n]]$ and the solution process $x(t,s,y)$ through (s,y) is continuous with respect to (s,y), it is obvious that $W(s,\omega)$ is sample continuous for $s \ge t_0$. For small $h > 0$ such that $s + h < t$ we have

$$\begin{aligned} W(s+h,\omega) - W(s,\omega) &= V(s+h,x(t,s+h,y(s+h,\omega)),\omega) \\ &\quad -V(s,x(t,s,y(s,\omega)),\omega) \\ &= V(s+h,x(t,s+h,y(s+h,\omega)),\omega) \\ &\quad -V(s+h,x(t,s+h,y(s,\omega) \\ &\quad +hf(s,y(s,\omega),Ty(s,\omega),\omega)),\omega) \\ &\quad +V(s+h,x(t,s+h,y(s,\omega) \\ &\quad +hf(s,y(s,\omega),Ty(s,\omega),\omega)),\omega) \\ &\quad -V(s,x(t,s,y(s,\omega)),\omega). \end{aligned}$$

This together with the local Lipschitzian property of $x(t,s,y)$ in y and $V(t,x,\omega)$ in x implies

$$\begin{aligned} W(s+h,\omega) - W(s,\omega) &\le K||y(s+h,\omega) - y(s,\omega) - hf(s,y(s,\omega),Ty(s,\omega),\omega)||e \\ &\quad +V(s+h,x(t,s+h,y(s,\omega) \\ &\quad +hf(s,y(s,\omega),Ty(s,\omega),\omega)),\omega) \\ &\quad -V(s,x(t,s,y(s,\omega)),\omega), \end{aligned}$$

where K is the local Lipschitz constant relative to $x(t,s,y)$ and $V(t,x,\omega)$ and $e = (1,1,1,\ldots,1) \in \mathbb{R}^n$. This together with the definition of $D^+V(s,x(t,s,y),\omega)$, (H_3) and the sample continuity of $W(s,\omega)$ yields the inequalities $D^+W(s,\omega) \le g(s,W(s,\omega),\omega)$, $t_0 \le s \le t$, and

$$W(t_0,\omega) \le u_0(\omega).$$

Using Lemma 3, [11], we get $D_-W(s,\omega) \leq g(s, W(s,\omega),\omega)$, $t_0 \leq s \leq t$, and $W(t_0,\omega) \leq u_0(\omega)$ on Ω. Using Lemma 3.1,

$$W(s,\omega) \leq r(s, t_0, u_0(\omega), \omega), \quad t_0 \leq s < t, \tag{3.4}$$

since

$$W(s,\omega) \to V(t, x(t,t,y(t,\omega)),\omega) = V(t, y(t,\omega),\omega), \tag{3.5}$$

with probability 1, $t \geq t_0$.

Remark 3.1: If $u_0 = V(t_0, x(t, t_0, y_0(\omega)), \omega)$, $t \geq t_0$, then (3.5) becomes

$$V(t, y(t,\omega),\omega) \leq r(t, t_0, V(t_0, x(t,\omega),\omega),\omega), \quad t \geq t_0. \tag{3.6}$$

This theorem relates the solution process of three kinds of initial value problems, namely (2.1), (2.2) and (2.3). On the other hand, the usual comparison theorems [4, 9] relate the solutions of two kinds of initial value problems.

Example 3.1: Consider the following (SIDE)

$$y' = A(t,\omega)y + \int_{t_0}^{t} K(t, s, y(s,\omega),\omega)\, ds, \quad y(t_0,\omega) = y_0(\omega), \tag{3.7}$$

where $A(t,\omega)$ and $K(t,s,y(s,\omega),\omega)$ are measurable and finite everywhere. The corresponding (DIVP) is given by

$$m' = \hat{A}(t)m + \int_{t_0}^{t} \hat{K}(t, s, m(s))\, ds, \quad m(t_0) = m_0 \tag{3.8}$$

and the random initial value problem (RIVP),

$$x' = \hat{A}(t)x + \int_{t_0}^{t} \hat{K}(t, s, x(s))\, ds, \quad x(t_0,\omega) = x_0(\omega). \tag{3.9}$$

Consider the Lyapunov function, $V(t,x,\omega) = e^{\beta t}||x||$.

We compute $D^+_{3.9}V(s, x(t,s,y),\omega)$. For this purpose we consider

$$\begin{aligned} &V(s+h, x(t, s+h, y+hf(s,y,Ty,\omega)),\omega) - V(s, x(t,s,y),\omega) \\ &\quad = e^{\beta(s+h)}||x(t, s+h, y+hf(s,y,Ty,\omega))|| - e^{\beta s}||x(t,s,y)||. \end{aligned}$$

Using Taylor's expansion,

$$||x(t, s+h, y+hf(s, y, Ty, \omega))||$$

$$= ||x(t,s,y) + h\Phi(t,s,y)\left[-\hat{A}(s)y - \int_{t_0}^{s} \hat{a}(s,\xi,y(\xi))y(\xi)\,d\xi\right] \tag{3.10}$$

$$+h\Phi(t,s,y)\left[A(s,\omega)y(s) + \int_{t_0}^{s} a(s,\xi,y(\xi),\omega)y(\xi)\,d\xi\right] + O(h)||,$$

where $a(t,s,y,\omega)$ is the sample derivative of $K(t,s,y,\omega)$ exists, and it is sample continuous in x for each $s \le t \in \mathbb{R}^+$. Here

$$a(t,s,y,\omega) = \int_0^1 K_y(t,s,\theta y,\omega)\,d\theta \cdot y.$$

Rearranging (3.10), we obtain

$$\begin{aligned} &||x(t,s+h,y+hf(t,y,Ty,\omega))|| \\ &= ||x(t,s,y) + h\Phi(t,s,y)(A(s,\omega) - \hat{A}(s))y \\ &\quad +h\Phi(t,s,y)\left[\int_{t_0}^{s} [a(s,\xi,y(\xi),\omega) - \hat{a}(s,\xi,y(\xi))]y(\xi)\,d\xi\right] + O(h)||. \end{aligned} \tag{3.11}$$

Since $x(t,s,y) = \int_0^1 \Phi(t,s,\theta y)d\theta \cdot y = \Psi(t,s,y)y$, which gives $y = \Psi^{-1}(t,s,y)x(t,s,y)$. This together with (3.11) and (3.10) implies

$$\begin{aligned} D_+V \;\le\; &\left[\mu[\Phi(t,s,y)(A(s,\omega) - \hat{A}(s))\Psi^{-1}(t,s,y)] + \beta\right] V \\ &+||\Phi(t,s,y)|| \int_{t_0}^{t} e^{\beta(s-\xi)}||[a(s,\xi,y(\xi),\omega) - \hat{a}(s,\xi,y(\xi))]||\cdot \\ &\cdot e^{\beta\xi}||y(\xi)||\,d\xi, \end{aligned} \tag{3.12}$$

as long as $x \in \Omega = \{x : e^{\beta}||\Psi(t,r,y(r))||\,||y(r)|| \le e^{\beta s}||\Psi(t,s,y(s))||\,||y(s)||,$ $t_0 \le r \le s\}$. By assuming

$$\mu[\Phi(t,s,y)(A(s,\omega) - \hat{A}(s))\Psi^{-1}(t,s,y)] \le \gamma(s,\omega), \tag{3.13}$$

$$||a(s,\xi,y(\xi),\omega) - \hat{a}(s,\xi,y(\xi))||\,||\Phi(t,s,y)||\,||\Psi^{-1}(t,\xi,y(\xi))|| \le \nu(s,\xi,\omega), \tag{3.14}$$

the relation (3.12) reduces to

$$\begin{aligned} &D_+V(s,x(t,x,y),\omega) \\ &\le \left[\gamma(s,\omega) + \beta + \int_{t_0}^{t} e^{\beta(s-\xi)}\nu(s,\xi,\omega)\,d\xi\right] V(s,x(t,s,y),\omega) \end{aligned} \tag{3.15}$$

with the comparison equation

$$u' = \left[\gamma(s,\omega) + \beta + \int_{t_0}^{t} e^{\beta(s-\xi)}\nu(s,\xi,\omega)\,d\xi\right] u, \quad u(t_0,\omega) = u_0(\omega), \tag{3.16}$$

where γ and ν are sample integrable stochastic processes.

From (3.14), (3.15) and an application of Theorem 3.1, we obtain

$$V(t,y(t,\omega),\omega) \le V(t_0,x(t,\omega),\omega)\exp\left[\int_{t_0}^{t}[\gamma(s,\omega) + \beta + \int_{t_0}^{t} e^{\beta(s-\xi)}\nu(s,\xi,\omega)\,d\xi]\,ds\right].$$

This leads to the inequality

$$\|y(t,\omega)\| \le \|x(t,\omega)\|\exp\left[\int_{t_0}^{t}[\gamma(s,\omega) + \beta + \int_{t_0}^{s} e^{\beta(s-\xi)}\nu(s,\xi,\omega)\,d\xi]\,ds\right], \tag{3.17}$$

where $y(t,\omega)$, $x(t,\omega)$ are solution processes of (3.7), (3.9), respectively.

Remark 3.2: By adding and subtracting $E[a(s,\xi,y(\xi),\omega)]$ we can express the right hand side of (3.12) as functions of $a(s,\xi,y(\xi),\omega) - E[a(s,\xi,y(\xi),\omega)]$ and $E[a(s,\xi,y(\xi),\omega)] - \hat{a}(s,\xi,y(\xi))$. By assuming

$$\mu[\Phi(t,s,y)(A(s,\omega) - \hat{A}(s))\Psi^{-1}(t,s,y)] \le \alpha(s,\omega),$$

$$\|a(s,\xi,y(\xi),\omega) - E[a(s,\xi,y(\xi),\omega)]\|\,\|\Phi(t,s,y)\|\,\|\Psi^{-1}(t,\xi,y(\xi))\| \le \gamma(s,\xi,\omega),$$

$$\begin{aligned}\|E[a(s,\xi,y(\xi),\omega)] - \hat{a}(s,\xi,y(\xi))\|\,\|\Phi(t,s,y)\|\,\|\Psi^{-1}(t,\xi,y(\xi))\|&\\ \le \delta(s,\xi),&\end{aligned} \tag{3.18}$$

relation (3.15) reduces to

$$\begin{aligned}&D^+V(s,x(t,s,y),\omega)\\ &\le \left[\alpha(s,\omega) + \beta + \int_{t_0}^{t} e^{\beta(s-\xi)}[\gamma(s,\xi,\omega) + \delta(s,\xi)]\,d\xi\right] V(s,x(t,s,y),\omega),\end{aligned} \tag{3.19}$$

where α, γ and δ are sample integrable stochastic processes, which leads to

$$\begin{aligned}&\|y(t,\omega)\|\\ &\le \|x(t,\omega)\|\exp\left[\int_{t_0}^{t}[a(s,\omega) + \beta + \int_{t_0}^{t} e^{\beta(s-\xi)}[\gamma(s,\xi,\omega) + \delta(s,\xi)]\,d\xi]\,ds\right].\end{aligned} \tag{3.20}$$

We now state and prove a comparison theorem which has wide range of applications in the theory of error estimates and stability analysis of stochastic integro-differential equations.

Theorem 3.2. *Assume that all the hypotheses of Theorem 3.1 are satisfied except that relations (3.1) and (3.2) are replaced by*

$$D^+V(s, x(t, s, y) - x(t, s, z), \omega) \leq g(s, V(s, x(t, s, y) - z(t, s, z), \omega)) \tag{3.21}$$

on the set Ω, *where* $\Omega = \{x \in C[\mathbb{R}^+, \mathbb{R}^n] : V(r, x(t, r, y(r, \omega)) - \bar{x}(t, r, \bar{x}(r)), \omega) \leq V(s, x(t, s, y(s, \omega)) - \bar{x}(t, s, \bar{x}(s)), \omega),\ r \leq s \leq t\}$ *and*

$$V(t_0, x(t, \omega) - x(t, t_0, z_0), \omega) \leq u_0(\omega), \tag{3.22}$$

where $x(t, t_0, x_0) = \bar{x}(t)$ *is a solution process of either* (2.2) *or* (2.3) *depending on the choice of* z_0. *Then*

$$V(t, y(t, \omega) - \bar{x}(t), \omega) \leq r(t, t_0, u_0(\omega)), \quad t \geq t_0. \tag{3.23}$$

Proof. Let $y(t, \omega)$ be a solution process of (2.1) and let $x(t, \omega)$ be the solution process of (2.3) with $x(t_0, \omega) = y_0(\omega)$. Let $\bar{x}(t) = x(t, t_0, z_0)$ be the solution process of either (2.2) or (2.3) depending on the choice of z_0. Set, for $t_0 \leq s \leq t$,

$$W(s, \omega) = V(s, x(t, s, y(s, \omega)) - x(t, s, \bar{x}(s)), \omega),$$

$$W(t_0, \omega) = V(t_0, x(t, \omega) - \bar{x}(t), \omega).$$

By following the proof of Theorem 3.1, the proof of this theorem can be completed analogously.

To demonstrate the scope of the result, we present a few examples.

Example 3.2: We consider the initial value problems (3.7), (3.8) and (3.9). Let $V(t, x, \omega) = e^{\beta t}||x||$. By assuming

$$\mu[\Phi(t, s, y)(A(s, \omega) - \hat{A}(s))\Psi^{-1}(t, s, y)] \leq \alpha(s, \omega),$$

$$||a(s, \xi, y(\xi), \omega) - E[a(s, \xi, y(\xi), \omega)]||\,||\Phi(t, s, y)||\,||\Psi^{-1}(t, \xi, y(\xi), m(\xi))|| \leq \gamma(s, \xi, \omega),$$

$$||E[a(s, \xi, y(\xi), \omega)] - \hat{a}(s, \xi, y(\xi))||\ ||\Phi(t, s, y)||\ ||\Psi^{-1}(t, \xi, y(\xi), m(\xi))|| \leq \delta(s, \xi),$$

$$||a(s, \xi, y(\xi), \omega) - E[a(s, \xi, y(\xi), \omega)]||\ ||\Phi(t, s, y)|| \leq \nu(s, \xi, \omega),$$

$$||E[a(s, \xi, y(\xi), \omega)] - \hat{a}(s, \xi, y(\xi))||\ ||\Phi(t, s, y)|| \leq \theta(s, \xi) \tag{3.24}$$

and following the discussions in Example 3.2 and Remark 3.2, we obtain

$$
\begin{aligned}
&D^+V(s, x(t,s,y) - x(t,x,m), \omega) \\
&\leq \left[\alpha(s,\omega) + \beta + \int_{t_0}^{t} e^{\beta(s-\xi)}\left[\gamma(s,\xi,\omega) + \delta(s,\xi)\right] d\xi\right] \cdot \\
&\cdot V(s, x(t,s,y) - x(t,s,m), \omega) \\
&+ \int_{t_0}^{s} e^{\beta s}\nu(s,\xi)\|m(\xi)\|\, d\xi + \int_{t_0}^{s} e^{\beta s}\theta(s,\xi)\|m(\xi)\|\, d\xi
\end{aligned}
\tag{3.25}
$$

with a comparison equation

$$
\begin{aligned}
u' &= g(s, u(s), \omega) \\
&= \left[\alpha(s,\omega) + \beta + \int_{t_0}^{s} e^{\beta(s-\xi)}\left[\gamma(s,\xi,\omega) + \delta(s,\xi)\right] d\xi\right] u \\
&\quad + \int_{t_0}^{s} e^{\beta s}[\nu(s,\xi,\omega) + \theta(s,\xi)]\|m(\xi)\|\, d\xi,
\end{aligned}
\tag{3.26}
$$

where $\gamma(s,\xi,\omega) > 0$, $\delta(s,\xi) > 0$ and $\theta(s,\xi) > 0$ with probability 1 are sample integrable stochastic processes. Solving (3.26), we obtain

$$
\begin{aligned}
&u(t,\omega) = \\
&u(t_0,\omega)\exp\left[\int_{t_0}^{t}[\alpha(s,\omega) + \beta + \int_{t_0}^{s} e^{\beta(s-\xi)}\left[\gamma(s,\xi,\omega) + \delta(s,\xi)\right] d\xi]\, ds\right] \\
&+ \int_{t_0}^{t}\left[\int_{t_0}^{s} e^{\beta s}[\nu(s,\xi,\omega) + \theta(s,\xi)]\|m(\xi)\|\, d\xi\right] \cdot \\
&\cdot \exp\left[\int_{s}^{t}[\alpha(s,\omega) + \beta + \int_{t_0}^{s} e^{\beta(s-\xi)}\left[\gamma(s,\xi,\omega) + \delta(s,\xi)\right] d\xi]\, ds\right].
\end{aligned}
\tag{3.27}
$$

From (3.26), (3.27) and an application of Theorem 3.1 in the context of the particular case $u_0 = V(t_0, x(t, t_0, x_0(\omega)) - x(t, t_0, m_0), \omega)$ we obtain

$$
\begin{aligned}
&\|y(t,\omega) - m(t)\| \\
&\leq \|x(t,\omega) - x(t,t_0,m_0)\| \cdot \\
&\quad \exp\left[\int_{t_0}^{t} [\alpha(s,\omega) + \beta + \int_{t_0}^{s} e^{\beta(s-\xi)}[\gamma(s,\xi,\omega) + \delta(s,\xi)]\, d\xi]\, ds\right] \\
&\quad + \int_{t_0}^{t} \left[\int_{t_0}^{s} [\nu(s,\xi,\omega) + \theta(s,\xi)]\|m(\xi)\|\, d\xi\right] \cdot \\
&\quad \cdot \exp\left[\int_{s}^{t} [\alpha(z,\omega) + \beta + \int_{t_0}^{s} e^{\beta(z-\xi)}[\gamma(s,\xi,\omega) + \delta(s,\xi)]\, d\xi]\, ds\right].
\end{aligned} \tag{3.28}
$$

In the following example, we consider the linear version of (3.7) and its corresponding determinisitic system.

Example 3.3: Let

$$
y' = A(t,\omega)y + \int_{t_0}^{t} K(t,s,\omega)y(s,\omega)\, ds, \quad y(t_0,\omega) = y_0(\omega) \tag{3.29}
$$

and

$$
m' = \hat{A}(t)m + \int_{t_0}^{t} \hat{K}(t,s)m(s)\, ds, \quad m(t_0) = m_0 \tag{3.30}
$$

and the random initial value problem (RIVP)

$$
x' = \hat{A}(t)x + \int_{t_0}^{t} \hat{K}(t,s)x(s)\, ds, \quad x(t_0,\omega) = x_0(\omega), \tag{3.31}
$$

where $A(t,\omega)$ is an $n \times n$ random matrix function, $K(t,s,\omega) \in C[\mathbb{R}^+ \times \mathbb{R}^+, \mathbb{R}^n]$ and A, K satisfy the desired regularity conditions so that (3.29) has a solution process for $t \geq t_0$ with power 1, and $\hat{A}$, $\hat{K}$ are obtained by ignoring the randomness in $A(t,\omega)$ and $K(t,s,\omega)$. Consider the Lyapunov function

$$
V(t,x,\omega) = e^{\beta t}\|x\|, \quad \beta > 0.
$$

Then

$$
x(t,t_0,x_0(\omega)) = R(t,t_0)x_0(\omega), \tag{3.32}
$$

where $R(t, t_0)$ is the resolvent kernel and, in the linear case,

$$\Psi(t, s, y(s, \omega), m(s)) = R(t, s). \tag{3.33}$$

Moreover, $x(t, t_0, z_0)$ is sample continuously differentiable with respect to (t_0, z_0) and

$$\frac{\partial x(t, t_0, z_0)}{\partial x_0} = R(t, t_0)$$

is the solution process of the adjoint equation

$$R_s(t, s) + R(t, s)\hat{A}(s) + \int_s^t R(t, r)\hat{K}(r, s)\, dr = 0, \tag{3.34}$$

associated with (3.31). With regard to equations (3.29)–(3.31), estimates corresponding to (3.17), (3.18), (3.28) can be obtained analogously.

(i) The estimate corresponding to (3.17) with regard to (3.29)–(3.31) is given by

$$\begin{aligned} &D^+V(s, x(t, s, y), \omega) \\ &\le \left[\gamma(s, \omega) + \beta + \int_{t_0}^s e^{\beta(s-\xi)}\nu(s, \xi, \omega)\, d\xi\right] V(s, x(t, s, y), \omega), \end{aligned} \tag{3.35}$$

where

$$\mu[R(t, s)(A(s, \omega) - \hat{A}(s))R^{-1}(t, s)] \le \gamma_1(s, \omega),$$

$$\|K(s, \xi, \omega) - \hat{K}(s, \xi)\|\, \|R(t, s)R^{-1}(t, \xi)\| \le \nu_1(s, \xi, \omega). \tag{3.36}$$

γ_1 and $\nu_1 > 0$ are sample integrable stochastic processes. Hence the comparison equation is

$$u'(s, \omega) = \left[\gamma_1(s, \omega) + \beta + \int_{t_0}^s e^{\beta(s-\xi)}\nu_1(s, \xi, \omega)\, d\xi\right] u(s, \omega).$$

By following the argument used in Example 3.2, one can arrive at

$$\|y(t, \omega)\| \le \|x(t, \omega)\| \exp\left[\int_{t_0}^t [\gamma_1(s, \omega) + \beta + \int_{t_0}^s e^{\beta(s-\xi)}\nu_1(s, \xi, \omega)\, d\xi]\, ds\right], \tag{3.37}$$

where $y(t, \omega)$, $x(t, \omega)$ are solution processes of (3.29), (3.31) respectively.

(ii) Estimates corresponding to (3.20) for the function $V(s, x(t,s,y), \omega)$ with regard to equations (3.29)–(3.31) are obtained from

$$D^+V(s, x(t,s,y), \omega)$$
$$\leq \left[\alpha_1(s,\omega) + \beta_1(s,\omega) + \beta + \int_{t_0}^{s} e^{\beta(s-\xi)}[\gamma(s,\xi,\omega) + \delta(s,\xi)]\, d\xi\right] \cdot$$
$$\cdot V(s, x(t,s,y), \omega),$$

where

$$\mu[R(t,s)(A(s,\omega) - E[A(s,\omega)])R^{-1}(t,s)] \leq \alpha_1(s,\omega),$$
$$\mu[R(t,s)(E[A(s,\omega)] - \hat{A}(s))R^{-1}(t,s)] \leq \beta_1(s,\omega),$$
$$||K(s,\xi,\omega) - EK(s,\xi,\omega)||\ ||R(t,s)R^{-1}(t,\xi)|| \leq \gamma_1(s,\xi,\omega),$$

$$||EK(s,\xi,\omega) - \hat{K}(s,\xi)||\ ||R(t,s)R^{-1}(t,\xi)|| \leq \delta_1(s,\xi), \tag{3.38}$$

where $\alpha_1, \beta_1, \gamma_1 > 0$ are sample integrable stochastic processes. Hence the comparison equation is

$$u'(s,\omega) =$$
$$\left[\alpha_1(s,\omega) + \beta_1(s,\omega) + \beta + \int_{t_0}^{s} e^{\beta(s-\xi)}[\gamma(s,\xi,\omega) + \delta(s,\xi)]\, d\xi\right] u(s,\omega). \tag{3.39}$$

By following the analysis of Example 3.1, one can arrive at

$$\begin{aligned} ||y(t,\omega)|| \ &\leq\ ||x(t,\omega)|| \exp\left[\int_{t_0}^{t} [\alpha_1(s,\omega) + \beta_1(s,\omega) + \beta \right. \\ &\left. + \int_{t_0}^{s} e^{\beta(s-\xi)}[\gamma(s,\xi,\omega) + \delta(s,\xi)]\, d\xi]\, ds\right]. \end{aligned} \tag{3.40}$$

(iii) The estimate corresponding to (3.29) with regard to (3.30)–(3.31) is given by

$$D^+V(s, x(t,s,y) - x(t,s,m), \omega)$$
$$\leq \left[\alpha_1(s,\omega) + \beta_1(s,\omega) + \beta + \int_{t_0}^{s} e^{\beta(s-\xi)}[\gamma_1(s,\xi,\omega) + \delta_1(s,\xi)]\, d\xi\right] \cdot \tag{3.41}$$
$$\cdot V(s, x(t,s,y) - x(t,s,m), \omega),$$

$$+ \textstyle\int_{t_0}^{s} e^{\beta s}\nu_1(s,\xi,\omega)||m(\xi)||\, d\xi + \int_{t_0}^{s} e^{\beta s}\theta_1(s,\xi)||m(\xi)||\, d\xi$$

where

$$\mu[R(t,s)(A(s,\omega) - E[A(s,\omega)])R^{-1}(t,s)] \le \alpha_1(s,\omega),$$

$$\mu[R(t,s)(E[A(s,\omega)] - \hat{A}(s))R^{-1}(t,s)] \le \beta_1(s,\omega),$$

$$||K(s,\xi,\omega) - EK(s,\xi,\omega)||\ ||R(t,s)R^{-1}(t,\xi)|| \le \gamma_1(s,\xi,\omega),$$

$$||EK(s,\xi,\omega) - \hat{K}(s,\xi)||\ ||R(t,s)R^{-1}(t,\xi)|| \le \delta_1(s,\xi),$$

$$||K(s,\xi,\omega) - EK(s,\xi,\omega)||\ ||R(t,s)|| \le \nu_1(s,\xi,\omega),$$

$$||EK(s,\xi,\omega) - \hat{K}(s,\xi)||\ ||R(t,s)|| \le \theta_1(s,\xi). \tag{3.42}$$

Here $\alpha_1, \beta_1, \gamma_1 > 0$, $\delta > 0$, $\nu_1 > 0$, $\theta_1 > 0$ with probability 1 are sample integrable stochastic processes. Hence the comparison equation is

$$u'(s,\omega) =$$

$$\left[\alpha_1(s,\omega) + \beta_1(s,\omega) + \beta + \int_{t_0}^{s} e^{\beta(s-\xi)}[\gamma_1(s,\xi,\omega) + \delta_1(s,\xi)]\, d\xi\right] u(s,\omega) \tag{3.43}$$

$$+ \int_{t_0}^{s} e^{\beta s}[\nu_1(s,\xi,\omega) + \theta_1(s,\xi)]||m(\xi)||\, d\xi.$$

By following the analysis of Example 3.1, one can arrive at

$$||y(t,\omega) - m(t)|| \le$$

$$||x(t,\omega) - x(t,t_0,m_0)||\cdot$$

$$\cdot \exp\left[\int_{t_0}^{t} [\alpha_1(s,\omega) + \beta_1(s,\omega) + \int_{t_0}^{s} e^{\beta(s-\xi)}[\gamma_1(s,\xi,\omega) + \delta_1(s,\xi)]\, d\xi]\, ds\right]$$

$$+ \int_{t_0}^{t}\left[\int_{t_0}^{s} [\nu_1(s,\xi,\omega) + \theta_1(s,\xi)]||m(\xi)||\, d\xi\right]\cdot \tag{3.44}$$

$$\cdot \exp\left[\int_{s}^{t} [\alpha_1(z,\omega) + \beta_1(z,\omega)\right.$$

$$\left. + \int_{t_0}^{z} e^{\beta(z-\xi)}[\gamma_1(z,\xi,\omega) + \delta_1(s,\xi)]\, d\xi]\, dz\right]\, ds.$$

4. Stability analysis

In this section, we employ the comparison theorem 3.1 to study the p-th moment stability of the trivial solution process of (2.1).

Theorem 4.1. *Suppose the following holds:*

(i) *The conditions* (H_0–H_4) *of Theorem* 3.1 *are satisfied.*

(ii) $g(t,0,\omega) \equiv 0$, $f(t,0,0) \equiv 0$ *with probability* 1, $t \in \mathbb{R}^+$.

(iii)

$$b(||y(t,\omega)||^p) \leq \sum_{i=1}^{m} V_i(t, y(t,\omega), \omega) \leq a(t, ||y(t,\omega)||^p), \tag{4.1}$$

where $a \in CK$, $b \in \nu K$ *and* $p \geq 1$.

If the trivial solution of

$$u' = g(t, u(t,\omega), \omega), \qquad u(t_0, \omega) = u_0(\omega),$$

is stable in the p-th mean, then the trivial solution process of (2.1) *is stable in the p-th mean.*

The *proof* of this theorem is left to the reader. As an application of Theorem 4.1 we give the following example.

Example 4.1: We use equation (3.36) from Example 3.3 and $V(t,x) = e^{\beta t}||x||$. Further, we assume that there exists some Lebesgue integrable function Λ defined on $\mathbb{R}^+$ into $\mathbb{R}$ such that

$$||x(t,\omega)|| \leq x_0(\omega) \exp\left[\int_{t_0}^{t} \Lambda(u)\, du\right] \quad \text{with probability 1,}$$

where $x(t,\omega)$ is the solution process of (3.31). Using estimat (3.28), we get

$$\begin{aligned} &E[||y(t,\omega)||] \\ &\leq E[||x_0(\omega)||] E\left[\exp\left[\int_{t_0}^{t} \gamma_1(s,\omega) + \beta + \int_{t_0}^{s} \nu_1(s,\xi,\omega)\, d\xi\right] ds\right] \cdot \\ &\cdot \exp\left[\int_{t_0}^{t} \Lambda(s)\, ds\right]. \end{aligned} \tag{4.2}$$

As an application of Theorem 3.2 we conclude that (2.1) is asymptotically stable whenever

(4.3)
$$E\left[\exp\left(\int_{t_0}^{t}\left[\gamma_1(s,\omega)+\int_{t_0}^{s}e^{\beta(s-\xi)}\nu_1(s,\xi,\omega)\,d\xi\right]\,ds\right)\right] \text{ is finite for all } t\geq t_0$$

and

$$\limsup_{t\to\infty}\left[\frac{1}{t-t_0}\int_{t_0}^{t}[\Lambda(u)+\beta]\,du\right]=-a(t_0), \quad \text{where } a(t_0)>0. \tag{4.4}$$

5. Error estimate results

In this section we shall employ the comparison method to derive an error estimate on the p-th moment deviation of a solution process of (2.1) from the solution process of either (2.2) or (2.3).

Theorem 5.1. *Let the hypotheses of Theorem* 3.1 *be satisfied. Further, assume that*

$$b(\|x\|^p)\leq\sum_{i=1}^{n}V_i(t,x,\omega), \tag{5.1}$$

where $b\in CK$ is a convex function and $p\geq 1$. Then

$$b(E[\|y(t,\omega)-\bar{x}(t)\|^p])\leq\sum_{i=1}^{n}E[r_i(t,t_0,V(t_0,x(t,\omega)-\bar{x}(t),\omega),\omega)]. \tag{5.2}$$

Proof. By the choice of $u_0=V(t_0,x(t,\omega)-\bar{x}(t),\omega)$, (3.4) reduces to

$$\sum_{i=1}^{n}V_i(t,y(t,\omega)-\bar{x}(t),\omega)\leq\sum_{i=1}^{n}r_i(t,t_0,V(t_0,x(t,\omega)-\bar{x}(t),\omega),\omega). \tag{5.3}$$

This together with (5.1) and the convexity of b yields

$$b(E[\|y(t,\omega)-\bar{x}(t)\|^p])\leq\sum_{i=1}^{n}E[r_i(t,t_0,V(t_0,x(t,\omega)-\bar{x}(t),\omega),\omega)]. \tag{5.4}$$

This completes the proof of the theorem.

Example 5.1: We use $V(t,x,\omega) = e^{\beta t}||x||$ and $b(||x||) = ||x||$. By an application of Theorem 5.1, (3.28) becomes

$$
\begin{aligned}
&E||y(t,\omega) - m(t)||\\
&\le 2E[||x(t,\omega) - x(t,t_0,m_0)||]\cdot\\
&\quad\cdot\exp\left[\int_{t_0}^{t}[\alpha(s,\omega) + \beta + \int_{t_0}^{s} e^{\beta(s-\xi)}[\gamma(s,,\xi,\omega) + \delta(s,\xi)]\,d\xi]\,ds\right]\\
(5.5)\qquad&\quad+2E\left[\int_{t_0}^{t}\left[\int_{t_0}^{s}[\nu(s,\xi,\omega) + \theta(s,\xi)]||m(\xi)||\,d\xi\right]\cdot\right.\\
&\quad\left.\cdot\exp\left[\int_{s}^{t}[\alpha(z,\omega) + \beta + \int_{t_0}^{s} e^{\beta(z-\xi)}[\gamma(s,\xi,\omega) + \delta(s,\xi)]\,d\xi]\,dz\right]\,ds\right].
\end{aligned}
$$

Using Schwarz's inequality and after simplification, the above equation reduces to

$$
\begin{aligned}
&E||y(t,\omega) - m(t)|| \le\\
&2E[||x(t,\omega) - x(t,t_0,m_0)||]\cdot\\
&\cdot E^{1/2}\left[\exp\left[\int_{t_0}^{t}\left[\alpha(s,\omega) + \beta + \int_{t_0}^{s} e^{\beta(s-\xi)}[\gamma(s,,\xi,\omega) + \delta(s,\xi)]\,d\xi\right]\,ds\right]\right]\\
(5.6)\qquad&+2\int_{t_0}^{t}\left[\int_{t_0}^{s}||m(\xi)||\,d\xi\right]^2 E\left[\left[\int_{t_0}^{s}[\nu(s,\xi,\omega) + \theta(s,\xi)]\,d\xi\right]^2\cdot\right.\\
&\left.\cdot\exp\left[2\int_{s}^{t}\left[\alpha(z,\omega) + \beta + \int_{t_0}^{s} e^{\beta(z-\xi)}[\gamma(s,\xi,\omega) + \delta(s,\xi)]\,d\xi\right]\,dz\right]\,ds\right]^{1/2}.
\end{aligned}
$$

6. Numerical results

To show the scope and usefulness of the various results in this paper, we present some numerical examples. By an application of Newton's method we developed a computer code to generate sample solutions of random integro-differential equations. We used IMSL subroutine to generate random numbers and modified those numbers to our need. We consider the following equations

$$
\begin{aligned}
(6.1)\qquad y'(t,\omega) &= \tfrac{1}{2}a_1(t,\omega)y(t,\omega) - 9a_2(t,\omega)\int_0^t e^{-7(t-s)}y(s,\omega)\,ds,\\
y(t_0,\omega) &= y_0(\omega),
\end{aligned}
$$

$$m(t) = \tfrac{1}{2}E(a_1(t,\omega))m(t) - 9E(a_2(t,\omega))\int_0^t e^{-7(t-s)}m(s)\,ds, \qquad E(y_0) = m_0, \tag{6.2}$$

and

$$y'(t,\omega) = \left(e^{-t_0 b_1(\omega)} - e^{-tb_1(\omega)} + \frac{1}{t}\right)y(t,\omega) - b_1(\omega)t\int_{t_0}^t e^{-y(s,\omega)}ds, \qquad y(t_0,\omega) = y_0(\omega), \quad t > 0, \tag{6.3}$$

$$m(t) = \left(e^{-t_0 m_0} - e^{-tm_0} + \frac{1}{t}\right)m(t) - m_0 t\int_{t_0}^t e^{-m(s)}ds, \qquad E(y_0) = m_0, \quad t > 0, \tag{6.4}$$

where a_1, a_2, b_1 are normal random variables.

Example 6.1: (6.1), (6.2) with $a_1, a_2, b_1 \in N(1, 0.1^2)$, $t \in [0,1]$.

TABLE I

t	$E(y(t,\omega))$	$m(t)$	$E\lvert y(t,\omega) - m(t)\rvert$	Error estimate
0.1	1.015915	1.016368	0.000453	0.0025
0.2	0.982032	0.981930	0.000102	0.0100
0.3	0.922842	0.922454	0.000386	0.0100
0.4	0.852956	0.852436	0.000520	0.0240
0.5	0.780449	0.779881	0.000568	0.0450
0.6	0.709639	0.709071	0.000568	0.7400
0.7	0.642709	0.642165	0.000544	0.1110
0.8	0.580632	0.580123	0.000509	0.1560
0.9	0.523711	0.523242	0.000469	0.2120
1.0	0.471884	0.471456	0.000428	0.2750

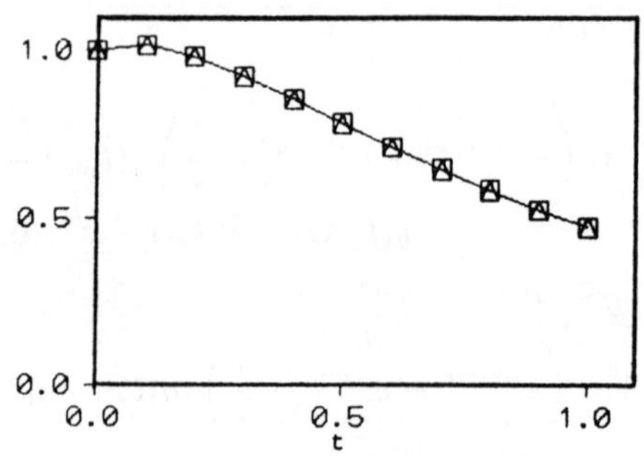

Figure 1. (Example 6.1)

□ E(y(t,w)), △ m(t)

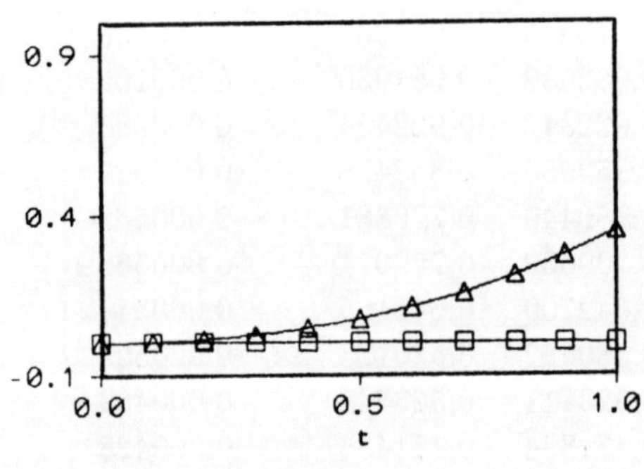

Figure 2. (Example 6.1)

□ E|y(t,w)-(m(t)|, △ Error Estimates

Example 6.2: (6.3), (6.4) with $b_1 \in N(1, 0.01^2)$, $t \in [1, 2]$.

TABLE II

t	$E(y(t,\omega))$	$m(t)$	$E\|y(t,\omega) - m(t)\|$	Error estimate
1.1	1.09813	1.09990	0.00177	0.04103
1.2	1.19746	1.19939	0.00193	0.08428
1.3	1.29592	1.29801	0.00209	0.12997
1.4	1.39301	1.39525	0.00224	0.17836
1.5	1.48817	1.49057	0.00240	0.22974
1.6	1.58081	1.58337	0.00256	0.28442
1.7	1.67032	1.67302	0.00270	0.34275
1.8	1.75602	1.75887	0.00285	0.40511
1.9	1.83721	1.84020	0.00299	0.47192
2.0	1.91312	1.91625	0.00313	0.54366

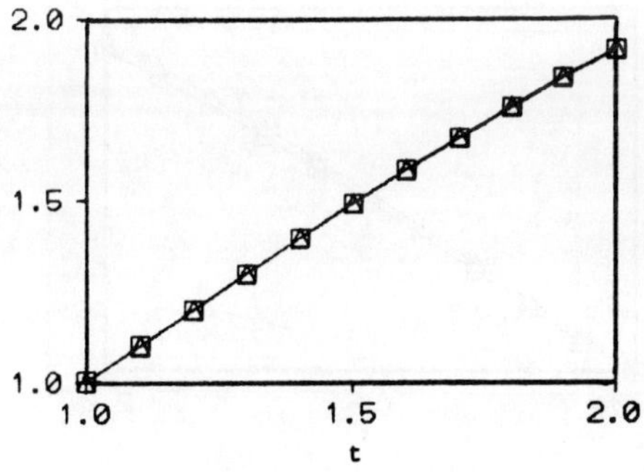

Figure 3. (Example 6.2)

□ E(y(t,w)), △ m(t)

Example 6.3: (6.3), (6.4) with $b_1 \in N(1, 0.1^2)$, $t \in [1, 2]$.

TABLE III

t	$E(y(t,\omega))$	$m(t)$	$E\|y(t,\omega) - m(t)\|$	Error estimate
1.1	1.08221	1.09990	0.01769	0.04103
1.2	1.18010	1.19939	0.01929	0.08428
1.3	1.27714	1.29801	0.02087	0.12997
1.4	1.37283	1.39525	0.02242	0.17836
1.5	1.46662	1.49057	0.02395	0.22974
1.6	1.55794	1.58337	0.02543	0.28442
1.7	1.64618	1.67302	0.02684	0.34275
1.8	1.73067	1.75887	0.02820	0.40511
1.9	1.81071	1.84020	0.02949	0.47192
2.0	1.88554	1.91625	0.03071	0.54366

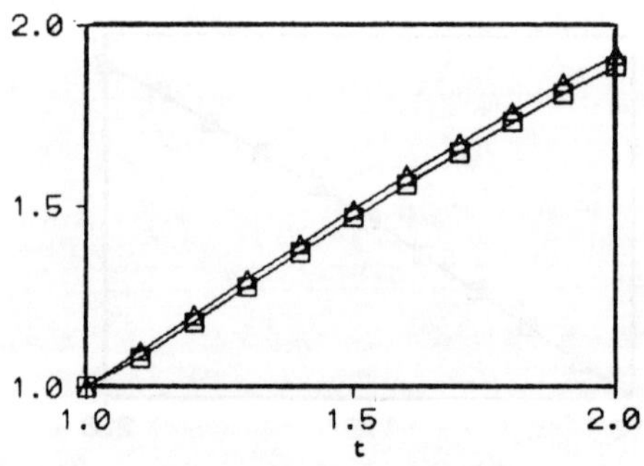

Figure 4. (Example 6.3)

□ E(y(t,w)), △ m(t)

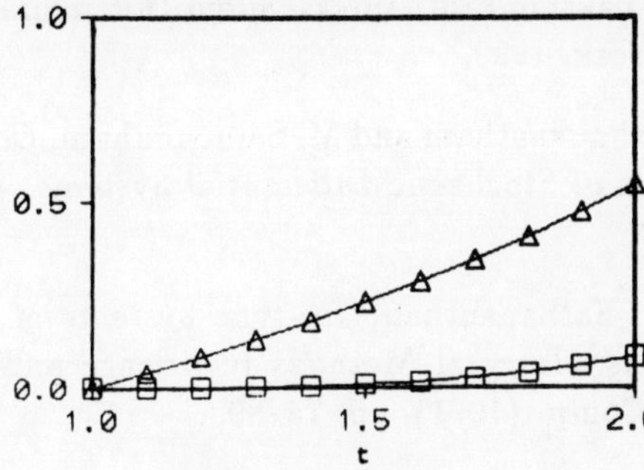

Figure 5. (Example 6.3)

□ E|y(t,w)-(m(t)|, △ Error Estimates

All the results presented in this section are from a sample of size 200. Table 1 contains $E(y(t,\omega))$, $m(t)$, $E|y(t,\omega)-m(t)|$ and error estimates (3.44) for Example 1.

Figure 1 contains $E(y(t,\omega))$ and $m(t)$ and Figure 2 contains $E|y(t,\omega) - m(t)|$ and error estimates (3.28) for Example 6.1. Tables II and III contain the same information as Table I, but for Examples 6.2 and 6.3 respectively. Figures 3 and 4 contain $E(y(t,\omega))$ and $m(t)$ for Example 6.2 and 6.3 respectively. Figure 5 contains $E|y(t,\omega) - m(t)|$ and error estimates for Example 6.3.

Acknowledgement. This research was partially supported by the U.S. Army Research Grant No. DAAL03-90-G-0180.

References

1. A.T. Bharucha-Reid, *Random Integral Equations.* Academic Press, New York, 1972.

2. J. Chandra, G.S. Ladde and V. Lakshmikantham, *On the fundamental theory of nonlinear second order stochastic boundary value problems.* Stoch. Anal. Appl. **1** (1983), 1–19.

3. J. Chandra, G.S. Ladde and V. Lakshmikantham, *Stochastic analysis of compressible gas lubricated slider bearing problem.* SIAM J. Appl. Math. **43** (1983), 1174–1186.

4. G.S. Ladde and V. Lakshmikantham, *Random Differential Inequalities*. Academic Press, New York, 1980.

5. G.S. Ladde, V. Lakshmikantham and M. Sambandham, *Comparison Theorem and Error Estimates of Stochastic Differential Systems*. Stoch. Anal. Appl. **3** (1) (1985), 23–62.

6. G.S. Ladde and S. Sathananthan, *Itô-type systems of stochastic integro-differential equations*. Integral Methods in Science and Engineering - 90, Hemisphere Publ. Comp. (1991), pp. 75–89.

7. G.S. Ladde and S. Sathananthan, *Error Estimates and Stability of Itô-type systems of Nonlinear Stochastic Integro-Differential Equations*. Applicable Analysis. To appear.

8. V. Lakshmikantham, *Some problems in integro-differential equations of Volterra type*. J. Int. Eqs. **10** (1985), 137–146.

9. V. Lakshmikantham and S. Leela, *Differential and Integral Inequalities*. Academic Press, New York, 1968.

10. V. Lakshmikantham, S. Leela and A.A. Martynuk, *Stability Analysis of Nonlinear Systems*. Marcel Dekker, New York and Basel, 1989.

11. H.L. Royden, *Real Analysis*. Macmillan, 1968.

12. G.R. Shendge, *A new approach to the stability theory of functional differential equations*. J. Math. Anal. Appl. **95** (1983), 319–334.

G.S. Ladde, Department of Mathematics, University of Texas at Arlington, Box 19408, Arlington, TX 76012, USA

M. Sambandham, Morehouse College, Atlanta, GA 30314, USA

S. Sathananthan, Division of Science and Mathematics, Jarvis Christian College, Hawkins, TX 75765, USA

International Series of Numerical Mathematics, Vol. 103, © 1992 Birkhäuser Verlag Basel

Nagumo theory without the Nagumo point

Raymond M. Redheffer

Abstract

In the Nagumo-Westphal theory of parabolic inequalities the first step typically involves a strict inequality, and the result for weak inequality is then obtained by introduction of a suitable perturbation. As a rule, it is the latter result that is really wanted. By a double use of the perturbation, bypassing the Nagumo theory, we reach the goal in a single step. Avoiding the Nagumo point is one novelty of the presentation. Another is the use of a discontinuous comparison function in the main theorem.

Ever since the pioneering work of Nagumo in 1939 and Westphal in 1949, parabolic problems have been analyzed by means of a *Nagumo point.* This is a point (x^*, t^*) with smallest t such that the desired conclusion, for example $u < 0$, fails. The principal reason for introducing a Nagumo point is to account for the fact that the region is open-ended at the top; as a rule, we have $0 \le t < \infty$ or $0 \le t < T$ where no boundary condition is given at $t = T$. A different method of dealing with the open-ended region is suggested here. It takes us in one step from weak inequality in the hypothesis to weak inequality in the conclusion. Without striving for maximum generality, we introduce only enough complication to illustrate the main ideas.

1. Notation

Let B denote a connected open subset of $\mathbb{R}^n$. As usual, the parabolic interior G and parabolic boundary Γ are subsets of $\mathbb{R}^{n+1}$ defined by

$$G = \{(x,t) : x \in B, t > 0\}, \qquad \Gamma = \overline{G} - G.$$

It is assumed that u is a real-valued function which is continuous in $\overline{G}$ and twice differentiable in G. The expression u_x is the gradient and u_{xx} is the Hessian. All our differential inequalities for u hold in G.

The letters a, b, c denote functions from G to $\mathbb{R}$, $\mathbb{R}^n$, and S^n respectively, where S^n is the set of real n by n matrices. We use the contractions

$$bu_x = \sum_{j=1}^{n} b_j \frac{\partial u}{\partial x_j}, \quad cu_{xx} = \sum_{i,j=1}^{n} c_{ij} \frac{\partial^2 u}{\partial x_i \partial x_j}$$

and an inequality such as $c \geq 0$ is defined by quadratic forms. The lemma of Schur and Féjer ensures that $cu_{xx} \leq 0$ at any point where $c \geq 0$ and $u_{xx} \leq 0$.

By f we mean a function

$$G \times \mathbb{R} \times \mathbb{R}^n \times S^n \to \mathbb{R}.$$

The letters ρ, J and K denote positive functions on G with ρ twice differentiable. To ensure that our various functions are defined at (x, t) conditions such as $|x| = r$, $|x| < \alpha$, $|x| \geq \alpha$, $t \leq T$ are interpreted respectively as

$$x \in B,\ |x| = r, \quad x \in B,\ |x| < \alpha, \quad x \in B,\ |x| \geq \alpha, \quad 0 \leq t \leq T.$$

When B is unbounded, $u \leq o(\rho)$ means

$$\limsup_{r \to \infty} \sup_{|x|=r} \frac{u(x,t)}{\rho(x)} \leq 0$$

uniformly in t. It will be clear from the proofs that the lim sup can sometimes be replaced by lim inf. However, this refinement will not be emphasized.

2. Infinite interval

We denote by I the class of functions defined on G and admitting a monotone increasing majorant $M(t)$. Thus I is identical with the class of functions $G \to \mathbb{R}$ such that $g(x, t)$ admits a finite upper bound $M(T)$ whenever $t \leq T$. If B is bounded, I contains all real-valued functions continuous in $\overline{G}$. Hence $u \leq o(\rho)$ implies $u/\rho \in I$. The class I is closed under formation of sums and products.

Lemma 1. *Let $g_j \in I$. Then there exists a function $R(t)$ such that*

$$R(t) > 1, \quad R(t) > g_1(x,t), \quad R'(t) > g_2(x,t)R(t), \qquad (x,t) \in G.$$

For *proof* let $g_j(x, t) \leq M_j(t)$ where M_j are positive and increasing and set $M = M_1 + M_2$. Now look at

$$P(t) = 1 + \int_0^{t+1} M(\tau)\, d\tau, \quad Q(t) = \int_0^{t+1} P(\tau)\, d\tau, \quad R(t) = e^{Q(t)}.$$

Example 1. Let $a \in I$ and $c \geq 0$. Suppose further that B is bounded and that

$$u_t \leq au + bu_x + cu_{xx}.$$

Then $u(\Gamma) \leq 0 \Rightarrow u(G) \leq 0$. To see why, choose R so that for $(x,t) \in G$

$$R(t) > 1, \quad R(t) > e^t u(x,t), \quad R'(t) > a(x,t)R(t).$$

If the conclusion fails the function $w = u/R$ has a positive maximum interior to G. At this point

$$u = wR, \quad u_t = wR', \quad u_x = 0, \quad u_{xx} \leq 0$$

and the differential inequality leads to a contradiction.

Example 2. Let B be unbounded, let $u \leq o(\rho)$, let $c \geq 0$, and let the condition $a \in I$ of Example 1 be replaced by $\lambda \in I$ where $\lambda = \lambda(x,t)$ is defined by

$$a\rho + b\rho_x + c\rho_{xx} = \lambda\rho.$$

Then $u(\Gamma) \leq 0 \Rightarrow u(G) \leq 0$. For proof, choose R so that

$$R > 1, \quad R(t) > \frac{u(x,t)}{\rho(x)}e^t, \quad R'(t) > \lambda(x,t)R(t).$$

If the conclusion fails the function the function $w = u/(\rho R)$ has a positive maximum in G. A this point

$$u = w\rho R, \quad u_t = w\rho R', \quad u_x = w\rho_x R, \quad u_{xx} \leq w\rho_{xx} R$$

and the differential inequality leads to a contradiction.

Example 3. We now consider the inequality $u_t \leq f(x,t,u,u_x,u_{xx})$ under the hypothesis that B is bounded and that

$$f(x,t,z,0,q) \leq f(x,t,z,0,0) \quad \text{for} \quad z > 0,\ q \leq 0.$$

This holds, in particular, if $f(x,t,z,0,q)$ is weakly increasing with respect to its matrix argument q. We assume also that there is a positive increasing function $\beta = \beta(t)$, depending on u, such that

(1) $$f(x,t,z,0,0) \leq \beta(t)z \quad \text{when} \quad z = u(x,t) > 0.$$

Taking

$$R(t) > 1, \quad R(t) > e^t u(x,t), \quad R'(t) > \beta(t)R(t)$$

we find again that $u(\Gamma) \leq 0 \Rightarrow u(G) \leq 0$. The proof is virtually identical to that in Example 1.

Example 4. Suppose next that (1) is replaced by

$$f(x,t,z,0,0) \leq \beta(t)z \quad \text{when} \quad 0 < z \leq \epsilon(t) \tag{2}$$

where β and ϵ are positive increasing and decreasing functions, respectively. Suppose also that $f(x,t,z,0,0)$ is continuous. If the remaining hypotheses of Example 3 hold, the conclusion still holds. To see why, note that $f(x,t,u(x,t),0,0)$ is continuous. Hence it admits a positive increasing majorant $\gamma(t)$. This yields

$$f(x,t,u(x,t),0,0) \leq \frac{\gamma(t)}{\epsilon(t)} u(x,t) \quad \text{when} \quad u(x,t) \geq \epsilon(t).$$

Hence (1) holds with β replaced by $\beta + (\gamma/\epsilon)$, and this completes the proof.

Example 5. If B is unbounded, we assume a condition of the form

$$f(x,t,z,p,q) \leq \beta(t)(z + J(x)|p| + K(x)|s|) \tag{3}$$

whenever $z = u(x,t) > 0$ and $q \leq s$. Here β is a positive increasing function and $|\cdot|$ denotes any convenient norm. In particular, (3) holds if $f(x,t,z,p,q)$ is weakly increasing with respect to its matrix argument q and the inequality is satisfied with $q = s$.

Let B be unbounded, let $u \leq o(\rho)$, and let (3) hold. Suppose further that $J|\rho_x|/\rho$ and $K|\rho_{xx}|/\rho$ are bounded in B. Then $u(\Gamma) \leq 0 \Rightarrow u(G) \leq 0$. For proof proceed as in Example 2. There will be a contradiction at the maximum of $w = u/(R\rho)$ if

$$wR'\rho > f(x,t,wR\rho,wR\rho_x,u_{xx})$$

when $u_{xx} \leq wR\rho_{xx}$. This holds if

$$wR'\rho > \beta(t)(wR\rho + JwR|\rho_x| + KwR|\rho_{xx}|).$$

Dividing through by $w\rho$ we find that R can be constructed and the result follows.

Example 6. Let $\alpha > 0$. Suppose the hypothesis of Example 5 holds, except that (3) is assumed only for $|x| \geq \alpha$ and (1) for $|x| < \alpha$. Then the conclusion still holds. The proof is postponed to Section 3, where the result is given in a sharper form.

3. Finite interval

We begin with some preliminary observations that are well known to all who work in parabolic inequalities. If B is bounded and $u(x,0) \leq 0$ we can choose $T > 0$ so small that $u(x,t) < \epsilon$ for $0 \leq t \leq T$. Since $z = u$ in our proofs, this will give $0 < z < \epsilon$. To extend to larger t, let U denote the set of values T such that $u(x,t) \leq 0$ for $t \leq T$. Then U is closed. On the other hand if $T_0 \in U$ we can choose $T > T_0$ so close to T_0 that $u(x,t) < \epsilon$ for $t < T$. Letting T_0 take the role formerly taken by 0 we shall find that $u(x,t) \leq 0$ for $t < T$ and hence U is open. This shows that U must be the whole interval $0 \leq t < \infty$.

When $t \leq T$ any monotone majorant $M(t)$ can be replaced by the constant $M = M(T)$. The following lemma is proved by inspection:

Lemma 2. *If M and T are positive constants the function R defined for $t < T$ on the left below satisfies the inequalities on the right:*

$$P = \frac{T}{T-t}, \qquad R = e^{(1+MT)P}; \qquad R > P, \quad R' > MR.$$

Instead of discussing Examples 1–5 in this setting, we confine attention to Example 6. The letters α, β, ϵ denote positive constants corresponding to α, $\beta(t)$, $\epsilon(t)$ in the previous discussion. Relations involving these constants are required to hold on each interval $t \leq T_1$, and all three constants can depend on T_1. In a like fashion, the inequality defining $u \leq o(\rho)$ is required to hold uniformly when $t \leq T_1$, for each fixed T_1, and not for $0 \leq t < \infty$ as in Section 2.

Besides the basic hypotheses

$$\text{(A)} \qquad u_t \leq f(x,t,u,u_x,u_{xx}), \qquad u \leq o(\rho)$$

we assume

$$\text{(B)} \qquad \begin{cases} f(x,t,z,0,q) \leq \beta z & \text{when } |x| < \alpha, \\ z = u(x,t), \quad 0 < z < \epsilon, & q \leq 0. \end{cases}$$

We also need

$$\text{(C)} \qquad \begin{cases} f(x,t,z,p,q) \leq \beta(z + J(x)|p| + K(x)|s|) \quad \text{when} \quad |x| \geq \alpha, \\ z = u(x,t) > 0, \quad 0 < z < \varepsilon\rho(x), \quad q \leq s, \quad |p| \leq \epsilon|\rho_x|, \quad |s| \leq \epsilon|\rho_{xx}|. \end{cases}$$

Finally, we assume the following for $x \in B$:

$$\text{(D)} \qquad J|\rho_x| \leq 1 + \rho, \quad K|\rho_{xx}| \leq 1 + \rho.$$

Inclusion of the term 1 is desirable because it is not assumed that ρ is bounded away from zero. No constant factor β is introduced on the right-hand side of (D) because β multiplies J and K in (C).

Theorem 1. *If conditions* (ABCD) *hold then* $u(\Gamma) \leq 0 \Rightarrow u(G) \leq 0$.

By the above remarks, it can be assumed in the proof that $0 \leq t \leq T$ where T is positive but arbitrarily small. Condition (C) with $p = q = 0$ implies (B) for $|x| \geq \alpha$ with ϵ replaced by $\epsilon\rho$. If B is bounded we can ensure $0 < z < \epsilon$ when $z = u$ by taking T small. The conclusion then follows as in Example 3.

Assuming now that B is unbounded let $0 < \delta < 1$ and, without loss of generality, $\beta > 1$. Choose R by Lemma 2 so that

$$R > \frac{T}{T-t}, \qquad R' > 3\beta R.$$

Since (B) holds for all α we can increase α at will. The condition $u \leq o(\rho)$ yields

$$u(x,t) \leq \phi(x)\rho(x)$$

where $\phi(x)$ tends to zero as $|x| \to \infty$. We increase α, if necessary, so that $\phi(x) < \epsilon$ holds for $|x| \geq \alpha$. Now that α is fixed, we have (B) with a possibly smaller ϵ. (The same letters (ϵ, α) are used for the new values.) We now diminish T enough to ensure $u(x,t) < \epsilon$ for $|x| \leq \alpha$, $0 \leq t \leq T$.

After these preliminaries let

$$\sigma(x) = \rho(x) \text{ when } |x| \geq \alpha; \qquad \sigma(x) = 0 \text{ when } |x| < \alpha$$

and define

$$w = \frac{u(x,t)}{R(t)(1+\delta\sigma(x))}.$$

If the conclusion $u(G) \leq 0$ fails there are two cases.

Case 1. Suppose there is some value x with $|x| > \alpha$ such that $w(x,t) > 0$ and

$$w(x,t) > \sup_{|\xi| \leq \alpha, \tau < T} w(\xi, \tau).$$

Then w has a positive maximum interior to the part of G where $|x| > \alpha$ and $t < T$. At the maximum

$$u = wR(1+\delta\rho), \quad u_t = wR'(1+\delta\rho), \quad u_x = wR\delta\rho_x, \quad u_x x \leq wR\delta\rho_{xx}.$$

This fits the pattern (C) with

$$z = u > 0, \quad p = \frac{\delta u}{1+\delta\rho}\rho_x, \quad s = \frac{\delta u}{1+\delta\rho}\rho_{xx}.$$

Since $u \leq \epsilon\rho$ by choice of α, we have

$$|p| \leq \frac{\epsilon\delta\rho}{1+\delta\rho}|\rho_x| \leq \epsilon|\rho_x|, \qquad |s| \leq \frac{\epsilon\delta\rho}{1+\delta\rho}|\rho_{xx}| \leq \epsilon|\rho_{xx}|.$$

Conditions (C) and (A) now yield

$$R'(1+\delta\rho) \leq \beta R(1+\delta\rho + J\delta|\rho_x| + K\delta|\rho_{xx}|).$$

In view of (D), this contradicts the inequality $R' > 3\beta R$.

Case 2. By Case 1, every value x with $|x| > \alpha$ yields

$$w(x,t) \leq \sup_{|\xi|\leq\alpha, \tau<T} w(\xi,\tau).$$

Since this is true for each $\delta > 0$ a moment's thought shows that it holds also for $\delta = 0$. This means that u/R has a maximum in the part of G where $|x| \leq \alpha$ and $t < T$. Since $u < \epsilon$ in this region, by choice of T, Condition (B) leads to a contradiction as in Example 3. This completes the proof.

As is clear from the proof, it does not matter how $\rho(x)$ is defined, or even whether it is defined, for small $|x|$.

If $\rho(x) = \ln|x|$ the inequality (C) is

$$f(x,y,z,p,q) \leq \beta(z + |p||x|\ln|x| + |s||x|^2\ln|x|)$$

and the side conditions for $|x| \geq \alpha \geq e$ are

$$z = u(x,t), \quad 0 < z < \epsilon\ln|x|, \quad |p| < \frac{\epsilon}{|x|}, \quad |s| < \frac{\epsilon}{|x|^2}, \quad q \leq s.$$

This sharpens [1], Theorem 1. Further results can be deduced from the general theorem with equal ease.

It would be of interest to look systematically at other applications of the Nagumo procedure from the point of view of this paper. Among these are parabolic inequalities with functionals, monotonicity and comparison theorems for parabolic operators, and parabolic or parabolic-functional systems. However, as was said long ago:

To adorn oneself with one's own rag is better than to ask for the loan of a robe. If it should not touch anyone's ear of desire, the messenger has told his tale; it is enough.

—from the Gulistan of Sa'di

References

1. R. M. Redheffer and W. Walter, *Das Maximumprinzip in unbeschränkten Gebieten für parabolische Ungleichungen mit Funktionalen.* Math. Ann. **226** [1977] 155-170. Further references can be found here.

Raymond M. Redheffer, UCLA, Mathematics Department, Los Angeles, CA 90024, USA

International Series of Numerical Mathematics, Vol. 103, © 1992 Birkhäuser Verlag Basel

Inequalities for the solutions of the heat equation

Saburou Saitoh

Abstract

For the solutions $u(t,x)$ of the heat equation

$$\partial_t u = \Delta u \quad \text{on} \quad \mathbb{R}^+ \times \mathbb{R}^n$$

satisfying the initial condition

$$u(0,x) = F(x) \quad \text{on} \quad \mathbb{R}^n$$

for $L_2(\mathbb{R}^n, dx)$ functions F, inequalities for the functions $D^\alpha u(t,x)$ for any fixed x are derived from both points of view of analyticity in t and integral transforms by the heat kernel.

1. Introduction and results

As an application of the general method [8, 10] for integral transforms using the theory of reproducing kernels, the author [9] obtained a curious isometrical identity in connection with the space dependence of the analyticity of the solutions of the heat equation. N. Hayashi found very important applications of the isometrical identity from the point of view of the theory of partial differential equations and developed the investigations of the analyticity of the solutions of nonlinear partial differential equations. See, for example, [4, 5, 6].

In this paper, we shall examine the time dependence of $D^\alpha u(t,x)$ for the solutions $u(t,x)$ of the heat equation

$$\partial_t u = \Delta u \quad \text{on} \quad \mathbb{R}^+ \times \mathbb{R}^n \tag{1.1}$$

satisfying the initial condition

$$u(0,x) = F(x) \quad \text{on} \quad \mathbb{R}^n \tag{1.2}$$

This paper is in final form and no version of it will be submitted for publication elsewhere.

for $L_2(\mathbb{R}^n, dx)$ functions F. Here,

$$D^\alpha = \frac{\partial^{|\alpha|}}{\partial x_1^{\alpha_1} \cdots \partial x_n^{\alpha_n}}, \qquad |\alpha| = \sum_{j=1}^{n} \alpha_j.$$

For the functions $D^\alpha u(t,x)$, we obtain naturally the following inequalities.

Theorem 1.1. *For the solutions $u(t,x)$ of the heat equation (1.1) satisfying (1.2), the derivatives $D^\alpha u(t,x)$ at any fixed x are extensible analytically onto the right half plane $R^+ = \{\operatorname{Re}\tau = t > 0\}$ on the complex $\tau = t + i\hat{t}$ plane in the form $D^\alpha u(\tau, x)$.*

If $2|\alpha| + n > 2$, then we have the inequality

$$\frac{\pi^{n-1} 2^{|\alpha| + \frac{3}{2}n - 2} (|\alpha| + \frac{n}{2} - 1)}{\prod_{j=1}^{n} \Gamma(\alpha_j + \frac{1}{2})} \cdot \iint_{R^+} |D^\alpha u(\tau, x)|^2 t^{|\alpha| + \frac{n}{2} - 2}\, dt\, d\hat{t} \leq \int_{\mathbb{R}^n} |F(\xi)|^2\, d\xi. \tag{1.3}$$

If $n = 2$ and $|\alpha| = 0$, then we have the inequality

$$2 \int_{\mathbb{R}} |u(i\hat{t}, x)|^2\, d\hat{t} \leq \int_{\mathbb{R}^n} |F(\xi)|^2\, d\xi, \tag{1.4}$$

in the sense of Fatou's nontangential boundary values $u(i\hat{t}, x)$ from R^+.

If $n = 1$ and $|\alpha| = 0$, then we have the inequality

$$2 \int_{\mathbb{R}} |u(i\hat{t}, x)|^4\, d\hat{t} \leq \left\{ \int_{\mathbb{R}} |F(\xi)|^2\, d\xi \right\}^2, \tag{1.5}$$

in the sense of Fatou's nontangential boundary values $u(i\hat{t}, x)$ from R^+.

Compare our Theorem 1.1 with Theorem 2.1 in [7]. See also [3, Proposition 4.4].

In particular, from $\Delta u(t,x) = \partial_t u(t,x)$ for the solutions $u(t,x)$ and from Theorem 1.1 we have

Corollary 1.1. *For the solutions $u(t,x)$ of the heat equation (1.1) satisfying (1.2), the derivatives $\partial_t u(t,x)$ at any fixed x are extensible analytically onto R^+ in the form $\partial_\tau u(\tau, x)$, and we have the inequality*

$$\frac{\pi^{n-\frac{3}{2}}}{n^2} 2^{\frac{3}{2}n+1} \left(\frac{1}{2}n + 1\right) \iint_{R^+} |\partial_\tau u(\tau, x)|^2 t^{\frac{n}{2}}\, dt\, d\hat{t} \leq \int_{\mathbb{R}^n} |F(\xi)|^2\, d\xi. \tag{1.6}$$

More generally, we obtain

Theorem 1.2. *In Theorem 1.1, we have the inequality, for any positive integer m $(m \geq 2)$,*

$$\frac{\pi^{mn-1}2^{m|\alpha|+\frac{3}{2}mn-2}}{\prod_{j=1}^{n}\Gamma(a_j+\frac{1}{2})^m}\left(m|\alpha|+\frac{mn}{2}-1\right) \tag{1.7}$$
$$\cdot\iint_{R^+}|D^\alpha u(\tau,x)|^{2m}t^{m|\alpha|+\frac{mn}{2}-2}\,dt\,d\hat{t} \leq \left\{\int_{\mathbb{R}^n}|F(\xi)|^2\,d\xi\right\}^m .$$

As in Corollary 1.1, from Theorem 1.2 we have

Corollary 1.2. *In Corollary 1.1, we have the inequality, for any positive integer m $(m \geq 2)$,*

$$\frac{\pi^{mn-\frac{m}{2}-1}2^{3m+\frac{3}{2}mn-2}(2m+\frac{mn}{2}-1)}{n^{2m}} \tag{1.8}$$
$$\cdot\iint_{R^+}|\partial_\tau u(\tau,x)|^{2m}t^{2m+\frac{mn}{2}-2}\,dt\,d\hat{t} \leq \left\{\int_{\mathbb{R}^n}|F(\xi)|^2\,d\xi\right\}^m .$$

For the time dependence of $u(t,x)$ for $L_2(\mathbb{R}^n, e^{a|x|^2}dx)$ initial functions F, we have

Theorem 1.3. *For any fixed x, the solutions $u(t,x)$ of the heat equation (1.1) with the initial condition (1.2) for $L_2(\mathbb{R}^n, e^{a|x|^2}dx)$ functions F are extensible analytically in the form $u(\tau,x)$ to the following domain D_a on the τ plane:*

(i) *for* $a>0,\quad D_a=\left\{\left|\tau+\frac{1}{4a}\right|>\frac{1}{4a}\right\}\setminus\left\{t\leq-\frac{1}{2a},\ \hat{t}=0\right\},$

(ii) *for* $a=0,\quad D_a=R^+,$

and

(iii) *for* $a<0,\quad D_a=\left\{\left|\tau+\frac{1}{4a}\right|<\frac{1}{4|a|}\right\}.$

In particular, for $n = 1$ we have the inequality

$$(1.9)\quad \frac{1}{2\pi}\int_0^{\infty}\left|\int_{\mathbb{R}} u\left(\frac{1}{t+i\hat{t}}, x\right)\frac{e^{i\xi\hat{t}}}{(t+i\hat{t})^{\frac{1}{2}}}d\hat{t}\right|^2 \frac{\sqrt{\xi}e^{2\xi(t+2a)}}{\cosh(4ax\sqrt{\xi})}d\xi \le \int_{\mathbb{R}} |F(\xi)|^2 e^{a\xi^2}d\xi$$

for any $t > -2a$ and for any point x.

Note that our inequalities are sharp in the sense that the equalities hold for some functions, as we see from our arguments.

2. Proof of Theorem 1.1

Recall the integral representation of the solutions $u(t,x)$ of (1.1) satisfying (1.2):

$$(2.1)\qquad u(t,x) = \int_{\mathbb{R}^n} F(\xi)\frac{1}{(4\pi t)^{\frac{n}{2}}}\exp\left\{-\frac{1}{4t}|x-\xi|^2\right\}d\xi.$$

We shall consider the identities

$$(2.2)\qquad D^{\alpha}u(\tau,x) = \int_{\mathbb{R}^n} F(\xi)D^{\alpha}\frac{1}{(4\pi\tau)^{\frac{n}{2}}}\exp\left\{-\frac{1}{4\tau}|x-\xi|^2\right\}d\xi$$

and

$$(2.3)\qquad \frac{1}{(4\pi\tau)^{\frac{n}{2}}}\exp\left\{-\frac{1}{4\tau}|x-\xi|^2\right\} = \prod_{j=1}^{n}\frac{1}{2\pi}\int_{\mathbb{R}}\exp\{-p_j^2\tau + ip_j(x_j-\xi_j)\}dp_j,$$

for $\mathrm{Re}\,\tau > 0$. Following the general method [8, 10] for integral transforms, we shall compute the kernel form in the complex form, in order to investigate the integral transform (2.2),

$$K_{D_\alpha}(\tau,\overline{\tau'}) := \int_{\mathbb{R}^n} D^\alpha \frac{1}{(4\pi\tau)^{\frac{n}{2}}} \exp\left\{-\frac{1}{4\tau}|x-\xi|^2\right\} \cdot D^\alpha \frac{1}{(4\pi\tau')^{\frac{n}{2}}} \exp\left\{-\frac{1}{4\tau'}|x-\xi|^2\right\} d\xi$$

(2.4)

$$\begin{aligned} &= \int_{\mathbb{R}^n} \left[\prod_{j=1}^n \frac{1}{2\pi} \int_{\mathbb{R}} \exp\{-p_j^2\tau + ip_j(x_j-\xi_j)\}(ip_j)^{\alpha_j} dp_j\right] \\ &\quad \cdot \left[\prod_{j=1}^n \frac{1}{2\pi} \int_{\mathbb{R}} \exp\left\{-\hat{p}_j^2\overline{\tau'} - i\hat{p}_j(x_j-\xi_j)\right\}\overline{(i\hat{p}_j)^{\alpha_j} d\hat{p}_j}\right] d\xi \\ &= \prod_{j=1}^n \frac{1}{2\pi} \int_{\mathbb{R}} p_j^{2\alpha_j} \exp\left\{-p_j^2(\tau+\overline{\tau'})\right\} dp_j \\ &= \frac{1}{(2\pi)^n} \prod_{j=1}^n \Gamma\left(a_j + \frac{1}{2}\right) \frac{1}{(\tau+\overline{\tau'})^{|\alpha|+\frac{n}{2}}}, \end{aligned}$$

for $\tau, \overline{\tau'} \in R^+$.

Let $H_{K_{D_\alpha}}$ denote the Hilbert space composed of analytic functions $f(\tau)$ on R^+ admitting the reproducing kernel $K_{D_\alpha}(\tau,\overline{\tau'})$. In particular, note that the Hilbert space $H_{K_{D_\alpha}}$ is independent of the point x.

Since the family of functions

$$\left\{D^\alpha \exp\left(-\frac{1}{4\tau}|x-\xi|^2\right);\ \tau \in R^+\right\}$$

is not complete in $L_2(\mathbb{R}^n, d\xi)$ in the integral transform (2.2), we obtain the inequality, by (2.2), (2.4) and [8] or [10, p. 82, Theorem 3.1]

$$||D^\alpha u(\cdot,x)||^2_{H_{K_{D_\alpha}}} \le \int_{\mathbb{R}^n} |F(\xi)|^2\, d\xi. \tag{2.5}$$

In general, for $q > \frac{1}{2}$,

$$K_q(\tau,\overline{\tau'}) = \frac{\Gamma(2q)}{(\tau+\overline{\tau'})^{2q}} \quad \text{on} \quad R^+ \times \overline{R^+}$$

is the Bergman reproducing kernel for the Bergman-Selberg space H_{K_q} composed of analytic functions $f(\tau)$ on R^+ with finite norms

$$||f||^2_{H_{K_q}} = \frac{2^{2q-2}}{\pi\Gamma(2q-1)} \iint_{R^+} |f(\tau)|^2 t^{2q-2}\, dt\, d\hat{t}. \tag{2.6}$$

For $q = \frac{1}{2}$, $K_{\frac{1}{2}}(\tau, \overline{\tau'})$ is the Szegö reproducing kernel on R^+ composed of analytic functions $f(\tau)$ on R^+ satisfying

$$\sup_{t>0} \int_{\mathbb{R}} |f(t + i\hat{t})|^2 \, d\hat{t} < \infty.$$

The norm in $H_{K_{\frac{1}{2}}}$ is given by

$$\left\{ \frac{1}{2\pi} \int_{\mathbb{R}} |f(i\hat{t})|^2 \, d\hat{t} \right\}^{\frac{1}{2}} < \infty. \tag{2.7}$$

Furthermore,

$$\lim_{t \to +0} \left\{ \frac{1}{2\pi} \int_{\mathbb{R}} |f(t + i\hat{t}) - f(i\hat{t})|^2 \, d\hat{t} \right\}^{\frac{1}{2}} = 0.$$

Here, $f(i\hat{t})$ means Fatou's nontangential boundary values from R^+ (cf. [2, pp. 113–116]).

If $2|\alpha| + n > 2$, then from (2.4), (2.5) and (2.6) we have (1.3).

If $2|\alpha| + n = 2$ and so if $n = 2$ and $|\alpha| = 0$, then from (2.4), (2.5) and (2.7) we have (1.4).

If $n = 1$ and $|\alpha| = 0$, then

$$K_{D_0}(\tau, \overline{\tau'})^2 = \frac{1}{4\pi(\tau + \overline{\tau'})}.$$

Hence, from Theorem II ([1, p. 361, see also p. 393 (4)]), we have (1.5).

3. Proof of Theorem 1.2

From the identity

$$K_{D_\alpha}(\tau, \overline{\tau'})^m = \frac{1}{(2\pi)^{mn}} \prod_{j=1}^{n} \Gamma\left(\alpha_j + \frac{1}{2}\right)^m \frac{1}{(\tau + \overline{\tau'})^{m|\alpha| + \frac{mn}{2}}}$$

and the theory of Aronszajn [1, p. 361], we have Theorem 1.2, using the representation (2.6) of the norm in H_{K_q} ($q > \frac{1}{2}$) space.

4. Proof of Theorem 1.3

We consider the kernel form

$$
\begin{aligned}
K_a(\tau,\overline{\tau'};x) &= \int_{\mathbb{R}^n} h(x-\xi,\tau)h(x-\xi,\overline{\tau'})e^{-a|\xi|^2}d\xi \\
&= \frac{1}{2^n\pi^{\frac{n}{2}}(\tau+\overline{\tau'}+4a\tau\overline{\tau'})^{\frac{n}{2}}}\exp\left[\frac{-a|x|^2(\tau+\overline{\tau'})}{\tau+\overline{\tau'}+4a\tau\overline{\tau'}}\right]
\end{aligned}
\tag{4.1}
$$

on some subdomain of D_a stated in Theorem 1.3. However, the last term in (4.1) is analytic on $D_a \times \overline{D_a}$. The identity (4.1) implies that for any functions G satisfying

$$\int_{\mathbb{R}^n} |G(\xi)|^2 e^{-a|\xi|^2}\, d\xi < \infty,$$

the functions

$$g(\tau) = \int_{\mathbb{R}^n} G(\xi)h(x-\xi,\tau)e^{-a|\xi|^2}\, d\xi$$

are analytic on D_a and belong to the Hilbert space H_{K_a} admitting the reproducing kernel $K_a(\tau,\overline{\tau'};x)$ on D_a. Furthermore, we have the identity

$$||g||^2_{H_{K_a}} = \int_{\mathbb{R}^n} |G(\xi)|^2 e^{-a|\xi|^2}\, d\xi$$

([8, Theorem 3.1] and [10, Theorem 3.1]). Hence, we obtain the desired result for the analyticity of $u(\tau,x)$ by setting $F(\xi) = G(\xi)e^{-a|\xi|^2}$.

Next, for $n=1$, we consider the kernel form

$$
\begin{aligned}
K_a^*(\tau,\overline{\tau'};x) &= K_a\left(\frac{1}{\tau},\frac{1}{\overline{\tau'}};x\right) \\
&= \frac{\tau^{\frac{1}{2}}\overline{\tau'}^{\frac{1}{2}}}{2\pi}\int_0^\infty e^{\tau\xi}e^{-\overline{\tau'}\xi}\frac{\cosh(4ax\sqrt{\xi})}{\sqrt{\xi}}e^{-4a\xi}\, d\xi
\end{aligned}
\tag{4.2}
$$

for $\operatorname{Re}\tau, \operatorname{Re}\tau' > -2a$. In order to avoid the multiply-valuedness of the function $\tau^{\frac{1}{2}}$, we shall consider the kernel form (4.2) on the halfplane $R^+_{-2a} = \{\operatorname{Re}\tau > -2a\} \setminus \{-2a \le \operatorname{Re}\tau \le 0\}$ cut by the slit $[-2a, 0]$ for $a > 0$. Let $H_{K_a^*}$ denote the Hilbert space admitting the reproducing kernel (4.2). Then, from the representation (4.2) we see that the members $\varphi(\tau)$ in $H_{K_a^*}$ are analytic on R^+_{-2a} (see also (4.1)) and are expressible in the form

$$\varphi(\tau) = \frac{\tau^{\frac{1}{2}}}{2\pi}\int_0^\infty \psi(\xi)e^{-\tau\xi}\frac{\cosh(4ax\sqrt{\xi})}{\sqrt{\xi}}e^{-4a\xi}\, d\xi \tag{4.3}$$

for some uniquely determined functions ψ satisfying

$$\int_0^\infty |\psi(\xi)|^2 \frac{\cosh(4ax\sqrt{\xi})}{\sqrt{\xi}} e^{-4a\xi}\, d\xi < \infty. \tag{4.4}$$

Furthermore, the norms in $H_{K_a^*}$ are given by

$$||\varphi||^2_{H_{K_a^*}} = \frac{1}{2\pi} \int_0^\infty |\psi(\xi)|^2 \frac{\cosh(4ax\sqrt{\xi})}{\sqrt{\xi}}\, e^{-4a\xi}\, d\xi. \tag{4.5}$$

By using the Fourier transform in (4.3) in the framework of the L_2 space we have from (4.5) the representation of the norms in $H_{K_a^*}$

$$||\varphi||^2_{H_{K_a^*}} = \frac{1}{2\pi} \int_0^\infty \left| \int_{-\infty}^\infty \frac{\varphi(t+i\hat{t})}{(t+i\hat{t})^{\frac{1}{2}}}\, e^{i\xi\hat{t}}\, d\hat{t} \right|^2 \frac{\sqrt{\xi}}{\cosh(4ax\sqrt{\xi})}\, e^{2t(\xi+2a)}\, d\xi, \tag{4.6}$$

for any $t > -2a$. Hence, from the expression

$$u\left(\frac{1}{t}, x\right) = \frac{t^{\frac{1}{2}}}{\sqrt{4\pi}} \int_{\mathbb{R}} \exp\left[-\frac{(x-\xi)^2 t}{4}\right] G(\xi) e^{-a\xi^2}\, d\xi$$

for $L_2(\mathbb{R}^n, e^{-a\xi^2} d\xi)$ functions G and from (4.2), (4.6), we obtain (1.9) by setting $F(\xi) = G(\xi)e^{-a\xi^2}$.

Acknowledgments. The author wishes to express his sincere thanks to Professor Nakao Hayashi for his valuable advice and suggestions.

The travel expenses of the author for attending to the 6th International Conference on General Inequalities, Oberwolfach, December 9–15, 1990 were partially supported by the Gunma University Foundation for Science & Technology.

References

1. N. Aronszajn, *Theory of reproducing kernels.* Trans. Amer. Math. Soc. **68** (1950), 337–404.

2. J. Burbea, *Total positivity of certain reproducing kernels.* Pacific J. Math. **67** (1976), 101–130.

3. J. Ginibre and G. Velo, *Scattering theory in the energy space for a class of nonlinear Schrödinger equations.* J. Math. Pures et Appl. **64** (1985), 363–401.

4. N. Hayashi and S. Saitoh, *Analyticity and smoothing effect for the Schrödinger equation.* Ann. Inst. Henri Poincaré, Physique théorique **52** (1990), 163–173.

5. N. Hayashi and S. Saitoh, *Analyticity and global existence of small solutions to some nonlinear Schrödinger equations.* Commun. Math. Phys. **129** (1990), 27–41.

6. N. Hayashi, *Global existence of small analytic solutions to nonlinear Schrödinger equations.* Duke Math. J. **60** (1990), 717–727.

7. C.E. Kenig, G. Ponce and L. Vega, *Oscillatory integrals and regularity of dispersive equations.* Preprint.

8. S. Saitoh, *Hilbert spaces induced by Hilbert space valued functions.* Proc. Amer. Math. Soc. **89** (1983), 74–78.

9. S. Saitoh, *The Weierstrass transform and an isometry in the heat equation.* Applicable Analysis **16** (1988), 1–6.

10. S. Saitoh, *Theory of Reproducing Kernels and its Applications.* Pitman Res. Notes in Math. Series 189, Longman Science & Technical, England 1988.

Saburou Saitoh, Department of Mathematics, Faculty of Engineering, Gunma University, Kiryu 376, Japan

International Series of Numerical Mathematics, Vol. 103, © 1992 Birkhäuser Verlag Basel

An isoperimetric inequality for the first eigenvalue of the Laplacian under Robin boundary conditions

René P. Sperb

Abstract

It is proven that among all regions with given value for the maximum of the torsion function, the first eigenvalue of the Laplacian under Robin boundary condition is a minimum for an infinite slab. This holds provided the mean curvature of the boundary is nonnegative.

1. Introduction

The eigenvalue problem

$$(1)\qquad \begin{cases} \Delta u + \lambda u = 0 & \text{in } \Omega \subseteq R^N \\ \dfrac{\partial u}{\partial n} + \alpha u = 0 & \text{on } \partial\Omega \end{cases}$$

arises in many applications in various areas. Here $\frac{\partial}{\partial n}$ denotes the outward normal derivative. It is also often referred to as the problem of the "elastically supported membrane". In that context the parameter α (a positive constant here) is proportional to a spring constant. The limiting cases are then the "free membrane" ($\alpha = 0$) and the "fixed membrane" ($\alpha = \infty$).

One of the most important informations in such an eigenvalue problem is a lower bound for the first eigenvalue.

In the limiting case $\alpha = \infty$, i.e. in the fixed membrane problem

$$(2)\qquad \begin{cases} \Delta u + \lambda u = 0 & \text{in } \Omega \\ u = 0 & \text{on } \partial\Omega \end{cases}$$

This paper is in final form and no version of it will be submitted for publication elsewhere.

a famous result of Faber [3] and Krahn [5] states that for a given N-volume of Ω the ball yields the lowest eigenvalue λ_1^*. It is said that Krahn had tried in vain to extend this result to the eigenvalue $\lambda_1(\alpha)$ of (1). Quite recently M.H. Bossel [1] was able to extend this result to $\lambda_1(\alpha)$ for plane domains Ω and also for domains on a surface [2].

Protter [10] and Hersch [4] proved an inequality which is optimal in a different sense. They showed that if Ω is a convex domain then one has

$$\lambda_1(\infty) \geq \frac{\pi^2}{4\rho^2}. \tag{3}$$

Here ρ is the radius of the largest ball contained in Ω, and equality is attained if Ω degenerates into an infinite slab of thickness 2ρ.

By a completely different method Payne and Stakgold [9] showed that inequality (3) holds provided the mean curvature H of $\partial\Omega$ is nonnegative. Note that this assumption is much less restrictive than convexity. For example if Ω is the torus in R^3 obtained by rotating the circle

$$x = R_1 + R_2 \cos\varphi, \quad y = R_2 \sin\varphi, \quad 0 \leq \varphi \leq 2\pi,$$

around the y-axis, then $H \geq 0$ provided $R_2 \leq 1/2\, R_1$.

Later Payne [8] showed that if $H \geq 0$ one even has

$$\lambda_1(\infty) \geq \frac{\pi^2}{8t_m}. \tag{4}$$

Here $t_m = \max_\Omega t$ and t is the "torsion function", i.e. the solution of

$$\begin{cases} \Delta t + 1 = 0 & \text{in } \Omega \\ t = 0 & \text{on } \partial\Omega. \end{cases} \tag{5}$$

Various bounds for t_m are known (see e.g. [7, 6, 11]). In particular if one uses the bound of Payne [7]

$$t_m \leq \frac{\rho^2}{2} \tag{6}$$

one is led back to (3).

Inequality (4) can also be read as follows:

> "*Among all regions with given value of t_m $\lambda_1(\infty)$ is a minimum for the slab.*"

In the next section we will see that this last statement is true for all values of $\alpha > 0$.

2. An Isoperimetric Inequality for $\lambda_1(\alpha)$

We now prove the main result stated as

Theorem 1. *Assume that Ω is a finite domain of R^N with smooth ($C^{2+\epsilon}$) boundary $\partial\Omega$, such that the mean curvature H of $\partial\Omega$ is nonnegative.*

Then for any $\alpha > 0$ one has

$$\lambda_1(\alpha) \geq \nu_1(\alpha), \tag{7}$$

where $\nu_1(\alpha)$ is the first positive root of

$$\tan(\sqrt{2t_m\nu}) = \frac{\alpha}{\sqrt{\nu}} \tag{8}$$

and t_m is the maximum of the torsion function.

Equality holds in (7) if Ω degenerates into an infinite slab.

It is convenient to state the following well known fact first as

Lemma 2. *Suppose the nonnegative function v satisfies*

$$\Delta v + \nu v \leq 0 \text{ in } \Omega \quad \text{and} \quad \frac{\partial v}{\partial n} + \alpha v \geq 0 \text{ on } \partial\Omega. \tag{9}$$

Then

$$\nu \leq \lambda_1(\alpha). \tag{10}$$

In order to keep this paper self contained the simple proof is given:

It is well known that the first eigenfunction u_1 of (1) does not change sign and can therefore be taken as positive. Multiplying inequality (9) by u_1 and integrating over Ω gives (with dx = volume element)

$$\int_\Omega (v\,\Delta u_1 + \nu u_1 v)\,dx + \oint_{\partial\Omega} (u_1 \frac{\partial v}{\partial n} - \nu\frac{\partial u_1}{\partial n})\,ds \leq 0$$

that is

$$\int_\Omega (\nu - \lambda_1(\alpha))u_1 v\,dx + \oint_{\partial\Omega} (\frac{\partial v}{\partial n} + \alpha v)u_1\,ds \leq 0.$$

But since $v \geq 0$ in Ω and $\frac{\partial v}{\partial n} + \alpha v \leq 0$ the positivity of u_1 implies inequality (10).

In order to prove inequality (7) we construct a function v using an idea of Payne [8].

Let x be a generic point of Ω and let $t(x)$ be the torsion function of Ω. We then set

$$v(x) = X(s(x)), \tag{11}$$

where

$$s(x) = \sqrt{2(t_m - t(x))}, \tag{12}$$

and $X(s)$ is the solution of the one-dimensional problem

$$\begin{cases} \dfrac{d^2X}{ds^2} + \nu X = 0 \quad \text{in } (0, s_0) \\ \dfrac{dX}{ds} = 0 \quad \text{at } s = 0, \qquad \dfrac{dX}{ds} + \alpha X = 0 \quad \text{at } s = s_0 = \sqrt{2t_m} \end{cases} \tag{13}$$

Of course the solution of (13) can be written down immediately. But it seems more convenient in view of extensions to nonlinear problems to leave it in the present form. We now use the relations (prime denotes $\frac{d}{ds}$)

$$\Delta v = X' \, \Delta s + X'' \, |\nabla s|^2, \tag{14}$$

and

$$\nabla s = \frac{-\nabla t}{s}, \tag{15}$$

$$\Delta s = \frac{-\Delta t}{s} + \frac{-\nabla t \, \nabla s}{s^2} = \frac{1}{s} - \frac{|\nabla t|^2}{s^3}, \tag{16}$$

to find

$$\Delta v + \nu v = \underbrace{(\frac{X'}{s} + \nu X)}_{*}(1 - \frac{|\nabla t|^2}{s^2}). \tag{17}$$

It was shown in Payne [7] that for domains with smooth boundary $\partial\Omega$ of nonnegative mean curvature one has

$$|\nabla t|^2 + 2t(x) \leq 2t_m \quad \text{in } \Omega, \tag{18}$$

that is

$$\frac{|\nabla t|}{s} \leq 1 \tag{19}$$

holds in $\overline{\Omega}$. To check the sign of the term $*$ in (17) we consider the function

$$g(s) = X'(s) + \nu s X$$

which satisfies

$$g'(0) = 0 \quad \text{and} \quad g'(s) = X'' + \nu X + \nu s X' = \nu s X' \leq 0,$$

since $X(s) = \cos(\sqrt{\nu_1} s)$ and $\nu_1(\alpha)$ is the first positive root of

$$\tan(\sqrt{\nu} s_0) = \frac{\alpha}{\sqrt{\nu}}. \tag{20}$$

It remains to check the boundary condition for v. Now

$$\begin{aligned} \frac{\partial v}{\partial n} + \alpha v &= -X'(s_0)\frac{\partial t}{\partial n}\frac{1}{s_0} + \alpha X(s_0) = \frac{|\nabla t|}{s_0} X'(s_0) + \alpha X(s_0) \\ &= \alpha X(s_0)(1 - \frac{|\nabla t|}{s_0}) \geq 0, \end{aligned} \tag{21}$$

by again applying (19). It is easy to see that $\nu_1(\alpha)$ is a decreasing function of s_0. We may therefore use any upper bound $\bar{t}_m$ for t, that is we can take

$$s_0 = \sqrt{2\bar{t}_m}$$

in our theorem.

3. Remarks

(a) Comparison of the lower bounds.

Since the optimal domain is a strip (in two dimensions) we can expect that our lower bound is sharper than the bound of M.H. Bossel if the domain is long and thin. To this end consider an ellipse of semiaxes $a = 3$, $b = 1$. Then the value t_m can be given explicitly: $t_m = \frac{1}{2(1/9+1)}$ and (7) gives in this case $\lambda_1(1) \geq \nu(1) \cong 0.791$.

Comparison with the circle of the same area gives $\lambda_1(\alpha) \geq \frac{x_1^2}{R^2}$, where $R = \sqrt{3}$ and $x_1 =$ first positive solution of $J_1(x) = \alpha R \frac{J_0(x)}{x}$, J_0, J_1 = Besselfunctions. One finds $\lambda_1(1) \geq 0.779$.

(b) Necessity of the assumption $H \geq 0$.

Consider the domain $\Omega = \{x \in R^3,\ R_1 < |x| < R_2\}$ and take for simplicity $\alpha = \infty$. The torsion function for Ω can be calculated explicitly and one attains

$$t_m = \frac{R_2^2}{6}\left(\gamma^2 + \gamma + 1 - \frac{4}{3}(\frac{1}{2}(1+\gamma))^{\frac{2}{3}}\right), \quad \gamma = \frac{R_1}{R_2} < 1.$$

On the other hand, one has $\lambda_1(\infty) = \frac{\pi^2}{(1-\gamma)^2}\frac{1}{R_2^2}$, and it is easy to check that the inequality $\lambda_1(\infty) \geq \frac{\pi^2}{8t_m}$ does not hold if $\gamma < 0.34$. This shows that the assumption $H \geq 0$ can in general not be omitted.

(c) In [12] a similar technique was applied in the nonlinear problem

$$\Delta u + \lambda f(u) = 0 \text{ in } \Omega \subset M, \quad u = 0 \text{ on } \partial\Omega.$$

Here M is a surface with Gaussian curvature K_G, $\partial\Omega$ has geodesic curvature k_g and it is assumed that $K_G|\Omega| + k_g|\partial\Omega| \geq 0$ holds, with $|\cdot|$ denoting area and arc-length respectively.

It is quite clear that the arguments of [12] can be used exactly the same way to prove an extension of the Theorem 1 to the case that Ω is a domain on a surface.

References

1. M.H. Bossel, *Longeurs extrémales et fonctionelles de domaine.* Complex Variables **6** (1986), 203–234.

2. M.H. Bossel, *Membranes élastiquement liées inhomogènes ou sur une surface: une nouvelle extension du théorème isopérimetrique de Rayleigh-Faber-Krahn.* ZAMP **39** (1988), 733–742.

3. G. Faber, *Beweis, dass unter allen homogenen Membranen von gleicher Fläche und gleicher Spannung die kreisförmige den tiefsten Grundton hat.* Sitz.ber. bayer. Akad. Wiss. (1923), 169–172.

4. J. Hersch, *Sur la fréquence fondamental d'une membrane vibrante; évaluation par défaut et principe de maximum.* J. Math. Phys. Appl. **11** (1960), 387–413.

5. E. Krahn, *Über Minimaleigenschaften der Kugel in drei und mehr Dimensionen.* Acta. Comm. Univ. Tartu (Dorpat) **A9** (1926), 1–144.

6. L.E. Payne, *Isoperimetric inequalities and their applications.* SIAM Rev. **9** (1967), 453–488.

7. L.E. Payne, *Bounds for the maximum stress in the Saint Venant torsion problem.* Indian J. Mech. Math., special issue (1968), 51–59.

8. L.E. Payne, *Bounds for solutions of a class of quasilinear elliptic boundary value problems in terms of the torsion function.* Proc. Roy. Soc. Edinborough **88A** (1981), 251–265.

9. L.E. Payne and I. Stakgold, *On the mean value of fundamental mode in the fixed membrane problem.* Applicable Anal. **3** (1971), 295–303.

10. M.H. Protter, *Lower bounds for the first eigenvalue of elliptic equations.* Ann. of Math. **71** (1960), 423–444.

11. R.P. Sperb, *Maximum principles and their applications.* Mathematics in Science and Engineering **157**, Academic Press (1981.)

12. R.P. Sperb, *Optimal bounds for the critical value in a semilinear boundary value problem on a surface.* General ineq. 5, ISNM **80** (1987), 391–400.

René P. Sperb, Seminar für Angewandte Mathematik, FLI, ETHZ, CH-8092 Zürich, Switzerland

International Series of Numerical Mathematics, Vol. 103,

Minimum principles for weak solutions of second order ordinary differential equations

Wolfgang Walter

Abstract

We shall give theorems of the following type under weak conditions on u and the coefficients of the differential operator:

$$(p(x)u'(x))' + c(x)u(x) \leq 0 \quad \text{a.e. in } J^\circ = (a,b), \quad u(a), u(b) \geq 0$$

implies "$u \geq 0$ in J" (minimum principle) or "$u \equiv 0$ in J or $u > 0$ in J°" (strong minimum principle). The relation between these results and the "first eigenvalue case" are investigated, and operators in non-selfadjoint form are also considered.

1. Introduction

Let J denote a compact interval $[a,b]$ on the real line and $J^\circ = (a,b)$ the interior of J. We shall consider the differential inequality

$$(pu')' + cu \leq 0 \quad \text{a.e. in } J^\circ \tag{1}$$

together with an assumption on the boundary

$$u(a) \geq 0, \quad u(b) \geq 0. \tag{2}$$

Our aim is to show under minimal conditions on u, p and c that a minimum principle

(MP) (1) and (2) imply $u \geq 0$ in J

or a strong minimum principle

(SMP) (1) and (2) imply $u \equiv 0$ or $u > 0$ in J°

holds. The problem is closely related to the "first eigenvalue case". This will be discussed in Section 3.

This paper is in final form and no version of it will be submitted for publication elsewhere.

2. The main theorems

The class Z of 'admissible' functions consists of all functions u which are continuous in J and such that u and pu' are locally absolutely continuous in J°. The functions p and c are defined in J°, and p is positive a.e.

The first theorem uses the classical assumption $c \le 0$ and no other conditions on p and c, while in the second theorem the general case is handled by an auxiliary function h.

Theorem 1. *Assume that $p > 0$ and $c \le 0$ a.e. in J°. Then* (MP) *holds, i.e., every function $u \in Z$ satisfying* (1) *and* (2) *is nonnegative in J.*

Proof. Assume that $\min_J u = u(d) < 0$ and that $u < 0$ in (α, β), $u(\alpha) = u(\beta) = 0$, where $\alpha < d < \beta$. Then $u' < 0$ on a set of positive measure in (α, d) and $u' > 0$ on a set of positive measure in (d, β). Since p is positive a.e., there are points γ, δ such that

$$\alpha < \gamma < d < \delta < \beta \quad \text{and} \quad p(\gamma)u'(\gamma) < 0 < p(\delta)u'(\delta).$$

It follows that $(pu')' > 0$ on a set of positive measure in (γ, δ). Since $cu \ge 0$ in (γ, δ), a contradiction to inequality (1) is obtained. □

Let us introduce the following assumption (P) on the coefficients:

(P) The functions p and c are measurable, $p > 0$ a.e. in J°, for any $\alpha \in J^\circ$ there exists $\varepsilon > 0$ such that

$$C(\alpha, x) = \int_\alpha^x \frac{1}{p(t)} \int_\alpha^t c^-(s)\, ds\, dt < \infty \quad \text{for} \quad |x - \alpha| < \varepsilon.$$

Here, $c^-(x) = \max\{-c(x), 0\} \ge 0$.

Note that the integral is nonnegative for $x > \alpha$ and for $x < \alpha$. It clear that the finiteness of the integral implies $C(\alpha, x) \to 0$ as $x \to \alpha$.

Theorem 2. *Assume that* (P) *holds. If $u \in Z$ satisfies* (1)(2) *and if there exists a function $h \in Z$ with the properties*

(3) $$(ph')' + ch \le 0 \text{ a.e. in } J^\circ \text{ and } h > 0 \text{ in } J,$$

then $u \equiv 0$ in J or $u > 0$ in J°.

Note that in the case $c \le 0$ assumption (3) holds with $h(x) \equiv 1$.

Proof. Assume that $\min u < 0$. Since $h(x) \geq \delta > 0$ in J, the function $u + \mu h$ is positive for large values of μ. Let μ_0 be the smallest number such that $v := u + \mu_0 h$ is nonnegative in J. Obviously, $\mu_0 > 0$,

$$(pv')' + cv \leq 0 \quad \text{and} \quad v(a), v(b) > 0 \tag{4}$$

and hence $v(\alpha) = 0$ for some $\alpha \in J^\circ$. We may assume that α is the largest number with this property, i.e., that $v > 0$ in $(\alpha, b]$. In every neighborhood of α there are points to the right of α where pv' is positive and to the left of α where pv' is nonpositive. Since pv' is continuous, we get $p(\alpha)v'(\alpha) = 0$. According to (4) $(pv')' \leq -cv \leq c^- v$ and hence, for $x > \alpha$,

$$p(x)v'(x) = \int_\alpha^x (pv')'\,dt \leq \int_\alpha^x c^- v\,dt.$$

Dividing by p, we get

$$v(x) \leq \int_\alpha^x \frac{1}{p(s)} \int_\alpha^x c^-(t)v(t)\,dt\,ds.$$

Now let $x_m > \alpha$ be a point with the property $v(x_m) = \max\,\{v(x) : \alpha \leq x \leq x_m\}$. Then the last inequality implies

$$v(x_m) \leq v(x_m)C(\alpha, x_m).$$

Since $C(\alpha, x) < 1$ for x close to α and since there are points x_m arbitrarily close to α, a contradiction is obtained. It follows that $u \geq 0$ in J.

Now assume that $\max u > 0$ and that u is not positive in J°. Then there exists a subinterval $[\alpha, \beta]$ of J° such that u is positve in (α, β) and $u(\alpha) = 0$ or $u(\beta) = 0$. We consider the first case only. Using the same reasoning as above, an inequality $u(x_m) \leq u(x_m)C(\alpha, x_m)$ with $x_m \in (\alpha, \beta)$ is obtained. It leads to a contradiction as before, which proves the theorem. □

Our next theorem deals with the non-selfadjoint case, that is, with the differential inequality

$$au'' + bu' + cu \leq 0 \quad \text{a.e. in } J^\circ. \tag{5}$$

A function u is admissible, in short $u \in Z^*$, if it is continuous in J and differentiable in J° and if u' is locally absolutely continuous in J°. The coefficients are assumed to satisfy assumption

(P*) a, b, c are defined and measurable in J°, $a > 0$ a.e. in J°, the functions b/a and c/a belong to $L^1_{\text{loc}}(J^\circ)$.

Theorem 3. *Assume that* (P*) *holds and that there exists a function* $h \in Z^*$ *with the properties*

$$ah'' + bh' + ch \le 0 \text{ a.e. in } J^\circ \text{ and } h > 0 \text{ in } J. \tag{6}$$

Then any function $u \in Z^*$ *satisfying* (5) *and* (2) *is either identically zero or positive in* J°.

Proof. Let $\alpha \in J$. The functions $p(x) = \exp\left(\int_\alpha^x \frac{b}{a}\,dt\right)$ is locally absolutely continuous and positive in J°, hence pu' is also locally absolutely continuous. The function u satisfies (1) with c replaced by $c^* := c/ap$, and h satisfies (3) with this coefficent c^*. It is easily seen that (P) holds with c^* since p is strictly positive in J°. Hence Theorem 2 is applicable, and the theorem follows. □

Remarks. We shall discuss some conditions which imply assumption (P).

1. If $1/p$ and c^- belong to $L^1_{\text{loc}}(J^\circ)$, then obviously (P) holds.

2. If c is locally bounded below in J° and if to every $\alpha \in J^\circ$ there exist positive constants γ, δ, ε such that

$$p(x) \ge \gamma|x-\alpha|^{2-\delta} \quad \text{for} \quad |x-\alpha| < \varepsilon,$$

then again (P) holds. In this case $\left|\int_\alpha^t c^-\,ds\right| \le C|t-\alpha|$ and hence $\frac{1}{p(t)}\left|\int_\alpha^t c^-\,ds\right| \le C_1|t-\alpha|^{\delta-1}$. This example shows that p may have isolated zeros in J°. But it is not allowed that p behaves near a zero at $\alpha \in J^\circ$ like $(x-\alpha)^2$ as long as $c < 0$ near α. For this case a simple counterexample to Theorem 2 is obtained by putting $J = [-1,1]$, $p = x^2$, $c = -6$. The function $h \equiv 1$ satisfies (3), and $u = x^2$ is a solution to $(pu')' + cu = 0$, but u is not positive in J°.

3. If $c \ge 0$, then (P) holds without further conditions on p.

4. The preceding counterexample can also be used in connection with Theorem 1. It shows that under the weak assumptions of Theorem 1 the stronger proposition (SMP) does not hold in general.

3. Near the first eigenvalue

Theorem 1 shows that $c \le 0$ implies (MP). According to Theorem 2 (SMP) holds also in cases where $\sup c > 0$ as long as there exists an auxiliary function h with the properties of (3). On the other hand, (MP) is certainly false if $c(x) = \lambda_1$, where λ_1 is the first eigenvalue of the problem

$$(pu')' + \lambda u = 0 \text{ in } J^\circ, \quad u(a) = u(b) = 0, \tag{7}$$

or if the first eigenvalue of the problem

$$(8) \qquad (pu')' + cu + \lambda u = 0 \text{ in } J^\circ, \quad u(a) = u(b) = 0,$$

is zero. In both cases $\pm$ the first eigenfunction gives a counterexample.

But if (p and c can be extended beyond J in a proper way and) (7) or (8) is considered in an slightly larger interval $J_\varepsilon = [a - \varepsilon, b + \varepsilon]$ with u vanishing at the endpoints, then the first eigenfunction is positive in the closed interval J. Hence it may be used as an auxiliary function h in Theorem 2 (and, mutatis mutandis, also in Theorem 3). In this way the following results are obtained.

(a) If $c(x) \le \lambda_1^\varepsilon$, where λ_1^ε is the first eigenvalue of problem (7) in J_ε for some $\varepsilon > 0$, then Theorem 2 applies.

(b) If the first eigenvalue to problem (8) in J_ε is nonnegative, then again Theorem 2 applies.

Our next theorem shows that under some additional assumptions one may relax condition (3) and assume only that h is positive in J°. The theorem is a best theorem of this type where the "first eigenvalue case" is included in the proposition. For this result stronger assumptions on u and the coefficients p, c are needed. Let $Z_1 \subset Z$ denote the class of functions u such that u and pu' are absolutely continuous in J. We require assumption

(P_1) p and c satisfy assumption (P), where the inequality $C(\alpha, x) < \infty$ holds also at the endpoints $\alpha = a$ and $\alpha = b$; furthermore p is bounded near a and near b.

Theorem 4. *Assume that* (P_1) *holds. Let* $u, h \in Z_1$, *where* u *satisfies* (1)(2) *and* h *satisfies*

$$(9) \qquad (ph')' + ch \le 0 \text{ a.e. in } J^\circ \text{ and } h > 0 \text{ in } J^\circ,$$

and furthermore $u(x) \ge -M(x - a)$ *near* a *and* $u(x) \ge -M(b - x)$ *near* b *with a positive constant* M. *Then*

(i) $u \equiv 0$ *in* J *or* (ii) $u > 0$ *in* J° *or* (iii) $u = -\beta h$, *where* $\beta > 0$.

Note that in the third case u (and h) is an eigenfunction to problem (8) corresponding to the first eigenvalue $\lambda_1 = 0$.

The proof is based on the following

Lemma. *Assume that* (P_1) *holds and that* $u \in Z_1$ *satisfies* (1)(2) *and* $u \geq 0$, $u \not\equiv 0$. *Then there exists* $\gamma > 0$ *such that*

$$u(x) \geq \gamma d(x) \text{ in } J, \text{ where } d(x) = \operatorname{dist}\{x, \partial J\} = \min\{x - a, b - x\}.$$

Proof of the lemma. If we assume that $u > 0$ in (α, β) and $u(\alpha) = 0$, where $\alpha \in J^\circ$, then a contradiction is obtained as in the proof of Theorem 2. The case where $u(\beta) = 0$ is handled in the same way. Hence $u > 0$ in J°. Now assume that $(pu')(0) = 0$. Then the reasoning in the proof of Theorem 2 leads again to a contradiction. Hence $(pu')(0) > 0$. Since pu' is continuous in J and p is bounded near a, say, $p \leq K$ in $[a, a + \varepsilon)$, there exists $\delta > 0$ such that

$$0 < \delta < p(x)u'(x) \leq Ku'(x) \quad \text{in} \quad [a, a + \varepsilon).$$

This inequality shows that $u(x) > \delta_1(x - a)$ near a with $\delta_1 = \delta/K$. A similar estimate near the right endpoint b leads to the desired inequality $u(x) \geq \gamma d(x)$. □

Proof of Theorem 4. If $u \geq 0$, then it follows from the lemma that either (i) or (ii) holds. Now let $\min u < 0$. According to the lemma, h satisfies an inequality $h(x) \geq \gamma d(x)$ with $\gamma > 0$. Our assumptions on u imply that $u(x) \geq -Cd(x)$ for some $C > 0$. Hence $u + \mu h$ is certainly positive for $\mu > C/\gamma$. As in the proof of Theorem 2 we consider the function $v = u + \mu_0 h$, where $\mu_0 > 0$ is the smallest number such that $v \geq 0$. If $v \equiv 0$ then case (iii) obtains. If $\max v > 0$, then the lemma implies $v(x) \geq \alpha d(x)$, $\alpha > 0$. We choose a positive number ε satisfying $\varepsilon < \gamma(\mu_0 - \varepsilon)/C$. According to the minimality of μ_0, we have

$$(\star) \qquad u + (\mu_0 - \varepsilon)h = v - \varepsilon h < 0 \quad \text{for some } x = x_\varepsilon \in J^\circ.$$

At the point x_ε we have the inequalities $u \geq -Cd$, $v \geq \gamma d$ and $(\mu_0 - \varepsilon)h \leq -u \leq Cd$, hence

$$v - \varepsilon h \geq \gamma d - \frac{\varepsilon C}{\mu_0 - \varepsilon} d > 0$$

due to our choice of ε. This inequality contradicts $(\star)$. Hence the case $\min u < 0$ implies (iii) □

Remark. 1. If case (ii) of Theorem 4 applies, then an inequality $u(x) \geq \gamma d(x)$ with $\gamma > 0$ holds. In particular, if $u(a) = 0$ or $u(b) = 0$, then $u'(a) > 0$ or $u'(b) < 0$, resp.

(if the derivative does not exist, then the inequality holds with the corresponding Dini derivative).

2. If (λ_1, φ_1) is the first eigenpair of problem (7), then $h = \varphi_1$ satisfies (9). In this case Theorem 4 applies if $c(x) \leq \lambda_1$. We illustrate this by a simple example:

$$u'' + cu \leq 0 \text{ a.e. in } (0, \pi), \quad u(0) \geq 0, \quad u(\pi) \geq 0$$

implies $u \equiv 0$ or $u > 0$ in $(0, \pi)$ or $u(x) = -\beta \sin x$ $(\beta > 0)$ if $c(x) \leq 1$.

3. Theorem 4 applies if the first eigenvalue of problem (8) is nonnegative. Again $h = \varphi_1$ satisfies (9).

4. In applications of Theorem 4 one often has additional information on u or h which excludes the eigenvalue case (iii), i.e., which leads to (SMP). For example, case (iii) is excluded if one of the following conditions holds:

$$u(a) > 0; \quad u(b) > 0; \quad (pu')' + cu \not\equiv 0 \text{ in } J$$

or one of these conditions for h.

5. Example: $J = [0, \pi]$, $p(x) = 1/\sqrt{\sin x}$, $c = 3/(2\sqrt{\sin x})$. The function $h(x) = (\sin x)^{3/2}$ is the first eigenfunction to problem (8), the first eigenvalue is 0. In this example c and $1/p$ belong to $L^1(J)$, but p is not bounded near 0 and π, and (P_1) does not hold. Obviously, there is no $\gamma > 0$ such that $h(x) \geq \gamma d(x)$.

Final remarks. 1. Consider an elliptic second order operator L defined in a bounded domain $\Omega \subset \mathbb{R}^n$. Then the strong maximum principle states that

$$Lu + cu \leq 0 \text{ in } \Omega, \quad u \geq 0 \text{ on } \partial\Omega \tag{10}$$

implies $u \equiv 0$ or $u > 0$ in Ω if $c(x) \leq 0$. The proof for classical solutions $u \in C(\overline{\Omega}) \cap C^2(\Omega)$ is simple. One deals first with strict inequalities and reduces the $\leq$ case to this case by means of a limiting argument which involves also an auxiliary function. Theorems of this type hold also for weak solutions where L is in divergence form or in general form, but the proof is rather difficult; see [1], Sections 3.1–2, 8.1 and 9.1. In contrast to this situation the case $n = 1$ can be handled in a simple straightforward way under very general assumptions; cf. Theorem 1 and its proof. It is one of the purposes of the present article to make this (possibly new) method of proof more widely known.

2. Theorem 4 describes the strong maximum principle in its most general form where the first eigenvalue case is included. A similar theorem for elliptic differential inequalities of the form (10) was first proved in [2].

References

1. D. Gilbarg and N.S. Trudinger, *Elliptic Partial Differential Equations of Second Order*, 2nd ed. Grundlehren d. math. Wiss. 224, Springer Verlag 1983.

2. W. Walter, *A theorem on elliptic differential inequalities with an application to gradient bounds.* Math. Z. **200** (1989), 293–299.

Wolfgang Walter, Mathematisches Institut I, Universität Karlsruhe, Postfach 6980, D-7500 Karlsruhe, Germany

International Series of Numerical Mathematics, Vol. 103, © 1992 Birkhäuser Verlag Basel

Uniqueness and existence for viscosity solutions of quasimonotone systems

Volkmar Weckesser

Abstract

In this paper we expand the notion of viscosity solutions to fully nonlinear quasimonotone elliptic systems. The analytical heart of the theory of viscosity solutions is formulated for systems and we prove uniqueness for the Dirichlet problem. Finally we show the existence of a viscosity solution by the Perron method.

1. Introduction

In recent years the notion of viscosity solutions for fully nonlinear, degenerate elliptic pde's

$$F(x, u(x), \mathrm{D}u(x), \mathrm{D}^2u(x)) = 0 \tag{1}$$

was developed. It was first introduced in 1983 by M. G. Crandall and P.-L. Lions for Hamilton–Jacobi equations ([3], [1]), and then P.-L. Lions extended it to elliptic equations of second order [8]. While Lions used methods of optimal control theory to prove the uniqueness of viscosity solutions, R. Jensen was able to prove this with purely analytical methods in 1987 [7]. His method to work with regularizations of the viscosity subsolution resp. supersolution was simplified in subsequent papers.

In this work we shall expand the notion of viscosity solution to weakly coupled, quasimonotone systems and prove uniqueness and existence.

We consider the following system of differential equations:

$$\begin{aligned} F_1(x, u_1, \ldots, u_N, \mathrm{D}u_1, \mathrm{D}^2u_1) &= 0 \\ \vdots \quad & \vdots \quad \vdots \\ F_N(x, u_1, \ldots, u_N, \mathrm{D}u_N, \mathrm{D}^2u_N) &= 0. \end{aligned} \tag{2}$$

This paper is in final form and no version of it will be submitted for publication elsewhere.

The functions u_i are defined on Ω ($\Omega \subseteq \mathbb{R}^n$ a bounded, open domain) and real valued; $\mathrm{D}u_i = \left(\frac{\partial u_i}{\partial x_j}\right)$ is the vector of the first derivatives and $\mathrm{D}^2 u_i = \left(\frac{\partial^2 u_i}{\partial x_j \partial x_k}\right)$ the Hessian of the second derivatives.

The system (2) is weakly coupled, i. e. the function F_i depends only on $\mathrm{D}u_i$ and $\mathrm{D}^2 u_i$, but not on $\mathrm{D}u_j$ or $\mathrm{D}^2 u_j$ for $i \neq j$.

As usually we write $u := (u_1, \ldots, u_N)$ and $F := (F_1, \ldots, F_N)$. Then (2) can be written as

$$F(x, u, \mathrm{D}u, \mathrm{D}^2 u) = 0. \tag{3}$$

We assume $F_i : \Omega \times \mathbb{R}^N \times \mathbb{R}^n \times S^n \to \mathbb{R}^N$ to be continuous ($i = 1, \ldots, N$). We use the following notations: S^n is the space of symmetric $n \times n$ matices provided with the usual order relation, i. e. $r \leq s$ in S^n means that $s - r$ is positive semidefinite. For $r \in S^n$ we define the norm of r as $\|r\| := \left(\sum_{i,j=1}^n r_{ij}^2\right)^{1/2}$ and the trace of r as $\operatorname{tr} r := \sum_{i=1}^n r_{ii}$. For $p, q \in \mathbb{R}^n$ the inequality $p \leq q$ means $p_i \leq q_i$ is valid for all $i = 1, \ldots, n$ and the vector e is defined as $e := (1, \ldots, 1)$. With $B_r(\xi)$ we denote the open ball with centre ξ and radius r, i. e. $B_r(\xi) := \{x \in \mathbb{R}^n \mid |x - \xi| < r\}$, where $|.|$ denotes the Euclid norm. For a bounded real valued function u on Ω we define $\|u\|_\Omega := \sup_{x \in \Omega} |u(x)|$ and for a bounded vector valued function $\|u\|_\Omega := \max_{i=1,\ldots,N} \|u_i\|_\Omega$.

The system (2) is called degenerate elliptic, if all components are so. This means that all component functions are decreasing in $\mathrm{D}^2 u$ (the terms decreasing and increasing are always used in the weak sense), i. e. for $i = 1, \ldots, N$, $(x, z, p) \in \Omega \times \mathbb{R}^N \times \mathbb{R}^n$ and $r, s \in S^n$

$$r \leq s \ \text{ implies } \ F_i(x, z, p, r) \geq F_i(x, z, p, s). \tag{4}$$

We make this assumption throughout the paper.

It is also provided that F is quasimonotone decreasing.

Definition 1. F is called quasimonotone decreasing in z, if all F_i are decreasing in z_j for $i \neq j$. This means that for all $i = 1, \ldots, N$, $x \in \Omega$, $(p, r) \in \mathbb{R}^n \times S^n$ and $z, w \in \mathbb{R}^n$ with $z_i = w_i$ and $z \leq w$ the inequality

$$F_i(x, w, p, r) \leq F_i(x, z, p, r) \tag{5}$$

holds.

Furthermore F is supposed to be monotone increasing in the following sense:

Definition 2. F is called monotone increasing in z, if for all $(x, z, p, r) \in \Omega \times \mathbb{R}^N \times (\mathbb{R}^n)^N \times (S^n)^N$ and all $\lambda \in \mathbb{R}$ with $\lambda > 0$ the following inequality is valid

(6) $$F(x, z, p, r) \leq F(x, z + \lambda e, p, r).$$

F is called strictly increasing in z, if (6) holds with the strict "$<$" sign.

If F has continuous first derivatives quasimonotone decreasing in z means $\frac{\partial F_i}{\partial z_j}(x, z, p, r) \leq 0$ for $i \neq j$ and monotone increasing in z means $\sum_{j=1}^{N} \frac{\partial F_i}{\partial z_j}(x, z, p, r) \geq 0$ for $i = 1, \ldots, N$.

2. Definition of the viscosity solution for systems

The superdifferential of a function $u : \Omega \to \mathbb{R}$ at $\xi \in \Omega$ is defined as the set

(7) $$D^+_{1,2}u(\xi) := \left\{ (p, r) \in \mathbb{R}^n \times S^n \,\middle|\, \limsup_{h \to 0} \frac{u(\xi + h) - u(\xi) - p \cdot h - \frac{1}{2} h^T r h}{h^2} \leq 0 \right\}$$

and the subdifferential of a function $u : \Omega \to \mathbb{R}$ as

(8) $$D^-_{1,2}u(\xi) := \left\{ (p, r) \in \mathbb{R}^n \times S^n \,\middle|\, \liminf_{h \to 0} \frac{u(\xi + h) - u(\xi) - p \cdot h - \frac{1}{2} h^T r h}{h^2} \geq 0 \right\}.$$

(see [1], [8]).

With these we define the viscosity solution for systems:

Definition 3. a) An upper semi-continuous function $u : \Omega \to \mathbb{R}^N$ is called viscosity subsolution of (2), if we have for all $x \in \Omega$ and $i = 1, \ldots, N$

(9) $$F_i(x, u(x), p, r) \leq 0 \quad \text{for all } (p, r) \in D^+_{1,2}u_i(x).$$

b) A lower semi-continuous function $u : \Omega \to \mathbb{R}^N$ is called viscosity supersolution of (2), if we have for all $x \in \Omega$ and $i = 1, \ldots, N$

(10) $$F_i(x, u(x), p, r) \geq 0 \quad \text{for all } (p, r) \in D^-_{1,2}u_i(x).$$

c) $u : \Omega \to \mathbb{R}^N$ is called viscosity solution of (2), if u is viscosity subsolution and viscosity supersolution of (2).

Proposition 1. a) *Let* $u : \Omega \to \mathbb{R}^N$ *be upper semi-continuous. Then:*

u is a viscosity subsolution of (2), if and only if for all components $i = 1, \ldots, N$, all $\psi \in C^2(\Omega, \mathbb{R})$ and for each local maximum point ξ of $u_i - \psi$ we have

$$F_i(\xi, u(\xi), \mathrm{D}\psi(\xi), \mathrm{D}^2\psi(\xi)) \leq 0. \tag{11}$$

b) *Let* $u : \Omega \to \mathbb{R}^N$ *be lower semi-continuous. Then:*

u is a viscosity supersolution of (2), if and only if for all components $i = 1, \ldots, N$, all $\psi \in C^2(\Omega, \mathbb{R})$ and for each local minimum point ξ of $u_i - \psi$ we have

$$F_i(\xi, u(\xi), \mathrm{D}\psi(\xi), \mathrm{D}^2\psi(\xi)) \geq 0. \tag{12}$$

Proof. The proof is exactly the same as for $N = 1$. It is only to be done for $i = 1, \ldots, N$ (see [8]).

Remark. Proposition 1 remains valid, if we replace local extrema by global extrema or $\psi \in C^2(\Omega, \mathbb{R})$ by $\psi \in C^\infty(\Omega, \mathbb{R})$.

For functions $u, v : \Omega \to \mathbb{R}^N$ we define $w := \max(u, v)$ componentwise, i. e. $w_i := \max(u_i, v_i)$, and correspondingly $\sup u^k$ for a sequence of functions $u^k : \Omega \to \mathbb{R}^N$.

Proposition 2. a) *Let* $u, v : \Omega \to \mathbb{R}^N$ *be viscosity subsolutions (resp. viscosity supersolutions) of (2). Then also* $\max(u, v)$ *is a viscosity subsolution (resp.* $\min(u, v)$ *is a viscosity supersolution) of (2).*

b) *Let (F^k) be a sequence of functions from $C(\Omega \times \mathbb{R}^N \times (\mathbb{R}^n)^N \times (S^n)^N)$ that converges uniformly on compact subsets of $\Omega \times \mathbb{R}^N \times (\mathbb{R}^n)^N \times (S^n)^N$ to a function F. Let (u^k) be a sequence of viscosity subsolutions (viscosity supersolutions, viscosity solutions) of*

$$F^k(x, u^k, Du^k, D^2u^k) = 0,$$

that converges uniformly on compact subsets of Ω to u.

Then u is a viscosity subsolution (viscosity supersolution, viscosity solution) of (2).

c) *Let (u^k) be a locally bounded sequence of viscosity subsolutions (viscosity supersolutions) of (2). Then $w := (\sup u^k)^*$ is also a viscosity subsolution ($w := (\inf u^k)_*$ a viscosity supersolution) of (2).*

Here u^* denotes the upper semi-continuous envelope of u (defined in each component), i. e.

$$(u^*)_i(x) := \lim_{r\downarrow 0} \sup\{u_i(y) \mid y \in \Omega \text{ and } |x-y| \leq r\},$$

and the lower semi-continuous envelope u_* is defined analogously.

Proof. a) Upper semicontinuity of u and v carries over to $\max(u,v)$. Now the assertion follows from

$$\mathrm{D}^+_{1,2} \max(u_i, v_i)(\xi) \subseteq \mathrm{D}^+_{1,2} u_i(\xi) \cup \mathrm{D}^+_{1,2} v_i(\xi)$$

for $i = 1, \ldots, N$.

b) Because of Propositon 1 the proof of Propositon I.3 in [8] can be adopted.

c) The proof of b) must be only slightly modified. If we use the sequence $w^k := \sup_{1\leq j\leq k} u^j$, we have $w^k \nearrow \sup u^k$ and for all compact sets $K \subset \Omega$ on which the ith component $(w^*)_i$ has an inner maximum the following convergence assertion holds:

$$\max_K (w^k)_i \to \max_K (w^*)_i \quad \text{for } k \to \infty. \qquad \square$$

3. ε–Envelopes

In [6] the upper resp. lower ε–envelopes for bounded functions $u, v : \Omega \to \mathbb{R}$ are defined by

$$u^\varepsilon(x) := \sup\{u(y) - \frac{1}{\varepsilon}|x-y|^2 \mid y \in \Omega\} \tag{13}$$

$$v_\varepsilon(x) := \inf\{v(y) + \frac{1}{\varepsilon}|x-y|^2 \mid y \in \Omega\}. \tag{14}$$

This definition carries over to vector valued functions by using our general agreement about taking suprema componentwise, i. e.

$$u^\varepsilon := ((u_1)^\varepsilon, \ldots, (u_N)^\varepsilon), \qquad v_\varepsilon := ((v_1)_\varepsilon, \ldots, (v_N)_\varepsilon).$$

So the properties of the ε–envelopes remain valid for the components of u^ε and v_ε. In detail: u^ε is bounded above on Ω and Lipschitz continuous, it suffices to take the supremum in the ith component of u^ε over those $y \in \Omega$ with $|x-y| \leq \sqrt{2\varepsilon\|u_i\|_\Omega}$ and $u_i^\varepsilon + \frac{1}{\varepsilon}|x|^2$ is convex on Ω ($i = 1, \ldots, N$).

For $\varrho > 0$ we define the set $\Omega_\varrho := \{x \in \Omega \mid \text{dist}(x, \partial\Omega) > \varrho\}$.

Proposition 3. *Let $F \in C(\Omega \times \mathbb{R}^N \times (\mathbb{R}^n)^N \times (S^n)^N)$ be quasimonotone decreasing and monotone increasing in z and $\varepsilon > 0$.*

Let furthermore $u : \Omega \to \mathbb{R}^N$ be a bounded viscosity subsolution of (2) and $\varrho := \sqrt{2\varepsilon\|u\|_\Omega}$. Define $F_\varepsilon : \Omega_\varrho \times \mathbb{R}^N \times (\mathbb{R}^n)^N \times (S^n)^N \to \mathbb{R}^N$ in the following way

$$F_\varepsilon(\xi, z, p, r) := \min\{F(x, z, p, r) \mid x \in \overline{B_\varrho(\xi)}\}. \tag{15}$$

Then u^ε is a viscosity subsolution of $F_\varepsilon(x, u, Du, D^2u) = 0$.

A corresponding statement holds for bounded viscosity supersolution with respect to the function F^ε defined by

$$F^\varepsilon(\xi, z, p, r) := \max\{F(x, z, p, r) \mid x \in \overline{B_\varrho(\xi)}\}. \tag{16}$$

Proof. The proof of Proposition 4.2 in [5] must be modified slightly.

Let $\psi \in C^2(\Omega)$ be real valued, $i \in \{1, \dots, N\}$ and let $\xi \in \Omega_\varrho$ satisfy

$$\max_{x \in \Omega_\varrho} (u_i^\varepsilon - \psi)(x) = (u_i^\varepsilon - \psi)(\xi).$$

Then there is an $\eta \in B_\varrho(\xi)$ with

$$u_i^\varepsilon(\xi) = u_i(\eta) - \frac{1}{\varepsilon}|\xi - \eta|^2$$

and the definition of the ε–envelope gives for $i \neq j$:

$$u_j^\varepsilon(\xi) \geq u_j(\eta) - \frac{1}{\varepsilon}|\xi - \eta|^2.$$

So the function $y \mapsto u_i(y) - \psi(y - \eta + \xi)$ attains a maximum at $y = \eta$. If we now use the definiton of F_ε, the monotone increasing in z of F, the condition for quasimonotone decreasing in z and Proposition 1, we obtain:

$$\begin{aligned} (F_\varepsilon)_i(\xi, u^\varepsilon(\xi), D\psi(\xi), D^2\psi(\xi)) &\leq F_i(\eta, u^\varepsilon(\xi) + \frac{1}{\varepsilon}|\xi - \eta|^2 e, D\psi(\xi), D^2\psi(\xi)) \\ &\leq F_i(\eta, u(\eta), D\psi(\xi), D^2\psi(\xi)) \leq 0. \end{aligned}$$

Now the assertion follows from Proposition 1. □

Basic Lemma 1. *Let $u : \Omega \to \mathbb{R}^N$ be a bounded viscosity subsolution of*

$$F(x, u, Du, D^2u) = 0$$

and $v : \Omega \to \mathbb{R}^N$ *a bounded viscosity supersolution of*

$$G(x, v, Dv, D^2v) = 0.$$

Define $\varrho := (2\varepsilon \max\{\|u\|_\Omega, \|v\|_\Omega\})^{1/2}$.

Let $F, G : \Omega \times \mathbb{R}^N \times (\mathbb{R}^n)^N \times (S^n)^N \to \mathbb{R}^N$ *be continuous, degenerate elliptic, quasimonotone decreasing and monotone increasing in* z. *Let* $\varepsilon > 0$, $\phi \in C^2(\Omega \times \Omega, \mathbb{R})$ *and* $i \in \{1, \ldots, N\}$.

Put $w(x, y) := u_i^\varepsilon(x) - v_{i\varepsilon}(y)$ *for* $(x, y) \in \Omega_\varrho \times \Omega_\varrho$ *and assume that the function* $w - \phi$ *attains its maximum at* $(\bar{x}, \bar{y}) \in \Omega_\varrho \times \Omega_\varrho$. *Then there are matrices* $r, s \in S^n$, *so that*

$$(17) \qquad -\frac{2}{\varepsilon} I \le \begin{pmatrix} r & 0 \\ 0 & s \end{pmatrix} \le D^2\phi(\bar{x}, \bar{y})$$

$$(18) \qquad (F_\varepsilon)_i(\bar{x}, u^\varepsilon(\bar{x}), D_x\phi(\bar{x}, \bar{y}), r) \le 0$$

$$(19) \qquad \text{and} \ \ (G^\varepsilon)_i(\bar{y}, v_\varepsilon(\bar{y}), -D_y\phi(\bar{x}, \bar{y}), -s) \ge 0.$$

Proof. Because of Proposition 3 the proof of Proposition 5.1 in [5] can completely be adopted. □

4. Uniqueness

For a moment we will assume that $u, v : \Omega \to \mathbb{R}^N$ are twice differentiable, classical sub– resp. supersolutions of (2) and that F is strongly increasing in z.

Then it follows from $u \le v$ on $\partial\Omega$ that $u \le v$ on Ω.

Otherwise there would exist an $i \in \{1, \ldots, N\}$ and a $\xi \in \Omega$ with the property:

$$u_i(\xi) - v_i(\xi) = \max_{j \in \{1, \ldots, N\}} \max_{x \in \Omega} (u_j(x) - v_j(x)) =: \lambda > 0.$$

Using the ellipticity of F and the fact that F is strongly increasing and quasimonotone decreasing in z, we get

$$\begin{aligned} 0 &\ge F_i(\xi, u, \mathrm{D}u_i, \mathrm{D}^2u_i) \\ &> F_i(\xi, u - \lambda e, \mathrm{D}u_i, \mathrm{D}^2u_i) \\ &\ge F_i(\xi, v, \mathrm{D}v_i, \mathrm{D}^2v_i) \ge 0. \end{aligned}$$

This is a contradiction.

Corresponding results for viscosity supersolutions and subsolutions will be presented now. The case $N = 1$ of the first theorem is treated in [5].

Theorem 1. *Let* $F : \mathbb{R}^N \times (\mathbb{R}^n)^N \times (S^n)^N \to \mathbb{R}^N$ *be continuous, degenerate elliptic, quasimonotone decreasing and monotone increasing in* z.

Let u *and* v *be bounded viscosity subsolution resp. supersolution of*

$$u + F(u, Du, D^2u) = 0. \tag{20}$$

Then $u \le v$ *on* $\partial\Omega$ *implies*

$$u \le v \quad in\ \Omega. \tag{21}$$

Proof. We assume that the statement is false. Then a point ξ exists in the interior of Ω and an $i \in \{1, \ldots, N\}$ with

$$\delta := u_i(\xi) - v_i(\xi) > 0.$$

We put all those indices from $\{1, \ldots, N\}$ in the set I, for which a $x \in \Omega$ exists with $u_i(x) - v_i(x) \ge \delta$. Because of the convergence properties of the ε–envelopes there is an ε with $\sqrt{\delta} > \varepsilon > 0$, so that for all $i \in I$ points $(\bar{x}_i, \bar{y}_i) \in \Omega_\varrho \times \Omega_\varrho$ (ϱ defined in the Basic Lemma) exist with $|\bar{x}_i - \bar{y}_i| < \varepsilon$ and

$$u_i^\varepsilon(\bar{x}_i) - v_{i\varepsilon}(\bar{y}_i) \ge \delta.$$

Define ϕ by $\phi(x, y) := \frac{1}{2}|x - y|^2$. If we chose ε small enough, then for all $i \in I$ there are points $(\hat{x}_i, \hat{y}_i)$ in the interior of $\Omega_\varrho \times \Omega_\varrho$ with

$$u_i^\varepsilon(\hat{x}_i) - v_{i\varepsilon}(\hat{y}_i) - \phi(\hat{x}_i, \hat{y}_i) = \max_{(x,y)\in\Omega_\varrho\times\Omega_\varrho} (u_i^\varepsilon(x) - v_{i\varepsilon}(y) - \phi(x, y)) \ge \frac{\delta}{2}.$$

We choose now a component $i \in I$ with

$$u_i^\varepsilon(\hat{x}_i) - v_{i\varepsilon}(\hat{y}_i) = \max_{j\in I} [u_j^\varepsilon(\hat{x}_j) - (v_{j\varepsilon}(\hat{y}_j)] =: \lambda > 0 \tag{22}$$

and we put $\hat{x} := \hat{x}_i$, $\hat{y} := \hat{y}_i$.

We apply the Basic Lemma and we get matrices $r, s \in S^n$ with

$$\begin{pmatrix} r & 0 \\ 0 & s \end{pmatrix} \le \mathrm{D}^2\phi(\hat{x}, \hat{y}) \tag{23}$$

(24) $$u_i^\epsilon(\hat{x}) + F_i(u^\epsilon(\hat{x}), \mathrm{D}_x\phi(\hat{x},\hat{y}), r) \le 0$$

(25) $$\text{and}\quad v_{i\epsilon}(\hat{y}) + G_i(\hat{y}, v_\epsilon(\hat{y}), -\mathrm{D}_y\phi(\hat{x},\hat{y}), -s) \ge 0.$$

(We have $F \equiv F^\epsilon \equiv F_\epsilon$.)

Because we have $\phi(x,y) = \varphi(x-y)$, (23) implies $r + s \le 0$ and $\mathrm{D}_x\phi(x,y) = -\mathrm{D}_y\phi(x,y)$ is valid. So the inequality (24) combined with the ellipticity condition gives

(26) $$u_i^\epsilon(\hat{x}) + F_i(u^\epsilon(\hat{x}), -\mathrm{D}_y\phi(\hat{x},\hat{y}), -s) \le 0.$$

If we look at the choice of the index i and the assumptions of monotony, we get

$$\begin{aligned} F_i(u^\epsilon(\hat{x}), -\mathrm{D}_y\phi(\hat{x},\hat{y}), -s) &\ge F_i(u^\epsilon(\hat{x}) - \lambda e, -\mathrm{D}_y\phi(\hat{x},\hat{y}), -s) \\ &\ge F_i(v_\epsilon(\hat{y}), -\mathrm{D}_y\phi(\hat{x},\hat{y}), -s). \end{aligned}$$

Combining this with the inequalities (25) and (26) gives us

$$u_i^\epsilon(\hat{x}) \le v_{i\epsilon}(\hat{y}).$$

This is obviously a contradiction to (22). □

We will establish now a more general uniqueness result, which can be found in [6] for the case $N = 1$. Further restrictions are made to the equation (2): The equation (2) is called strictly elliptic, if and only if for all $R > 0$ there is a $\nu_R > 0$, so that for all $i = 1, \ldots, N$ and for $r, s \in S^n$ with $r \le s$

(27) $$F_i(x,z,p,r) - F_i(x,z,p,s) \ge \nu_R \operatorname{tr}(s - r)$$

holds for all $x \in \Omega, z \in \mathbb{R}^N$ with $|z| \le R$ and $p \in \mathbb{R}^n$. We will also use a weaker version of (27), where p just varies in $B_R(0)$ — we refer to it by (27w). It is obvious that the inequality (27) is stronger than degenerate ellipticity. We also have to restrict the x-dependence of F. We assume that we have for all $i = 1, \ldots, N$ and $R > 0$

(28) $$\begin{aligned} &|F_i(x,z,p,r) - F_i(y,z,p,r)| \\ &\quad \le C_R + \omega_R(|x-y|)\,|x-y|^{\tau_i}\,|p|^{2+\tau_i} + \mu_R(|x-y|)\|r\| \end{aligned}$$

for all $x, y \in \Omega, z \in \mathbb{R}^N$ with $|z| < R, p \in \mathbb{R}^n, r \in S^n$. Here is $\tau_i \in [0,1], C_R$ a positive constant and ω_R, μ_R are non negative functions defined on $[0,\infty)$ with $\omega_R(\sigma) \to 0$ for $\sigma \to 0+$ and $\int_{0+} (\mu_R(\sigma)/\sigma)\, \mathrm{d}\sigma < \infty$ $(i = 1, \ldots, N)$.

Another condition: there is a $\vartheta > \frac{1}{2}$ with

$$|F_i(x,z,p,r) - F_i(y,z,p,r)| \leq \omega_R\left(|x-y|^\vartheta(1+\|r\|)\right) \tag{29}$$

for all $x, y \in \Omega$, $z \in \mathbb{R}^N$ with $|z| \leq R$, $p \in B_R$ and $r \in S^n$. Here $\omega_R(\sigma) \to 0$ for $\sigma \to 0+$ and $\omega_R(\sigma)(1+\sigma)$ remains bounded for all $\sigma \geq 0$.

Finally we assume that for every $R > 0$ there is a $\gamma_R > 0$, so that we have for all $z \in \mathbb{R}^N$ and $\lambda > 0$ with $-Re \leq z \leq z + \lambda e \leq Re$

$$F(x,z,p,r) - F(x,z-\lambda e,p,r) \geq \gamma_R \lambda e \tag{30}$$

for all $x \in \Omega$, $p \in B_R$ and $r \in S^n$.

Theorem 2. a) *Assume* (27), (28), (29), *and* (30) *for the equation* (2). *Let* u *and* v *be a bounded viscosity subsolution resp. supersolution that satisfy* $u \leq v$ *on* $\partial\Omega$. *Then we have* $u \leq v$ *on* Ω.

b) *Assume* (27w), (29), *and* (30) *for the equation* (2). *Let* u *and* v *be a bounded viscosity subsolution resp. supersolution and one of them be locally Lipschitz continuous. Then* $u \leq v$ *on* $\partial\Omega$ *implies* $u \leq v$ *on* Ω.

Proof. b) We assume again that the assumption is false and that there is a point ξ in the interior of Ω and an index $i \in \{1, \ldots, N\}$ with

$$\delta := u_i(\xi) - v_i(\xi) > 0.$$

In the following we work with the set I of those indices, for which an $x \in \Omega$ with $u_i(x) - v_i(x) > 0$ exists. For $\delta > 0$ we define the function $\phi(x,y) := \frac{1}{2\delta}|x-y|^2$. Then we can choose δ so small that for all $i \in I$ the functions

$$u_i(x) - v_i(y) - \phi(x,y)$$

attain a global maximum over $\Omega \times \Omega$ and all maxima are contained in a fixed compact subset of $\Omega \times \Omega$.

Now we make all conclusions in [6] following the inequality (3.4) in the proof of Theorem III.1(2) for all $i \in I$. So we get points of global maximum $(\bar{x}_i, \bar{y}_i)$ of $u_i(x) - v_i(y) - \phi(x,y)$ for all $i \in I$ with the same properties as those mentioned in [6].

In the following we only look at an $i \in I$ with

$$\lambda := u_i(\bar{x}_i) - v_i(\bar{y}_i) = \max_{j \in I}\left(u_j(\bar{x}_j) - v_j(\bar{y}_j)\right)$$

and write for short $\bar{x} := \bar{x}_i$, $\bar{y} := \bar{y}_i$.

Then the proof of [6] goes through for the component function F_i — we just have to use the above Basic Lemma. The only thing that remains to do is to establish the inequality (3.14). From the definition of λ we have:

$$u(\bar{x}) - \lambda e \leq v(\bar{y}).$$

And this combined with the quasimonotony and (30) gives us:

$$\begin{aligned} & F_i(\bar{y}, u(\bar{x}), \frac{1}{\delta}(\bar{x}-\bar{y}), r) - F_i(\bar{y}, v(\bar{y}), \frac{1}{\delta}(\bar{x}-\bar{y}), r) \\ \geq\ & F_i(\bar{y}, u(\bar{x}), \frac{1}{\delta}(\bar{x}-\bar{y}), r) - F_i(\bar{y}, u(\bar{x}) - \lambda e, \frac{1}{\delta}(\bar{x}-\bar{y}), r) \\ \geq\ & \gamma_R \lambda = \gamma_R(u_i(\bar{x}) - v_i(\bar{y})). \end{aligned}$$

a) The proof of Theorem III.1(1) will be done for F_i — i the index we fixed above. In the proof of Proposition III.1 we have to use again the above argument to show the inequality (3.22). □

5. The Perron method

In this section we will formulate Perron's method for a proof of existence of viscosity solutions. This was first done by H. Ishii for Hamilton-Jacobi-equations (see [4], [5]).

Therorem 3 (Perron's method). *Let $\mathcal{S}$ be a nonempty set of viscosity subsolutions of* (2) *with the following two properties:*

(I) *If $v \in \mathcal{S}$ is no viscosity solution of* (2)*, then there is a $w \in \mathcal{S}$, an $i \in \{1, \ldots, N\}$ and a $\xi \in \Omega$ with $v_i(\xi) < w_i(\xi)$.*

(II) *The function $u : \Omega \to \mathbb{R}^N$ defined by*

$$u(x) := \sup\{v(x) \mid v \in \mathcal{S}\} \tag{31}$$

belongs to $\mathcal{S}$.

Then u – defined by (31) *– is a viscosity solution of* (2).

Proof. The proof is obvious.

Proposition 4. *Let g be a viscosity supersolution of* (2). *Define the set $\mathcal{S}$ as*

$$\mathcal{S} := \{v \mid v \leq g \text{ auf } \Omega \text{ and } v \text{ is viscosity subsolution of } (2)\}.$$

Then S has the property (I) *of Theorem* 3.

Proof. Because of Proposition 1 the corresponding part of the proof of Theorem 3.1 in [4] goes through.

For the case $N = 1$ the following theorem can be found in [2].

Theorem 4. *We consider the Dirichlet Problem*

$$F(x, u, Du, D^2u) = 0 \text{ in } \Omega, \qquad u(x) = f(x) \text{ on } \partial\Omega \tag{32}$$

($f : \partial\Omega \to \mathbb{R}^N$ continuous) and assume that the comparison theorem holds, i. e. if u is a viscosity subsolution and v a viscosity supersolution with $u \leq v$ on $\partial\Omega$, then we have $u \leq v$ in Ω. Furthermore we assume that there is a viscosity subsolution u and a viscosity supersolution v, that satisfy the boundary condition in the following sense: $u_*(x) = v^*(x) = f(x)$ on $\partial\Omega$.

Then the function

$$W(x) := \sup\{w(x) \mid w \leq v \text{ on } \Omega \text{ and } w \text{ is a viscosity subsolution of } (2)\}.$$

is a viscosity solution of (32) *that satisfies the boundary conditions in the classical sense.*

Proof. We have $u_* \leq W_* \leq W \leq W^* \leq v^*$ on $\overline{\Omega}$ and so in particular $W(x) = f(x)$ on $\partial\Omega$. So the boundary conditions are satisfied.

Put

$$S := \{w \mid w \leq v \text{ on } \Omega \text{ and is a viscosity subsolution of (2) }\}.$$

The condition (I) in Theorem 3 is satisfied by S because of Proposition 4.

According to Proposition 2c the function W^* is a viscosity subsolution of (2). So from the comparison theorem follows $W^* \leq v$ in Ω and so $W^* \leq W$. This gives us $W = W^* \in S$ and also condition (II) in Theorem 3 is true.

Now the statement follows from Theorem 3. □

Acknowledgement. I thank Prof. Dr. W. Walter who gave me the idea to expand the notion of viscosity solution to systems.

References

1. M. G. Crandall, L. C. Evans, and P.-L. Lions, *Some properties of viscosity solutions of Hamilton-Jacobi equations.* Trans. Amer. Math. Soc. **282** (1984), 487–502.

2. M. G. Crandall, H. Ishii, and P.-L. Lions, *User's Guide to viscosity solutions of second order partial differential equations.* Bull. Amer. Math. Soc., to appear.

3. M. G. Crandall and P.-L. Lions, *Viscosity solutions of Hamilton-Jacobi equations.* Trans. Amer. Math. Soc.**277** (1983), 1–42.

4. H. Ishii, *Perron's method for Hamilton-Jacobi equations.* Duke Math. J. **55(2)** (1986), 369–384.

5. H. Ishii, *On uniqueness and existence of viscosity solutions of fully nonlinear second-order elliptic pde's.* Comm. Pure Appl. Math. **42** (1989), 15–45.

6. H. Ishii and P.-L. Lions, *Viscosity solutions of fully nonlinear second-order elliptic partial differential equations.* J. Diff. Equs. **83** (1990), 26–78.

7. R. Jensen, *The maximum principle for viscosity solutions of fully nonlinear second order partial differential equations.* Arch. Rational Mech. Anal. **101** (1988), 1–29.

8. P.-L. Lions, *Optimal control of diffusion processes and Hamilton-Jacobi-Bellman equations. Part 2. Viscosity solutions and uniqueness.* Comm. Part. Diff. Equs. **8** (1983), 1229–1276.

9. P. J. McKenna and W. Walter, *On the Dirichlet problem for elliptic systems.* Applicable Analysis **21** (1986), 207–224.

Volkmar Weckesser, Universität Karlsruhe, Mathematisches Institut I, Kaiserstr. 12, D-7500 Karlsruhe 1, Germany

Inequalities in Economics, Optimization and Applications

In the snow: Jogger 1990

International Series of Numerical Mathematics, Vol. 103, © 1992 Birkhäuser Verlag Basel

On some problems concerning best constants for the midpoint and trapezoidal rule

Bernhard Büttgenbach, Gerald Lüttgens* and Rolf J. Nessel

Abstract

It is the purpose of this paper to discuss some questions on best constants in connection with Jackson-type inequalities for the compound midpoint and trapezoidal rule. To this end representations of the remainders are employed which explicitly exhibit differences of that order which is characteristic for the exactness of the rule. The matter is considered for continuous and Riemann integrable functions, using ordinary moduli of continuity and τ-moduli, respectively.

1. Introduction

The problem we have in mind in connection with some commonly used compound quadrature rules is very popular in approximation theory and originates, at least to our knowledge, in work of the Russian school on the approximation by trigonometric convolution processes. To be a little bit more precise, let $C_{2\pi}$ be the space of 2π-periodic, continuous functions on the real axis $\mathbb{R}$, endowed with the usual sup-norm $\|f\|_{C_{2\pi}} := \{|f(u)| : u \in \mathbb{R}\}$. Consider the n-th partial sum

$$(S_n f)(x) := \sum_{k=-n}^{n} f\hat{}(k)\, e^{ikx}, \qquad f\hat{}(k) := \frac{1}{2\pi}\int_{-\pi}^{\pi} f(u)\, e^{-iku}\, du,$$

of the Fourier series of f with complex Fourier coefficients $f\hat{}$. In terms of the first order modulus of continuity

$$\omega_1(\delta, f) := \sup\{|f(u+h) - f(u)| : u \in \mathbb{R}, |h| \le \delta\} \tag{1.1}$$

This paper is in final form and no version of it will be submitted for publication elsewhere.

*Supported by Deutsche Forschungsgemeinschaft Grant No. Ne 171/11-1

one then has that for every $f \in C_{2\pi}$ there holds true the familiar Lebesgue (or Jackson-type) inequality

$$\|S_n f - f\|_{C_{2\pi}} \leq c\left(\|S_n\|_{[C_{2\pi}]} + 1\right)\omega_1(1/n, f) \qquad (f \in C_{2\pi}),$$

where $\|S_n\|_{[C_{2\pi}]}$ denotes the norm of the partial sum operator S_n, considered as a mapping from $C_{2\pi}$ into itself. It was already shown by Lebesgue (see [8]) that this estimate is sharp with regard to the order. Concerning constants a more refined analysis finally delivers (see [6] and the literature cited there) that for each $n \in \mathbb{N}$ (= set of natural numbers)

$$(1.2) \qquad \sup_{f \in C_{2\pi}, f \neq const} \frac{\|S_n f - f\|_{C_{2\pi}}}{\omega_1\left(\dfrac{2\pi}{2n+1}, f\right)} = \frac{1}{2}\left(\|S_n\|_{[C_{2\pi}]} + 1\right) \qquad (n \in \mathbb{N}).$$

In many papers, including work of e.g. V.V. Arestov, N.I. Chernikh, V.T. Gavriljuk, S.B. Stečkin, results of a similar nature are derived, for example, in different metrics or for various linear means of the Fourier series.

The present paper is concerned with corresponding questions in connection with some compound quadrature rules. Indeed, it may be worthwhile to mention that the problems to be discussed have already been posed at several places in the literature. See e.g. [3], [7], [9, Chapter 3]. The starting point for the treatment here are certain representations of the remainders (see [2]) which explicitly exhibit differences of that order which is characteristic for the exactness of the rule. After introducing some preliminary facts in Section 2, Section 3 deals with the compound midpoint rule, where the situation is particularly simple and the results are already complete. This is followed by Section 4 on the compound trapezoidal rule. The matter is considered for continuous and Riemann integrable functions, employing ordinary moduli of continuity (cf. (1.1)) and τ-moduli, respectively. The results obtained also contribute to the question whether there exist extremal functions. The paper concludes with some remarks concerning relevant problems for the Simpson rule.

2. Preliminaries

Let $C = C[0,1]$, R or B be the spaces of real-valued functions, defined on the compact interval $[0,1]$ which are continuous, Riemann integrable or bounded, respectively. Obviously, $C \subset R \subset B$, and under the familiar sup-norm $\|f\|_B := \sup\{|f(u)| : 0 \leq u \leq 1\}$ each of these spaces is a Banach space. For $r \in \mathbb{N}$ and $f \in B$ let r-th differences be introduced via

$$\Delta_h^r f(u) := \sum_{k=0}^{r} (-1)^{r-k} \binom{r}{k} f(u + kh).$$

In connection with continuous functions the appropriate measure of smoothness will of course be the usual r-th modulus of continuity (cf. (1.1))

$$\omega_r(\delta, f) := \omega_r(\delta, f)_{[0,1]} := \sup\{|\Delta_h^r f(u)| : u, u + rh \in [0,1], |h| \leq \delta\}.$$

Indeed, for $f \in B$ one has

$$f \in C \iff \omega_1(\delta, f) = o(1) \qquad (\delta \to 0+).$$

In connection with the space of Riemann integrable functions, which naturally is of particular interest for quadrature, the appropriate measure of smoothness is given by the so-called r-th τ-modulus $\tau_r(\delta, f)$, which for any bounded function f may be defined via

$$\tau_r(\delta, f) := \overline{\int_0^1} \omega_r(\delta, f, x)\, dx,$$

$$\omega_r(\delta, f, x) := \sup\{|\Delta_h^r f(u)| : u, u + rh \in [x - \delta, x + \delta] \cap [0,1]\},$$

as the upper Riemann integral $\overline{\int}$ over the local modulus of continuity $\omega_r(\delta, f, x)$. This notion, independently introduced by P.P. Korovkin and Bl. Sendov around 1968, has turned out to be very useful in approximation theory. Parallel to the continuous situation one has that for $f \in B$

$$f \in R \iff \tau_1(\delta, f) = o(1) \qquad (\delta \to 0+).$$

For this and further elementary properties of the τ-modulus one may consult the treatment in [9] (and the literature cited there). Thus it is an immediate consequence of the definitions that

$$\tau_r(\delta, f) \leq \|\omega_r(\delta, f, x)\|_B = \omega_r(2\delta/r, f) \qquad (f \in B, \delta > 0). \tag{2.1}$$

Let us also recall the familiar fact that in the special case of affine transformations the usual integration by substitution remains valid for upper Riemann integrals. Thus, if $A(u) := a + (u - c)(b - a)/(d - c)$ for $u \in [c, d]$ is the affine transformation of the interval $[c, d]$ onto $[a, b]$, then for every $g \in B[a, b]$

$$\overline{\int_a^b} g(x)\, dx = \overline{\int_c^d} g(A(u))A'(u)\, du. \tag{2.2}$$

For $f \in R[a, b]$ let $Q_n f$ be the quadrature formula for the approximate calculation of the integral $If := \int_a^b f(u)\, du$, given by

$$Q_n f := \sum_{k=1}^n a_{kn} f(x_{kn})$$

with weights $0 \neq a_{kn} \in \mathbb{R}$ and knots $a \leq x_{1n} < \ldots < x_{nn} \leq b$, $1 \leq k \leq n \in \mathbb{N}$. Considered as a bounded linear functional on $C[a,b]$ it is well-known (cf. [1, p. 32]) that for its operatornorm one has the representation

$$\|Q_n\|_{C^*} := \sup\left\{|Q_n f| : \|f\|_B = 1, f \in C[a,b]\right\} = \sum_{k=1}^{n} |a_{kn}|. \tag{2.3}$$

Based upon an application of the Riesz representation theorem, it was shown in [5] that for the remainder $R_n := Q_n - I$ there holds true the representation

$$\|R_n\|_{C^*} = \sum_{k=1}^{n} |a_{kn}| + (b-a). \tag{2.4}$$

In this connection it may be worthwhile to mention that a proof of (2.4) may also be given by the aid of those elementary arguments, usually employed for the proof of (2.3). Indeed, the upper estimate being obvious, to show that the right hand side of (2.4) is also a lower bound for $\|R_n\|_{C^*}$, consider the piecewise linear function $f_\varepsilon \in C[a,b]$, given for sufficiently small $\varepsilon > 0$ via

$$f_\varepsilon(x) := \begin{cases} \operatorname{sgn} a_{kn}, & x = x_{kn} \\ -1, & x \notin [x_{kn} - \varepsilon, x_{kn} + \varepsilon] \cap [a,b], 1 \leq k \leq n. \end{cases}$$

Then $\|f_\varepsilon\|_B = 1$, $Q_n f_\varepsilon = \sum_{k=1}^{n} |a_{kn}|$, and

$$I f_\varepsilon = -(b-a) + \int_a^b [f_\varepsilon(u) + 1]\, du \leq -(b-a) + 2n\varepsilon < 0.$$

Therefore in view of the definition of $\|R_n\|_{C^*}$

$$\|R_n\|_{C^*} \geq |R_n f_\varepsilon| = \sum_{k=1}^{n} |a_{kn}| - I f_\varepsilon \geq \sum_{k=1}^{n} |a_{kn}| + (b-a) - 2n\varepsilon,$$

which already completes the proof of (2.4) since $n \in \mathbb{N}$ is fixed and ε may be arbitrarily small. Note that (2.4) together with (2.3) implies

$$\|Q_n - I\|_{C^*} = \|Q_n\|_{C^*} + \|I\|_{C^*}, \tag{2.5}$$

which nicely corresponds to the fact that for the identity operator id and for every compact operator A from $C[a,b]$ into itself (see [4])

$$\|A + id\|_{[C]} = \|A\|_{[C]} + \|id\|_{[C]},$$

the latter equality being frequently used in the trigonometric context of results like (1.2).

Let us conclude with some elementary manipulations, valid for compound quadrature procedures. Consider the (elementary) rule

$$Q^{el} f := \sum_{i=1}^{j} a_i f(x_i)$$

with $0 \neq a_i \in \mathbb{R}$, $0 \leq x_1 < \ldots < x_j \leq 1$ and the remainder

$$R^{[a,b]} f := Q^{[a,b]} f - If := (b-a) \sum_{i=1}^{j} a_i f(a + (b-a)x_i) - \int_a^b f(u)\, du$$

of its affine transformation. In these terms the remainder of the compound rule

$$R_n f := Q_n f - If := \frac{1}{n} \sum_{k=1}^{n} \sum_{i=1}^{j} a_i f\Big(\frac{k-1+x_i}{n}\Big) - \int_0^1 f(u)\, du$$

may be represented in the form

$$R_n f = \sum_{k=1}^{n} R^{[(k-1)/n, k/n]} f. \tag{2.6}$$

If for the elementary rule Q^{el} there holds true the Jackson-type inequality

$$|R^{el} g| \leq c\, \omega_r(\delta, g) \qquad (g \in R[0,1]),$$

then via the substitution $g(x) = f(a+(b-a)x)$ one has for the affine transformation that for every $f \in R[a,b]$

$$|R^{[a,b]} f| = (b-a)|R^{el} g| \leq (b-a)\, c\, \omega_r(\delta, g) \leq c\,(b-a)\, \omega_r(\delta(b-a), f)_{[a,b]}. \tag{2.7}$$

3. Midpoint rule

Let us consider the compound midpoint rule

$$Q_n^{Mi} f := \frac{1}{n} \sum_{k=1}^{n} f\Big(\frac{2k-1}{2n}\Big), \qquad R_n^{Mi} := Q_n^{Mi} - I,$$

for the approximate calculation of the integral $If := \int_0^1 f(u)\, du$. As already mentioned, the situation here is particularly simple, and the results are already complete. Indeed, for the remainder it is well-known (cf. [2], [9, p. 41]) that there holds true the explicit representation

$$R_n^{Mi} f = -\sum_{k=1}^{n} \int_0^{1/2n} \Delta_h^2 f\Big(\frac{2k-1}{2n} - h\Big)\, dh \qquad (f \in R[0,1]), \tag{3.1}$$

the second difference precisely corresponding to the fact that the midpoint rule is exact for polynomials of degree 1. Consequently for every $f \in R[0,1]$ and $n \in \mathbb{N}$ there holds true the Jackson-type inequality

$$|R_n^{Mi} f| \le \frac{1}{2}\tau_2\left(\frac{1}{2n}, f\right) \le \frac{1}{2}\omega_2\left(\frac{1}{2n}, f\right). \tag{3.2}$$

To include a proof for the sake of completeness, in view of the identity (3.1) and (2.2) one has that (see [7], [9, p. 42]SP)

$$\begin{aligned}
|R_n^{Mi} f| &\le \sum_{k=1}^{n} \frac{1}{2}\int_0^{1/2n}\left[\left|\Delta_h^2 f\left(\frac{2k-1}{2n}-h\right)\right| + \left|\Delta_{-h}^2 f\left(\frac{2k-1}{2n}+h\right)\right|\right] dh \\
&\le \frac{1}{2}\sum_{k=1}^{n}\int_0^{1/2n}\left[\omega_2\left(\frac{1}{2n}, f, \frac{2k-1}{2n}-h\right) + \omega_2\left(\frac{1}{2n}, f, \frac{2k-1}{2n}+h\right)\right] dh \\
&= \frac{1}{2}\sum_{k=1}^{n}\left[\int_{\frac{2k-2}{2n}}^{\frac{2k-1}{2n}} \omega_2\left(\frac{1}{2n}, f, u\right) du + \int_{\frac{2k-1}{2n}}^{\frac{2k}{2n}} \omega_2\left(\frac{1}{2n}, f, u\right) du\right],
\end{aligned}$$

which together with (2.1) establishes the inequalities (3.2).

For the compound midpoint rule one may then introduce the following constants

$$\begin{aligned}
A_n^{Mi}\binom{C}{R} &:= \sup\left\{\frac{|R_n^{Mi} f|}{\omega_2(1/2n, f)} : f \in \binom{C}{R} \setminus \mathcal{P}_1\right\}, \\
B_n^{Mi}\binom{C}{R} &:= \sup\left\{\frac{|R_n^{Mi} f|}{\tau_2(1/2n, f)} : f \in \binom{C}{R} \setminus \mathcal{P}_1\right\},
\end{aligned} \tag{3.3}$$

where $\mathcal{P}_r$ denotes the set of all algebraic polynomials of degree less than or equal to r. Note that in view of (2.1)

$$A_n^{Mi}\binom{C}{R} \le B_n^{Mi}\binom{C}{R}, \qquad \binom{A}{B}_n^{Mi}(C) \le \binom{A}{B}_n^{Mi}(R). \tag{3.4}$$

Moreover, (3.2) shows that for all these constants

$$\binom{A}{B}_n^{Mi}\binom{C}{R} \le \frac{1}{2} \qquad (n \in \mathbb{N}). \tag{3.5}$$

Concerning their actual value, a first result was given by Ivanov (see [7]), namely

$$B_n^{Mi}(C) = B_n^{Mi}(R) = 1/2 \qquad (n \in \mathbb{N}). \tag{3.6}$$

Indicating the proof, for the upper bound Ivanov proceeds via (3.2), for the lower bound he employs some extremal sequence. Following the arrangement in the book of Sendov–Popov (see [9, p. 52]SP) one may consider the bell-shaped C^∞-function

$$\varphi(x) := \begin{cases} \exp\Big\{ -\dfrac{x^2}{1-x^2}\Big\} & , \quad |x| \le 1 \\ 0 & , \quad |x| > 1 \end{cases}$$

to construct the extremal sequence (for fixed $n \in \mathbb{N}$ and arbitrary $j \in \mathbb{N}$)

$$f_j(x) := \varphi\Big(2jn\big(x - \frac{2k-1}{2n}\big)\Big) \quad \text{for } \Big|x - \frac{2k-1}{2n}\Big| < \frac{1}{2n}, \qquad 1 \le k \le n.$$

In this connection our first assertion completes Ivanov's result and states that $1/2$ is also the value for the smaller constants A.

Theorem 3.1. *For each $n \in \mathbb{N}$ one has*

$$A_n^{Mi}(C) = A_n^{Mi}(R) = B_n^{Mi}(C) = B_n^{Mi}(R) = 1/2.$$

Proof. In view of (3.4-6) the assertion follows once it is shown that $1/2$ is also a lower bound for $A_n^{Mi}(C)$. But for the midpoint rule this is an immediate consequence of (2.5) and the fact that $\omega_2(\delta, f) \le 4\,\|f\|_B$:

$$\begin{aligned} A_n^{Mi}(C) \;&\ge\; \sup_{f \in C\setminus \mathcal{P}_1} \frac{|R_n^{Mi} f|}{4\,\|f\|_B} \;=\; \frac{1}{4} \sup_{0 \ne f \in C} \frac{|R_n^{Mi} f|}{\|f\|_B} \\ &=\; \frac{1}{4}\Big\|Q_n^{Mi} - I\Big\|_{C^*} \;=\; \frac{1}{4}\Big(\Big\|Q_n^{Mi}\Big\|_{C^*} + 1\Big) \;=\; \frac{1}{2}. \end{aligned}$$

■

It may be mentioned that the present argument in particular shows that

$$A_n^{Mi}(C) = \frac{1}{4}\Big(\Big\|Q_n^{Mi}\Big\|_{C^*} + 1\Big) \qquad (n \in \mathbb{N}),$$

quite parallel to the result (1.2) on Fourier partial sums.

Let us finally contribute to the question whether there in fact exist extremal elements.

Theorem 3.2. (a) *Given $n \in \mathbb{N}$, the function $f^* \in R[0,1] \setminus \mathcal{P}_1$, defined by*

$$f^*(x) := \begin{cases} 1, & x = \dfrac{2k-1}{2n},\ 1 \le k \le n \\ 0, & \textit{elsewhere on } [0,1] \end{cases},$$

is an extremal element for the constants $A_n^{Mi}(R)$, $B_n^{Mi}(R)$.
(b) *Given* $n \in \mathbb{N}$, *the function* $g^* \in C[0,1] \setminus \mathcal{P}_1$, *defined by*

$$g^*(x) := \begin{cases} 1-2nx, & 0 \le x \le 1/2n \\ 0 \quad , & 1/2n < x \le 1 \end{cases},$$

is an extremal element for the constant $B_n^{Mi}(C)$.
(c) *For each* $f \in C[0,1] \setminus \mathcal{P}_1$ *one has*

$$|R_1^{Mi} f| < \frac{1}{2}\,\omega_2(1/2, f).$$

Thus for the constant $A_1^{Mi}(C)$ *there cannot exist an extremal function.*

Proof. (a). Obvious, since $R_n^{Mi} f^* = 1$ and $\omega_2(1/2n, f^*) = \tau_2(1/2n, f^*) = 2$.

(b). Again it is immediate that $|R_n^{Mi} g^*| = I g^* = 1/4n$. Moreover, considering cases it follows that (note that g^* is piecewise linear)

$$\omega_2(\frac{1}{2n}, g^*, x) = \begin{cases} 2nx \quad , & 0 \le x \le 1/2n \\ 2\,(1-nx), & 1/2n \le x \le 1/n \\ 0 \quad , & 1/n \le x \le 1 \end{cases}.$$

Therefore $\tau_2(1/2n, g^*) = 1/2n$, establishing the assertion.

(c). Suppose there exists some $f_0 \in C \setminus \mathcal{P}_1$ such that, without loss of generality, $R_1^{Mi} f_0 = -(1/2)\,\omega_2(1/2, f_0)$. Then for $g(x) := \left[f_0(x) - f_0(1/2)\right]/\omega_2(1/2, f_0) \in C \setminus \mathcal{P}_1$ which (otherwise consider $[g(x) + g(1-x)]/2$) may be assumed to be symmetric to $1/2$, one has that

$$\omega_2(1/2, g) \le 1, \qquad g(1/2) = 0, \qquad Ig = -R_1^{Mi} g = 1/2.$$

It follows that for $0 \le x \le 1/2$

$$2\,g(\frac{1}{2} + x) = 2\,g(\frac{1}{2} - x) = \Delta_x^2\, g(\frac{1}{2} - x) \le \omega_2(1/2, g) \le 1.$$

This implies $g(x) \le 1/2$ on $[0,1]$, in fact $g(x) = 1/2$ for $0 \le x \le 1$, since $g \in C$ and $Ig = 1/2$. But this is a contradiction to $g(1/2) = 0$. ∎

4. Trapezoidal rule

Clearly, the results for the midpoint rule now outline a program for compound rules of higher order. And indeed, the starting point for the present treatment of problems of this kind is the fact that, parallel to (3.1), there are available explicit representations of the remainders in some other cases (see [2]), too. But the results concerning the various constants already split in the case of the compound trapezoidal rule

$$Q_n^{Tr} f := \frac{1}{n}\Big[\frac{1}{2}f(0) + \sum_{k=1}^{n-1} f(\frac{k}{n}) + \frac{1}{2}f(1)\Big], \qquad R_n^{Tr} := Q_n^{Tr} - I.$$

Again for the remainder there holds true an explicit representation (see [2], [9, p. 54]SP and the literature cited there)

$$R_n^{Tr} f = \sum_{k=1}^{n} \int_0^{1/2n} \Big[\Delta_h^2 f(\frac{k-1}{n}) + \Delta_{-h}^2 f(\frac{k}{n})\Big]\, dh, \tag{4.1}$$

the trapezoidal rule being exact on $\mathcal{P}_1$. Hence for every $f \in R[0,1]$ and $n \in \mathbb{N}$ there holds true the Jackson-type inequality

$$|R_n^{Tr} f| \le \tau_2(1/2n, f) \le \omega_2(1/2n, f). \tag{4.2}$$

To this end, following again [9, p. 54]SP, since for $x \in [0, 1/2n]$

$$\frac{k-1}{n} + x - \frac{1}{2n} \le \frac{k-1}{n} \le \frac{k-1}{n} + 2x \le \frac{k-1}{n} + x + \frac{1}{2n},$$

one has, with an analogous argument for the second estimate

$$|\Delta_x^2 f(\frac{k-1}{n})| \le \omega_2(\frac{1}{2n}, f, \frac{k-1}{n} + x), \qquad |\Delta_{-x}^2 f(\frac{k}{n})| \le \omega_2(\frac{1}{2n}, f, \frac{k}{n} - x).$$

Therefore, in view of the identity (4.1) and (2.2) one has that

$$\begin{aligned}
|R_n^{Tr} f| &\le \sum_{k=1}^{n} \int_0^{1/2n} \Big[|\Delta_h^2 f(\frac{k-1}{n})| + |\Delta_{-h}^2 f(\frac{k}{n})|\Big]\, dh \\
&\le \sum_{k=1}^{n} \Big[\bar{\int_0^{1/2n}} \omega_2(\frac{1}{2n}, f, \frac{k-1}{n} + x)\, dx + \bar{\int_0^{1/2n}} \omega_2(\frac{1}{2n}, f, \frac{k}{n} - x)\, dx\Big] \\
&= \sum_{k=1}^{n} \Big[\bar{\int_{\frac{2k-2}{2n}}^{\frac{2k-1}{2n}}} + \bar{\int_{\frac{2k-1}{2n}}^{\frac{2k}{2n}}}\Big] \omega_2(\frac{1}{2n}, f, u)\, du \;=\; \tau_2(\frac{1}{2n}, f),
\end{aligned}$$

which together with (2.1) establishes the inequalities (4.2).

If for the trapezoidal rule one introduces best constants A and B as for the midpoint rule (cf. (3.3)) it follows from (4.2) that for all these constants

(4.3) $$\binom{A}{B}_n^{Tr}\binom{C}{R} \le 1 \qquad (n \in \mathbb{N}).$$

Theorem 4.1. *For each $n \in \mathbb{N}$ one has*

$$B_n^{Tr}(R) = B_n^{Tr}(C) = 1.$$

Moreover, given $n \in \mathbb{N}$, the function $f^* \in R[0,1] \setminus \mathcal{P}_1$, defined by

$$f^*(x) := \begin{cases} 1, & x = 0 \\ 0, & 0 < x \le 1 \end{cases},$$

is an extremal element for the constant $B_n^{Tr}(R)$. On the other hand, for each $f \in C[0,1] \setminus \mathcal{P}_1$ one has

$$|R_1^{Tr} f| < \tau_2(1/2, f).$$

Thus for the constant $B_1^{Tr}(C)$ there cannot exist an extremal function.

Proof. A straightforward calculation for the candidate f^* shows that

$$\omega_2(1/2n, f^*, x) = \begin{cases} 1 & , \quad 0 \le x \le 1/2n \\ 0 & , \quad 1/2n < x \le 1 \end{cases},$$

and therefore $\tau_2(1/2n, f^*) = 1/2n = R_n^{Tr} f^*$. Together with (4.3) this implies $B_n^{Tr}(R) = 1$. The case C is dealt with by smoothing f^* appropriately. Thus, given $n \in \mathbb{N}$, for arbitrary $0 < \varepsilon < 1/2n$ consider the piecewise linear function $f_\varepsilon \in C \setminus \mathcal{P}_1$, defined via

$$f_\varepsilon(x) := \begin{cases} 1 & , \quad x = 0 \\ 0 & , \quad \varepsilon \le x \le 1 \end{cases}.$$

Again, straightforward calculations establish the estimates

$$\tau_2(1/2n, f_\varepsilon) \le (1/2n) + \varepsilon, \qquad |R_n^{Tr} f_\varepsilon| \ge (1/2n) - (\varepsilon/2),$$

which together with (4.3) imply

$$1 \ge \frac{|R_n^{Tr} f_\varepsilon|}{\tau_2(1/2n, f_\varepsilon)} \ge \frac{(1/2n) - (\varepsilon/2)}{(1/2n) + \varepsilon}.$$

Since ε may be arbitrarily small, this shows $B_n^{Tr}(C) = 1$.

On the other hand, suppose there exists some $f_0 \in C \setminus \mathcal{P}_1$ such that $|R_1^{Tr} f_0| = \tau_2(1/2, f_0)$. Without loss of generality one may normalize f_0 according to $f_0(0) = f_0(1) = 0$ and $f_0(1/2) \geq 0$. The definition of f_0 as well as a review of the proof of (4.2) for f_0 and $n = 1$ shows that there holds true equality at each step of the proof so that the continuity of the expressions involved implies that for all $0 < h \leq 1/2$

$$|\Delta_h^2 f_0(0)| = \omega_2(\frac{1}{2}, f_0, h), \qquad |\Delta_{-h}^2 f_0(1)| = \omega_2(\frac{1}{2}, f_0, 1-h).$$

Hence by the definition of the local modulus of continuity

$$\begin{aligned} 0 &= \lim_{h\to 0} |\Delta_h^2 f_0(0)| = \lim_{h\to 0} \omega_2(\frac{1}{2}, f_0, h) = \omega_2(\frac{1}{2}, f_0, 0) \\ &= \sup\left\{|\Delta_\mu^2 f_0(u)| : u, u+2\mu \in [0, 1/2]\right\}, \end{aligned}$$

which implies $\Delta_\mu^2 f_0(u) = 0$ for all $u, u + 2\mu \in [0, 1/2]$. Thus f_0 is linear on $[0, 1/2]$, and analogously on $[1/2, 1]$ so that, due to the normalizations imposed, f_0 necessarily is of the form

$$f_0(x) = \begin{cases} 2\alpha x &, \quad 0 \leq x \leq 1/2 \\ 2\alpha(1-x) &, \quad 1/2 < x \leq 1 \end{cases}$$

for some $\alpha \geq 0$. But then on the one hand side $|R_1^{Tr} f_0| = \alpha/2$, on the other hand for $0 \leq x \leq 1/2$

$$\omega_2(1/2, f_0, x) \geq |\Delta_x^2 f_0(\frac{1}{2} - x)| = 4\alpha x,$$

and analogously for $1/2 \leq x \leq 1$. Therefore

$$\tau_2(1/2, f_0) \geq 2 \int_0^{1/2} 4\alpha x \, dx = \alpha.$$

Thus $\alpha/2 = |R_1^{Tr} f_0| = \tau_2(1/2, f_0) \geq \alpha$ which necessarily implies $\alpha = 0$, a contradiction to $f \notin \mathcal{P}_1$. ∎

The assertions of Theorem 4.1 solve the problem with regard to the constants B, that is to say, with regard to the τ-modulus. What can be said for the smaller constants A, in other words, for those constants, connected with the usual (L^∞–)modulus $\omega_2(\delta, f)$ (cf. (3.3))? Apart from the fact that 1 is an upper bound (cf. (4.3)), the result is that 1 is the optimal constant only for the elementary trapezoidal rule. For all the genuine compound cases, however, the optimal constant is smaller than 1 so that for $n > 1$ our candidate 1 for the optimal value, suggested by the Jackson-type inequality (4.2), is in fact too large.

Theorem 4.2. *For the constants $A_1^{T_r}$ (cf. (3.3)) there holds true*

$$A_1^{T_r}(R) = A_1^{T_r}(C) = 1.$$

Whereas there exists an extremal element (see (4.5)) in connection with the assertion for $A_1^{T_r}(R)$, for each $f \in C[0,1] \setminus \mathcal{P}_1$ one has

$$|R_1^{T_r} f| < \omega_2(1/2, f).$$

Proof. Concerning $A_1^{T_r}(C)$, in order to show that 1 is also a lower bound (cf. (4.3)), let us first normalize the problem via

$$\begin{aligned} A_1^{T_r}(C) &= \sup_{f \in C \setminus \mathcal{P}_1} \frac{-R_1^{T_r} f}{\omega_2(1/2, f)} \\ &= \sup\Big\{\frac{If}{\omega_2(1/2, f)} : f \in C \setminus \mathcal{P}_1, f(0) = f(1) = 0\Big\} \qquad (4.4) \\ &= \sup\big\{If : f \in D := \{g \in C \setminus \mathcal{P}_1 : g(0) = g(1) = 0, \\ &\qquad\qquad \omega_2(1/2, g) = 1\}\big\}. \end{aligned}$$

For a dyadic decomposition of the interval $[0,1]$ with step size 2^{-p}, $p \in \mathbb{N}$, consider the piecewise linear function $f_p \in C \setminus \mathcal{P}_1$, determined via

$$f_p(x) := \begin{cases} 1 - 2^{-m} & , \quad x = j\,2^{-m} \text{ with } m \le p \text{ and } j \text{ odd } \ (m, j \in \mathbb{N}) \\ 0 & , \quad x = 0,\ x = 1 \end{cases}.$$

It follows that $f_p \in D$ for all $p \in \mathbb{N}$. Moreover,

$$\begin{aligned} If_p &:= \int_0^1 f_p(u)\,du = \sum_{k=1}^{2^p} 2^{-(p+1)}\big[f_p((k-1)2^{-p}) + f_p(k\,2^{-p})\big] \\ &= 2^{-p} \sum_{k=1}^{2^p} f_p((k-1)2^{-p}) = 2^{-p} \sum_{k=1}^{p} 2^{k-1}(1 - 2^{-k}) = 1 - 2^{-p} - p2^{-p-1}. \end{aligned}$$

Therefore $\lim_{p \to \infty} If_p = 1$, establishing $A_1^{T_r}(C) = 1$.

To determine $A_1^{Tr}(R)$, the preceding argument suggests to consider the following candidate

$$(4.5) \qquad f^*(x) := \begin{cases} 0 & , \quad x = 0,\ x = 1 \\ 1 - 2^{-n} & , \quad x = j\,2^{-n} \text{ for } j \text{ odd} \\ 1 & , \quad \text{elsewhere on } [0,1] \end{cases}$$

for an extremal element. Obviously, the set

$$E := \{x \in [0,1] : x = 0, x = 1 \text{ or } x = j\,2^{-n} \text{ for some } j, n \in \mathbb{N}, j \text{ odd}\}$$

is denumerable, and f^* is continuous in x if and only if $x \notin E$ since $\lim_{n\to\infty}(1 - 2^{-n}) = 1$. Therefore $f^* \in R$ by a theorem of Lebesgue. Moreover, considering cases, it follows that $\omega_2(1/2, f^*) = 1$. Since $If^* = 1$ and since the analogue to (4.4) also holds true for $A_1^{Tr}(R)$, one has $A_1^{Tr}(R) = 1$ with f^*, being an extremal element. On the other hand, in view of (2.1) and Theorem 4.1 there cannot exist an extremal element for $A_1^{Tr}(C)$. ∎

Let us finally show that the actual value of the constants $A_n^{Tr}(C)$, $A_n^{Tr}(R)$ for all the other cases $n > 1$ is strictly less than 1.

Theorem 4.3. *For each* $1 < n \in \mathbb{N}$ *one has* $1/2 \le A_n^{Tr}(C) \le A_n^{Tr}(R) \le 179/180$.

Proof. Proceeding as for the proof of Theorem 3.1, in view of (2.5)

$$A_n^{Tr}(C) \ge \frac{1}{4} \sup_{0 \ne f \in C} \frac{|R_n^{Tr} f|}{\|f\|_B} = \frac{1}{4}\left(\|Q_n^{Tr}\|_{C^*} + \|I\|_{C^*}\right) = \frac{1}{2}.$$

Concerning the upper bound let us first show that (cf. (3.4))

$$(4.6) \qquad A_2^{Tr}(R) := \sup_{f \in R \setminus \mathcal{P}_1} |R_2^{Tr} f| / \omega_2(1/4, f) \le \frac{119}{120}.$$

To this end, analogously to (4.4) one may now normalize the problem for $A_2^{Tr}(R)$ via

$$A_2^{Tr}(R) = \sup\big\{ -R_2^{Tr} f : f \in D := \{g \in R \setminus \mathcal{P}_1 : g(0) = g(1) = 0, \omega_2(1/4, g) = 1\}\big\}.$$

It then follows that with $\alpha := f(1/2)$

$$-R_2^{Tr} f = \Big(\int_0^{1/4} + \int_{1/4}^{3/4} + \int_{3/4}^{1}\Big) f(u)\,du - \frac{\alpha}{2} \qquad (f \in D).$$

To estimate the integrals, for $f \in D$

$$\begin{aligned}\int_0^{1/4} f(u)\,du &= \int_0^{1/4}\Big[-\frac{1}{2}f(0)+f(u)-\frac{1}{2}f(2u)\Big]\,du+\int_0^{1/4}\frac{1}{2}f(2u)\,du\\ &\le \frac{1}{8}\omega_2(1/4,f)+\frac{1}{4}\Big(\int_0^{1/4}+\int_{1/4}^{1/2}\Big)f(t)\,dt,\end{aligned}$$

and therefore

$$\int_0^{1/4} f(u)\,du \le \frac{4}{3}\Big[\frac{1}{8}+\frac{1}{4}\int_{1/4}^{1/2} f(u)\,du\Big]. \tag{4.7}$$

Analogously,

$$\int_{3/4}^{1} f(u)\,du \le \frac{4}{3}\Big[\frac{1}{8}+\frac{1}{4}\int_{1/2}^{3/4} f(u)\,du\Big]. \tag{4.8}$$

Moreover,

$$\begin{aligned}\int_{1/4}^{3/4} f(u)\,du &= \int_0^{1/4}\Big[f(\frac{1}{2}-h)-2f(\frac{1}{2})+f(\frac{1}{2}+h)\Big]\,dh+\frac{1}{2}f(\frac{1}{2})\\ &\le \frac{1}{4}\omega_2(1/4,f)+\frac{\alpha}{2}=\frac{1}{4}+\frac{\alpha}{2}.\end{aligned} \tag{4.9}$$

Altogether these estimates imply

$$-R_2^{T_r}f \le \frac{1}{3}+\frac{4}{3}\int_{1/2}^{3/4} f(u)\,du-\frac{\alpha}{2}\le\frac{2}{3}+\frac{\alpha}{6}. \tag{4.10}$$

On the other hand, since $-\Delta_{1/4}^2 f(0)\le 1$ and $-\Delta_h^2 f(1/4)\le 1$ for $h\in[0,1/4]$ and $f\in D$, one also has (cf. (4.9))

$$\int_{1/4}^{3/4} f(u)\,du \le \frac{3}{4}+\frac{\alpha}{4},$$

and therefore

$$-R_2^{T_r}f \le \frac{1}{3}+\frac{4}{3}\int_{1/2}^{3/4} f(u)\,du-\frac{\alpha}{2}\le\frac{4}{3}-\frac{\alpha}{6}. \tag{4.11}$$

As a consequence of (4.10,11) one may therefore note that

$$\Big|\frac{1}{6}f(\frac{1}{2})-\frac{1}{3}\Big| \le 1+R_2^{T_r}f \qquad (f\in D). \tag{4.12}$$

Suppose now that for $\varepsilon \in (0,1]$ there exists $f_\varepsilon \in D$ such that

$$-R_2^{Tr} f_\varepsilon \geq 1 - \varepsilon. \tag{4.13}$$

Without loss of generality one may assume f_ε to be symmetric to $1/2$. By (4.12) it then follows that

$$\left|f_\varepsilon(\frac{1}{2}) - 2\right| \leq 6\,(1 + R_2^{Tr} f_\varepsilon) \leq 6\,\varepsilon. \tag{4.14}$$

Since $\Delta_h^2 f_\varepsilon\big((1/2)-h\big) \leq 1$ for $h \in [0,1/4]$, this and the symmetry of f_ε imply $f_\varepsilon\big((1/2)-x\big) \leq (5/2) + 6\,\varepsilon$ for $h \in [0,1/4]$, thus

$$f_\varepsilon(x) \leq \frac{5}{2} + 6\,\varepsilon \text{ for } x \in [1/4, 3/4]. \tag{4.15}$$

The symmetry of f_ε also implies $I = 2\int_0^{1/2} f(u)\,du$. Moreover, by (4.7–9,14)

$$\int_0^{1/4} f_\varepsilon(u)\,du \leq \frac{1}{6} + \frac{1}{6}\int_{1/4}^{3/4} f_\varepsilon(u)\,du \leq \frac{1}{6} + \frac{1}{6}\Big(\frac{1}{4} + \frac{1}{2} f_\varepsilon(\frac{1}{2})\Big) \leq \frac{3}{8} + \frac{\varepsilon}{2}. \tag{4.16}$$

Since $-\Delta_{-h}^2 f_\varepsilon(1/2) \leq 1$ for $h \in [1/8, 1/4]$, one has by (4.14) that

$$\int_{1/4}^{3/8} f_\varepsilon(u)\,du = \int_{1/8}^{1/4} f_\varepsilon(\frac{1}{2} - u)\,du \leq \frac{3}{16} + \frac{3}{8}\,\varepsilon + \frac{1}{4}\int_0^{1/4} f_\varepsilon(u)\,du. \tag{4.17}$$

Therefore by (4.15–17)

$$I f_\varepsilon = 2\Big(\int_0^{1/4} + \int_{1/4}^{3/8} + \int_{3/8}^{1/2}\Big) f_\varepsilon(u)\,du \leq \frac{31}{16} + \frac{7}{2}\,\varepsilon.$$

Thus in view of (4.13,14)

$$1 - \varepsilon \leq -R_2^{Tr} f_\varepsilon = I f_\varepsilon - \frac{1}{2} f_\varepsilon(\frac{1}{2}) \leq \frac{31}{16} + \frac{7}{2}\,\varepsilon - \frac{1}{2}(2 - 6\,\varepsilon) = \frac{15}{16} + \frac{13}{2}\,\varepsilon,$$

which necessarily implies $\varepsilon \geq 1/120$. This establishes (4.6).

To show that $A_n^{Tr}(R) \leq 179/180$ for $n \geq 3$, if $n = 2m$, then in view of (2.6–7) and (4.6)

$$\begin{aligned} |R_{2m}^{Tr} f| &= \Big|\sum_{k=1}^{2m} R^{Tr[(k-1)/2m,\,k/2m]} f\Big| \leq \sum_{k=1}^{m} \big|R_2^{Tr[(k-1)/m,\,k/m]} f\big| \\ &\leq \frac{119}{120}\sum_{k=1}^{m} \frac{1}{m}\,\omega_2(\frac{1}{4m}, f)_{[\frac{k-1}{m},\frac{k}{m}]} \leq \frac{119}{120}\,\omega_2(\frac{1}{2n}, f). \end{aligned}$$

If $n = 2m+1$ for some $m \in \mathbb{N}$, then by (2.6–7), (4.6) and Theorem 4.2

$$|R^{Tr}_{2m+1}f| = \Big|\sum_{k=1}^{2m+1} R^{Tr[\frac{k-1}{2m+1},\frac{k}{2m+1}]}f\Big| \le \sum_{k=1}^{m}\Big|R_2^{Tr[\frac{2k-2}{2m+1},\frac{2k}{2m+1}]}f\Big| + \Big|R_1^{Tr[\frac{2m}{2m+1},1]}f\Big|$$

$$\le \frac{119}{120}\sum_{k=1}^{m}\frac{2}{2m+1}\,\omega_2(\frac{1}{4}\,\frac{2}{2m+1},f)_{[\frac{2k-2}{2m+1},\frac{2k}{2m+1}]}$$

$$+\frac{1}{2m+1}\,\omega_2(\frac{1}{2}\,\frac{1}{2m+1},f)_{[\frac{2m}{2m+1},1]}$$

$$\le \omega_2(\frac{1}{2n},f)\Big[\frac{119}{120}\,\frac{2m}{2m+1}+\frac{1}{2m+1}\Big] \le \frac{179}{180}\,\omega_2(\frac{1}{2n},f). \qquad \blacksquare$$

5. Concluding remarks

Let us conclude with some remarks concerning the compound Simpson rule

$$Q^{Si}_n f := \frac{1}{6n}\sum_{k=1}^{n}\Big[f(\frac{k-1}{n}) + 4\,f(\frac{2k-1}{2n}) + f(\frac{k}{n})\Big], \qquad R^{Si}_n := Q^{Si}_n - I.$$

Again for the remainder there holds true the explicit representation (see [2])

$$R^{Si}_n f := \frac{2}{3}\sum_{k=1}^{n}\int_0^{1/4n}\Big[\Delta^4_h f(\frac{k-1}{n}) + \frac{2}{3}\,\Delta^4_h f(\frac{2k-1}{2n} - 2h) + \Delta^4_{-h} f(\frac{k}{n})\Big]\,dh,$$

the fourth differences precisely corresponding to the fact that the Simpson rule is exact on $\mathcal{P}_3$. Based upon this representation and (2.1) it can then be shown that for every $f \in R[0,1]$ and $n \in \mathbb{N}$ one has the Jackson-type inequality

$$|R^{Si}_n f| \le \frac{4}{9}\,\tau_4(1/2n, f) \le \frac{4}{9}\,\omega_4(1/4n, f).$$

See also [3], [9, p. 46], where these estimates are obtained via general interpolation (K-functional) methods, but with unspecified constants. Analogously to (3.3), one may now introduce optimal constants A^{Si}_n, B^{Si}_n for the Simpson rule and pose the question whether $4/9$ is a candidate for the actual value of at least some of the relevant quantities. This remains an open problem. For example, concerning

$$A^{Si}_n(C) := \sup\Big\{\frac{|R^{Si}_n f|}{\omega_4(1/4n, f)} : f \in C[0,1]\setminus \mathcal{P}_3\Big\},$$

apart from the upper bound $4/9$, we are at the moment only able to show (by some tedious calculations) that $10/27$ performs a lower bound for $A^{Si}_1(C)$.

References

1. H. Brass, *Quadraturverfahren.* Vandenhoeck und Ruprecht, Göttingen 1977.

2. B. Büttgenbach, G. Lüttgens, R.J. Nessel, *An explicit representation of the remainder of some Newton–Cotes formulas in terms of higher order differences.* To appear.

3. P.L. Butzer, K. Scherer, U. Westphal, *On the Banach–Steinhaus theorem and approximation in locally convex spaces.* Acta Sci. Math. (Szeged) **34** (1973), 25–34.

4. I.K. Daugavet, *A property of completely continuous operators in the space* C (Russian). Uspekhi Mat. Nauk **18** (1963), 157–158.

5. S. Filippi, H. Esser, *Eine Reihe neuer Sätze und Ergebnisse zur numerischen Quadratur.* Elektron. Datenverarbeitung **11** (1969), 166–180.

6. V.T. Gavriljuk, S.B. Stečkin, *Approximation of continuous periodic functions by Fourier sums.* Soviet Math. Dokl. **19** (1978), 882–884.

7. K.G. Ivanov, *New estimates of errors of quadrature formulae, formulae of numerical differentiation and interpolation.* Anal. Math. **6** (1980), 281–303.

8. H. Lebesgue, *Sur la représentation trigonométrique approchée des fonctions satisfaisant à une condition de Lipschitz.* Bull. Soc. Math. France **38** (1910), 184–210.

9. Bl. Sendov, V.A. Popov, *The Averaged Moduli of Smoothness.* Wiley, Chichester 1988.

B. Büttgenbach, G. Lüttgens, and R.J. Nessel, Lehrstuhl A für Mathematik, RWTH Aachen, Templergraben 55, D-5100 Aachen, Germany

References

1. [illegible]

2. [illegible]

8. [illegible]

9. [illegible] Math. Appl. [illegible]

10. [illegible]

International Series of Numerical Mathematics, Vol. 103, © 1992 Birkhäuser Verlag Basel

The uncertainty inequality in quantum holography, Bohr's indeterminacy principle, and synchronized neural networks

Chen Han-lin and Walter Schempp

Some ten years after writing his fundamental papers on optics, Hamilton made a startling observation: that the same formalism applies to mechanics of point particles. Replace the optical axis by the time. Then the transformation from initial position and momenta to final position and momenta is always symplectic. This discovery led to remarkable progress in the nineteenth century. In the 1920s — almost a century later — Hamilton's analogy between optics and mechanics served as one of the major clues in the discovery of quantum mechanics.

Shlomo Sternberg (1988)

Great progress was possible thanks to the wonderful tool in atomic physics which is the laser.

Alain Aspect (1986)

Abstract

A rigorous proof of quantum parallelism cannot be based on the Heisenberg inequality because the standard deviation is insensitive to fine structures of the interference pattern generated by a Mach-Zehnder interferometer. Actually Niels Bohr's indeterminacy principle cannot be based on any of the known uncertainty principles.

It is shown how the holographic transform allows to circumvent the difficulties with the standard deviation by using a group theoretical implementation of the canonical commutation relation of quantum electrodynamics. The geometric quantization approach makes the heuristic arguments concerning

This paper is in final form and no version of it will be submitted for publication elsewhere.

quantum parallelism rigorous by considering wave packets as symplectic spinors on the hologram plane. It includes the standard uncertainty inequality as a special case.

As an application to the area of quantum computers, we study synchronized neural networks and their photonic implementations as suggested by recent experiments in brain research. As another result of the geometric quantization approach to holography, the uncertainty minimizing Gabor wavelets arise which provide useful wavelet expansions for image analysis, segmentation, and compression.

The Schrödinger equation preserves the norm of the wave function and thus the number of particles. Now it is well known that in the presence of sources, photons can be absorbed or emitted. Thus one cannot introduce a Schrödinger equation for a single photon in the presence of sources. In fact, the electromagnetic field itself must be quantized, and photons then occur as elementary excitations of the quantized field.

Claude Cohen-Tannoudji (1989)

Each photon bears information about the entire system.

H. John Caulfield (1991)

1. Holographic imaging

Optical holography is a two-step coherent imaging process. In the first processing step, the writing step, the three-dimensional image is spatially encoded in both amplitude and phase by an interference pattern in the hologram plane. The interference pattern arises by coherently mixing the diffuse reference beam of frequency ν and the beam coherently scattered by the object. Thus the hologram forms a linear superposition of interference fringe patterns which are generated by phase-shifted beam splitter interferometer experiments.

The result of these experiments are photonic excitation distributions recording the raw optical data that are transmitted by coherent optical channels. In terms of geometric quantization, the symplectic hologram plane $(\mathbb{R} \oplus \mathbb{R}, \Omega_\nu)$ carries the structure of a metaplectic manifold.

In the second processing step, the readout step, the raw optical data encoded by the hologram as an excitation distribution are spatially decoded by coherent diffuse light illumination. After slice selection by the Bragg angle condition similar to the spin slice selection procedure in magnetic resonance imaging MRI [6], the three-dimensional image of the object is reconstructed by coherent diffraction in both amplitude and phase. According to the geometric quantization procedure, the reconstruction is implemented by the Hilbert space bundle associated with the metaplectic manifold structure of the symplectic hologram plane $(\mathbb{R} \oplus \mathbb{R}, \Omega_\nu)$.

Both the writing and readout steps are of a quantum theoretical character. The quantum parallelism according to which different alternatives at the quantum level are allowed to coexist in linear superposition, irrespective of how different from one another the states might be, is equivalent to Niels Bohr's indeterminacy principle which says that in the phase shifted beam splitter interferometer experiment the interference pattern appears if and only if we cannot determine the paths of the optical photons. Bohr established his conclusion by a heuristic application of the uncertainty principle.

The uncertainty principle occupies a peculiar position in physics and in signal processing. Since the fundamental 1927 paper by Werner Heisenberg [7], it is often considered as the hallmark of quantum mechanics. In his Chicago lectures of spring 1929, he regarded the inequality

$$(\Delta q)_\psi \cdot (\Delta p)_\psi \geq \frac{1}{2}\,\hbar$$

as the precise mathematical expression of the uncertainty principle within the formalism of quantum mechanics [8]. On the other hand, although it is easy to derive it in a rigorous mathematical manner, there is still a great deal of discussion about its interpretation in physics and information theory [9], [10], [17], [18]. It turns out that the preceding inequality has a number of weaknesses, however, particularly related to the fact that the standard deviations $(\Delta q)_\psi$ and $(\Delta p)_\psi$ only give very general information about the spreads of the probability density functions of position and momentum, respectively. Moreover, in quantum holography we find that $(\Delta q)_\psi$ or $(\Delta p)_\psi$ are divergent. Therefore the inequality is not appropriate for the purposes of quantum holography. In fact, it does not conclusively establish Niels Bohr's undeterminacy principle which states that in the Mach-Zehnder interferometer "the interference pattern appears if and only if we cannot determine the paths of the optical photons."

> *The correct way to understand the uncertainty relations is to see that they represent an inherent limitation on the sort of states which can be produced for quantum mechanical systems, but that they cannot be 'explained' or 'deduced' by the naive disturbance argument.*
>
> Michael Redhead (1987)

2. The uncertainty inequality

Let ψ denote the state vector of a one-dimensional quantum-mechanical system. Thus $\psi \in L^2(\mathbb{R}; dx)$ satisfies the normalization condition

$$||\psi||_2 = \left(\int_{\mathbb{R}} |\psi(x)|^2 \, dx \right)^{1/2} = 1.$$

The position and momentum translation operators are defined by the $\mathbb{C}$-linear mappings

$$q : \psi \mapsto x \cdot \psi,$$

and

$$p : \psi \mapsto i\hbar\psi',$$

where ψ' denotes the distributional derivative of ψ, h denotes Planck's constant, and as usual in quantum theory, $\hbar = h/2\pi$. Suppose

$$\int_{\mathbb{R}} |x|^2 |\psi(x)|^2 \, dx < +\infty$$

and

$$\int_{\mathbb{R}} |\xi|^2 |\mathcal{F}\psi(\xi)|^2 \, d\xi < +\infty,$$

where $\mathcal{F}\psi$ denotes the Fourier transform of ψ in the real variable ξ which is dual to the real variable x. Hence

$$\mathcal{F}\psi(\xi) = \int_{\mathbb{R}} \psi(x) e^{-2\pi i \xi x} \, dx$$

in the dual state space $L^2(\mathbb{R}; dx)$. The expected values of position and momentum are given by

$$\langle q \rangle_\psi = \int_{\mathbb{R}} x |\psi(x)|^2 \, dx,$$

$$\langle p \rangle_\psi = \int_{\mathbb{R}} \frac{\xi}{h} \left| \mathcal{F}\psi\left(\frac{\xi}{h}\right) \right|^2 d\xi,$$

and their standard deviations are the expressions

$$(\Delta q)_\psi = \left(\int_{\mathbb{R}} |x - \langle q \rangle_\psi|^2 |\psi(x)|^2 \, dx \right)^{1/2},$$

$$(\Delta p)_\psi = \left(\int_{\mathbb{R}} |\xi - \langle p \rangle_\psi|^2 \, \frac{1}{h} \left| \mathcal{F}\psi\left(\frac{\xi}{h}\right) \right|^2 d\xi \right)^{1/2}.$$

The Heisenberg inequality

$$\int_{\mathbb{R}} x^2|\psi(x)|^2\, dx \cdot \int_{\mathbb{R}} \xi^2|\mathcal{F}\psi(\xi)|^2\, d\xi \geq \frac{1}{16\pi^2}\left(\int_{\mathbb{R}} |\psi(x)|^2\, dx\right)^{1/2},$$

which is a direct consequence of the Plancherel theorem combined with the Cauchy-Schwarz inequality implies the uncertainty inequality

$$(\Delta q)_\psi \cdot (\Delta p)_\psi \geq \frac{1}{2}\hbar,$$

mentioned in Section 1 supra. It follows from the Cauchy-Schwarz inequality that in the preceding inequality identity holds if and only if

$$\psi'(x) = c \cdot x\bar{\psi}(x) \qquad (c \in \mathbb{C})$$

holds in $L^2(\mathbb{R}; dx)$, i.e., if and only if $\psi \in L^2(\mathbb{R}; dx)$ is a vacuum coherent state vector

$$\psi(x) = C \cdot e^{-|c|x^2/2} \qquad (x \in \mathbb{R}),$$

where the constant $C \in \mathbb{C}$ in front of the Gaussian wave packet is determined by the condition $||\psi||_2 = 1$ up to a phase factor. The uncertainty minimizing property of the coherent states is an important ingredient of the Gabor wavelet expansions [3] for image analysis, segmentation and compression; see Section 5 infra.

> *An advantage of the Heisenberg point of view is that it allows one to define 'two-time averages', that is to say, mean values in the state of the system of a product of two operators taken at two different times t and t'. Important examples of two-time averages are the symmetric correlation functions and the linear response functions.*
>
> Claude Cohen-Tannoudji (1989)

> *Something stronger is needed to uphold Bohr's conclusion.*
>
> John F. Price (1985)

3. Quantum holography

The main idea to avoid the difficulties involved in the standard deviation Δ is to start off with the quantum mechanical commutation relation

$$q \circ p - p \circ q = i\hbar$$

and to put it in a group theoretical context. Let G denote the connected Lie group [20]

$$G = \left\{ \begin{pmatrix} 1 & x & z \\ 0 & 1 & \xi \\ 0 & 0 & 1 \end{pmatrix} \mid x, \xi, z \in \mathbb{R} \right\}$$

with the real Lie algebra $\mathbf{g}$ spanned by the set $\{P, Q, Z\}$ of matrices

$$P = \begin{pmatrix} 0 & 1 & 0 \\ 0 & 0 & 0 \\ 0 & 0 & 0 \end{pmatrix}, \quad Q = \begin{pmatrix} 0 & 0 & 0 \\ 0 & 0 & 1 \\ 0 & 0 & 0 \end{pmatrix}, \quad Z = \begin{pmatrix} 0 & 0 & 1 \\ 0 & 0 & 0 \\ 0 & 0 & 0 \end{pmatrix},$$

which satisfy the commutator condition

$$[P, Q] = PQ - QP = Z.$$

For any frequency $\nu \neq 0$, the Heisenberg nilpotent Lie group G admits an irreducible unitary linear representation U_ν, unique up to unitary isomorphy, acting on $\psi \in L^2(\mathbb{R}; dx)$ according to the prescription

$$U_\nu \left(\begin{pmatrix} 1 & x & z \\ 0 & 1 & \xi \\ 0 & 0 & 1 \end{pmatrix} \right) \psi(t) = e^{2\pi i \nu (z + \xi t)} \psi(t - x) \qquad (t \in \mathbb{R}).$$

Then the Heisenberg inequality takes the form of a Robertson relation [19], [18]

$$\Delta U_\nu(P) \cdot \Delta U_\nu(Q) \geq \frac{1}{2} |U_\nu(Z)|.$$

The mapping

$$\psi(t')\, dt' \otimes \varphi(t)\, dt \mapsto H_\nu(\psi, \varphi; x, \xi)\, dx \wedge d\xi,$$

which is called the holographic transform of the two-mixing wave packet densities $\psi(t')\, dt'$ and $\varphi(t)\, dt$ describes the interference pattern in the symplectic hologram plane $(\mathbb{R} \oplus \mathbb{R}, \Omega_\nu)$. Directional derivatives identify the plane $(\mathbb{R} \oplus \mathbb{R}, \Omega_\nu)$ with each tangent plane to the natural linear symplectic manifold modelled on $(\mathbb{R} \oplus \mathbb{R}, \Omega_\nu)$. In the preceding formula, the standard symplectic form

$$\Omega_\nu = \nu \cdot dx \wedge d\xi = \nu \cdot d(x \cdot d\xi) \qquad (\nu \neq 0)$$

carries the frequency ν as its scaling factor. The coefficient $H_\nu(\psi, \varphi; x, \xi)$ of Ω_ν on the hologram plane $(\mathbb{R} \oplus \mathbb{R}, \Omega_\nu)$ represents a cross-correlation function and takes the form of an entry function of the representation U_ν:

$$H_\nu(\psi, \varphi; x, \xi) = \int_{\mathbb{R}} \psi(t - x) \bar{\varphi}(t) e^{2\pi i \nu \xi t}\, dt = \left\langle U_\nu \left(\begin{pmatrix} 1 & x & 0 \\ 0 & 1 & \xi \\ 0 & 0 & 1 \end{pmatrix} \right) \psi \middle| \varphi \right\rangle$$

The integration along the line $\mathbb{R}$ supposes temporal coherence of the mixing wave packet densities. The Poisson summation type operator

$$w_1 : \psi \mapsto \left(\begin{pmatrix} 1 & x & z \\ 0 & 1 & \xi \\ 0 & 0 & 1 \end{pmatrix} \mapsto e^{2\pi i z} \sum_{m \in \mathbb{Z}} e^{2\pi i \xi m} \psi(m - x) \right)$$

maps the linear Schrödinger representation U_1 of G onto the right translations δ_1 of G such that

$$H_1(\psi, \varphi; x, \xi) = \left\langle \delta_1 \left(\begin{pmatrix} 1 & x & 0 \\ 0 & 1 & \xi \\ 0 & 0 & 1 \end{pmatrix} \right) w_1(\psi) \middle| w_1(\varphi) \right\rangle$$

holds. This identity gives rise to a "distributed" form of the holographic transform $\psi(t')\, dt' \otimes \varphi(t)\, dt \mapsto H_1(\psi, \varphi; x, \xi) \cdot \nu\, dx \wedge d\xi$. It allows to introduce the notion of holographic fractal [21] and to rigorously establish Niels Bohr's indeterminacy principle. As a consequence, it is impossile to construct a position operator for the optical photon. However, more than Bohr's conclusion about quantum parallelism in transmission channels actually is true.

Wir haben entdeckt, daß die Antworten von Merkmalsdetektoren rhythmisch sind und mit einer mittleren Frequenz von etwa 40 Hertz oszillieren. Wir haben ferner beobachtet, daß räumlich verteilt liegende Gruppen von Merkmalsdetektoren ihre rhythmischen Aktivitäten synchronisieren können und dann in Phase schwingen. Solche Synchronisationen traten besonders häufig auf zwischen Zellgruppen, die ähnliche Merkmale codieren, also zum Beispiel zwischen Neuronen, die ähnliche Richtungs- und Orientierungspräferenzen aufweisen. Die oszillierenden Antworten räumlich verteilter Merkmalsdetektoren beginnen in Phase zu schwingen, wenn im Bereich ihrer rezeptiven Felder Konturen angeboten werden, die sich mit gleicher Geschwindigkeit in die gleiche Richtung bewegen.

Wolf Singer (1990)

Multilayer optical neural network architectures based on storing weights as holographic gratings in photorefractive crystals have been proposed which are capable of implementing such neural network paradigms as backward propagation and simulated annealing.

Yuri Owechko (1989)

4. Synchronized neural networks

The preceding identity shows that the information distributed by the holographic encoding procedure on the natural linear symplectic manifold structure of the symplectic hologram plane $(\mathbb{R}\oplus\mathbb{R}, \Omega_\nu)$ is a non-local one. Similar experimental results in brain research concerning synchronized connections in neural oscillator layers are supporting the holographic hypothesis of memory structure in brain function and perception [1], [15], [16]: Cell assemblies coding for coherent features in visual scenes may not be distinguished by the fact that the constituting neurons are particularly active. Rather it appears that such cell assemblies are characterized by synchronization of oscillatory responses over considerable tangential distances across spatially separate functional columns and even between spatially separate regions of the cortex. In particular, these studies present experimental evidence that temporal coherence is a basic principle of cortical self-organization [5], [22]. From the theory of symplectic spinors [13] it is known that the symplectic hologram plane $(\mathbb{R} \oplus \mathbb{R}, \Omega_\nu)$ forms the base of a fiber bundle, the fiber at each of its points (x_0, ξ_0) being a copy of the three-dimensional Heisenberg nilpotent group G. More precisely, a choice of a symplectic frame in a tangent plane at (x_0, ξ_0) to the natural symplectic manifold modelled on $(\mathbb{R} \oplus \mathbb{R}, \Omega_\nu)$ determines a definite Lie group isomorphism of a central extension [20] of the tangent plane onto G. This definite isomorphism models the information redundancy of the holographic encoding procedure: Each part of the optical hologram contains all the information stored by the whole optical hologram. In view of Niels Bohr's indeterminacy principle, the excitation distribution spatially encoded by the hologram generates a synchronized neural network modelled by the associated Hilbert space bundle. The associative holographic memory of this adaptive optical neural network can be decoded by the fundamental read out formula

$$\iint_{\mathbb{R}\oplus\mathbb{R}} H_\nu(\psi, \varphi; x, \xi) e^{-2\pi i\nu\xi t} \bar{\psi}(t - x) \cdot \nu \, dx \, d\xi = \bar{\varphi}(t) \cdot \nu \, dt \qquad (\|\psi\|_2 = 1).$$

The read out formula which is written for an arbitrary fiber represents the synchronized response of the neural network and therefore the associative recall [2]. The phase conjugation involved allows to implement the backpropagating learning mode of the synchronized neural network by degenerate four-wave mixing. In this way, the Owechko-Soffer resonator [14], [23], [24], for instance, efficiently uses the associative holographic memory function of volume holograms stored in photorefractive crystals [4] in the fields of pattern recognition and pattern completion. Other completely different recording media for optical holograms are certain specific dyes at extremely low temperatures of 2°–3°K. They allow to implement molecular computers by the spectral hole-burning procedure [25], [6]. Because such processors can be considered as quantum computers, quantum holography provides a new approach to the highly active area of quantum computation.

> *Ein Computer ist eine Organisationsform elementarer funktionaler Komponenten, an denen für das Verhalten des ganzen Systems nur ihre Funktion relevant ist.*
>
> Herbert A. Simon (1990)

5. Gabor wavelets

If V and W denote two transversal Lagrangian lines of the tangent space at the point (x_0, ξ_0) to the natural symplectic manifold modelled on $(\mathbb{R} \oplus \mathbb{R}, \Omega_\nu)$, the one-dimensional complex vector space of tempered distributions on the plane $\mathbb{R} \oplus \mathbb{R}$ annihilated by the vaccum state $U_1|V \oplus W$ for the Lagrangian subspace $V \oplus W$ of $\mathfrak{g}$ is spanned by the Gaussian envelope at (x_0, ξ_0):

$$(x, \xi) \mapsto e^{-\pi(a^2(x-x_0)^2 + b^2(\xi-\xi_0)^2)},$$

where the real coefficients a, b determine a positive semi-definite quadratic form at the point (x_0, ξ_0). Lifting from the polarized cross-section to the central extension [20] at $(x_0, \xi_0) \in \mathbb{R} \oplus \mathbb{R}$ yields the resolution generator or analyzing Gabor wavelet

$$(x, \xi) \mapsto e^{2\pi i(u(x-x_0)+v(\xi-\xi_0))} \cdot e^{-\pi(a^2(x-x_0)^2 + b(\xi-\xi_0)^2)},$$

where the real modulation parameters u, v determine a phase factor at the position (x_0, ξ_0). This family of complex-valued functions on $\mathbb{R} \oplus \mathbb{R}$ for which identity holds in the Heisenberg inequality for $L^2(\mathbb{R} \oplus \mathbb{R})$ give rise to expansions which provide excellent fits to the empirical simple cell neural receptive field profiles [11], [12].

Acknowledgments. The first author would like to thank the Max-Planck-Gesellschaft for awarding him with a stipendium for studying High Tech Math at the Lehrstuhl für Mathematik I der Universität Siegen. The second author would like to thank the Rheinische Friedrich-Wilhelms-Universität at Bonn and Professor Dr. Michael Clausen (Informatik V), whose invitation to initiate the Informatik-Kolloquium in Bonn and to teach an advanced course on high resolution imaging at the Institute of Computer Science provided the occasion for writing a preliminary version of this article. Finally, valuable discussions on optical holography with Professors H. John Caulfield, Alex T. Granik, and Emmett N. Leith, and on molecular computers with Professor Urs P. Wild are gratefully acknowledged. It is a pleasure to thank all of them for their advice and tutorial comments.

References

1. R.J. Baron, M. Nuwer and K.H. Pribram, *The holographic hypothesis of memory structure in brain function and perception.* Contemporary Developments in Mathematical Psychology, Vol. II: Measurement, Psychophysics, and Neural Information Processing (R.C. Atkinson, D.H. Krantz, R.D. Luce, P. Suppes, Eds.), pp. 416–457. W.H. Freeman, San Francisco, 1974.

2. H.J. Caulfield, *Optical neural networks.* Real-Time Signal Processing for Industrial Applications, Critical Reviews of Optical Science and Technology (Bahram Javidi, Ed.), SPIE Vol. 960, pp. 242–252 (1989).

3. J.G. Daugman, *Entropy reduction and decorrelation in visual coding by oriented neural receptive fields.* IEEE Trans. Biomedical Eng. **36** (1989), 107–114.

4. J. Feinberg and K.R. MacDonald, *Phase-conjugate mirrors and resonators with photorefractive materials.* Photorefractive Materials and Their Applications II (P. Günther, J.-P. Huignard, Eds.), pp. 151–203. Springer, Berlin, Heidelberg, New York, London, Paris, Tokyo, 1989.

5. C.M. Gray and W. Singer, *Stimulus-specific neuronal oscillations in orientation columns of cat visual cortex.* Proc. Natl. Acad. Sci. USA **86** (1989), 1698–1702.

6. E.L. Hahn, *NMR and MRI in retrospect.* Phil. Trans. R. Soc. London A **333** (1990), 403–411.

7. W. Heisenberg, *Über den anschaulichen Inhalt der quantenmechanischen Kinematik und Mechanik.* Z. Phys. **43** (1927), 172-198.

8. W. Heisenberg, *Physikalische Prinzipien der Quantentheorie.* BI-Wissenschaftsverlag, Mannheim, Wien, Zürich, 1958.

9. J. Hilgevoord and J.B.M. Uffink, *The mathematical expression of the uncertainty principle.* Microphysical Reality and Quantum Formalism (A. van der Merwe, F. Selleri, G. Tarozzi, Eds.), pp. 91–114. Kluwer Academic Publishers, Dordrecht, Boston, London, 1988.

10. J. Hilgevoord and J.B.M. Uffink, *A new view on the uncertainty principle.* International School of History of Science, Second Course: Sixty-Two Years of Uncertainty: Historical, Philosophical and Physical Inquiries into the Foundations of Quantum Mechanics, Erice, Sicily, 1989.

11. J.P. Jones and L.A. Palmer, *The two-dimensional spatial structure of simple receptive fields in cat striate cortex.* J. Neurophys. **58** (1987), 1187–1211.

12. J.P. Jones and L.A. Palmer, *An evaluation of the two-dimensional Gabor filter model of simple receptive fields in cat striate cortex.* J. Neurophys. **58** (1987), 1233–1258.

13. B. Kostant, *Symplectic spinors.* Symposia Math., Vol. XIV, pp. 139–152. Academic Press, New York, London, 1974.

14. Y. Owechko, *Nonlinear holographic associative memories.* IEEE J. Quantum Electronics **25** (1989), 619–634.

15. K.H. Pribram, *Languages of the Brain.* 5th ed. Brandon House, New York, 1982.

16. K.H. Pribram, *Holography and brain function.* Encyclopedia of Neuroscience, Vol. I (G. Adelman, Ed.), pp. 499–500. Birkhäuser, Basel, Boston, Stuttgart, 1987.

17. J.F. Price, *Uncertainty principles and interference patterns.* Proc. Centre for Mathematical Analysis, Australian National University, Vol. 9 (1985), 241–258.

18. M. Redhead, *Incompleteness, Nonlocality, and Realism.* Oxford Univ. Press, Oxford, New York, Toronto, 1987.

19. H.P. Robertson, *The uncertainty principle.* Phys. Rev. **34** (1929), 163–164.

20. W. Schempp, *Harmonic Analysis on the Heisenberg Nilpotent Lie Group, with Applications to Signal Theory.* Pitman Research Notes in Math., Vol. 147. Longman Scientific and Technical, Harlow, Essex, 1986.

21. W. Schempp, *Holographic fractals.* Signal Theory and Image Processing (W. Schempp, D.C. Struppa, Eds.). Mediterranean Press, 1992. To appear.

22. W. Singer, *Search for coherence: A basic principle of cortical self-organization.* Concepts in Neuroscience **1** (1990), 1–26.

23. B.H. Soffer, G.J. Dunning, Y. Owechko and E. Marom, *Associative holographic memory with feedback using phase-conjugate mirros.* Opt. Lett. **11** (1986), 118-120.

24. B.H. Soffer, Y. Owechko and G.J. Dunning, *A photorefractive optical neural network.* Optics in Complex Systems (F. Lanzl, H.-J. Preuss, G. Weigelt, Eds.). Proc. SPIE **1319** (1990), 196–197.

25. U.P. Wild, *Spectral hole-burning: Applications to data storage and processing — the molecular computer.* Gordon Research Conference on Holography and Optical Information Processing, University of Rhode Island, Kingston, RI, 1991.

Chen Han-lin and Walter Schempp, Lehrstuhl für Mathematik I, Universität Siegen, P.O. Box 10 12 40, D-5900 Siegen

International Series of Numerical Mathematics, Vol. 103, © 1992 Birkhäuser Verlag Basel

Optimization and inequalities on product spaces

Don-lin Mon and Chung-lie Wang

Abstract

Basic inequalities, such as the Hölder inequality, Minkowski inequalities, and the weighted mean inequality, etc., concerning mixed versions of discrete and continuous models, are established. By these newly formed inequalities, abundant optimization problems can be readily solved.

1. Introduction

In view of the facts gradually gathering in the literature (see, e.g. [1–3, 7–11, 14, 18–34] we can surely say that an inequality can be used to establish many optimization problems while an optimization problem may be solved by various inequalities. In other words, the development of the theory of inequalities has been closely related to that of finding techniques for solving optimization problems.

In this regard, we should recall that in 1966, one of the most interesting generalization of the discrete Hölder inequality due to E.F. Beckenbach was proved as Theorem 1 of [1, p. 24] by an optimization technique. We now cite it from [1] (see also Mitrinović [14, p. 52]) as follows.

Theorem 1. *Let positive reals c_i, $1 \le i \le m$, and k_j, $1 \le j \le n$, with $0 \le m \le n$ be given and let p and q satisfy $p^{-1} + q^{-1} = 1$. Then for $p > 1$, the Beckenbach inequality*

$$B(x) \ge B(d), \tag{1}$$

where

$$B(x) = \left(\sum_{j=1}^{m} c_j^p + \sum_{j=m+1}^{n} x_j^p \right)^{1/p} \left(\sum_{j=1}^{m} c_j k_j + \sum_{j=m+1}^{n} x_j k_j \right)^{-1} \tag{2}$$

This paper is in final form and no version of it will be submitted for publication elsewhere.

and

$$d_i = \left(\sum_{j=1}^{m} c_j^p k_i / \sum_{j=1}^{m} c_j k_j \right)^{q/p}, \quad m+1 \le i \le n,$$

holds for all nonnegative $x_{m+1}, \ldots, x_n$. *The sign of inequality in* (1) *is reversed for* $0 < p < 1$. *In either case, the sign of equality holds iff*

$$x_i = d_i, \quad m+1 \le i \le n.$$

In 1979, Wang proved Theorem 1 by the method of applying the convexity of a real function of several variables [18] and the functional equation approach of dynamic programming [25] respectively. In 1983, Wang extended the discrete model (2) to its continuous model and successively developed techniques to further establish both cases and their variants (see, e.g. Iwamoto and Wang [9], Wang [28, 30]).

In this paper, we shall consider optimization problems of the type $B(x)$ given in (2), but with a mixture of discrete and continuous arguments. The setting of the mixture of discrete and continuous models concerning basic inequalities appears to be new and useful. This is the motivation of this paper.

To this end, in Section 2, we shall summarize notation, symbols and definitions that will be used. In Section 3, we shall present several theorems concerning the characteristics of nonlinear positive functionals on product spaces. In the following sections, we shall successively establish the mixed versions of some basic inequalities, such as the Hölder inequality, Minkowski inequalities, etc. and conclude with some examples and remarks.

2. Notation, symbols and definitions

Let us begin by displaying some notation and symbols that we shall need:

$\mathbb{R}$ = the field of real numbers,

$\mathbb{R}^+ = \{t \in \mathbb{R} \mid t > 0\}$,

$\mathbb{R}^n = \{x = (x_1, \ldots, x_n)^\top \mid x_j \in \mathbb{R},\ j = 1, \ldots, n\}$,

$E = \pi_{j=1}^m B_j$ = product of a finite number of Banach spaces (see, e.g. Cullen [5, pp. 65–72] for definitions and characteristics of product spaces), where B_j may be a general Banach space, $j = 1, \ldots, m$. However, we only consider $B_j = E_p^{(j)}$, where $E_p^{(j)}$ may be $\mathbb{R}^n$, l^p (p-th power summable sequence spaces) and L^p (p-th power integrable function spaces), $j = 1, \ldots, m$ (see, e.g. Kreyszig [13, p. 11, p. 65] for definitions of l^p, L^p, etc. and their properties). For each $E_p^{(j)}$, we use $|b_j|_{j,p}$ to indicate its p-norm. For each $z \in E_p$ with $z = (b_1, \dots, b_m)^\top$, we define a p-norm on E_p as follows

$$\|z\|_p = \left(\sum_{j=1}^{m} |b_j|_{j,p}^p \right)^{1/p}, \quad p > 1. \tag{3}$$

Consequently, E_p is itself a Banach space in association with the norm $\|z\|_p$ given in (3).

U = a suitable (such as open convex, compact, etc.) subset of E_p,

$\mathcal{F} = \{F \mid F : U \to \mathbb{R}^+\}$ = the set of positive functionals on U,

$\varphi(t) = F(z + th)$, $F \in \mathcal{F}$, $z \in U$, $h \in E_p$, $t \in \mathbb{R}$, $z + th \in U$,

$$\varphi'(0) = \frac{d}{dt} F(z + th)|_{t=0} = \langle \partial_z F, h \rangle = \langle \partial F, h \rangle,$$

$$\varphi''(0) = \frac{d^2}{dt^2} F(z + th)|_{t=0} = \langle \partial_z^2 F, h^2 \rangle = \langle \partial^2 F, h^2 \rangle = Q,$$

where $\varphi'(0)$ and $\varphi''(0)$ are said to be the first and second Gateaux differentials (see, e.g. Vainberg [16, p. 35]) respectively and $\langle , \rangle$ is a usual bilinear form (or inner product).

Remark. As a demonstration of the preciseness of Gateaux differentials $\varphi'(0)$ and $\varphi''(0)$, we consider the functional $F : z = (x, f) \in \mathbb{R}^n \times L(a, b) \to \mathbb{R}^+$, where

$$F = F(z) = F(x, f).$$

Setting $\varphi(t) = F(z_0 + th)$, with $h = z - z_0$, we have

$$\varphi'(0) = \langle \partial_z F(z_0), h \rangle = \langle \partial_x F(z_0), x - x_0 \rangle + \langle \partial_f F(z_0), f - f_0 \rangle,$$

$$\varphi''(0) = \langle \partial_z^2 F(z_0), h^2 \rangle = \langle \partial_x^2 F(z_0), (x - x_0)^2 \rangle + \langle \partial_f^2 F(z_0), (f - f_0)^2 \rangle.$$

Definition 1. The functional $A : U \to \mathbb{R}$ is affine if for every $z \in U$, $A(z) = L(z) + c$, where $L : U \to \mathbb{R}$ is linear and c is a constant in $\mathbb{R}$.

Definition 2. A functional $F : U \to \mathbb{R}^+$ has support at $w \in U$ if there is an affine functional $A : U \to \mathbb{R}$ such that $A(w) = F(w)$, and $A(z) \le F(z)$ (resp. $A(z) \ge F(z)$) for every $z \in U$.

We also refer to Roberts and Varberg [15] and Vainberg [16] for the properties and definitions of the Gateaux differentials used throughout the paper (unless specified otherwise).

For the classical inequalities cited here without mentioning any source, one should refer to Beckenbach et al. [2], Hardy et al. [7], or Mitrinović [14] for details.

3. Theorems concerning optimization and convexity

In correspondence with the definition and characteristics of the Gateaux differential of a nonlinear operator given in Vainberg [16, pp. 35–53], we can present the same for the Gateaux differential $\langle \partial_z F, w \rangle$, $z, w \in U$, of a nonlinear positive functional F of $\mathcal{F}$. However, we omit the almost duplicated detail and refer to Vainberg [16] for it.

We now adopt the notion of Vainberg [16] and produce several theorems (see also Courant and John [4]; Roberts and Varberg [15] for comparison) concerning nonlinear positive functionals on product spaces.

Theorem 2. *Let F have continuous second Gateaux differential on U. Then a point $w \in U$ at which $\partial F(w)$ vanishes is a maximum point of F if $\partial^2 F(w) < 0$ throughout U, and a minimum point of F if $\partial^2 F(w) > 0$ throughout U.*

Theorem 3. *Let F have continuous second Gateaux differential on U. Then F is convex (resp. concave) iff the quadratic form Q is nonnegative (resp. nonpositive) for all $z \in U$, $h \in E$. Moreover, if Q is positive (resp. negative) on U, then F is strictly convex (resp. concave).*

Theorem 4. *The nonlinear positive functional F is convex (resp. concave) on U iff F has support $A(z)$ at each point $w \in U$, $A(w) = F(w)$ and $A(z) \le F(z)$ (resp. $A(z) \ge F(z)$) for every $z \in U$.*

In view of the Taylor formula of F

$$F(z) = F(z_0) + \langle \partial F(z_0), h \rangle + \frac{1}{2} \langle \partial^2 F(z_0 + \theta h), h^2 \rangle, \tag{4}$$

where $z, z_0 \in U$, $h = z - z_0$, $0 < \theta < 1$, proofs of Theorems 2–4 can be readily given by transplantation from those for ordinary functions (see, e.g. Courant and John [4, p. 243]; Roberts and Varberg [15, pp. 103, 108]) and a routine duplication is thus omitted.

On the other hand, for solving an optimization opt F without appealing to the differentiability of functions, after we establish

$$F(z) = F(z_0) + \Phi(z, z_0) \tag{5}$$

it is easy to see that (i) if $\Phi(z, z_0) \le 0$, we establish $\max F$ or (ii) if $\Phi(z, z_0) \ge 0$, we establish $\min F$. Apparently, the optimum opt $F(z) = F(z_0)$ is attained at $z = z_0$ with $\Phi(z, z_0) = 0$.

Certainly, we might have from (4) and (5)

$$\Phi(z, z_0) = \langle \partial F(z_0), h \rangle + \frac{1}{2}\langle \partial^2 F(z_0 + \theta h), h^2 \rangle, \quad 0 < \theta < 1,$$

if the differentiability of corresponding functions is assumed.

4. Basic inequalities

In this section we deal with some basic inequalities, such as the Hölder inequality and Minkowski inequalities concerning the p-norm on product spaces defined in (3) as follows.

Theorem 5. *If $p > 1$, then*

$$\langle z, \frac{w}{||w||_q} \rangle \leq ||z||_p, \quad z \in U \cap E_p, \quad w \in U \cap E_q, \tag{6}$$

where w is given and $q = p(p-1)^{-1}$. The inequality sign in (6) is reversed for $0 < p < 1$. In either case the sign of equality holds iff $z^p = cw^q$, where $c \in \mathbb{R}^+$.

Theorem 6. *If $p > 1$, then*

$$||z + w||_p \leq ||z||_p + ||w||_p, \quad z, w \in U \cap E_p. \tag{7}$$

The inequality sign in (7) is reversed for $0 < p < 1$. In either case, the sign of equality holds iff $z = cw$, where $c \in \mathbb{R}^+$.

Theorem 7. *Let $M_t(z)$ be the weighted mean of order t of the argument z on U with weight $\alpha = (\alpha_1, \ldots, \alpha_m)^\top$, $\alpha_j \in \mathbb{R}^+$, $j = 1, \ldots, m$ defined as follows:*

$$\begin{aligned} M_t(z) &= ||\alpha^{1/t} z||_t, \quad t \neq 0, \quad t \in \mathbb{R}, \quad \textstyle\sum \alpha_j = 1, \\ M_0(z) &= G(z), \quad \textit{geometric mean}. \end{aligned} \tag{8}$$

Then it satisfies Minkowski inequality

$$M_t(z + w) \leq M_t(z) + M_t(w), \quad z, w \in U \tag{9}$$

for $t > 1$. The inequality sign is reversed in (9) for $t \leq 1$. In either case the sign of equality holds iff $z = cw$, where $c \in \mathbb{R}^+$.

In order to prove Theorems 5, 6 and 7, we simply apply the appropriate one of Theorems 2, 3 and 4, respectively, to establish the inequality in question. These turn out to reissue the same arguments for the discrete model in [18] or the continuous model in [25]. So, we omit the details which are repetitious.

5. Derivations from the Hölder inequality

Since there is one and only one Hölder inequality (see, e.g. Wang [28]), the other forms are simply it variations or variants regardless of discrete models or continuous models. Although we are now dealing with the mixture of discrete and continuous models, this new situation leads to the same consequence without any exception. Consequently, we reestablish here three theorems in the previous section by means of a combination of the usual Hölder inequality for discrete model and its continuous model. For the purpose of removing the entanglement of notation and symbols, we simply consider

$$E = E_p = E_p^{(1)} \times E_p^{(2)}$$

for $m = 2$, where $E_p^{(1)}$ denotes $\mathbb{R}^n$, l^p, etc. and $E_p^{(2)}$ denotes $L^p(U, S, \mu)$, etc.

Letting $q = p(p-1)^{-1}$, we have for $p > 1$

$$(10) \quad \sum x_j y_j \leq \left(\sum x_j^p\right)^{1/p} \left(\sum y_j^q\right)^{1/q}, \quad x \in E_p^{(1)} \cap l^p, \quad y \in E_p^{(2)} \cap l^q$$

and

$$(11) \quad \int fg \, d\mu \leq \left(\int f^p \, d\mu\right)^{1/p} \left(\int g^q \, d\mu\right)^{1/q}, \quad f \in E_p^{(2)} \cap L^p, \quad g \in E_p^{(2)} \cap L^q.$$

Here and in what follows $\sum$ is used to indicate $\sum_{j=1}^n$ ($n = \infty$ is also allowed), while $\int$ and $d\mu$ are used to indicate $\int_U$ and $d\mu(x) = dx_1 \cdots dx_n$ respectively, whenever confusion is unlikely to occur.

The inequality sign in (10) as well as in (11) is reversed for $0 < p < 1$. In either case, the sign of equality holds in (10) iff $x = cy$, while the sign of equality holds in (11) iff $f = cg$, where $c \in \mathbb{R}^+$.

Adding (11) to (10) and applying the simple Hölder inequality to the sum of the right hand side, we obtain the mixed Hölder inequality for $p > 1$

$$(12) \quad \sum x_j y_j + \int fg \, d\mu \leq \left(\sum x_j^p + \int f^p \, d\mu\right)^{1/p} \left(\sum y_j^q + \int g^q \, d\mu\right)^{1/q}.$$

Evidently, the inequality sign in (12) is reversed for $0 < p < 1$. In either case, the sign of equality holds in (12) iff $(x, f) = c(y, g)$, where $c \in \mathbb{R}^+$.

We now state and prove a generalization of inequality (12) as follows.

Theorem 8. *For* $0 < s < t$,

$$(13)\qquad \sum c_j x_j^s + \int df^s\, d\mu \le \left[\sum (c_j^t a_j^{-s})^{1/(t-s)} + \int (d^t b^{-s})^{1/(t-s)} d\mu\right]^{(t-s)/t} \left[\sum a_j x_j^t + \int b f^t d\mu\right]^{s/t},$$

where $a_j, c_j, x_j \in \mathbb{R}^+$, $j = 1, 2, \ldots$ *and* b, d, f *are suitable positive functions. The inequality sign in (13) is reversed for* $s < t < 0$ *or* $s < 0 < t$. *In either case, the sign of equality holds in (13) iff*

$$(14)\qquad \frac{x_1^t}{(c_1 a_1^{-1})^{t/(t-s)}} = \frac{x_2^t}{(c_2 a_2^{-1})^{t/(t-s)}} = \cdots = \frac{f^t}{(db^{-1})^{t/(t-s)}}.$$

Proof. Noting

$$(15)\qquad \sum c_j x_j^s + \int df^s\, d\mu = \sum (c_j a_j^{-s/t})(a_j^{s/t} x_j^s) + \int (db^{-s/t})(b^{s/t} f^s) d\mu$$

and setting $t/(t-s) = p$, $t/s = q$, (13) follows from (12) via (15). Naturally, the sign of equality holds iff (14) holds.

Theorem 9. *For* $s < t$, *the weighted mean inequality*

$$(16)\qquad M(s) \le M(t),$$

where

$$M(t) = \left(\frac{\sum a_j x_j^t + \int b f^t d\mu}{\sum a_j + \int b\, d\mu}\right)^{1/t}, \quad t \ne 0,$$

$$M(0) = G(x, f) = \left(\pi x_j^{a_j} \exp \int b \log f\, d\mu\right)^{1/(\sum a_j + \int b d\mu)},$$

holds with equality holding iff $x_1 = x_2 = \cdots = f$.

Proof. In fact, inequality (16) is a special case of inequality (13) by setting $c_j = a_j$, $j = 1, 2, \ldots, d = b$ and following a simple transposition.

Remark. Inequality (12) can be readily extended to a general m, so are Theorems 8 and 9. In turn, Theorem 8 can be regarded as a generalization of Theorems 5 and 9 for a general m. As Minkowski inequalities usually follow from the Hölder inequalty, Theorems 6 and 7 can be derived from Theorem 5 by routine inequality arguments.

6. Examples on optimization

As applications of the results given above and a demonstration of solving optimization problems by inequalities, we first consider the problem

$$\text{opt } F(x, f),$$

where

$$F(x,f) = \frac{(\sum a_j x_j + \int bf\, d\mu)^p}{\sum x_j^p + \int f^p\, d\mu}, \tag{17}$$

and a_j, $j = 1, 2, \ldots, b$ are given. From (12), it follows that the optimum

$$\text{opt } F(x,f) = \left(\sum a_j^q + \int b^q\, d\mu\right)^{p/q}$$

is attained at

$$\frac{x_1^p}{a_j^q} = \frac{x_2^p}{a_2^q} = \cdots = \frac{f^p}{b^q},$$

where opt $=$ min if $p > 1$, opt $=$ max if $0 < p < 1$.

For an instructive purpose, three examples for $p = 2$ are in order.

Example 1. Consider $\max F$ where

$$F = F(x,f) = \frac{(\sum_{j=1}^n jx_j + \int_0^1 f(t)\, dt)^2}{\sum_{j=1}^n x_j^2 + \int_0^1 f^2\, dt}.$$

The maximum $\max f = \frac{n(n+1)(2n+1)}{6} + 1$ is attained at $x_1 = \frac{x_2}{2} = \cdots = \frac{x_n}{n} = f(t)$.

Example 2. Consider $\min G$ where

$$G = G(f,g) = \frac{(\int_0^1 f^2(t)dt + \int_0^\infty g^2(t)dt)}{[\int_0^1 \sqrt{3t} f(t)dt + \int_0^\infty e^{-t/2}t^{-1/2}g(t)dt]^2}.$$

The minimum $\min G = 1/2$ is attained at

$$\frac{f(t)}{\sqrt{3t}} = \frac{g(t)}{e^{-t/2}t^{-1/2}}.$$

Example 3. Consider $\max H$ where

$$H = H(x,f) = \frac{[\sum_{j=1}^\infty j^{-1}x_j + \int_0^1 [t(1-t)]^{-1/4} f(t)dt]^2}{\sum_{j=1}^\infty x_j^2 + \int_0^1 f^2(t)dt}.$$

The maximum $\max H = \frac{\pi^2}{6} + \pi$ is attained at

$$x_1 = 2x_2 = \cdots = nx_n = \cdots = [t(1-t)]^{-1/4} f(t).$$

A straight forward manipulation in conjunction with standard formulas for summations and integrals (see, e.g. Kreyszig [12]) yields the above results.

7. Concluding remarks

In the above we have established some basic inequalities, such as the Hölder inequality, Minkowski inequalities, etc., in conjunction with the mixture of discrete and continuous models. With these newly found inequalities, certain optimization problems (see, e.g. [11, 26, 27]) can be readily solved without appealing to the differentiability of concerned functions. On the other hand, the discrete versions of the basic inequalities or their continuous versions may simply be regarded as the special cases of the mixed ones. In this way, more inequalities and their variants (see, e.g. [2, 7, 14]) can be reformed into the mixed versions if necessary. In probability and statistics, mixtures of discrete and continuous density functions have been considered in practice and experiment (see, e.g. Dwass [6, pp. 207–210]). Our results may be useful in this connection. Furthermore, some optimization problems, such as the type of problem (17) and a fortiori basic inequalities — Theorems 5–7 can be established by a dynamic programming argument as introduced in [3, 9, 20]. However, this will not be explored here.

In view of the results given here and in most selected references below, a further development of the idea for forming a close linkage of optimization problems and the theory of inequalities appears to be promising.

Acknowledgement. This work was done when the second author was on sabbatical at Chung Cheng Institute of Technology for the year 1990, supported (in part) by the National Science Council of the Repbulic of China.

References

1. E.F. Beckenbach, *On Hölder inequality.* J. Math. Anal. Appl. **15** (1966), 21–29.

2. E.F. Beckenbach and R. Bellman, *Inequalities.* 3rd rev. ed., Springer-Verlag, Berlin, 1965.

3. R. Bellman, *Dynamic programming.* Princeton Univ. Press, Princeton, N.J., 1967.

4. R. Courant and F. John, *Calculus and Analysis.* Vol. 1, Wiley-Interscience, New York, 1965.

5. H.F. Cullen, *Introduction to General Topology.* D.C. Heath and Co., Boston, Ms., 1968.

6. M. Dwass, *Probability and Statistics: An Undergraduate Course.* W.A. Benjamin, Inc., Menlo Park, Calif., 1970.

7. G.H. Hardy, J.E. Littlewood and G. Pólya, *Inequalities.* 2nd ed., Cambridge Univ. Press, Cambridge, 1952.

8. S. Iwamoto and Chung-lie Wang, *Continuous dynamic programming approach to inequalities.* J. Math. Anal. Appl. **96** (1983), 119–129.

9. S. Iwamoto and Chung-lie Wang, *Continuous dynamic programmming approach to inequalities, II.* J. Math. Anal. Appl. **118** (1986), 279–286.

10. S. Iwamoto, R.J. Tomkins and Chung-lie Wang, *Some theorems on reverse inequalities.* J. Math. Anal. Appl. **119** (1986), 282–299.

11. S. Iwamoto, R.J. Tomkins and Chung-lie Wang, *Inequalities and mathematical programming, III.* In: "General Inequalities 5 (Proceedings, Fifth International Conference on General Inequalities, Oberwolfach)" (W. Walter, Ed.), pp. 419–432, Birkhäuser, Basel–Stuttgart, 1987.

12. E. Kreyszig, *Advanced Engineering Mathematics.* John Wiley, New York, 7th Printing, 1966.

13. E. Kreyszig, *Introductory Functional Analysis with Applications.* John Wiley, New York, 1978.

14. D.S. Mitrinović, *Analytic Inequalities.* Springer-Verlag, Berlin, 1970.

15. A.W. Roberts and D.E. Varberg, *Convex Functions.* Academic Press, New York, 1973.

16. M.M. Vainberg, *Variational Methods for the Study of Nonlinear Operators* (translated by A. Feinstein). Holden-Day, San Francisco, 1964.

17. Chung-lie Wang, *Convexity and inequalities.* J. Math. Anal. Appl. **72** (1979), 355–361.

18. Chung-lie Wang, *On development of inverses of the Cauchy and Hölder inequalities.* SIAM Review, Vol. 21, No. 4 (1979), 550–557.

19. Chung-lie Wang, *Functional equation approach to inequalities.* J. Math. Anal. Appl. **71** (1979), 423–430.

20. Chung-lie Wang, *Functional equation approach to inequalities, II.* J. Math. Anal. Appl. **78** (1980), 522–530.

21. Chung-lie Wang, *Functional equation approach to inequalities, III.* J. Math. Anal. Appl. **80** (1981), 31–35.

22. Chung-lie Wang, *Functional equation approach to inequalities, IV.* J. Math. Anal. Appl. **86** (1982), 96–98.

23. Chung-lie Wang, *A generalization of the HGA inequalities.* Soochow J. Math. **6** (1980), 149–152.

24. Chung-lie Wang, *Functional equation approach to inequalities, VI.* J. Math. Anal. Appl. **104** (1984), 95–102.

25. Chung-lie Wang, *Characteristics of nonlinear positive functionals and their applications.* J. Math. Anal. Appl. **95** (1983), 564–574.

26. Chung-lie Wang, *Inequalities and mathematics programming.* In: "General Inequalities 3 (Proceedings, Third International Conference on General Inequalities, Oberwolfach)" (E.F. Beckenbach, Ed.), pp. 149–164, Birkhäuser, Basel–Stuttgart, 1983.

27. Chung-lie Wang, *Inequalities and mathematics programming, II.* In: "General Inequalities 4 (Proceedings, Fourth International Conference on General Inequalities, Oberwolfach)" (W. Walter, Ed.), pp. 381–392, Birkhäuser, Basel–Stuttgart, 1984.

28. Chung-lie Wang, *Beckenbach inequality and its variants.* J. Math. Anal. Appl. **130** (1988), 252–256.

29. Chung-lie Wang, *The principle and models of dynamic programming.* J. Math. Anal. Appl. **118** (1986), 287–308.

30. Chung-lie Wang, *The principle and models of dynamic programming, II.* J. Math. Anal. Appl. **135** (1988), 268–283.

31. Chung-lie Wang, *The principle and models of dynamic programming, III.* J. Math. Anal. Appl. **135** (1988), 284–296.

32. Chung-lie Wang, *The principle and models of dynamic programming, IV.* J. Math. Anal. Appl. **137** (1989), 148–160.

33. Chung-lie Wang, *The principle and models of dynamic programming, V.* J. Math. Anal. Appl. **137** (1989), 161–167.

34. Chung-lie Wang, *Dynamic programming and inequalities.* J. Math. Anal. Appl. **150** (1990), 528–550.

Don-lin Mon, Department of System Engineering, Chung-Cheng Institute of Technology, Taiwan, Republic of China

Chung-lie Wang, Department of Mathematics and Statistics, University of Regina, Regina, Saskatchewan, Canada S4S 0A2

Notes, Problems and Remarks

Gasthof Hirschen, Oberwolfach-Walke

International Series of Numerical Mathematics, Vol. 103, © 1992 Birkhäuser Verlag Basel

A note on Henrici's inequalities

J.L. Brenner and Horst Alzer

In 1956 P. Henrici [2] published for the special case $p_1 = \cdots = p_n$ the following interesting inequalities:

If $a_i \geq 1$ $(i = 1, \ldots, n)$, then

$$\sum_{i=1}^{n} \frac{p_i}{1+a_i} \geq \frac{P_n}{1+G_n}, \tag{1}$$

and if $0 < a_i \leq 1$ $(i = 1, \ldots, n)$, then

$$\sum_{i=1}^{n} \frac{p_i}{1+a_i} \leq \frac{P_n}{1+G_n}, \tag{2}$$

where G_n denotes the weighted geometric mean of $a_1, \ldots, a_n$, i.e. $G_n = \prod_{i=1}^{n} a_i^{p_i/P_n}$ with positive weights p_i $(i = 1, \ldots, n)$ and $P_n = \sum_{i=1}^{n} p_i$. Equality holds in (1) and (2) if and only if $a_1 = \cdots = a_n$.

(See also [1], [3] and the references therein.)

The geometric mean belongs to the family of power mean values defined by

$$M_r = \begin{cases} \left(\dfrac{1}{P_n} \displaystyle\sum_{i=1}^{n} p_i a_i^r \right)^{1/r}, & r \neq 0, \\ G_n, & r = 0. \end{cases}$$

Since $r \mapsto M_r$ is increasing on the real line it is natural to ask whether inequalities (1) and (2) can be refined by replacing G_n by a power mean. The answer is "no"!

Indeed, setting $a_1 = \cdots = a_{n-1} = 1$, $a_n = 1 + x$ (where $x \neq 0$ is sufficiently small) we obtain

$$1 + M_r - \frac{P_n}{\sum_{i=1}^{n} \frac{p_i}{1+a_i}} = \frac{p_n P_{n-1} r}{2P_n^2} x^2 + \cdots,$$

which immediately implies that neither (1) nor (2) can be sharpened by power means.

Now we consider the following related problems: What is the smallest number r such that

$$\sum_{i=1}^{n} \frac{p_i}{1-a_i} \leq \frac{P_n}{1-M_r}$$

holds for all real $a_i > 1$ $(i = 1, \ldots, n)$; and what is the greatest number s such that

$$\sum_{i=1}^{n} \frac{p_i}{1-a_i} \geq \frac{P_n}{1-M_s}$$

is valid for all real $a_i \in (0,1)$ $(i = 1, \ldots, n)$?

It might be interesting that the answer to these questions leads to the two other classical mean values — to the arithmetic and harmonic means. In what follows we denote by A_n and H_n the weighted arithmetic and harmonic means of $a_1, \ldots, a_n$ (i.e. $A_n = M_1$ and $H_n = M_{-1}$). Further we define:

$$R_n = \sum_{i=1}^{n} \frac{p_i}{1-a_i}, \quad S_n = \frac{P_n}{1-H_n}, \quad T_n = \frac{P_n}{1-A_n}.$$

Then we have

Theorem 1. *If $a_i > 1$ $(i = 1, \ldots, n)$, then*

$$R_n - S_n \leq R_{n-1} - S_{n-1}, \tag{3}$$

and if $a_i \in (0,1)$ $(i = 1, \ldots, n)$, then

$$R_n - T_n \geq R_{n-1} - T_{n-1}. \tag{4}$$

Equality holds in (3) if and only if $a_n = H_{n-1}$, and in (4) if and only if $a_n = A_{n-1}$.

Proof. We designate by f the function

$$f : (1, \infty) \to \mathbb{R}, \quad f(a_n) = R_n - S_n.$$

Using well-known techniques from calculus we obtain that f attains its absolute maximum only at $a_n = H_{n-1}$. Hence we get

$$f(a_n) \le f(H_{n-1}) = R_{n-1} - S_{n-1}$$

with equality holding if and only if $a_n = H_{n-1}$.

The proof of (4) is similar. □

Repeated applications of (3) and (4) lead to the following variant of Henrici's proposition.

Theorem 2. *If $a_i > 1$ $(i = 1, \ldots, n)$, then*

$$\sum_{i=1}^{n} \frac{p_i}{1 - a_i} \le \frac{P_n}{1 - H_n}, \tag{5}$$

and if $a_i \in (0,1)$ $(i = 1, \ldots, n)$, then

$$\sum_{i=1}^{n} \frac{p_i}{1 - a_i} \ge \frac{P_n}{1 - A_n}. \tag{6}$$

Equality holds in (5) *and in* (6) *if and only if $a_1 = \cdots = a_n$.*

Finally we show that the harmonic and the arithmetic means are the best possible power mean values in inequalities (5) and (6). We assume that there exists a number $r < -1$ such that

$$\sum_{i=1}^{n} \frac{p_i}{1 - a_i} \le \frac{P_n}{1 - M_r} \tag{7}$$

holds for all $a_i > 1$ $(i = 1, \ldots, n)$. If we let a_i $(i = 1, \ldots, n-1)$ tend to ∞, then we obtain from (7):

$$\frac{1}{a_n} + \frac{P_n}{p_n}\left(1 - \frac{1}{a_n}\right) \le \left(\frac{p_n}{P_n}\right)^{1/r};$$

and if a_n tends to ∞ we conclude $r \ge -1$. Now we suppose that there exists a number $s > 1$ such that

$$\sum_{i=1}^{n} \frac{p_i}{1 - a_i} \ge \frac{P_n}{1 - M_s} \tag{8}$$

is valid for all $a_i \in (0,1)$ $(i = 1, \ldots, n)$. We let a_i $(i = 1, \ldots, n-1)$ tend to 0, then we get from (8):

$$\frac{p_n}{P_n - P_{n-1}a_n} \ge \left(\frac{p_n}{P_n}\right)^{1/s};$$

and if a_n tends to 0 we obtain $s \le 1$.

References

1. P.S. Bullen, D.S. Mitrinović and P.M. Vasić, *Means and Their Inequalities.* Reidel, Dordrecht, 1988.

2. P. Henrici, *Aufgabe 245*, Elem. Math. **11** (1956), 112.

3. D.S. Mitrinović, *Analytic Inequalities.* Springer, New York, 1970.

J.L. Brenner, 10 Phillips Road, Palo Alto, California 94303, USA

Horst Alzer, Department of Mathematics, Applied Mathematics and Astronomy, University of South Africa, P.O. Box 392, 0001 Pretoria, South Africa

International Series of Numerical Mathematics, Vol. 103, © 1992 Birkhäuser Verlag Basel

A note on an inequality of G. Talenti

R. Lemmert and H. Alzer

Abstract

The aim of this note is to provide not only a short and simple proof for an extension of (1), but also to establish a converse inequality of this extension. Furthermore, we show that Talenti's inequality holds for a class of functions which includes the set of decreasing functions as a proper subset.

G. Talenti presented a proof for the following interesting integral inequality:

If f is a positive decreasing function on $[a,b]$ $(a > 0)$, then

$$\text{(1)} \qquad \ln\left(1+\int_a^b f(t)\,dt/(1+af(a))\right) \le \int_a^b f(t)/(1+tf(t))\,dt.$$

Theorem. *Let $f, g \in L^1[a,b]$ be two positive functions. If*

$$\text{(2)} \qquad g(x)-g(a) \le \int_a^x f(t)\,dt \text{ almost everywhere (a.e.) on } [a,b],$$

then

$$\text{(3)} \qquad \ln\left(1+\int_a^b f(t)\,dt/g(a)\right) \le \int_a^b f(t)/g(t)\,dt;$$

and if

$$\text{(4)} \qquad g(b)-g(x) \le \int_x^b f(t)\,dt \text{ a.e. on } [a,b],$$

then

$$\text{(5)} \qquad \int_a^b f(t)/g(t)\,dt \le -\ln\left(1-\int_a^b f(t)\,dt/g(b)\right).$$

(In the last term we assume $g(b) > \int_a^b f(t)\,dt$.*) Equality holds in* (3) *and* (5) *if and only if* $g' = f$ *a.e..*

Proof. The proofs of (3) and (5) are very similar such that it suffices to establish inequality (3). We define

$$F : [a,b] \to \mathbb{R},$$

$$F(x) = \int_a^x f(t)/g(t)\,dt - \ln\left(1 + \int_a^x f(t)\,dt/g(a)\right)$$

and obtain

$$F'(x) = f(x)/g(x) - f(x)/\left(g(a) + \int_a^x f(t)\,dt\right) \text{ a.e. on } [a,b]. \tag{6}$$

From (2) and the assumption that f and g are positive we conclude

$$F'(x) \geq 0 \text{ a.e. on } [a,b] \tag{7}$$

and

$$F(b) \geq F(a) = 0,$$

which proves inequality (3).

If $g' = f$ a.e. on $[a,b]$, then the sign of equality holds in (3). Now we suppose $F(b) = 0$; then equality holds in (7), and from (6) we get

$$g(x) - g(a) = \int_a^x f(t)\,dt \text{ a.e. on } [a,b],$$

which implies $g' = f$ a.e. on $[a,b]$.

Remark. If we assume that f is a positive decreasing function on $[a,b]$ $(a > 0)$ and if we set $g(t) = 1 + tf(t)$, then the assumptions (2) and (4) are fulfilled and we obtain (1) as well as a converse inequality where the sign of equality holds if and only if $f \equiv$ const. a.e. on $[a,b]$.

Finally we note that Talenti's inequality is valid not only for decreasing functions but also for some non-monotonic functions as the following example reveals:

If f is defined by

$$f(x) = \begin{cases} 1 - \dfrac{x}{2}, & \alpha \leq x \leq 1 \\ \dfrac{x}{2}, & 1 \leq x \leq 1 + \beta \end{cases} \qquad (\alpha, \beta > 0),$$

then it is easy to check that the inequality (2) (with $g(t) = 1 + tf(t)$, $a = \alpha$ and $b = 1 + \beta$) is valid for all sufficiently small positive numbers α and β.

Roland Lemmert, Mathematisches Institut I, Universität Karlsruhe, Postfach 6980, D-7500 Karlsruhe, Germany

Horst Alzer, Department of Mathematics, Applied Mathematics and Astronomy, University of South Africa, P.O. Box 392, 0001 Pretoria, South Africa

International Series of Numerical Mathematics, Vol. 103, © 1992 Birkhäuser Verlag Basel

A simple proof for a quadratic inequality

L. Losonczi and Zs. Páles

The aim of this note is to give a simple proof for a generalization of the following inequality of S. Saitoh [2, Corollary 1.1]:

Let $A_1, \dots, A_m$ be $N \times N$ positive definite Hermitian matrices. Then for all $x_1, \dots, x_n$ in $\mathbb{C}^N$ we have the inequality

$$\left(\sum_{j=1}^m x_j\right)^* \left(\sum_{j=1}^m A_j^{-1}\right)^{-1} \left(\sum_{j=1}^m x_j\right) \le \sum_{j=1}^m x_j^* A_j x_j, \tag{1}$$

where * stand for conjugate transpose.

In (1) equality holds if and only if

$$A_1 x_1 = \cdots = A_m x_m.$$

Our proof is based on the next

Lemma. *Let H be a complex Hilbert space with inner product $\langle \cdot, \cdot \rangle$, and let $Q : H \to H$ be a linear bounded selfadjoint and strongly positive operator (i.e., there exists a positive constant q such that $\langle x, Qx\rangle \ge q\langle x, x\rangle$ for all $x \in H$). Then for all $x, y \in H$ we have*

$$\langle x, y\rangle + \langle y, x\rangle \le \langle x, Qx\rangle + \langle y, Q^{-1}y\rangle. \tag{2}$$

Equality holds here if and only if $Qx = y$.

Proof. The operator $P = Q^{1/2}$ is also linear, bounded and selfadjoint, and $Q = PP$, $Q^{-1} = P^{-1}P^{-1}$, furthermore

$$\begin{aligned}
\langle x, Qx\rangle &= \langle Px, Px\rangle,\\
\langle y, Q^{-1}y\rangle &= \langle P^{-1}y, P^{-1}y\rangle,\\
\langle x, y\rangle &= \langle Px, P^{-1}y\rangle,\\
\langle y, x\rangle &= \langle P^{-1}y, Px\rangle.
\end{aligned}$$

Using these equations (2) can be written as

$$\|Px - P^{-1}y\|^2 \geq 0. \tag{3}$$

This is obviously true, and so is (2). Inequality (3) also shows that equality holds in (3) as well as in (2) if and only if $Px = P^{-1}y$, that is if $Qx = y$.

Remark. Applying Schwarz's inequality for the vectors Px and $P^{-1}y$, we easily obtain

$$|\langle x, y\rangle|^2 \leq \langle x, Qx\rangle\langle y, Q^{-1}y\rangle. \tag{4}$$

Using (4) and the arithmetic mean – geometric mean inequality, we have

$$\begin{aligned}\langle x, y\rangle + \langle y, x\rangle &\leq 2|\langle x, y\rangle| = 2\sqrt{|\langle x, y\rangle|^2}\\ &\leq 2\sqrt{\langle x, Qx\rangle\langle y, Q^{-1}y\rangle}\\ &\leq \langle x, Qx\rangle + \langle y, Q^{-1}y\rangle.\end{aligned}$$

Thus we have got (2) again.

We generalize inequality (1) for complex Hilbert spaces and strongly positive selfadjoint operators.

Theorem (cf. Saitoh [2]). *Let H be a complex Hilbert space and $D_1, \ldots, D_m$ be linear bounded selfadjoint and strongly positive operators on H and let $D = (\sum_{j=1}^m D_j^{-1})^{-1}$. Then for all $x_1, \ldots, x_m \in H$,*

$$\left\langle \sum_{j=1}^m x_j, D\left(\sum_{j=1}^m x_j\right)\right\rangle \leq \sum_{j=1}^m \langle x_j, D_j x_j\rangle. \tag{5}$$

Equality holds here if and only if

$$D_1 x_1 = \ldots = D_m x_m.$$

Proof. Substitute $Q = D_j$, $x = x_j$ and $y = D(\sum_{j=1}^m x_j)$ into (2) for $j = 1, \ldots, m$ and add up the inequalities obtained. Then we arrive at

$$\sum_{j=1}^m \langle x_j, y\rangle + \sum_{j=1}^m \langle y, x_j\rangle \leq \sum_{j=1}^m \langle x_j, D_j x_j\rangle + \sum_{j=1}^m \langle y, D_j^{-1} y\rangle. \tag{6}$$

However we have

$$\sum_{j=1}^{m}\langle x_j, y\rangle = \left\langle \sum_{j=1}^{m} x_j, y \right\rangle = \left\langle \sum_{j=1}^{m} x_j, D\left(\sum_{j=1}^{m} x_j\right)\right\rangle,$$

which gives exactly the left hand side of (5). To complete the proof observe that

$$\sum_{j=1}^{m}\langle y, x_j\rangle = \left\langle y, \sum_{j=1}^{m} x_j \right\rangle = \langle y, D^{-1}y\rangle = \left\langle y, \sum_{j=1}^{m} D_j^{-1}y \right\rangle = \sum_{j=1}^{m}\langle y, D_j^{-1}y\rangle.$$

Thus the second terms on both sides of inequality (6) are equal, therefore they can be omitted.

By the lemma, equality holds if and only if $D_j x_j = y$ for all $j = 1, \ldots, m$, that is if $D_1 x_1 = \cdots = D_m x_m$.

Concerning generalizations of inequality (1) see S. Saitoh [3, 4], for applications see I. Fazekas [1].

Acknowledgement: This research was supported by the Hungarian National Science Foundation Operating Grant No. OTKA 1652.

References

1. I. Fazekas, Manuscript.
2. S. Saitoh, *Positive definite Hermitian matrices and reproducing kernels.* Linear Algebra Appl. **48** (1982), 119–130.
3. S. Saitoh, *Quadratic inequalites deduced from theory of reproducing kernels.* Linear Algebra Appl. **93** (1987), 171–178.
4. S. Saitoh, *Quadratic inequalities associated with integrals of reproducing kernels.* Linear Algebra Appl. **101** (1988), 269–280.

L. Losonczi and Zs. Páles, Institute of Mathematics, L. Kossuth University, H-4010 Debrecen, Pf. 12, Hungary

International Series of Numerical Mathematics, Vol. 103, © 1992 Birkhäuser Verlag Basel

Notes on convex functions

Josip E. Pečarić

Abstract

This paper contains some new results about the well-known Jensen inequality for Jensen-convex functions and for functions with nondecreasing increments as well as a result about Hadamard inequalities.

1. Introduction

Let f be a real-valued function defined on a convex set I from an arbitrary real linear space L. If for all points $x, y \in I$ the inequality

$$f\left(\frac{x+y}{2}\right) \leq \frac{f(x)+f(y)}{2} \tag{1}$$

is valid, f is a mid-convex function (a J-convex function or a convex function in Jensen sense). It is well-known that for mid-convex functions the Jensen inequality

$$f\left(\frac{1}{n}\sum_{i=1}^{n} x_i\right) \leq \frac{1}{n}\sum_{i=1}^{n} f(x_i), \tag{2}$$

is valid, where $x_i \in I$, $i = 1, \ldots, n$.

Many books bring Jensen's (i.e., Cauchy's) inductive proof of (2) (see, e.g., [1–3]). It is not generally known that two simpler proofs exist (see [4, 5]).

We shall sketch the proof from [4].

Of course, (2) for $n = 2$ is just (1). Now suppose that the result is valid for every k, $2 \leq k \leq n$. Writing $x = \frac{1}{n+1}\sum_{i=1}^{n+1} x_i$, we have

$$\begin{aligned} f(x) &= f\left(\frac{1}{n+1}\sum_{i=1}^{n+1} x_i\right) \\ &= f\left(\frac{1}{2}\left(\frac{1}{n}\sum_{i=1}^{n} x_i + \frac{n-1}{n}x + \frac{1}{n}x_{n+1}\right)\right) \\ &\le \frac{1}{2}\left(f\left(\frac{1}{n}\sum_{i=1}^{n} x_i\right) + f\left(\frac{n-1}{n}x + \frac{1}{n}x_{n+1}\right)\right) \\ &\le \frac{1}{2}\left(\frac{1}{n}\sum_{i=1}^{n} f(x_i) + \frac{1}{n}((n-1)f(x) + f(x_{n+1}))\right) \end{aligned}$$

which is just

$$f(x) \le \frac{1}{n+1}\sum_{i=1}^{n+1} f(x_i).$$

Now let $\mathbb{R}^k$ denote the k-dimensional vector lattice of points $x = (x_1, \ldots, x_k)$, x_i real for $i = 1, \ldots, k$, with the partial ordering

$$x = (x_1, \ldots, x_k) \le y = (y_1, \ldots, y_k)$$

if and only if $x_i \le y_i$ for $i = 1, \ldots, k$. We shall write $ax + by = (ax_1 + by_1, \ldots, ax_k + by_k)$, where $a, b \in \mathbb{R}$.

A real-valued function f on an interval $I \subseteq \mathbb{R}^k$ will be said to have nondecreasing increments if

$$f(a+h) - f(a) \le f(b+h) - f(b) \tag{3}$$

whenever $a \in I$, $b + h \in I$, $0 = (0, \ldots, 0) \le h \in \mathbb{R}^k$, $a \le b$.

For some properties of these functions see [6]. We only note that the functions $f(x_1, \ldots, x_k) = x_1 \cdots x_k$ ($x_j \ge 0$, $j = 1, \ldots, k$) have nondecreasing increments.

If we put $a = x$, $b + h = y$, $b = a + h$, then (3) becomes (1) with $x \le y$.

Now, using a classical inductive proof of Cauchy ([1–3]), we can prove (2) where $x_1 \le x_2 \le \cdots \le x_n$ are points from $I \subseteq \mathbb{R}^k$.

Moreover, the above proof is also valid. Indeed, if $x_1 \le \cdots \le x_n \le x_{n+1}$ we also have

$$x = \frac{1}{n+1}\sum_{i=1}^{n+1} x_i \le x_{n+1} \quad \text{and} \quad \frac{1}{n}\sum_{i=1}^{n} x_i \le \frac{1}{n}((n-1)x + x_{n+1}),$$

and we can work as above.

2. An interpolation of Jensen's inequality

Let f be a function of nondecreasing increments defined on I, and let x_i ($i = 1, \dots, n$) be points in I chosen so that $x_1 \le \dots \le x_n$. Then for $1 \le m \le n$, define $f_{m,n} = f_{m,n}(x_1, \dots, x_n)$ by

$$f_{m,n} = \frac{1}{\binom{n}{m}} \sum_{1 \le i_1 < \cdots < i_m \le n} f\left[\frac{1}{m}(x_{i_1} + \cdots + x_{i_m})\right].$$

For instance, with $n = 3$, we have the three possibilities

$$f_{1,3}(x_1, x_2, x_3) = \frac{1}{3}\{f(x_1) + f(x_2) + f(x_3)\},$$

$$f_{2,3}(x_1, x_2, x_3) = \frac{1}{3}\left\{f\left(\frac{x_1 + x_2}{2}\right) + f\left(\frac{x_1 + x_3}{2}\right) + f\left(\frac{x_2 + x_3}{2}\right)\right\},$$

$$f_{3,3}(x_1, x_2, x_3) = f\left(\frac{x_1 + x_2 + x_3}{3}\right).$$

Inequality (2) says $f_{3,3} \le f_{1,3}$. More is true: $f_{3,3} \le f_{2,3} \le f_{1,3}$. The value of $f_{2,3}$ interpolates between $f_{3,3}$ and $f_{1,3}$ and illustrates the following theorem about interpolating values.

Theorem. $f_{n,n} \le \cdots \le f_{m,n} \le \cdots \le f_{1,n}, \quad 1 \le m \le n.$

Proof. The theorem will be established by showing that, for $1 \le m \le n - 1$, $f_{m+1,n} \le f_{m,n}$. From $x_{i_1} \le \cdots \le x_{i_{m+1}}$, we have

$$\frac{1}{m}(x_{i_1} + \cdots + x_{i_{m+1}} - x_{i_{m+1}}) \le \cdots \le \frac{1}{m}(x_{i_1} + \cdots + x_{i_{m+1}} - x_{i_1}),$$

so by (2) we have

$$\begin{aligned} f_{m+1,n} &= \frac{1}{\binom{n}{m+1}} \sum_{1 \le i_1 < \cdots < i_{m+1} \le n} f\left[\frac{1}{m+1}(x_{i_1} + \cdots + x_{i_{m+1}})\right] \\ &\le \frac{1}{\binom{n}{m+1}} \sum_{1 \le i_1 < \cdots < i_{m+1} \le n} \frac{1}{m+1}\left[\sum_{j=1}^{m+1} f\left(\frac{1}{m}(x_{i_1} + \cdots + x_{i_{m+1}} - x_{i_j})\right)\right] \\ &= \frac{1}{\binom{n}{m}} \sum_{1 \le i_1 < \cdots < i_m \le n} f\left(\frac{1}{m}(x_{i_1} + \cdots + x_{i_m})\right) = f_{m,n}. \end{aligned}$$

Remark 1. As a special case we can give interpolating inequalities for the well-known Čebyšev inequalities for monotone sequences.

Remark 2. Note that all previous results for functions with nondecreasing increments are valid for a wider class of functions, namely for functions that satisfy (1) for all x, y chosen from I so that $x \leq y$.

Remark 3. A result related to our theorem is also valid for mid-convex functions (see [7, 8]). All results of this type are generalizations of a result from [9].

3. Hermite-Hadamard's inequalities

The following inequalities are known as Hadamard's inequalities for convex functions $f : I \to \mathbb{R}$, where now $I \subset \mathbb{R}$:

$$f\left(\frac{a+b}{2}\right) \leq \frac{1}{b-a}\int_a^b f(x)\,dx \leq \frac{f(a)+f(b)}{2}. \tag{4}$$

However, these results are due to Ch. Hermite (see for example [10]). So, we shall call these inequalities the Hermite-Hadamard inequalities.

The following generalization of (4) is valid (see for example [11, 12]):

Let p, q be two given positive numbers and $a_1 \leq a < b \leq b_1$. If y is a positive real number such that

$$y \leq \frac{b-a}{p+q}\min(p,q) \tag{5}$$

is valid, then for an arbitrary convex function $f : [a_1, b_1] \to \mathbb{R}$, the following inequalities are valid:

$$f\left(\frac{pa+qb}{p+q}\right) \leq \frac{1}{2y}\int_{A-y}^{A+y} f(t)\,dt \leq \frac{pf(a)+qf(b)}{p+q}, \tag{6}$$

where $A = \dfrac{pa+qb}{p+q}$.

First we note that the first inequality in (4) is stronger than the second, i.e., the following inequality is valid:

$$\frac{1}{b-a}\int_a^b f(x)\,dx - f\left(\frac{a+b}{2}\right) \leq \frac{f(a)+f(b)}{2} - \frac{1}{b-a}\int_a^b f(x)\,dx. \tag{7}$$

Indeed, (7) becomes

$$\frac{2}{b-a}\int_a^b f(x)\,dx \leq \frac{1}{2}\left(f(a)+f(b)+2f\left(\frac{a+b}{2}\right)\right),$$

i.e.

$$\frac{2}{b-a}\int_a^{\frac{a+b}{2}} f(x)\,dx + \frac{2}{b-a}\int_{\frac{a+b}{2}}^b f(x)\,dx$$
$$\leq \frac{1}{2}\left(f(a) + f\left(\frac{a+b}{2}\right)\right) + \frac{1}{2}\left(f\left(\frac{a+b}{2}\right) + f(b)\right).$$

This immediately follows if we apply the second inequality in (4) twice (on $\left[a, \frac{a+b}{2}\right]$ and $\left[\frac{a+b}{2}, b\right]$).

For $a = -1$, $b = 1$ we get a result from [13]. In fact, we showed that inequality (7), which is related to (4), is a simple consequence of (4).

Now, we note that the same assertion for (6) is also valid, i.e., under the same conditions we have

$$(8)\qquad \frac{1}{2y}\int_{A-y}^{A+y} f(t)\,dt - f\left(\frac{pa+qb}{p+q}\right) \leq \frac{pf(a)+qf(b)}{p+q} - \frac{1}{2y}\int_{A-y}^{A+y} f(t)\,dt.$$

By inequality (4) with a, b replaced by $A-y$, $A+y$, we obtain

$$(9)\qquad \frac{1}{2y}\int_{A-y}^{A+y} f(t)\,dt - f(A) \leq \frac{f(A-y)+f(A+y)}{2} - \frac{1}{2y}\int_{A-y}^{A+y} f(t)\,dt.$$

(Note that from (5) $a \leq A-y < A+y \leq b$ follows.)

By the definition of convexity we have for $a \leq x_1 < x_2 < x_3 \leq b$,

$$f(x_2) \leq \frac{x_3-x_2}{x_3-x_1}\,f(x_1) + \frac{x_2-x_1}{x_3-x_1}\,f(x_3).$$

Hence, taking $x_1 = a$, $x_3 = b$ we obtain

$$f(A-y) \leq \frac{b-(A-y)}{b-a}\,f(a) + \frac{A-y-a}{b-a}\,f(b),$$

$$f(A+y) \leq \frac{b-(A+y)}{b-a}\,f(a) + \frac{A+y-a}{b-a}\,f(b),$$

i.e.,

$$(10)\qquad \frac{1}{2}(f(A-y)+f(A+y)) \leq \frac{b-A}{b-a}\,f(a) + \frac{A-a}{b-a}\,f(b) = \frac{pf(a)+qf(b)}{p+q}.$$

By combining (9) and (10) we get (8).

References

1. D.S. Mitrinović, *Analytic Inequalities*. Springer Verlag, 1970.

2. A.W. Roberts and D.E. Varberg, *Convex Functions*. New York — London, 1973.

3. M. Kuczman, *An Introduction to the Theory of Functional Equations and Inequalities — Cauchy's Equation and Jensen's Inequality*. Warszawa — Kraków — Katowice, 1985.

4. J. Aczél, *Ungleichungen und ihre Verwendung zur elementaren Lösung von Maximum- und Minimumaufgaben*. Enseignement Math. **7** (1961), 214-249.

5. J. Aczél, *Nejednakosti i njihova primena u elementarnom rešavanju zadataka sa maksimumom i minimumom*. Mat. Biblioteka 18. Beograd, 1961, 111–138.

6. H.D. Brunk, *Integral inequalities for functions with nondecreasing increments*. Pacific J. Math. **14** (1964), 783–793.

7. D.S. Mitrinović and J.E. Pečarić, *Unified Treatment of Some Inequalities for Mixed Means*. Sb. Österr. Akad. Wiss. **197** (1988), 391–397.

8. J.E. Pečarić and V. Volenec, *Interpolation of the Jensen Inequality with some Applications*. Ibid. **197** (1988), 463–467.

9. S. Gabler, *Folgenkonvexe Funktionen*. Manuscripta math. **29** (1979), 29-47.

10. D.S. Mitrinović and I.B. Lacković, *Hermite and convexity*. Aequationes Math. **28** (1985), 229–232.

11. P.M. Vasić and I.B. Lacković, *Some complements to the paper: "On an inequality for convex functions"*. Univ. Beograd Publ. Elektrotehn. Fak. Ser. Mat. Fiz. No. **544–576** (1976), 59–62.

12. A. Lupaş, *A generalization of Hadamard's inequalities for convex functions*. Ibid. No. **544–576** (1976), 115–121.

13. P.S. Bullen, *Error estimates for some elementary quadrature rules*. Ibid. No. **602–633** (1978), 97–103.

Josip E. Pečarić, Faculty of Technology, Ive Lole Ribara 126, 41000 Zagreb, Croatia

International Series of Numerical Mathematics, Vol. 103, © 1992 Birkhäuser Verlag Basel

On special pairs of polygons with minimal area sum

Jürg Rätz

Abstract

Let n be a positive integer, $n \geq 3$ and C the unit circle line in $\mathbb{R}^2$. Let $a_1, \ldots, a_n$ be n pairwise distinct points of C, lying on C in this cyclic order, and such that the origin O is an interior point of $A := \operatorname{conv}\{a_1, \ldots, a_n\}$; put $a_{n+1} = a_1$. Finally, let B be the n-gon circumscribed to C which is obtained as the intersection of the closed supporting half-planes of C in $a_1, \ldots, a_n$. When is the area sum $F(A) + F(B)$ minimal as n and $a_1, \ldots, a_n$ vary? It is known that the pair of squares and only it realizes the minimum (cf. [1], [3], [5]). It is the purpose of this note to complement this result in several respects.

§ 1

If $x_1, \ldots, x_n$ denote the halves of the central angles $\angle a_i O a_{i+1}$ we have $0 < x_i < \pi/2$ $(i = 1, \ldots, n)$, $x_1 + \cdots + x_n = \pi$ and $F(A) + F(B) = f(x_1) + \cdots + f(x_n)$, where

$$(1) \qquad f(x) := \frac{1}{2} \sin 2x + \tan x \quad \left(0 \leq x < \frac{\pi}{2}\right).$$

Lemma 1. *For* $\vartheta := \arccos 2^{-1/4} \approx 0.572$, *$f$ is strictly concave on* $[0, \vartheta]$ *and strictly convex on* $[\vartheta, \frac{\pi}{2}[$.

This is proved in [1, p. 62].

Proof. For the reader's convenience the proof is sketched. The derivative

$$(2) \qquad f'(x) = \cos 2x + \frac{1}{\cos^2 x}$$

is positive in $\left[0, \frac{\pi}{2}\right[$, hence f is strictly increasing. The second derivative

$$f''(x) = \sin 2x \left(\frac{1}{\cos^4 x} - 2\right) \tag{3}$$

is negative in $]0, \vartheta[$ and positive in $\left]\vartheta, \frac{\pi}{2}\right[$, where ϑ satisfies $\cos^4 \vartheta = 1/2$. □

The main tool for our further procedure are the divided differences

$$q(\cdot, x') : x \mapsto \frac{f(x) - f(x')}{x - x'} \quad \left(x \in \left[0, \frac{\pi}{2}\right[\setminus \{x'\};\ x' \in \left[0, \frac{\pi}{2}\right[\right). \tag{4}$$

Lemma 2. a) *For $x' \in [0, \vartheta]$, $q(\cdot, x')$ is strictly decreasing on $[0, \vartheta] \setminus \{x'\}$.*

b) *For $x' \in \left[\vartheta, \frac{\pi}{2}\right[$, $q(\cdot, x')$ is strictly increasing on $\left[\vartheta, \frac{\pi}{2}\right[\setminus \{x'\}$.*

c) *For $0 \leq x_1 < x_2 < \pi/2$, we have*

$$f'(x_1) \leq f'(x_2) \Longleftrightarrow \cos x_1 \cdot \cos x_2 \leq \frac{\sqrt{2}}{2} \quad [1, \text{p. } 65].$$

Proof. a) and b) are well-known properties of strictly convex (or concave) functions.

c) Using (2) and $(\cos x_1)^2 - (\cos x_2)^2 > 0$, $\cos x_1 > 0$, $\cos x_2 > 0$, the assertion easily follows. □

§ 2

Theorem 3. *Among all pairs A, B of triangles related as described in the abstract, the pair of regular ones and only it minimizes $F(A) + F(B)$. The minimum value is $\frac{15\sqrt{3}}{4}$.*

Proof. Let $x_1, x_2, x_3 \in \left]0, \frac{\pi}{2}\right[$, $x_1 + x_2 + x_3 = \pi$. By definition, $\vartheta < \pi/4$, so at most one of the x_i's is less than ϑ.

Case 1: $x_1, x_2, x_3 \in \left[\vartheta, \frac{\pi}{2}\right[$. Lemma 1 ensures $f(x_1) + f(x_2) + f(x_3) \geq 3f\left(\frac{\pi}{3}\right) = \frac{15\sqrt{3}}{4}$ with equality if and only if $x_1 = x_2 = x_3 = \pi/3$.

Case 2: Let $0 < x_3 < \vartheta$; $x_1, x_2 \in \left[\vartheta, \frac{\pi}{2}\right[$. By Lemma 1, $f\left(\frac{1}{2}(x_1 + x_2)\right) < \frac{1}{2}[f(x_1 + f(x_2)]$ for $x_1 \neq x_2$ whence it is sufficient to consider the situation $x_1 = x_2 =: x$, and then $x_3 = \pi - 2x$, $0 < \pi - 2x < \vartheta \leq x < \pi/2$. This may immediately be sharpened to

$$0 < \pi - 2x < \vartheta < \frac{\pi}{4} < \frac{\pi - \vartheta}{2} < x < \frac{\pi}{2}. \tag{5}$$

By Lemma 2 b) $f'\left(\frac{\pi}{4}\right) < q\left(\frac{\pi}{4}, \frac{\pi-\vartheta}{2}\right) < f'\left(\frac{\pi-\vartheta}{2}\right)$, and by Lemma 2 c) $f'(0) = f'\left(\frac{\pi}{4}\right)$. Now let $\frac{\pi-\vartheta}{2} \le x < y < \frac{\pi}{2}$, i.e., $0 < \pi - 2y < \pi - 2x \le \vartheta$. From Lemma 2 a), b) we obtain

$$q(\pi - 2x, \pi - 2y) = q(\pi - 2y, \pi - 2x) < q(0, \pi - 2x) = q(\pi - 2x, 0) < f'(0)$$

$$= f'\left(\tfrac{\pi}{4}\right) < f'\left(\tfrac{\pi-\vartheta}{2}\right) < q\left(y, \tfrac{\pi-\vartheta}{2}\right) = q\left(\tfrac{\pi-\vartheta}{2}, y\right) \le q(x, y) = q(y, x),$$

thus $\frac{f(\pi-2x)-f(\pi-2y)}{2(y-x)} < \frac{f(y)-f(x)}{y-x}$, $2f(x) + f(\pi - 2x) < 2f(y) + f(\pi - 2y)$, i.e.,

(6) $\qquad 2f(\cdot) + f(\pi - 2\cdot)$ strictly increasing on $\left[\frac{\pi-\vartheta}{2}, \frac{\pi}{2}\right[$.

From (5), (6) and Lemma 1 we conclude $f(x_1)+f(x_2)+f(x_3) \ge 2f(x)+f(\pi-2x) \ge 2f\left(\frac{\pi-\vartheta}{2}\right) + f(\vartheta) > 3f\left(\frac{\pi}{3}\right)$, the last inequality by Case 1. □

Remark 4. The sequence $\left(nf\left(\frac{\pi}{n}\right)\right)$ $(n \ge 4)$ is strictly increasing and converges to 2π as $n \to \infty$. In fact: Since $\pi/6 < \vartheta$, Lemma 2 a) guarantees $q\left(\frac{\pi}{n}, 0\right) < q\left(\frac{\pi}{n+1}, 0\right)$ for $n \ge 6$, i.e., $nf\left(\frac{\pi}{n}\right) = \pi q\left(\frac{\pi}{n}, 0\right) < \pi q\left(\frac{\pi}{n+1}, 0\right) = (n+1)f\left(\frac{\pi}{n+1}\right)$ for $n \ge 6$. This is complemented by $4f\left(\frac{\pi}{4}\right) = 6$, $5f\left(\frac{\pi}{5}\right) = \frac{5}{8}\sqrt{\frac{5-\sqrt{5}}{2}}\,(5\sqrt{5} - 3) \approx 6.01$, $6f\left(\frac{\pi}{6}\right) = \frac{7\sqrt{3}}{2} \approx 6.06$, so the monotonicity assertion is established. Moreover, $nf\left(\frac{\pi}{n}\right) = \pi q\left(\frac{\pi}{n}, 0\right) \to \pi f'(0) = 2\pi$ $(n \to \infty)$.

Corollary 5. *Let S_n be the set*

$$S_n := \left\{ \sum_{i=1}^{n} f(x_i);\ 0 < x_i < \frac{\pi}{2}\ (i = 1, \ldots, n),\ \ \sum_{i=1}^{n} x_i = \pi \right\}, \text{ where } n \ge 3.$$

We have $\inf S_3 = \frac{15\sqrt{3}}{4} \approx 6.495$ *and* $\inf S_n = 6$ *for* $n \ge 4$.

Proof. Theorem 3 implies the first statement. By the theorem stated in the abstract above, the number 6 is a lower bound of each S_n $(n \ge 3)$, and $\inf S_4 = 6$. Let $n \ge 5$ and choose $z \in \left]0, \frac{\pi}{4(n-4)}\right[$ arbitrary. Then $0 < (n-4)z < \pi/4$. We now put $x_i = z$ $(i = 1, \ldots, n-4)$, $x_{n-3} = \frac{\pi}{4} - (n-4)z$, $x_{n-2} = x_{n-1} = x_n = \pi/4$. Then $0 < x_i < \pi/2$ $(i = 1, \ldots, n)$ and $x_1 + \cdots + x_n = \pi$. For $z \to 0+$ we get

$$\sum_{i=1}^{n} f(x_i) = (n-4)f(z) + f\left(\frac{\pi}{4} - (n-4)z\right) + 3f\left(\frac{\pi}{4}\right) \to (n-4)f(0) + 4f\left(\frac{\pi}{4}\right) = 6.$$

□

Remark 6. By Remark 4, $nf\left(\frac{\pi}{n}\right) > 4f\left(\frac{\pi}{4}\right) = 6$ for $n \geq 5$. Corollary 5 now says that, with respect to the area sum, the pair of regular n-gons is minimal among all pairs of n-gons if and only if $n = 3$ or $n = 4$.

References

1. J. Aczél and L. Fuchs, *A minimum problem on areas of inscribed and circumscribed polygons of a circle.* Compos. Math. **8** (1950), 61–67.

2. G.H. Hardy, J.E. Littlewood and G. Pólya, *Inequalities.* Cambridge Univ. Press 1934.

3. L. Kuipers, *On a minimum problem.* Nieuw Tijdschr. Wiskunde **38** (1951), 157–163.

4. J. Rätz, *A minimum property of the square.* In: General Inequalities 4, ISNM, Vol. 71, Birkhäuser Basel 1984, 415–416.

5. E. Trost, *Beweis einer Minimaleigenschaft des Quadrates.* Elemente Math. **6** (1951), 26–23.

Jürg Rätz, Mathematisches Institut, Universität Bern, Sidlerstr. 5, CH-3012 Bern, Switzerland

International Series of Numerical Mathematics, Vol. 103, © 1992 Birkhäuser Verlag Basel

On the Landau-Hadamard inequality

B. Saffari

1. Introduction

For any bounded real-valued function f defined on the whole real line $\mathbb{R}$ or on the half-line $(0, +\infty)$, denote by $\omega(f)$ the oscillation of f,

$$\omega(f) = \sup f - \inf f = \sup_{x,y} \, (f(x) - f(y)),$$

and put, as usual,

$$||f||_\infty = \sup_x \, |f(x)| = \max \, (|\sup f|, |\inf f|).$$

The classical Landau-Hadamard inequality says that if f is twice differentiable and if f'' is also bounded, then so is f', and moreover:

(1) $$||f'||_\infty^2 \leq 2||f||_\infty \cdot ||f''||_\infty \quad \text{(for } f \text{ defined on the whole line)},$$

(2) $$||f'||_\infty^2 \leq 4||f||_\infty \cdot ||f''||_\infty \quad \text{(for } f \text{ defined on the half-line)}.$$

These are proved by using a second order Taylor expansion for f. Actually, the same idea yields an inequality slightly sharper than (1), namely:

(3) $$||f'||_\infty^2 \leq \frac{1}{2}\omega(f)\,\omega(f'') \quad \text{(for } f \text{ defined on the whole line)}.$$

Indeed (3) is a slight refinement of (1) because $\omega(f) \leq 2||f||_\infty$ and $\omega(f'') \leq 2||f''||_\infty$.

Our purpose here is to point out that the *same* Taylor expansion method provides a further refinement of (3) (for f defined on the whole line). But it seems that in the "half-line case" much less can be said.

2. The case when f is defined on the whole line

Put

$$A = \sup f'' \quad \text{and} \quad -B = \inf f''.$$

Suppose henceforth that f is not identically constant on the whole line, so that necessarily

$$(4) \qquad A > 0 \ \underline{\text{and}} \ B > 0$$

(indeed, if we had $A \leq 0$, then f would be concave on $\mathbb{R}$; but the only functions that are simultaneously bounded and concave on the *whole* line are the identically constant functions; similarly we cannot have $B \leq 0$.)

Now $\omega(f'') = A + B$, so that inequality (3) can be rewritten as the equivalent inequality:

$$(3)^* \qquad ||f'||_\infty^2 \leq \omega(f) \cdot \frac{A+B}{2}.$$

We note that the term $(A+B)/2$ in $(3)^*$ is the *arithmetic* mean between A and B, and now we wish to point out that $(3)^*$ remains valid if this arithmetic mean $(A+B)/2$ is replaced by the *harmonic* mean $2AB/(A+B)$:

Theorem. *If f is not identically constant, then* (3) *and* $(3)^*$ *can be replaced by the sharper inequality*

$$(5) \qquad ||f'||_\infty^2 \leq \omega(f) \cdot \frac{2AB}{A+B}.$$

Proof. Let h and λ be any real numbers. From the two Taylor formulas

$$f(x+\lambda h) = f(x) + \lambda h f'(x) + \frac{1}{2}\lambda^2 h^2 f''(x+\alpha\lambda h) \quad (\text{with } 0 < \alpha < 1),$$

$$f(x-h) = f(x) - hf'(x) + \frac{1}{2}h^2 f''(x-\beta h) \quad (\text{with } 0 < \beta < 1)$$

we derive

$$f(x-h) - f(x+\lambda h) + (\lambda+1)hf'(x) + \frac{1}{2}\lambda^2 h^2 f''(x+\alpha\lambda h) - \frac{1}{2}h^2 f''(x-\beta h) = 0,$$

hence:

$$\omega(f) + (\lambda+1)hf'(x) + \frac{1}{2}h^2(\lambda^2 A + B) \geq 0.$$

Since, for any *fixed* λ, the above quadratic inequality holds for all h, it follows that the discriminant is ≤ 0. Thus,

$$(\lambda+1)^2(f'(x))^2 - 2(\lambda^2 A + B)\omega(f) \leq 0.$$

Hence, if $\lambda \neq -1$,

$$(f'(x))^2 \leq 2\omega(f) \cdot \frac{\lambda^2 A + B}{(\lambda+1)^2}.$$

As $\lambda \neq -1$ varies in $\mathbb{R}$, the minimum of the right-hand side (reached at $\lambda = B/A$) is $\omega(f) \cdot 2AB/(A+B)$. Thus (5) is proved.

Remark. It can be shown that the harmonic mean in (5) is best possible, that is, cannot be replaced by any power mean

$$M_p(A,B) = \left(\frac{A^p + B^p}{2}\right)^{1/p}$$

of order p less than -1.

3. The case when f is defined on the half-line

The "whole line inequality" (5), rewritten (equivalently) as

$$-\sqrt{\omega(f) \cdot \frac{2AB}{A+B}} \leq f'(x) \leq \sqrt{\omega(f) \cdot \frac{2AB}{A+B}}$$

provides an optimal *symmetric* interval (around the origin) in which $f'(x)$ is imprisoned. In contrast, in the "half-line case" (that is, if f is defined on the positive half-line only), we have the *asymmetric* (optimal) inequalities

$$-\sqrt{2A \cdot \omega(f)} \leq f'(x) \leq \sqrt{2B \cdot \omega(f)}, \tag{6}$$

where $A = \sup f''$ and $B = -\inf f''$. Not that in this "half-line case" we have $A \geq 0$ and $B \geq 0$; but when f is not identically constant, then we have *at least one* of the strict inequalities $A > 0$ and $B > 0$, but not necessarily both (contrary to (4) in the "whole line case").

The proof of (6) is straightforward: We have, for $h \neq 0$,

$$f'(x) = \frac{f(x+h) - f(x)}{h} - \frac{h}{2} f''(x + \theta H) \quad (\text{with } 0 < \theta < 1).$$

If $h > 0$, this implies

$$f'(x) \leq \frac{\omega(f)}{h} + \frac{h}{2} B,$$

whence the right side inequality (6) follows by optimizing over h. Similarly, if $h < 0$, we obtain the left side inequality (6).

Corollary to (6). *In the "half-line case" we have the following upper bound for* $\omega(f')$*:*

$$\omega(f') \leq \left(\sqrt{A} + \sqrt{B}\right) \sqrt{2\omega(f)} = 2\sqrt{2} \cdot \sqrt{\omega(f) \cdot M_{1/2}(A, B)}, \tag{7}$$

and it can be shown that the mean $M_{1/2}(A, B)$ *in* (7) *cannot be replaced by any power mean* $M_p(A, B)$ *of order less than* $1/2$.

4. Concluding remark

The observation (6) is certainly not new, and has probably been observed long ago (e.g., by Cavaretta?).

B. Saffari, Mathématiques, Bâtiment 425, Université de Paris-Orsay, F-91405 Orsay, France

International Series of Numerical Mathematics, Vol. 103, © 1992 Birkhäuser Verlag Basel

A lemma on the angles between a fixed line and the lines connecting a fixed point on it with the points of a convex arc

János Aczél and Che Tat Ng

The following result was observed by M. Kuczma in private correspondence. He supposed differentiability but conjectured that this condition can be dropped and asked for a simple proof. We give one here.

Lemma. *If f is convex (or concave) on an interval $I \subseteq]0,\infty[$ then $x \mapsto f(x)/x$ is either monotonic or unimodal (consists of two monotonic segments) there.*

Proof. Clearly (Figure 1), $f(x)/x$ is monotonic if, and only if, $\phi = \arctan(f(x)/x)$ is monotonic (ϕ is the varying angle in the title). Also, for an $f : I \to \mathbb{R}$, the ratio $f(x)/x$ (or the angle ϕ) is monotonic or unimodal if, and only if, it is monotonic or unimodal on every finite closed interval $J \subseteq I$. Let f be restricted to J and, say, convex. By compactness, there exists through 0 a "supporting straight line" (Stützgerade) from below to the graph of f (need not be a tangent). Let S be a common point of this line with the graph of f (over J). Figures 2 to 4 illustrate the three possible positions S may take on the graph. The last possibility is elaborated in Figure 5.

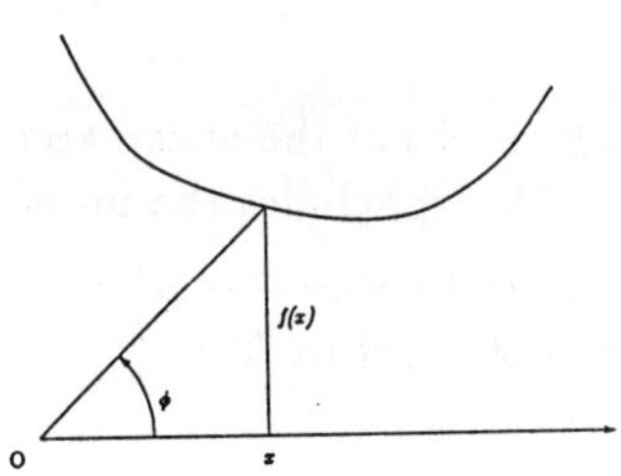

Figure 1

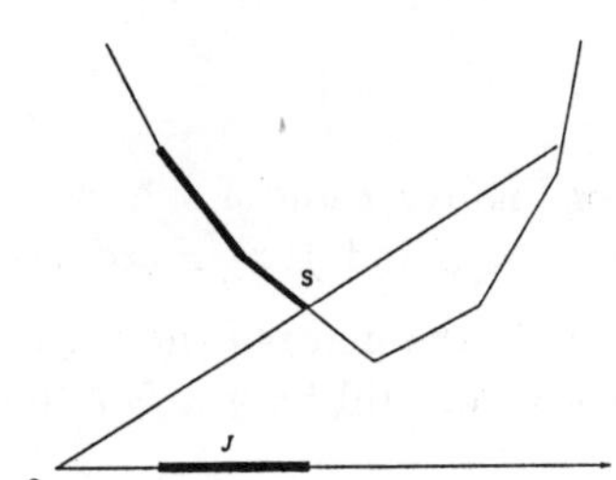

Figure 2

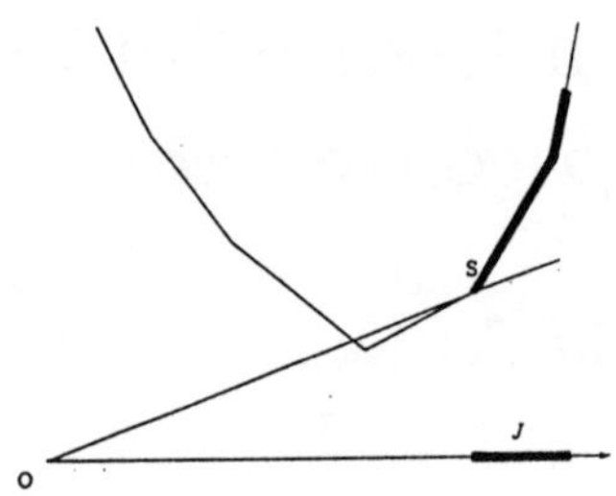

Figure 3

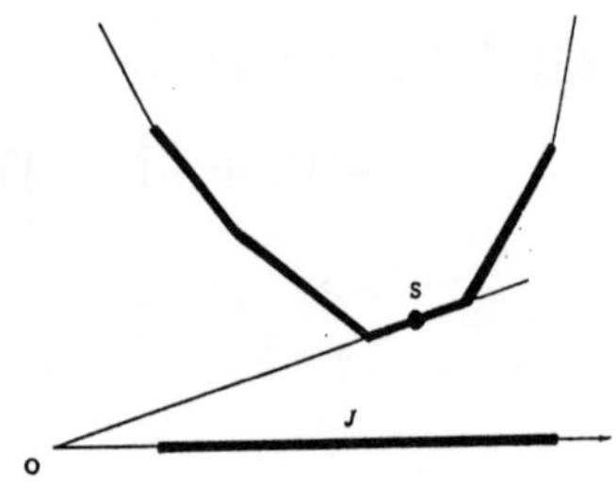

Figure 4

In Figure 5, by convexity, B is inside the triangle OAS (or *on* it if f is *not strictly* convex), so $\beta \leq \alpha$. Also, C is under SD, so much the more under OD, therefore $\delta \geq \gamma$. Thus *the angle decreases till* S, *increases after* S.

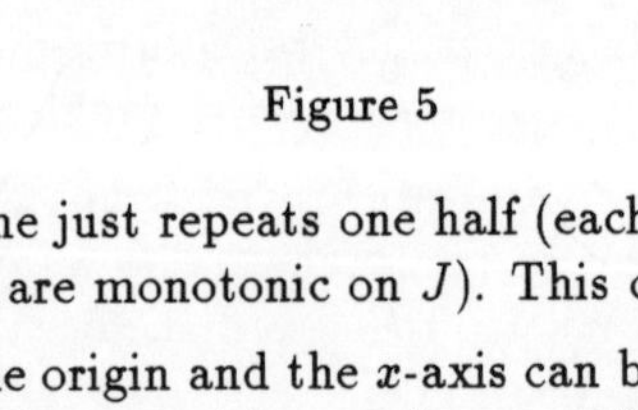

Figure 5

For Figures 2 and 3 one just repeats one half (each) of the above argument (in these cases ϕ and $f(x)/x$ are monotonic on J). This concludes the proof.

Clearly the place of the origin and the x-axis can be moved (shifted or rotated) as long as we still have a *function* (the arc does not "curl back").

János Aczél and Che Tat Ng, Department of Pure Mathematics, University of Waterloo, Waterloo, Ontario N2L 3G1, Canada

International Series of Numerical Mathematics, Vol. 103, © 1992 Birkhäuser Verlag Basel

Remarks

B.M. Brown, W.D. Evans and W.N. Everitt

Let Δ_0 denote the linear manifold of the space $L^2[0,\pi]$ defined by

$$\Delta_0 := \{f : [0,\pi] \to R \mid f, f' \in AC[0,\pi] \text{ and } f'' \in L^2[0,\pi]\}.$$

The inequality in question is

$$\text{(I)} \quad \left(\int_0^\pi (f'(x)^2 - f(x)^2)\,dx\right)^2 \le K \int_0^\pi f(x)^2 dx \int_0^\pi (f''(x) + f(x))^2 dx \quad (f \in \Delta),$$

where f is taken to belong to a linear manifold $\Delta \subseteq \Delta_0$. If (I) is a valid inequality we take K as the best (smallest) positive number, and then seek for all non-trivial cases of equality.

In the space $L^2[0,\pi]$ the set Δ_0 is the maximal linear manifold for which the inequality (I) can be considered.

The inequality (I) is a HELP integral inequality; references [1], [2] and [3].

We discuss three cases of this inequality on different choices of the domain Δ. Two of these cases require 'transcendental' proofs; the best possible numbers K can be characterised but are determined only from the roots of transcendental equations; also the non-trivial cases of equality appear as combinations of trigonometrical functions involving these roots.

The third case can be considered to have an 'elementary' proof in that only integration by parts, the Cauchy-Schwarz integral inequality, and a knowledge of the solutions of a simple Fourier boundary valve problem are required.

The use here of the terms 'elementary' and 'transcendental' is highly relative to the framework in which any particular inequality is considered. Such interest in the classification of proofs of inequalities was first raised by Hardy; see his address [4] and, in particular, [4, pages 63 to 66].

Let $\Delta_2 \subset \Delta_1 \subset \Delta_0$ be defined by

$$\Delta_1 := \{f \in \Delta_0 \mid f(\pi) = 0\}, \qquad \Delta_2 := \{f \in \Delta_0 \mid f(0) = f(\pi) = 0\},$$

where, clearly, the inclusion is strict.

As Bennewitz makes clear, see [1, Section 5] and [2, Sections 0 and 1] it is very exceptional that inequality (I) is valid when $\Delta = \Delta_0$ the maximal domain. From the detailed results in [2, Section 0] we have the result:

(I_0) if $\Delta = \Delta_0$ then the inequality (I) is valid with $K = K_0 \doteq 6.15$

with the number K_0 determined by consideration of a transcendental equation [1, Section 5] with a numerical approximation, quoted above in (I_0), given in [2, Section 0]. There are two distinct cases of non-trivial equality in (I_0) see [2],

(i) $f(x) = A\sin(x) + B\cos(x)$ ($x \in [0, \pi]$) with $A, B \in \mathbb{R}$ when both sides of the inequality reduce to zero.

(ii) a two-dimensional set of triogonometrically based functions, dependent upon the transcendental equation which determines K_0, for which both sides of inequality (I_0) are not zero.

In Evans and Everitt [3, Section 5.2] it is shown, independently of the inequality (I_0), that the inequality (I) is valid when $\Delta = \Delta_1$; we have:

(I_1) if $\Delta = \Delta_1$ then the inequality (I) is valid with $K = K_1 \doteq 4.64$.

with the number K_1 determined by consideration of a transcendental equation [3, Theorem 4.1] with a numerical approximation, quoted above in (I_1), given in [3, Section 5.2, page 389].

There are two distinct cases of non-trivial equality in (I_1), see [3, Theorem 4.1],

(i) $f(x) = A\sin(x)$ ($x \in [0, \pi]$) with $A \in \mathbb{R}$ when both sides of the inequality reduce to zero.

(ii) a one-dimensional set of trigonometrically based functions, dependent upon the transcendental equation which determines K_1, for which both sides of inequality (I_1) are not zero.

Both the inequalities (I_0) and (I_1) may be considered to require 'transcendental' proofs. On the other hand the inequality (I) on $\Delta = \Delta_2$ is valid but requires only an 'elementary' proof. We have

(I_2) if $\Delta = \Delta_2$ then the inequality (I) is valid with $K = K_2 = 1$.

All cases of equality can be described by

$$f(x) = A\sin(nx) \quad (x \in [0, \pi],\ n = 1, 2, ...) \text{ with } A \in \mathbb{R}.$$

When $n = 1$ both sides of the inequality (I_2) are zero; when $n \geq 2$ both sides are not zero.

The proof of (I_2) is obtained on integrating by parts the integral of f'^2 on the left-hand side, use of the boundary conditions for Δ_2, and then an application of the Cauchy-Schwarz inequality; this yields $K_2 \leq 1$. The cases of equality for $K_2 = 1$ are obtained by examing the case for equality in the application of the Cauchy-Schwarz inequality; the equalising functions appear as the non-trivial solutions of the boundary value problem

$$-y'' - y = \lambda y \quad \text{on} \quad [0, \pi], \qquad y(0) = y(\pi) = 0$$

which yields the result quoted above.

References

1. C. Bennewitz, *A general version of the Hardy-Littewood-Pólya-Everitt (HELP) inequality.* Proc. Roy. Soc. Edinb. (A) **97** (1984), 9–20.

2. C. Bennewitz, *The HELP inequality in the regular case.* In: General Inequalities 5 (Proc. Oberwolfach, 1986), edited by W. Walter, Birkhäuser-Verlag, Basel, 1987.

3. W.D. Evans and W.N. Everitt, *HELP inequalities for limit-circle and regular problems.* Proc. Roy. Soc. London A **432** (1991), 367–390.

4. G.H. Hardy, *Prolegomena to a chapter on inequalities.* J. London Math. Soc. **4** (1929), 61–78.

B.M. Brown, Department of Computing Mathematics, University of Wales, College of Cardiff, Senghennydd Road, Cardiff CF2 4YN, U.K.

W.D. Evans, School of Mathematics, University of Wales, College of Cardiff, Senghennydd Road, Cardiff CF2 4AG, U.K.

W.N. Everitt, School of Mathematics and Statistics, The University of Birmingham, Birmingham B15 2TT, U.K.

International Series of Numerical Mathematics, Vol. 103, © 1992 Birkhäuser Verlag Basel

A simple proof of the Minkowski inequality

Heinz König

The most common proof of the Minkowski inequality is via the Hölder inequality. Here is a short direct proof.

Lemma. *Let $1 \leq p < \infty$. Then*

$$(u+v)^p = \inf \left\{ t^{1-p}u^p + (1-t)^{1-p}v^p : 0 < t < 1 \right\} \quad \textit{for all real } u, v \geq 0.$$

Sketch of the *proof*: The nontrivial case is $p > 1$ and $u, v > 0$. Then the function $t \mapsto t^{1-p}u^p + (1-t)^{1-p}v^p$ in $t \in]0,1[$ tends to ∞ at the endpoints, and its minimum point and value can be determined by differential calculus.

Now let $f, g \in L^p$ with $1 \leq p < \infty$ on some measure space. From the lemma we obtain

$$|f+g|^p \leq (|f|+|g|)^p \leq t^{1-p}|f|^p + (1-t)^{1-p}|g|^p \quad \text{for } 0 < t < 1,$$

and hence by integration

$$||f+g||^p \leq t^{1-p}||f||^p + (1-t)^{1-p}||g||^p \quad \text{for } 0 < t < 1,$$

with $||\cdot||$ the L^p norm. Once more from the lemma we obtain $||f+g||^p \leq (||f|| + ||g||)^p$, and hence the Minkowski inequality.

Heinz König, Universität des Saarlandes, Fachbereich 9 Mathematik, D-6600 Saarbrücken, Germany

International Series of Numerical Mathematics, Vol. 103, © 1992 Birkhäuser Verlag Basel

On a property of convex functions

Heinz König

For an interval $I \subset \mathbb{R}$ define $M(I)$ to consist of the functions $\varphi : I \to \mathbb{R}$ which are either monotone or, for some $c \in I$, monotone decreasing on $\{x \in I : x \leq c\}$ and monotone increasing on $\{x \in I : x \geq c\}$. The property in question is that for a continuous convex function $f : I \to \mathbb{R}$ on an interval $I \subset]0, \infty[$ the function $x \mapsto f(x)/x$ is in $M(I)$. It has been presented by János Aczél as a problem attributed to Marek Kuczma, with mention of a solution due to himself and to Che Tat Ng. Here we want to present a simple proof.

Lemma. *For a function $\varphi : I \to \mathbb{R}$ on an interval $I \subset \mathbb{R}$ we have*

$$\varphi \in M(I) \Longrightarrow (\star) \quad \varphi(x) \leq \max(\varphi(u), \varphi(v)) \quad \text{for all } u < x < v \text{ in } I,$$

and $\Longleftarrow$ if φ is lower semicontinuous on I.

The *proof* is via appropriate manipulation with the subsets $I^\pm$ of I, defined to consist of the points $u \in I$ such that there exists an $x \in I$ with $x < u$ resp. $x > u$ and with $\varphi(x) < \varphi(u)$.

Proposition. *For a convex function $f : I \to \mathbb{R}$ on an interval $I \subset]0, \infty[$ the function $\varphi : \varphi(x) = f(x)/x$ for $x \in I$ satisfies $(\star)$, and hence is in $M(I)$ if f is continuous (= lower semicontinuous at the endpoints of I which are in I).*

Proof. Assume that $(\star)$ is violated for $u < x < v$ in I, that is $f(u) < \frac{u}{x} f(x)$ and $f(v) < \frac{v}{x} f(x)$. Then by the definition of convexity

$$\begin{aligned} (v-u)f(x) &\leq (v-x)f(u) + (x-u)f(v) \\ &< \left((v-x)\frac{u}{x} + (x-u)\frac{v}{x}\right) f(x) = (v-u)f(x), \end{aligned}$$

and hence a contradiction.

Heinz König, Universität des Saarlandes, Fachbereich 9 Mathematik, D-6600 Saarbrücken, Germany

International Series of Numerical Mathematics, Vol. 103, © 1992 Birkhäuser Verlag Basel

On a generalization of the plank problem

Zs. Páles

A *strip* or a *plank* S in $\mathbb{R}^n$ is a closed set bounded by two parallel hyperplanes. The distance of these hyperplanes is called the *width* of S. The *minimal width* of a convex closed set K is the minimal width of a strip containing K.

The following theorem was conjectured by A. Tarski in 1932 and proved by T. Bang [2] in 1951:

If a closed convex set K in $\mathbb{R}^n$ is covered by a finite number of strips, then the sum of their widths is greater than or equal to the minimal width of K.

This result has recently been generalized to Banach spaces by K. Ball [1].

If $n = 2$ and K is the unit disc then there is an extremely simple proof for the above result.

Assume that the unit disc in $\mathbb{R}^2$ is covered by strips $S_1, \ldots, S_k$ with widths $d_1, \ldots, d_k$. Without loss of generality we can also assume that both bounding lines of the strips intersect the unit circle. Now we consider the unit sphere in $\mathbb{R}^3$ and to each strip S_i in $\mathbb{R}^2$ we construct a three-dimensional strip S_i^* which is of width d_i and intersects the xy-plane in S_i. Since $S_1, \ldots, S_k$ cover the unit disc, hence $S_1^*, \ldots, S_k^*$ cover the unit sphere. The area of the intersection of the unit sphere and the strip S_i^* is $2\pi d_i$ independently of the position of the i-th strip. (This is a well knonw fact from calculus, already discovered by Archimedes.) Thus the sum of these areas exceeds the area of the unit sphere, i.e.,

$$\sum_{i=1}^{k} 2\pi d_i \geq 4\pi \quad \Longrightarrow \quad \sum_{i=1}^{k} d_i \geq 2,$$

which was to be proved.

We can interpret this proof in the following way: If S is a subset of the disc then we project it up to the sphere, measure the area of the projection and call this number the μ-measure of S. Then the μ-measure of the intersection of a strip and the disc is the width of the strip times 2π. Then the statement is a simple consequence of the subadditivity of μ. In what follows we generalize this idea and extend the result discussed above.

An *angular domain* in $\mathbb{R}^2$ is a closed convex set D bounded by two halflines. The *angle* of D is the angle closed by the bounding halflines. The *vertex* of D is the common endpoint of these two halflines.

Theorem 1. *Let two concentric circles k and K be given on the plane with radii r and R, $r < R$. Assume that the disc bounded by k is covered by angular domains whose vertices are within K. Then the sum of the angles of these angular domains is greater than or equal to the view angle of k from an arbitrary point of K.*

Remark. This result was proposed as a problem by the author on the 1985 M. Schweitzer competition (see [3]).

Proof. Denote by O the common center of the circles and by $D_1, \ldots, D_k$ the given angular domains with angles $\alpha_1, \ldots, \alpha_k$. An angular domain D will be called *regular* if the vertex of D is on K and both bounding halflines of D intersect k. Without loss of generality, we can assume that $D_1, \ldots, D_k$ are regular domains.

The idea of the proof is the following: We construct a rotation invariant non-negative measure μ on the closed disc T bounded by k such that the measure of the intersection of D and T is α, where D is an arbitrary regular angular domain with angle α. Having such a measure we can give a one line proof for the theorem:

$$\sum_{i=1}^{k} \alpha_i = \sum_{i=1}^{k} \mu(D_i \cap T) \geq \mu\left(\bigcup_{i=1}^{k} (D_i \cap T)\right) = \mu(T),$$

and observe that $\mu(T)$ is exactly the view angle of k from any point of K.

Now we construct the desired measure. Let $P \in T$ be an arbitrary point, and denote by ρ the distance of P and O. Then define

$$F(P) = f(\rho) = \frac{1}{\pi} \cdot \frac{1}{R^2 - \rho^2} \cdot \sqrt{\frac{R^2 - r^2}{r^2 - \rho^2}}.$$

If S is a Lebesgue measurable subset of T then let

$$\mu(S) = \int_S F(P)\, dP.$$

Obviously μ is a rotation invariant nonnegative measure on T. To prove the key property of μ, let D be a regular angular domain with vertex A and angle α. Then we want to show $\mu(D) = \alpha$. Without loss of generality we can assume that one of the bounding halflines of D is tangent to k at the point Q. (In the general case D can be obtained as the difference of two such angular domains.) Denote by ε the angle $OAQ <$ and by d the (signed) distance of the other bounding halfline from O. (This distance is positive if O is outside D, and negative if O is inside D.) Then $d = R\sin(\varepsilon - \alpha)$. Using successive integration, we obtain

$$\begin{aligned} \mu(D) &= \int_D F(P)\,dP = \int_d^r \int_{-\arccos(d/\rho)}^{\arccos(d/\rho)} f(\rho)\rho\,d\varphi\,d\rho \\ &= \int_d^r 2f(\rho)\rho \arccos(d/\rho)\,d\rho \\ &= 2f(\rho)\rho \arccos\left(\frac{R\sin(\varepsilon-\alpha)}{d}\right) d\rho. \end{aligned}$$

Thus we have to show that

$$\int_{R\sin(\varepsilon-\alpha)}^r 2f(\rho)\rho \arccos\left(\frac{R\sin(\varepsilon-\alpha)}{d}\right) d\rho = \alpha$$

for all $0 \le \alpha \le 2\varepsilon$. Substituting the new variable $t = R\sin(\varepsilon - \alpha)$ this reduces to

$$\int_t^r 2f(\rho)\rho \arccos(t/\rho)\,d\rho = \varepsilon - \arcsin(t/R)$$

for all $-r \le t \le r$. This latter equation is obviously valid for $t = r$, thus it suffices to show that the derivatives of both sides with respect to t are identical, i.e.,

$$\int_t^r \frac{2f(\rho)\rho}{\sqrt{\rho^2 - t^2}}\,d\rho = \frac{1}{\sqrt{R^2 - t^2}}, \qquad -r < t < r.$$

However,

$$\begin{aligned} \int_t^r \frac{2f(\rho)\rho}{\sqrt{\rho^2-t^2}} &= \int_t^r \frac{2}{\pi}\cdot\frac{\rho}{R^2-\rho^2}\cdot\sqrt{\frac{R^2-r^2}{(r^2-\rho^2)(\rho^2-t^2)}}\,d\rho \\ &= \left[\frac{2}{\pi}\cdot\frac{1}{\sqrt{R^2-t^2}}\cdot\arctan\sqrt{\frac{R^2-r^2}{R^2-t^2}\cdot\frac{\rho^2-t^2}{r^2-\rho^2}}\right]_{\rho=t}^{\rho=r} \\ &= \frac{1}{\sqrt{R^2-t^2}}. \end{aligned}$$

Thus the proof is complete.

Remark. When $n = 2$ and K is the unit disc, then the statement of the plank problem can easily be derived from our theorem. Denote the unit disc by T and assume that it is covered by strips $S_1, \dots, S_k$ (whose bounding lines intersect T). Take a concentric circle K with radius R, where R is sufficiently large. Assume that the two bounding lines of S_i intersect K in A_i, B_i and in C_i, D_i. We choose the notation such that S_i is covered by the two regular angular domains $A_iB_iD_i <$ and $B_iD_iC_i <$. Denote by α_i' and α_i'' their angle and by d_i the width of S_i. Then we have

$$\frac{d_i}{2R-2} \geq \tan \alpha_i' \geq \alpha_i', \qquad \frac{d_i}{2R-2} \geq \tan \alpha_i'' \geq \alpha_i''.$$

Thus the theorem yields

$$2\sum_{i=1}^{k} \frac{d_i}{2R-2} \geq \sum_{i=1}^{k} (\alpha_i' + \alpha_i'') \geq 2\varepsilon \geq 2\sin\varepsilon \geq \frac{2}{R}.$$

Now taking the limit $R \to \infty$ we obtain the statement.

Acknowledgement: This research was supported by the Hungarian National Science Foundation Operating Grant No. OTKA 1652.

References

1. K. Ball, *The plank problem in general normed spaces.* Manuscript.

2. T. Bang, *A solution of the "Plank problem".* Proc. Amer. Math. Soc. **2** (1951), 990–993.

3. B. Brindza and Zs. Páles, *Jelentés az 1985. évi Schweitzer Miklós emlékversenyről* (Report on the 1985 M. Schweitzer memorial competition, in Hungarian). Matematikai Lapok **13** (1982–1986), 149–169.

Zs. Páles, Institute of Mathematics, L. Kossuth University, H-4010 Debrecen, Pf. 12, Hungary

International Series of Numerical Mathematics, Vol. 103, © 1992 Birkhäuser Verlag Basel

Three problems

János Aczél

1. What are all linearly homogeneous functions

$$f(rx_1, \ldots, rx_n) = rf(x_1, \ldots, x_n),$$

($x_k, r \in \mathbb{R}$ or $x_k, r \in \mathbb{R}_+$ or $x_k \in \mathbb{R}$, $r \in \mathbb{R}_+$; $k = 1, 2, \ldots, n$), strictly increasing in each of $x_1, \ldots, x_n$? Here, $\mathbb{R}_+ =]0, \infty[$.

2. What are the solutions, strictly increasing in each of $x_1, \ldots, x_n$, of

$$f(rx_1 + s, \ldots, rx_n + s) = rf(x_1, \ldots, x_n) + s \quad (r > 0,\ x_1, \ldots, x_n, s \in \mathbb{R})?$$

(This question was asked by R.D. Luce, Irvine Research Unit Math. Behav. Sci., Univ. of Calif., Irvine, CA 92717, USA.)

3. K.A. Post (Nieuw Archief voor Wiskunde (4) 7 (1989), 117–119) shows that, given the arithmetic and harmonic means of n positive numbers, their geometric mean is extremal iff $(n - 1)$ of the numbers are equal. The proof is done by derivation and Lagrange multipliers. There ought to be an elementary proof.

János Aczél, Department of Pure Mathematics, University of Waterloo, Waterloo, Ontario, Canada N2L 3G1

International Series of Numerical Mathematics, Vol. 103, © 1992 Birkhäuser Verlag Basel

A problem

Claudi Alsina

According to the isoperimetric inequality ($4\pi A \leq P^2$, where A is the area and P the perimeter of a plane domain whose boundary is a closed Jordan curve) it is clear that $12A < P^2$, but it would be interesting to have an *explicit* construction of how taking 12 pairwise disjoint circles, all having radius r, and cutting them in a finite number of parts, one can effectively locate the collection of such pieces in a square of side $2\pi r$.

Claudi Alsina, Sec. Matemàtiques i Informàtica (ETSAB), Universitat Politècnica de Catalunya, Diagonal 649, E-08028 Barcelona, Spain

International Series of Numerical Mathematics, Vol. 103, © 1992 Birkhäuser Verlag Basel

Problem: The Grothendieck constant K_G

Hermann König

The Grothendieck inequality states that for any $m \in \mathbb{N}$, any $(m \times m)$ matrix (a_{jk}), any Hilbert space H and elements $x_j, y_k \in H$

$$\left| \sum_{i,j=1}^{m} a_{jk} \langle x_j, y_k \rangle \right| \leq K_G \cdot \left(\max_{1 \leq j,k \leq n} ||x_j|| \, ||y_k|| \right) \cdot \max_{|s_k|,|t_j| \leq 1} \left| \sum_{j,k=1}^{n} a_{jk} t_j s_k \right|$$

holds. The best constant K_G is the Grothendieck constant. In the real case, $K_G^{\mathbb{R}} \leq \dfrac{\pi}{2 \ln (1+\sqrt{2})}$ is known. This is conjectured to be the best value (Krivine, 1979). In the complex case, the conjecture is that $K_G^{\mathbb{C}} \leq 1 / \int_0^{\pi/2} \dfrac{\cos^2 \theta}{\sqrt{1+\sin^2 \theta}} \, d\theta$ $= 1.40458\ldots$ (Haagerup, 1987). The best known upper value, due to Haagerup, is $K_G^{\mathbb{C}} \leq \dfrac{8}{\pi(1+k_0)} = 1.40491\ldots$, where k_0 is determined from

$$k_0 \int_0^{\pi/2} \frac{\cos^2 \theta}{\sqrt{1-k_0^2 \sin^2 \theta}} \, d\theta = \frac{\pi}{2}(1+k_0).$$

The best known lower bounds (Krivine, Davie, unpublished) are

$$1.677 \leq K_G^{\mathbb{R}}, \qquad 1.338 \leq K_G^{\mathbb{C}}.$$

Hermann König, Mathematisches Seminar, Universität Kiel, Olshausenstr. 40, D-2300 Kiel, Germany

International Series of Numerical Mathematics, Vol. 103, © 1992 Birkhäuser Verlag Basel

A function and an associated inequality

S.H. Lehnigk

In connection with the problem of parameter estimation for the hypergamma probability distribution a function of one positive variable β arises which depends on a fixed parameter $x > 2$. This function has the form

$$(1) \qquad \alpha(\beta; x) = A^{1/2} C^{-1} (1 - A)^{-3/2} (1 - 3C + 2AC), \quad \beta > 0, \quad x > 2,$$

with

$$A(\beta; x) = \frac{\Gamma_2^2}{\Gamma_1 \Gamma_3} = \prod_{\nu=0}^{\infty} \frac{(y_\nu - 1)(y_\nu + 1)}{y_\nu^2},$$

$$C(\beta; x) = \frac{\Gamma_2 \Gamma_3}{\Gamma_1 \Gamma_4} = \prod_{\nu=0}^{\infty} \frac{(y_\nu - 1)(y_\nu + 2)}{y_\nu (y_\nu + 1)},$$

$\Gamma_\nu = \Gamma((x + \nu - 2)/\beta)$ (Γ = Gamma function), $y_\nu = x + \nu\beta$.

There are reasons to believe that the function $\alpha(\beta; x)$ is monotonically decreasing as a function of β (for every fixed $x > 2$). If this conjecture is true the inequality

$$(2) \qquad \frac{A - C}{1 - A} \le \frac{Q}{S}, \quad \beta > 0, \quad x \text{ fixed}, \quad x > 2,$$

must hold. It follows from the β-derivative of α after some manipulations. In (2)

$$Q(\beta; x) = \sum_{\nu=1}^{\infty} \frac{\nu y_\nu}{\rho_\nu}, \quad S(\beta; x) = \sum_{\nu=1}^{\infty} \frac{\nu(y_\nu + 2)}{\rho_\nu},$$

$\rho_\nu = (y_\nu - 1) y_\nu (y_\nu + 1)(y_\nu + 2)$. The infinite series S is defined by $A' = 2AS$ ($' = d/d\beta$), and $3Q \equiv 2U - S$ with U defined by $C' = 2CU$, $U(\beta; x) = \sum_{\nu=1}^{\infty} \frac{\nu(2y_\nu + 1)}{\rho_\nu}$.

A proof of inequality (2) is desired or an argument which shows that (2) does not hold for every $\beta > 0$. It certainly holds for $0 < \beta \le 1$ and for β large.

An inequality related to (2), namely

$$C < \frac{Q}{S}, \quad \beta > 1, \quad x \text{ fixed}, \quad x > 2, \tag{3}$$

has been presented by P. Volkmann at the conference. If (3) holds then so does (2).

S.H. Lehnigk, Amberger Straße 17, D-6748 Bad Bergzabern, Germany

International Series of Numerical Mathematics, Vol. 103, © 1992 Birkhäuser Verlag Basel

A problem

Harry I. Miller

Answering a question of Roman Ger, it has been shown that there exist compact sets $A, B \subseteq \mathbb{R}$ such that $A - B$ contains an interval, but $A + B$ does not. This result will appear in Acta Math. Hungarica and is written jointly with Miljenko Crnjac and Boris Guljaš.

I *conjecture* that:

If $f, g : \mathbb{R}^2 \to \mathbb{R}$ and f_x, f_y, g_x, g_y (partial derivatives) exist and are continuous in a neighborhood of the origin and

$$\frac{f_x(0,0)}{f_y(0,0)} < 0, \quad \frac{g_x(0,0)}{g_y(0,0)} > 0,$$

then there exist compact sets $A, B \subset \mathbb{R}$ such that $\{f(x,y) |\ x \in A,\ y \in B\}$ contains an interval, but $\{g(x,y) |\ x \in A,\ y \in B\}$ does not.

Harry I. Miller, Department of Mathematics, PM Faculty, University of Sarajevo, V. Putnika 43, 71000 Sarajevo, Yugoslavia

International Series of Numerical Mathematics, Vol. 103, © 1992 Birkhäuser Verlag Basel

A remark and problem for polynomials of several variables

Themistocles M. Rassias

David Hilbert [1] in his lecture that he delivered before the International Congress of Mathematicians at Paris in 1900 posed 23 problems; the 16th problem concerns the topology of algebraic curves and surfaces. In [2] I have proved the following result:

Theorem 1. *Let $f(x,y)$ be a polynomial of degree m in two real variables x, y. Then the set*

$$Z_f = \{(x,y) \in \mathbb{R}^2 : f(x,y) = 0\}$$

divides the plane $\mathbb{R}^2$ into at most $1 + [m(m+1)/2]$ connected components.

In [3] I have considered the problem for a homogeneous irreducible polynomial of degree k in three real variables x_0, x_1, x_2, which defines a nonsingular algebraic curve in $X = \mathbb{R}P^2 \simeq S^2/\{\pm 1\}$, and I proved the following

Theorem 2. *Let f be a homogeneous irreducible polynomial of degree k in three real variables x_0, x_1, x_2 which defines a nonsingular algebraic curve in $X = \mathbb{R}P^2 \simeq S^2/\{\pm 1\}$. Let $\overline{Z} = \{\underline{x} \in \mathbb{R}P^2 : f(\underline{x}) = 0\}$, $\underline{x} = [x_0, x_1, x_2]$. Define $\pi : S^2 \to X$ the universal covering map and $Z = \pi^{-1}(\overline{Z}) \subset S^2$. Let n be the number of connected components of $Y = S^2 - Z$. Then*

(a) *If k is even, $n \leq k^2 - 3k + 5$.*

(b) *If k is odd, $n \leq k^2 - 3k + 4$.*

Problem. Consider $f(x_1, x_2, \ldots, x_n)$ to be a polynomial of degree k, where $x_1, x_2, \ldots, x_n \in \mathbb{R}$. Let

$$Z_f = \{(x_1, x_2, \ldots, x_n) \in \mathbb{R}^n : f(x_1, x_2, \ldots, x_n) = 0\}.$$

Find the number of connected components in $\mathbb{R}^n$ of the set $\mathbb{R}^n - Z_f$.

Such a result is extremely useful as it gives some kind of a Fundamental Theorem of Algebra for polynomials in several variables.

References

1. D. Hilbert, *Mathematical Problems*, Göttinger Nachrichten (1900), 253–297.

2. Th.M. Rassias, *Remarks on the Hilbert's 16th problem.* Proc. of the Nat. Acad. of Athens, Greece, **54** (1979), 74–87.

3. Th.M. Rassias, *On the topology of algebraic curves.* Bull.of the Inst. Math., Academia Sinica **13** (1985), 273–278.

Themistocles M. Rassias, Department of Mathematics, University of La Verne, P.O. Box 51105, Kifissia, Athens, Greece 14510

International Series of Numerical Mathematics, Vol. 103, © 1992 Birkhäuser Verlag Basel

An eigenvalue problem

Themistocles M. Rassias

Consider the eigenvalue problem

$$r^2R'' + rR' + \lambda \cdot \frac{(r^4+1)^2}{r^2} \cdot R = (n^2 - 32)R,$$

where $R = R(r)$, $r \in [0, r_0]$, r_0 a fixed real number and $n = 1, 2, 3, \ldots$, with boundary condition $R(r_0) = 0$ and $R \in C_p^2[0, r_0]$ where C_p^2 stands for piecewise C^2 functions. Compute an upper bound of the sum of the dimensions of the eigenspaces corresponding to the negative eigenvalues of this eigenvalue problem as a function of r_0.

Note. The above estimate is useful in order to obtain the Morse-Smale index of Richmond's minimal surface in $\mathbb{R}^3$ (cf. [1, pp. 46–51], [2]).

References

1. Th.M. Rassias, *Foundations of Global Nonlinear Analysis*. Teubner-Texte zur Mathematik, Band 86, Leipzig, 1986.
2. Th.M. Rassias, *Query No. 257*, Notices Amer. Math. Soc. **29** (4) (1982), 326.

Themistocles M. Rassias, Department of Mathematics, University of La Verne, P.O. Box 51105, Kifissia, Athens, Greece 14510

International Series of Numerical Mathematics, Vol. 103, © 1992 Birkhäuser Verlag Basel

A problem for harmonic mappings

Themistocles M. Rassias

Does there exist a harmonic homeomorphism of the open unit ball B in $\mathbb{R}^n$ onto $\mathbb{R}^n$, where $n = 3, 4, \ldots$, i.e., do there exist harmonic functions $f_1, f_2, \ldots, f_n$, defined in B such that the mapping $f = (f_1, f_2, \ldots, f_n)$ is a homeomorphism of B onto all of $\mathbb{R}^n$?

Note. One can prove that there is no harmonic homeomorphism of the open unit disk in the plane onto the whole plane.

Themistocles M. Rassias, Department of Mathematics, University of La Verne, P.O. Box 51105, Kifissia, Athens, Greece 14510

International Series of Numerical Mathematics, Vol. 103, © 1992 Birkhäuser Verlag Basel

A problem

Dennis C. Russell

The following problem was told to me by A. Meir.

Let

$$F(p) := \sqrt{p} \int_0^\infty \left| \frac{\sin t}{t} \right|^p dt.$$

Then $F(1) = +\infty$, $F(2) = \pi/\sqrt{2}$.

Numerical evidence suggests that for $p > 1$, $F(p)$ decreases to a minimum and then increases to a limit less than or equal to $F(2)$, as $p \to +\infty$. For instance, $F(40)/F(2) = 0.9735\ldots$

Thus the problem is to prove:

(i) $F(p) < F(2)$ for $p > 2$;

(ii) There exists $\lim_{p \to +\infty} F(p) = L$, and find L.

Remark. Wolfgang Walter has shown me a proof that there exists

$$\lim_{p \to +\infty} F(p) = \sqrt{\frac{3\pi}{2}}.$$

Dennis C. Russell, Department of Mathematics, York University, North York (Toronto), Ontario M3J 1P3, Canada

International Series of Numerical Mathematics, Vol. 103, © 1992 Birkhäuser Verlag Basel

Problem: An integral inequality by Bushell and Okrasinski

Wolfgang Walter

In a recent paper Bushell and Okrasinski [1] prove the following inequality

$$(1)\qquad \int_0^x (x-s)^{\alpha-1} f(s)^\alpha \, ds \le \left(\int_0^x f(s)\, ds \right)^\alpha \quad \text{for } 0 < x \le b,$$

where $f \in C^0[0,b]$ is nonnegative and (weakly) increasing and $\alpha \ge 1$, $b \le 1$. The proof is done in three steps. First, for α an integer, the stronger inequality

$$(2)\qquad n \int_0^x (x-s)^{n-1} f^n(s)\, ds \le \left(\int_0^x f(s)\, ds \right)^n \quad (n = 1, 2, \ldots)$$

is proved inductively by showing that $I_n(x)$, the difference "right hand side minus left hand side", satisfies an inequality $I_n'(x) \ge n f(x) I_{n-1}(x)$. Then, by applying Hölder's inequality, (1) is deducted from (2) for rational values of α. Finally, irrational values of α are handled by a simple limiting argument.

Conjecture 1. The stronger inequality

$$(3)\qquad \alpha \int_0^x (x-s)^{\alpha-1} f(s)^\alpha \, ds \le \left(\int_0^x f(s)\, ds \right)^\alpha \quad \text{for } 0 < x \le b$$

holds for increasing functions $f : [0,b] \to [0,\infty)$ when $\alpha \ge 1$ ($b > 0$ arbitrary).

The proof described above in which first the stronger result (3) is established for $\alpha = n$, which is then lost in the second step, seems unnatural. Is the inequality known or can it easily be reduced to a known inequality? At any rate, a simpler proof for (1) or, even better, for (3) is in demand.

Conjecture 2. Inequality (3) holds also for $0 < \alpha < 1$ if the function f is nonnegative and decreasing.

References

1. P.J. Bushell and W. Okrasinski, *Nonlinear Volterra equations with convolution kernel*. J. London Math. Soc. (2) **41** (1990), 503–510.

Wolfgang Walter, Universität Karlsruhe, Mathematisches Institut I, Kaiserstr. 12, D-7500 Karlsruhe 1, Germany

Name Index

Subject Index